Lecture Notes in Artificial Intelligence   9244

Subseries of Lecture Notes in Computer Science

LNAI Series Editors

Randy Goebel
  University of Alberta, Edmonton, Canada
Yuzuru Tanaka
  Hokkaido University, Sapporo, Japan
Wolfgang Wahlster
  DFKI and Saarland University, Saarbrücken, Germany

LNAI Founding Series Editor

Joerg Siekmann
  DFKI and Saarland University, Saarbrücken, Germany

More information about this series at http://www.springer.com/series/1244

Honghai Liu · Naoyuki Kubota
Xiangyang Zhu · Rüdiger Dillmann
Dalin Zhou (Eds.)

Intelligent Robotics and Applications

8th International Conference, ICIRA 2015
Portsmouth, UK, August 24–27, 2015
Proceedings, Part I

 Springer

Editors
Honghai Liu
University of Portsmouth
Portsmouth
UK

Naoyuki Kubota
Tokyo Metropolitan University
Tokyo
Japan

Xiangyang Zhu
Shanghai Jiao Tong University
Shanghai
China

Rüdiger Dillmann
Karlsruhe Institute of Technology
Karlsruhe
Germany

Dalin Zhou
University of Portsmouth
Portsmouth
UK

ISSN 0302-9743   ISSN 1611-3349 (electronic)
Lecture Notes in Artificial Intelligence
ISBN 978-3-319-22878-5   ISBN 978-3-319-22879-2 (eBook)
DOI 10.1007/978-3-319-22879-2

Library of Congress Control Number: 2015946083

LNCS Sublibrary: SL7 – Artificial Intelligence

Springer Cham Heidelberg New York Dordrecht London

Springer International Publishing AG Switzerland is part of Springer Science+Business Media
(www.springer.com)

Preface

The Organizing Committee of the 8th International Conference on Intelligent Robotics and Applications aimed to facilitate interactions among active participants in the field of intelligent robotics, automation, and mechatronics. Through this conference, the committee intended to enhance the sharing of individual experiences and expertise in intelligent robotics with particular emphasis on technical challenges associated with varied applications such as biomedical application, industrial automations, surveillance, and sustainable mobility.

The 8th International Conference on Intelligent Robotics and Applications was most successful in attracting 228 submissions by researchers from 20 countries addressing the state-of-the art developments in robotics, automation, and mechatronics. Owing to the large number of valuable submissions, the committee was faced with the difficult challenge of selecting the most deserving papers for inclusion in these lecture notes and presentation at the conference. For this purpose, the committee undertook a rigorous review process. Despite the high quality of most of the submissions, a total of 172 papers were selected for publication in three volumes of Springer's *Lecture Notes in Artificial Intelligence* as subseries of *Lecture Notes in Computer Science*, with an acceptance rate of 75.4 %. The selected papers were presented at the 8th International Conference on Intelligent Robotics and Applications held during August 24–27, 2015, in Portsmouth, UK.

The contribution of the Technical Program Committee and the reviewers is deeply appreciated. Most of all, we would like to express our sincere thanks to the authors for submitting their most recent work and to the Organizing Committee for their enormous efforts to turn this event into a smoothly run meeting. Special thanks go to the University of Portsmouth for their generosity and direct support. Our particular thanks are due to Alfred Hofmann and Anna Kramer of Springer for enthusiastically supporting the project.

We sincerely hope that these volumes will prove to be an important resource for the scientific community.

June 2015

Honghai Liu
Naoyuki Kubota
Xiangyang Zhu
Rüdiger Dillmann
Dalin Zhou

Organization

International Advisory Committee

Jorge Angeles	McGill University, Canada
Tamio Arai	University of Tokyo, Japan
Hegao Cai	Harbin Institute of Technology, China
Tianyou Chai	Northeastern University, China
Jiansheng Dai	King's College London, UK
Han Ding	Huazhong University of Science and Technology, China
Toshio Fukuda	Nagoya University, Japan
Huosheng Hu	University of Essex, UK
Oussama Khatib	Stanford University, USA
Yinan Lai	National Natural Science Foundation of China, China
Zhongqin Lin	Shanghai Jiao Tong University, China
Guobiao Wang	National Natural Science Foundation of China, China
Kevin Warwick	University of Reading, UK
Bogdan M. Wilamowski	Auburn University, USA
Ming Xie	Nanyang Technological University, Singapore
Youlun Xiong	Huazhong University of Science and Technology, China
Huayong Yang	Zhejiang University, China

General Chair

Honghai Liu	University of Portsmouth, UK

General Co-chairs

Rüdiger Dillmann	Karlsruhe Institute of Technology, Germany
Jangmyung Lee	Pusan National University, Korea
Xiangyang Zhu	Shanghai Jiao Tong University, China

Program Chair

Naoyuki Kubota	Tokyo Metropolitan University, Japan

Program Co-chairs

Patrick Keogh	University of Bath, UK
Jeremy Wyatt	University of Birmingham, UK
Shengquan Xie	University of Auckland, New Zealand

Publicity Chairs

Darwin G. Caldwell   Istituto Italiano di Tecnologia, Italy
Yutaka Hata   University of Hyogo, Japan
Ning Xi   Michigan State University, USA

Award Chairs

Jianda Han   Chinese Academy of Sciences, China
Sabina Jesehke   RWTH Aachen University, Germany
Angelika Peer   University of the West of England, UK

Publication Chairs

William A. Gruver   Simon Fraser University, Canada
Jie Zhao   Harbin Institute of Technology, China
Tom Ziemke   University of Skövde, Sweden

Organized Session Chairs

Frank Guerin   University of Aberdeen, UK
Paolo Remagnino   Kingston University, UK
Bram Vanderborght   Vrije Universiteit Brussel, Belgium
Zhouping Yin   Huazhong University of Science and Technology, China

Organizing Committee Chairs

Kaspar Althoefer   King's College London, UK
Angelo Cangelosi   University of Plymouth, UK
Shengyong Chen   Zhejiang University of Technology, China
Feng Gao   Shanghai Jiao Tong University, China
Ping Li   Liaoning Shihua University, China
Robert Riener   ETH Zurich, Switzerland
Chunyi Su   Concordia University, Canada
Caihua Xiong   Huazhong University of Science and Technology, China

Technical Theme Committee

Chris Burbridge   University of Birmingham, UK
Jiangtao Cao   Liaoning Shihua University, China
Charles Fox   University of Sheffield, UK
Lars Kunze   University of Birmingham, UK
Takenori Obo   University of Malaya, Malaysia

Takahiro Takeda   Tokyo Metropolitan University, Japan
Xiaolong Zhou   Zhejiang University of Technology, China
Yahya Zweiri   Kingston University, UK

Local Arrangements Chairs

Zhaojie Ju   University of Portsmouth, UK
Hui Yu   University of Portsmouth, UK

Secretariat

Dalin Zhou   University of Portsmouth, UK

Contents – Part I

Drives and Actuators' Modeling

Biomechatronics in Bionic Dexterous Hand

Robot Actuators and Sensors

Intelligent Visual Systems

Contents – Part II

Navigation and Planning

Medical Robot

Prototyping

Manufacturing

Contents – Part III

Stiffening Mechanisms of Soft Robots

Robot Mechanism and Design

Robotic Vision, Recognition and Reconstruction

Active Control in Tunneling Boring Machine

Industrial Robot and Its Applications

Analysis and Control for Complex Systems

Observer-Based Robust Mixed H_2/H_∞ Control for Autonomous Spacecraft Rendezvous

Zhuqi Li$^{(\boxtimes)}$, Bolun Li, and Feng Xiao

Department of Control Science and Engineering,
School of Astronautics, Harbin Institute of Technology, Harbin, China
lizq900430@163.com

Abstract. This paper investigates the problem of observer-based robust mixed H_2/H_∞ control for a class of autonomous spacecraft rendezvous systems. The model of system is based on Clohessy-Wiltshire (C-W) equations, which contain non-circle uncertainty (NCU) of target track, white noise and bounded energy noise. Considering the unobservable states of system, the paper uses state-observer to implement state feedback control and separates the state-observer subsystem from the state-feedback subsystem with several sets of separated-parameters to reduce the complexity of derivation. In this paper, the aim of the control problem is to design a state-observer and a state feedback controller such that the closed-loop poles of two subsystems are placed within a given disc, the H_∞ norm of the transfer function from the bounded energy noise to output and the H_2 norm from the white noise to output is guaranteed to be less than two preset level. With Lyapunov theory, the observer and controller design problem can be transferred into solving a group of linear matrix inequalities (LMIs). An illustrative example is given to show the effectiveness of the proposed observer and the controller design approach.

Keywords: State observer · Mixed H_2/H_∞ · Spacecraft rendezvous

1 Introduction

Autonomous spacecraft rendezvous is the premise technology of many astronautic tasks. In order to describe the relative movement of spacecraft, the most famous model is C-W equations [1]. Based on C-W equations, many control problems of rendezvous have been studied in [2,3]. During past several years, there are many kinds of uncertainty to be studied in [4-6]. This paper uses C-W equations which contain the non-circle uncertainty (NCU) to build the model of system, The trust of chasing spacecraft is divided into three categories, impulsive trust, continuous trust and variable trust [7,8]. The robust controller for rendezvous was studied in [3,9,10]. However, in fact, the velocity states cannot be observed for state feedback, so the state-observer theory is applied to spacecraft rendezvous control [11-13].

According to above introduction, this paper provides a complete research of the approach on observer-based robust mixed H_2/H_∞ control for autonomous spacecraft rendezvous.

© Springer International Publishing Switzerland 2015
H. Liu et al. (Eds.): ICIRA 2015, Part I, LNAI 9244, pp. 3–14, 2015.
DOI: 10.1007/978-3-319-22879-2_1

2   Problem Formulation

In this section, the model of relative motion can be established based on C-W equations. What's more, NCU, bounded energy noise and white noise is considered. Finally, the observer-based robust mixed H_2/H_∞ control problem for autonomous spacecraft rendezvous is formulated.

2.1   Model of Relative Motion with NCU and Noise

Based on C-W equations, we assume that the orbit of target spacecraft is approximately circle and the distance between two spacecrafts is far less than the distance between spacecrafts and earth, such that the relative motion can be described as:

$$\ddot{x} - 2n\dot{y} - 3n^2 x = \frac{1}{m}\left(u_x + w_x + v_x\right)$$

$$\ddot{y} + 2n\dot{x} = \frac{1}{m}\left(u_y + w_y + v_y\right) \tag{1}$$

$$\ddot{z} + n^2 z = \frac{1}{m}\left(u_z + w_z + v_z\right)$$

where x, y, and z are the components of the chasing spacecraft's position relative to the target spacecraft's position, n is the angular velocity of the target spacecraft, m is the mass of the chasing spacecraft, u_i is the ith component of the control trust applied to the chasing spacecraft, w_i is the ith component of the bounded energy noise, v_i is the ith component of the white noise.

We can define the state vector $\mathbf{x}(t) = [x, y, z, \dot{x}, \dot{y}, \dot{z}]^T$, the control input vector $\mathbf{u}(t) = [u_x, u_y, u_z]^T$, the output vector $\mathbf{y}(t) = [x, y, z]^T$, the bounded energy noise vector $\mathbf{w}(t) = [w_x, w_y, w_z]^T$ and the white noise vector $\mathbf{v}(t) = [v_x, v_y, v_z]^T$, so we have

$$\mathbf{x}(t) = A\mathbf{x}(t) + B\mathbf{u}(t) + B_w\mathbf{w}(t) + B_v\mathbf{v}(t),$$
$$\mathbf{y}(t) = C\mathbf{x}(t), \tag{2}$$

where

$$A = \begin{bmatrix} 0 & 0 & 0 & 1 & 0 & 0 \\ 0 & 0 & 0 & 0 & 1 & 0 \\ 0 & 0 & 0 & 0 & 0 & 1 \\ 3n^2 & 0 & 0 & 0 & 2n & 0 \\ 0 & 0 & 0 & -2n & 0 & 0 \\ 0 & 0 & -n^2 & 0 & 0 & 0 \end{bmatrix}, B = \frac{1}{m}\begin{bmatrix} 0 & 0 & 0 \\ 0 & 0 & 0 \\ 0 & 0 & 0 \\ 1 & 0 & 0 \\ 0 & 1 & 0 \\ 0 & 0 & 1 \end{bmatrix}$$

$$C = \begin{bmatrix} 1 & 0 & 0 & 0 & 0 & 0 \\ 0 & 1 & 0 & 0 & 0 & 0 \\ 0 & 0 & 1 & 0 & 0 & 0 \end{bmatrix}, B_w = B_v = B$$

Considering the NCU, the matrix A is not accuracy. So we add the uncertainty matrix A_e to the matrix A and A_e can be described as [4]

$$A_e = E_1 \Delta E_2 \tag{3}$$

where

$$\Delta = diag\{-\sin M, \sin M, \cos M, \cos M, -0.5\cos M, \cos M\}$$

$$E_1 = \begin{bmatrix} 0 & 0 & 0 & 0 & 0 & 0 \\ 0 & 0 & 0 & 0 & 0 & 0 \\ 0 & 0 & 0 & 0 & 0 & 0 \\ 0 & 2\sigma & 4\sigma & 0 & 8\sigma & 0 \\ 2\sigma & 0 & 0 & 4\sigma & 0 & 0 \\ 0 & 0 & 0 & 0 & 6\sigma & 0 \end{bmatrix}, E_2 = \begin{bmatrix} n^2 & 0 & 0 & 0 & 0 & 0 \\ 0 & n^2 & 0 & 0 & 0 & 0 \\ 2.25n^2 & 0 & n^2 & 0 & -n & 0 \\ 0 & 0.25n^2 & 0 & n & 0 & 0 \\ 0 & 0 & n^2 & 0 & 0 & 0 \\ 0 & 0 & 0 & 0 & 0 & n^2 \end{bmatrix}$$

and M is the mean anomaly, σ is the upper bound of uncertain eccentricity. Obviously

$$\Delta^T \Delta \le I \tag{4}$$

So Eqs. (2) can be rewritten as

$$\mathbf{x}(t) = (A + A_e)\mathbf{x}(t) + B\mathbf{u}(t) + B_w \mathbf{w}(t) + B_v \mathbf{v}(t),$$
$$\mathbf{y}(t) = C\mathbf{x}(t), \tag{5}$$

2.2 Observer-Based Robust Mixed H_2/H_∞ Control Problem

We used state-observer to implement state feedback of the uncertain system and the equivalent of its state-space representation can be described as

$$\dot{\hat{\mathbf{x}}}(t) = GC\mathbf{x}(t) + (A + A_e - GC + BK)\hat{\mathbf{x}}(t)$$
$$\mathbf{u}(t) = K\hat{\mathbf{x}}(t) \tag{6}$$

where $\hat{\mathbf{x}}(t) \in \mathbb{R}^6$ is the state vector of the state-observer, $G \in \mathbb{R}^{6\times6}$ is the observer feedback matrix and $K \in \mathbb{R}^{3\times6}$ is the state feedback control law. By establishing an error vector $\mathbf{e}(t) = \mathbf{x}(t) - \hat{\mathbf{x}}(t)$ and two augmented vector $\overline{\mathbf{x}}(t)$ and $\overline{\mathbf{w}}(t)$, we can build an error system:

$$\dot{\bar{\mathbf{x}}}(t) = \bar{A}\bar{\mathbf{x}}(t) + \bar{B}\bar{\mathbf{w}}(t)$$
$$\bar{\mathbf{y}}(t) = \bar{C}\bar{\mathbf{x}}(t) \tag{7}$$

where

$$\bar{\mathbf{x}}(t) = \begin{bmatrix} \mathbf{x}(t)^T & \mathbf{e}(t)^T \end{bmatrix}^T, \quad \bar{\mathbf{w}}(t) = \begin{bmatrix} \mathbf{w}(t)^T & \mathbf{v}(t)^T \end{bmatrix}^T$$

$$\bar{A} = \begin{bmatrix} A + A_e + BK & -BK \\ 0 & A + A_e - GC \end{bmatrix}, \quad \bar{B} = \begin{bmatrix} B_w & B_v \\ B_w & B_v \end{bmatrix}, \quad \bar{C} = \begin{bmatrix} C & 0 \end{bmatrix}$$

The control problem to be investigated in this paper is to design a state-observer and a state feedback control law such that the system (7) satisfies the following requirements:

1. The closed-loop system (7) is asymptotically stable for the uncertainty A_e.
2. The closed-loop poles are placed in a disc $D(a,b)$
3. The transfer function from noise $\bar{\mathbf{w}}(t)$ to output $\mathbf{y}(t)$ is

$$\psi_{yw}(s) = \bar{C}(sI - \bar{A})^{-1}\bar{B} \tag{8}$$

and we must guarantee that the $\varPsi_{yw}(s)$ satisfies the H_∞ performance index, that is

$$\left\| \psi_{yw}(s) \right\|_\infty = \sup_\omega \sigma_{\max}(\psi_{yw}(jw)) < \gamma_\infty \tag{9}$$

4. We must guarantee that the $\varPsi_{yw}(s)$ satisfies the H_2 performance index, that is

$$\left\| \psi_{yw}(s) \right\|_2 = Trace(\frac{1}{2\pi} \int_{-\infty}^{\infty} \psi_{yw}(jw)\psi_{yw}^*(jw))^{1/2} < \gamma_2 \tag{10}$$

In the following proof, we need the following lemma.

Lemma 1. *Given a continuous system, if exists symmetrical positive matrix P so that*

$$\begin{bmatrix} -bP & aP + AP \\ * & -bP \end{bmatrix} < 0 \tag{11}$$

holds, the closed-loop poles of system is placed in $D(a,b)$.

2.3 Observer Parameter and Control Law Design Algorithm

First, the open-loop transfer function can be rewritten as

$$\det\begin{bmatrix} sI - \bar{A} \end{bmatrix} = \det\begin{bmatrix} sI - (A + A_e + BK) \end{bmatrix} \cdot \det\begin{bmatrix} sI - (A + A_e + BK) \end{bmatrix} \tag{12}$$

So we use separation design as the core to solve the control problem with several sets of separated-parameters η_i, $\mu_i \in (0,1)$, $i=1, 2,...$, where

$$\eta_i + \mu_i = 1$$

Theorem 1. For the closed-loop system (7), given constants $\varepsilon > 0$, $\gamma_2 > 0, \gamma_\infty > 0$, η_1, μ_1, η_2, μ_2, if exists symmetric positive matrixes P_1, P_2 such that the following LMIs

$$\begin{bmatrix} sym\left[P_1\left(A+A_e+BK\right)\right]+C^T C+\varepsilon P_1 P_1 & P_1 \tilde{B} \\ * & -\eta_1 \gamma_\infty^2 I \end{bmatrix} < 0 \tag{13}$$

$$\begin{bmatrix} sym\left[P_2\left(A+A_e-GC\right)\right]+\varepsilon^{-1}K^T B^T BK & P_2 \tilde{B} \\ * & -\mu_1 \gamma_\infty^2 I \end{bmatrix} < 0 \tag{14}$$

$$Trace\left(\tilde{B}^T P_1 \tilde{B}\right) < \eta_2 \gamma_2^2 \tag{15}$$

$$Trace\left(\tilde{B}^T P_2 \tilde{B}\right) < \mu_2 \gamma_2^2 \tag{16}$$

hold, where $\tilde{B} = \begin{bmatrix} B_w & B_v \end{bmatrix}$ and $sym(X)=X+X^T$(the $sym()$ will be used in following proof with the same meaning), then we get the conclusions that

1. System (7) is asymptotically stable;
2. $\|\Psi_{yw}(s)\|_\infty < \gamma_\infty$;
3. $\|\Psi_{yw}(s)\|_2 < \gamma_2$.

Proof. First, we assume that $B_v=0$, so that it's convenient to guarantee the H_∞ norm is bounded. Based on Lyapunov theory, we consider the Lyapunov function $V = \bar{x}^T P \bar{x}$, and we define the structure of P as

$$P = \begin{bmatrix} P_1 & 0 \\ 0 & P_2 \end{bmatrix} \tag{17}$$

where P_1 and P_2 are symmetric positive matrixes. We can obtain

$$\dot{V} + \bar{y}^T \bar{y} - \gamma_\infty^2 \bar{w}^T \bar{w} = \bar{x}^T \left(sym\left(\bar{A}^T P\right)+\bar{C}^T \bar{C}\right)\bar{x} + sym\left(\bar{w}^T \bar{B}^T P\bar{x}\right) - \gamma_\infty^2 \bar{w}^T w$$

$$= \begin{bmatrix} x \\ e \end{bmatrix}^T \begin{bmatrix} \Pi_K^T P_1 & 0 \\ -K^T B^T P_1 & \Pi_G^T P_2 \end{bmatrix} \begin{bmatrix} x \\ e \end{bmatrix} + \begin{bmatrix} x \\ e \end{bmatrix}^T \begin{bmatrix} P_1 \Pi_K & -P_1 BK \\ 0 & P_2 \Pi_G \end{bmatrix} \begin{bmatrix} x \\ e \end{bmatrix}$$

$$+ \begin{bmatrix} x \\ e \end{bmatrix}^T \begin{bmatrix} C^T C & 0 \\ 0 & 0 \end{bmatrix} \begin{bmatrix} x \\ e \end{bmatrix} \tag{18}$$

$$+ sym\left(\bar{w}^T \tilde{B}^T P_1 x\right) + sym\left(\bar{w}^T \tilde{B}^T P_2 e\right) - \gamma_\infty^2 \bar{w}^T \bar{w}$$

where, $\Pi_K = A + A_e + BK$, $\Pi_G = A + A_e - GC$ .Then we can obtain

$$-e^T K^T B^T P_1 x - x^T P_1 BKe \le \varepsilon x^T P_1 P_1 x + \varepsilon^{-1} e^T K^T B^T BKe \tag{19}$$

Combined with matrix inequality (19) and the first set of separated-parameters η_1 and μ_1, the Eq. (18) can be rewritten as

$$
\begin{aligned}
\dot{V} + \bar{y}^T \bar{y} - \gamma_\infty^2 \bar{w}^T \bar{w} \\
\leq x^T \left[sym(P_1 \Pi_K) + C^T C + \varepsilon P_1 P_1 \right] x + sym(\bar{w}^T \tilde{B}^T P_1 x) - \eta_1 \gamma_\infty^2 \bar{w}^T \bar{w} \\
+ e^T \left[sym(P_2 \Pi_G) + \varepsilon^{-1} K^T B^T BK \right] e + sym(\bar{w}^T \tilde{B}^T P_2 e) - \mu_1 \gamma_\infty^2 \bar{w}^T \bar{w}
\end{aligned} \tag{20}
$$

Now, we can separate the derivative of Lyapunov function and the definition of H_∞ norm into two parts, that is, direct state feedback part and state-observer part. When the two parts satisfy

$$
\dot{V}_1 < 0, \dot{V}_2 < 0 \tag{21}
$$

where

$$
\dot{V}_1 = \begin{bmatrix} x^T & w^T \end{bmatrix} \begin{bmatrix} sym\left[P_1(A + A_e + BK) \right] + C^T C + \varepsilon P_1 P_1 & P_1 \tilde{B} \\ * & -\eta_1 \gamma_\infty^2 I \end{bmatrix} \begin{bmatrix} x \\ w \end{bmatrix}
$$

$$
\dot{V}_2 = \begin{bmatrix} e^T & w^T \end{bmatrix} \begin{bmatrix} sym\left[P_2(A + A_e - GC) \right] + \varepsilon^{-1} K^T B^T BK & P_2 \tilde{B} \\ * & -\mu_1 \gamma_\infty^2 I \end{bmatrix} \begin{bmatrix} e \\ w \end{bmatrix}
$$

$$
\eta_1 \gamma_\infty^2 \bar{w}^T \bar{w} + \mu_1 \gamma_\infty^2 \bar{w}^T \bar{w} = \gamma_\infty^2 \bar{w}^T \bar{w}
$$

the system is asymptotically stable and $\|\Psi_{yw}(s)\|_\infty < \gamma_\infty$. Then, the following proof will guarantee the H_2 norm is bounded. We assume that $B_w = 0$. The necessary and sufficient condition [14] of bounded H_2 norm is

$$
\bar{A}^T P + P\bar{A} + \bar{C}^T \bar{C} < 0, \quad Trace\left(\bar{B}^T P \bar{B} \right) < \gamma_2^2 \tag{22}
$$

We expand the above first matrix inequality. So we can obtain

$$
\bar{A}^T P + P\bar{A} + \bar{C}^T \bar{C} \leq \left[sym(P_1 \Pi_K) + C^T C + \varepsilon P_1 P_1 \right] + \left[sym(P_2 \Pi_G) + \varepsilon^{-1} K^T B^T BK \right] \tag{23}
$$

Obviously, when matrix inequalities (13) and (14) hold, it satisfies the first inequality of matrix inequalities (22). Then, considering the special structure of matrix P, we obtain

$$
Trace\left(\bar{B}^T P \bar{B} \right) = Trace\left(\tilde{B}^T P_1 \tilde{B} \right) + Trace\left(\tilde{B}^T P_2 \tilde{B} \right) < \gamma_2^2 \tag{24}
$$

Therefore, with the second set of separated-parameters η_2 and μ_2, when matrix inequalities (15) and (16) hold, it satisfies the second inequality of matrix inequalities (22). This completes the proof.

Theorem 2. For the closed-loop system(7), given constants $\varepsilon > 0$, $\varepsilon_1 > 0$, $\varepsilon_2 > 0$, $\gamma_2 > 0$, $\gamma_\infty > 0$, η_1, μ_1, η_2, μ_2, if exists symmetric positive matrixes Q, P_2, M, N, and matrixes X, Y with appropriate dimensions, such that the following LMIs

$$
\begin{bmatrix}
sym(AQ) + sym(BX) + \varepsilon I & \tilde{B} & QC^T & E_1 & QE_2^T \\
* & -\eta_1\gamma_\infty^2 I & 0 & 0 & 0 \\
* & * & -I & 0 & 0 \\
* & * & * & -\varepsilon_1^{-1}I & 0 \\
* & * & * & * & -\varepsilon_1 I
\end{bmatrix} < 0 \quad (25)
$$

$$
\begin{bmatrix}
ym(P_2A) - sym(YC) & P_2\tilde{B} & K^T B^T & P_2 E_1 & E_2^T \\
* & -\mu_1\gamma_\infty^2 I & 0 & 0 & 0 \\
* & * & -\varepsilon I & 0 & 0 \\
* & * & * & -\varepsilon_2^{-1}I & 0 \\
* & * & * & * & -\varepsilon_2 I
\end{bmatrix} < 0 \quad (26)
$$

$$
Trace(M) < \eta_2\gamma_2^2 \quad (27)
$$

$$
Trace(N) < \mu_2\gamma_2^2 \quad (28)
$$

$$
\begin{bmatrix}
-Q & \tilde{B} \\
* & -M
\end{bmatrix} < 0 \quad (29)
$$

$$
\begin{bmatrix}
-P_2 & P_2\tilde{B} \\
* & -N
\end{bmatrix} < 0 \quad (30)
$$

hold, where $\tilde{B}$ is proposed in Theorem 1, then we get the conclusions that

1. System (7) is asymptotically stable;
2. $\|\Psi_{yw}(s)\|_\infty < \gamma_\infty$;
3. $\|\Psi_{yw}(s)\|_2 < \gamma_2$.

Proof. In this proof, we will use four LMIs proposed in the Theorem 1.
   For matrix inequality (13), we define

$$
Q = P_1^{-1}, X = KQ \quad (31)
$$

Then, post- and pre-multiplying the matrix inequality (13) by $diag\{Q, I\}$. So we obtain

$$
\begin{bmatrix}
sym(\Pi_K Q) + QC^T CQ + \varepsilon I & \tilde{B} \\
* & -\eta_1\gamma_\infty^2 I
\end{bmatrix} < 0 \quad (32)
$$

where Π_K is proposed in Theorem 1. By Eq. (3), the above matrix inequality holds if and only if there exists a positive constant ε_1 such that matrix inequality (25) is satisfied.

For matrix inequality (14), we define

$$Y = P_2 G \tag{33}$$

In the similar way, matrix inequality (26) should be satisfied.

For matrix inequality (15), we use a symmetric positive matrix M. Obviously, when the matrix inequality

$$Trace\left(\tilde{B}^T P_1 \tilde{B}\right) < Trace(M) < \eta_2 \gamma_2^2 \tag{34}$$

holds, matrix inequality (15) is satisfied, and the right half of matrix inequality (34) is matrix inequality (27). What's more, the left half of matrix inequality (34) can be satisfied by

$$\begin{bmatrix} -P_1 & P_1\tilde{B} \\ * & -M \end{bmatrix} < 0 \tag{35}$$

Post- and pre-multiplying the matrix inequality (35) by $diag\{Q, I\}$, and we obtain the matrix inequality (29)

For matrix inequality (16), we use a symmetric positive matrix N. Similarly, we can obtain the matrix inequality (28) and (30).

This completes the proof.

Theorem 3. For the closed-loop system (7), given a disc $D(a,b)$, if the following matrix inequalities

$$\begin{bmatrix} -bQ & aQ + AQ + BX & E_1 & 0 \\ * & -bQ & 0 & QE_2^T \\ * & * & -\varepsilon_1^{-1}I & 0 \\ * & * & * & -\varepsilon_1 I \end{bmatrix} < 0 \tag{36}$$

$$\begin{bmatrix} -bP_2 & aP_2 + P_2 A - YC & P_2 E_1 & 0 \\ * & -bP_2 & 0 & E_2^T \\ * & * & -\varepsilon_2^{-1}I & 0 \\ * & * & * & -\varepsilon_2 I \end{bmatrix} < 0 \tag{37}$$

hold, where ε_1, ε_2, Q, P_2, X, Y are proposed in Theorem 2, then we get the conclusion that the closed-loop poles are placed in $D(a,b)$.

Proof. We separate the transfer function into two parts. Using symmetric positive matrixes Q, $(P_2)^{-1}$ and Lemma 1, the poles constraint can be satisfied by

$$\begin{bmatrix} -bQ & aQ+AQ+A_eQ+BX \\ * & -bQ \end{bmatrix} < 0 \tag{38}$$

$$\begin{bmatrix} -bP_2^{-1} & aP_2^{-1}+AP_2^{-1}+A_eP_2^{-1}-GCP_2^{-1} \\ * & -bP_2^{-1} \end{bmatrix} < 0 \tag{39}$$

Post- and pre-multiplying the matrix inequality (39) by $diag\{P_2, P_2\}$, then the above two matrix inequalities can be rewritten as

$$\begin{bmatrix} -bQ & aQ+AQ+BX \\ * & -bQ \end{bmatrix} + \begin{bmatrix} 0 & E_1\Delta E_2 Q \\ * & 0 \end{bmatrix} < 0 \tag{40}$$

$$\begin{bmatrix} -bP_2 & aP_2+P_2A-GC \\ * & -bP_2^{-1} \end{bmatrix} + \begin{bmatrix} 0 & P_2E_1\Delta E_2 \\ * & 0 \end{bmatrix} < 0 \tag{41}$$

In order to deal with uncertainty problems, we use ε_1 and ε_2 proposed in Theorem 2, then we obtain

$$\begin{bmatrix} -bQ & aQ+AQ+BX \\ * & -bQ \end{bmatrix} + \varepsilon_1 \begin{bmatrix} E_1E_1^T & 0 \\ 0 & 0 \end{bmatrix} + \varepsilon_1^{-1} \begin{bmatrix} 0 & 0 \\ 0 & QE_2^TE_2Q \end{bmatrix} < 0 \tag{42}$$

$$\begin{bmatrix} -bP_2 & aP_2+P_2A-GC \\ * & -bP_2^{-1} \end{bmatrix} + \varepsilon_2 \begin{bmatrix} P_2E_1E_1^TP_2 & 0 \\ 0 & 0 \end{bmatrix} + \varepsilon_2^{-1} \begin{bmatrix} 0 & 0 \\ 0 & E_2^TE_2 \end{bmatrix} < 0 \tag{43}$$

The above two matrix inequalities are the equivalences of matrix inequalities (36) and (37).

This completes the proof.

Theorem 4. For the closed-loop system (7), given constants $\varepsilon > 0$, $\varepsilon_1 > 0$, $\varepsilon_2 > 0$, $\gamma_2 > 0$, $\gamma_\infty > 0$, η_1, μ_1, η_2, μ_2, and $D(a,b)$, if exists symmetric positive matrixes Q, P_2, M, N, and matrixes X, Y with appropriate dimensions, such that the LMIs (25) – (30), (38) and (39) holds, then we get the conclusion that:

1. System (7) is asymptotically stable;
2. $\|\Psi_{yw}(s)\|_\infty < \gamma_\infty$;
3. $\|\Psi_{yw}(s)\|_2 < \gamma_2$;
4. The closed-loop poles are placed in $D(a,b)$.

Proof. We follow from Theorem 2 and Theorem 3 and can get the above conclusions. This completes the proof

Remark 1. We can summarize the controller design approach and give several steps of solving the observer parameter and the state feedback control law:

1. First, according to the fact, decide the parameter of the system, that is, ε, ε_1, ε_2, γ_2, γ_∞, η_1, μ_1, η_2, μ_2, and $D(a,b)$;

2. Use the LMI tools of MATLAB to solve the LMIs (25), (27), (29) and (38). And then get the state feedback control law K by $K = XQ^{-1}$;

3. Use the LMI tools of MATLAB to solve the LMIs (26), (28), (30) and (39). Meanwhile, bring the K into the LMIs of this step as a known matrix. And then get the observer parameter G by $G = P_2^{-1}Y$.

3   Illustrative Example

In this section, we provide an example to illustrate the effectiveness of the approach proposed in the above section. Considering a target spacecraft and a chasing space-craft, we make some assumptions [4]., which is m= 400 kg, $n = 1.117 \times 10^{-3}$ rad/s, $x(0)$ = [400;50;-20;0;0.6702;0.2] with some random initial velocities, $\sigma = 0.001$, poles assignment area is $D(0.99, 0.9)$, $\gamma_2 = 3.5$ and $\gamma_\infty = 2$, $\varepsilon = 0.1$, $\varepsilon_1 = 2$, $\varepsilon_2 = 0.7$ and all of the separated-parameters are 0.5. This is the step 1 proposed in Remark 1.

Then, we execute step 2 proposed in Remark 1 and we can obtain

$$K = \begin{bmatrix} -550.6363 & -6.0330 \times 10^{-7} & -0.0015 & -788.5332 & -0.8936 & -0.0010 \\ 6.1354 \times 10^{-7} & -550.6429 & 2.4593 \times 10^{-7} & 0.8936 & -788.6318 & 8.7542 \times 10^{8} \\ -0.0015 & -2.4081 \times 10^{-7} & -550.6430 & -0.0010 & -2.3899 \times 10^{7} & -788.5321 \end{bmatrix}$$

Now, matrix K is a known matrix and we execute step 3. We can obtain

$$G = \begin{bmatrix} 1.3824 & 0.0013 & 0.0211 \\ -0.0013 & 1.3545 & -2.2085 \times 10^{-5} \\ 0.0211 & 1.1216 \times 10^{-5} & 1.3613 \\ 0.5703 & 0.0023 & 0.0173 \\ -0.0023 & 0.5475 & -6.3855 \times 10^{-5} \\ 0.0173 & 8.9067 \times 10^{-6} & 0.5530 \end{bmatrix}$$

So we can get the H_2 norm and H_∞ norm, that is, $\gamma_2 = 0.1931$ and $\gamma_\infty = 0.0217$. The assignment of closed-loop poles is described by Fig. 1. We can also draw the relative movement track of chasing spacecraft as Fig. 2. As shown in Fig. 3, the errors of three velocity states are all closed to zero in.

From the simulation result, we can see that the state-observer and the state feed-back controller solved by the approach in this paper can satisfy all the requirements of the system.

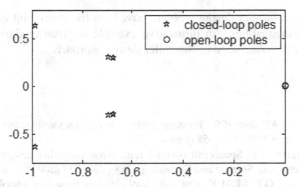

Fig. 1. Closed-loop poles and open-loop poles of system.

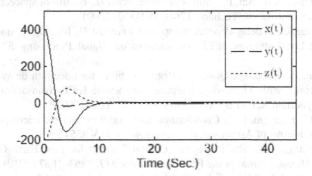

Fig. 2. Relative position along x-, y- and z- between chasing spacecraft and target spacecraft.

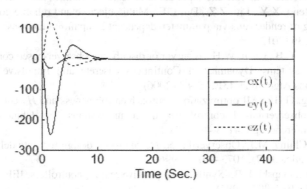

Fig. 3. Error of velocity between origin state and observer's state along x-, y- and z-.

4   Conclusions

This paper proposed an observer-based robust mixed H_2/H_∞ controller design approach for autonomous spacecraft rendezvous with some constraints, that is, NCU, bounded H_2 norm, bounded H_∞ norm and poles assignment. With Lyapunov theory, the observer and controller design problem have been transferred into solving a group of linear matrix inequalities. An illustrative example is given to show the effectiveness of the proposed observer and controller design approach.

References

1. Clohessy, W.H., Wiltshire, R.S.: Terminal guidance system for satellite rendezvous. Journal of Aerospace Science **27**(9), 653–658 (1960)
2. Luo, Y.Z., Tang, G.J.: Spacecraft optimal rendezvous controller design using simulated annealing. Aerospace Science and Technology **9**(8), 732–737 (2005)
3. Luo, Y.Z., Tang, G.J., Li, H.Y.: Optimal multi-objective linearized impulsive rendezvous. Journal of Guidance, Control and Dynamics **30**(2), 383–389 (2007)
4. Gao, H.J., Yang, X.B., Shi, P.: Multi-objective robust H_∞ control of spacecraft rendezvous. IEEE Trans. Control Syst. Technol. **17**(4), 794–802 (2009)
5. Gao, H., Wang, C.: A delay-dependent approach to robust H_∞ filtering for uncertain discrete-time state-delayed systems. IEEE Transactions on Signal Processing **52**(6), 1631–1640 (2004)
6. Li, H., Chen, B., Zhou, Q., Qian, W.: Robust stability for uncertain delay fuzzy hopfield neural networks with Markovian jumping parameters. IEEE Transactions on Systems, Man, and Cybernetics, Part B Cybernetics **39**(1), 94–102 (2009)
7. Wang, P.K.C., Hadaegh, F.Y.: Coordination and control of multiple microspacecraft moving in formation. Journal of Astronautical Sciences **44**(3), 315–355 (1996)
8. Gao, H.J., Yang, X.B., Shi, P.: Robust Orbital Transfer for Low Earth Orbit Spacecraft with Small-Thrust. Journal of the Franklin Institute **347**, 1863–1887 (2010)
9. Gao, X.Y., Teo, L.L., Duan, G.R.: Non-fragile robust H_∞ control for uncertain spacecraft rendezvous system with pole and input constraints. Int. J. Control **85**(7), 933–941 (2010)
10. Ma, L.C., Meng, X.Y., Liu, Z.Z., Du, L.F.: Multi-objective and reliable control for trajectory-tracking of rendezvous via parameter-dependent Lyapunov functions. Acta Astronaut. **81**(1), 122–136 (2012)
11. Guo, L., Feng, C.B., Chen, W.H.: A surver or disturbance-observer-based control for dynamic nonlinear systems. Dynamics of Continuous Discrete and Impulsive Systems-Series B-Applications Algorithms **13E**, 79–84 (2006)
12. Gu, Y., Yang, G.H.: LMI optimization approach on robustness and H_∞ control analysis via disturbance-observer-based control for uncertain systems. In: Chinese Control and Decision Conference, pp. 4158–4162 (2011)
13. Choi, H.H., Chung, M.J.: Observer-based H_∞ controller design for state delayed linear systems. Automatica **32**(7), 1073–1075 (1996)
14. Geromel, J.C., Gapski, P.B.: Synthesis of positive real H_2 controllers. IEEE Trans. Autom. Control **42**(7), 987–992 (1997)

Frequency Analysis of the In-Plane Rotating Hub-Beam System Considering Effects of the Hub

Jun-wei Chen, Le-tong Ma, Bo Zhang$^{(\boxtimes)}$, and Han Ding

State Key Laboratory of Mechanical System and Vibration, School of Mechanical Engineering,
Shanghai Jiao Tong University, Shanghai 200240, China
cjwbuaa@163.com, m835527262@126.com, {b_zhang,hding}@sjtu.edu.cn

Abstract. In this paper, the in-plane rotating hub-beam system is mainly investigated considering effects of the hub. By applying the extended Hamilton's principle and the Galerkin method, the governing equation of motion of the hub-beam system is derived. The hub-beam rotary inertia ratio is investigated to reveal its effect on the frequency characteristics of the rotating hub-beam system. Through the frequency analysis, it is shown that the natural frequencies of different orders vary between those of the clamped and the pinned boundary conditions at the connecting point between the hub and the beam. Finally the figure of connecting rigidity is plotted to show the variation.

Keywords: Hub-beam system · Dynamic stiffening · Rigid-flexible coupling

1   Introduction

The rotating beam is a fundamental element in mechanical engineering. Along with the high precision required in applications, such as rotating blades, antennas, robot arms, spinning space structures, etc., the dynamic characteristics of the rotating beam are of great importance for the mechanical design and evaluation due to dynamic stiffening.

Therefore many researchers have studied this phenomenon since it was pointed out in 1987 [1]. Yoo et al. [2-4] applied the Kane's method to model a rotating cantilever beam by assumed mode method. Utilizing the same discretization method, Hong & Liu developed first order approximating coupling (FOAC) model by the principle of virtual work [5,6]. In these papers, the dynamic stiffening due to the rotation of the single beam itself was mainly concerned and reasonable explanations about the natural frequencies were given and justified. Usually, it is regarded that it is clamped at the connecting point between the hub and the beam. This is reasonable when the rotary inertia of the hub is large. However, as the experimental and analytical work in [7-10] showed, the actual boundary condition is between the pinned and clamped boundary conditions at the connecting point, depending on the hub property.

This is meaningful for some engineering applications where a rotating beam is driven by a light hub. In such applications, the flexible connection between the hub and the beam has impact on the natural frequencies and dynamic response of the beam. So in this paper, considering the effect of dynamic stiffening, the effect of the

© Springer International Publishing Switzerland 2015
H. Liu et al. (Eds.): ICIRA 2015, Part I, LNAI 9244, pp. 15–23, 2015.
DOI: 10.1007/978-3-319-22879-2_2

hub property to the natural frequency of the rotating beam is investigated. First, the dynamic model of the rotating hub-beam system was developed by employing the extended Hamilton's principle and the Galerkin method. Then the governing equation for frequency analysis was formulated, and the variation of natural frequencies due to different values of rotary inertia of the hub was studied. Finally, conclusions are given.

2   Modeling

In this section, the rotating hub-beam system considering geometric nonlinearity is modelled based on the linear elasticity law and the Euler-Bernoulli model. Fig. 1 illustrates the configuration of the system. The rigid hub and the flexible beam are driven by the externally applied torque $M(t)$ at the hub center O in the global Cartesian coordinate system $O-X_0Y_0$. The body Cartesian coordinate system $O-x_by_b$, fixed to point O, is initially aligned to $O-X_0Y_0$.

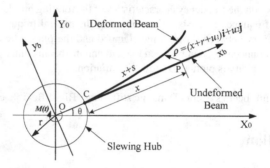

Fig. 1. Configuration of the rotating hub-beam system

2.1   Description of Deformation

As showed in Fig. 1, the displacement vector of a generic point P can be represented in $O-x_by_b$ as,

$$\boldsymbol{\rho} = (x + r + u_1)\boldsymbol{i} + u_2\boldsymbol{j} \tag{1}$$

where $\boldsymbol{i}, \boldsymbol{j}$ are unit vectors of $O-x_by_b$, and u_1, u_2 are the respective deformation along each unit vector. Introduce $\boldsymbol{A}$ to denote the transforming matrix from $O-x_by_b$ to $O-X_0Y_0$ as,

$$\boldsymbol{A} = \begin{bmatrix} cos\theta & -sin\theta \\ sin\theta & cos\theta \end{bmatrix} \tag{2}$$

The displacement vector in $O-X_0Y_0$ can be represented as follows.

$$\boldsymbol{r} = \boldsymbol{A}\boldsymbol{\rho} \tag{3}$$

Differentiating r yields,

$$\dot{r} = \dot{\theta} A \tilde{I} \rho + A \dot{\rho} \qquad (4)$$

where

$$\tilde{I} = \begin{bmatrix} 0 & -1 \\ 1 & 0 \end{bmatrix}, \quad \dot{\rho} = \begin{bmatrix} \dot{u}_1 \\ \dot{u}_2 \end{bmatrix}$$

Considering geometric nonlinearity, the axial displacements of Eulerian description is utilized instead of u_1, which is represented as in [3]

$$s = u_1 + \frac{1}{2} \int_0^x \left[\left(\frac{\partial u_2}{\partial x} \right)^2 \right] dx \qquad (5)$$

Then by the Galerkin method, one obtains.

$$s(x,t) = \phi_1^T(x) q_1(t) \qquad (6)$$

$$u_2(x,t) = \phi_2^T(x) q_2(t) \qquad (7)$$

where $\phi_i = [\phi_{i1} \ \phi_{i2} \ \cdots]^T$ and $q_i = [q_{i1} \ q_{i2} \ \cdots]^T$. Generally, any compact set of functions satisfying geometric boundary conditions can be applied as the spatial functions ϕ_i [11]. In this study, the orthonormal modes of a clamped-free beam [12] are utilized.

2.2   Potential Energy

The strain energy of the beam due to the axial inertia force and the transverse blending is expressed as in [3],

$$U = \frac{1}{2} \int_0^l EA \left(\frac{\partial s}{\partial x} \right)^2 dx + \frac{1}{2} \int_0^l EI \left(\frac{\partial^2 u_2}{\partial x^2} \right)^2 dx \qquad (8)$$

where E is the Young's modulus, A is the cross-section area of the beam, l the length of the beam, I the second area moment of inertia of the cross section.

2.3   Kinetic Energy

The kinetic energy of the system consists of those of the hub and the beam, i.e.,

$$T = \frac{1}{2} J_h \dot{\theta}^2 + \frac{1}{2} \int_0^l \rho A \dot{r} \cdot \dot{r} dx \qquad (9)$$

where J_h is the hub rotary inertia, and ρ mass density of the beam.

2.4   Governing Equations

Assuming $x + r \gg u_1$ and neglecting higher-order nonlinear terms, the governing equations of motion are formulated by the extended Hamilton's principle as,

$$\sum_{i=1}^{\mu_2} \left[\int_0^l \rho A(x+r)\phi_{2i}dx \right] \ddot{q}_{2i} + \left[J_h + \int_0^l \rho A(x+r)^2 dx \right] \ddot{\theta} = M(t) \tag{10}$$

$$\sum_{i=1}^{\mu_2} \ddot{q}_{2i} \int_0^l \rho A \phi_{2i}\phi_{2j} dx + \ddot{\theta} \int_0^l \rho A(x+r)\phi_{2j} dx + 2\dot{\theta} \sum_{i=1}^{\mu_1} \int_0^l \rho A \phi_{1i}\phi_{2j} dx \, \dot{q}_{1i} -$$

$$\dot{\theta}^2 \sum_{i=1}^{\mu_2} q_{2i} \left[\int_0^l \rho A \phi_{2i}\phi_{2j} dx + \frac{1}{2} \int_0^l \rho A(x^2 - l^2) \frac{\partial\phi_{2i}}{\partial x}\frac{\partial\phi_{2j}}{\partial x} dx + r \int_0^l \rho A(x -$$

$$l) \frac{\partial\phi_{2i}}{\partial x}\frac{\partial\phi_{2j}}{\partial x} dx \right] + \sum_{i=1}^{\mu_2} q_{2i} \int_0^l EI \frac{\partial^2\phi_{2i}}{\partial x^2}\frac{\partial^2\phi_{2j}}{\partial x^2} dx = 0 \qquad (j = 1,2,...\mu_2) \tag{11}$$

where μ_1 and μ_2 are the number of generalized coordinates for s and u_2, respectively. Obviously, Eqs. (10) and (11) are coupled due to the Coriolis effect. Since the stretching natural frequencies are much greater than the bending ones [13], it is sensible to ignore the coupling terms within the range concerned in the following.

For simplicity, introduce the following dimensionless variables.

$$\delta = \frac{r}{l}, \quad \beta = \frac{J_h}{\rho A l^3}, \quad \xi = \frac{x}{l}, \quad \tau = \sqrt{T_0}t, \quad \gamma(\tau) = \theta(t), \quad \dot{\gamma}(\tau) = \frac{\dot{\theta}(t)}{\sqrt{T_0}},$$

$$p(\tau) = \frac{q_2(t)}{l}, \quad \psi_2(\xi) = \phi_2(x), \quad \eta(\tau) = \frac{l}{EI}M(t), \quad T_0 = \frac{EI}{\rho A l^4}.$$

Then Eqs. (10) and (11) are simplified as

$$\begin{bmatrix} \sigma & C^T \\ C & \mathbf{M} \end{bmatrix} \begin{bmatrix} \ddot{\gamma} \\ \ddot{p} \end{bmatrix} + \left(\begin{bmatrix} 0 & 0 \\ 0 & K^s \end{bmatrix} - \dot{\gamma}^2 \begin{bmatrix} 0 & 0 \\ 0 & \Delta K \end{bmatrix} \right) \begin{bmatrix} \gamma \\ p \end{bmatrix} = \begin{bmatrix} \eta(\tau) \\ 0 \end{bmatrix} \tag{12}$$

where

$$\sigma = \beta + \frac{1}{3}(1 + 3\delta + 3\delta^2)$$

$$C = \int_0^1 (\xi + \delta)\psi_2 d\xi$$

$$\mathbf{M} = \int_0^1 \psi_2\psi_2^T d\xi$$

$$K^s = \int_0^1 \frac{\partial^2\psi_2}{\partial\xi^2}\frac{\partial^2\psi_2^T}{\partial\xi^2} d\xi$$

$$\Delta K = \mathbf{M} + \delta K^{d1} + \frac{1}{2}K^{d2}$$

$$\mathbf{K}^{d1} = \int_0^1 (\xi - 1) \frac{\partial \boldsymbol{\psi}_2}{\partial \xi} \frac{\partial \boldsymbol{\psi}_2^T}{\partial \xi} d\xi$$

$$\mathbf{K}^{d2} = \int_0^1 (\xi^2 - 1) \frac{\partial \boldsymbol{\psi}_2}{\partial \xi} \frac{\partial \boldsymbol{\psi}_2^T}{\partial \xi} d\xi$$

3   Frequency Analysis

In Eq. (12), rigid rotating motion and in-plane bending vibration are highly coupled, and hub radius ratio δ can affect the stiffness matrix obviously. Since this parameter has been studied a lot [14,11,15], rotary inertia of the hub is mainly focused on.

For frequency analysis, set $\eta(\tau)$ to zero. Let ω denote the dimensionless natural frequency, then it obtains,

$$\left[\omega^2 \left(\mathbf{M} - \frac{1}{\sigma} \boldsymbol{C} \cdot \boldsymbol{C}^T \right) - (\mathbf{K}^s - \dot{\gamma}^2 \Delta \mathbf{K}) \right] \boldsymbol{p} = 0 \qquad (13)$$

The curved surfaces of the first two order dimensionless natural frequencies are demonstrated in Fig. 2. It is showed that ω tends to increase as $\dot{\gamma}$ gets larger due to the effect of dynamic stiffening, which has been verified by many researchers. Meanwhile, hub-beam ratio of rotary inertia β has great effect on the varying trends of ω. In Fig. 2, the black solid lines represent the natural frequencies under the clamped boundary condition at the hub end, and the black solid lines with dots represent those of the pinned boundary condition (though they are plotted at $\beta = 5$). At different values of the angular velocity, the natural frequency of each order of the hub-beam system varies between those of the pinned-free and the clamped-free boundary conditions. That is, ω approaches the natural frequency of the rotating beam under the pinned-free boundary condition when β is close to zero, and that of the rotating beam under the clamped-free boundary condition when β is close to infinity. The above phenomenon was tested through a physical experiment using a high speed camera in Ref. [7].

To further reveal and verify the above phenomenon, Fig. 3 is plotted to show the curved lines of ω vs. γ. The black curved lines stand for the 1st order natural frequencies of different values of β, which clearly demonstrate the aforementioned changing rule in terms of ω about β. The same changing rule in terms of ω about β exists for other orders, so the corresponding curved lines are omitted. Besides, the blue curved line is plotted for comparison to represent the 2nd order natural frequency when β approximates to infinity. And the black asterisks and upper triangles represent the 1st and 2nd order dimensionless natural frequencies given in Ref. [3], respectively. Since the clamped-free boundary condition was applied in Ref. [3], its results agree with the curved lines of $\beta \to \infty$.

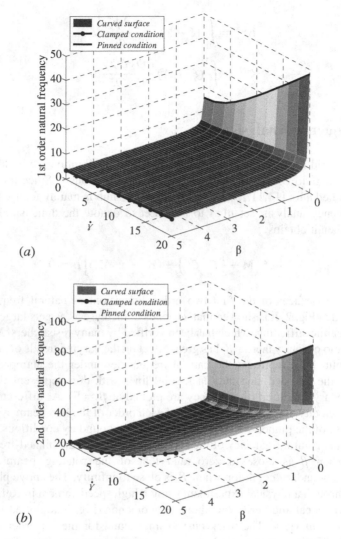

Fig. 2. Curved surfaces of ω in terms of β and $\dot{\gamma}$: (a) 1st order; (b) 2nd order

Fig. 4 shows the varying tendency of ω vs. β of different orders, where some specified points are compared with those in Ref. [8]. The corresponding results agree well. Besides, it can be observed that ω declines very fast when β is close to zero, and very slow after β passes 1. In other words, the greater β is, the less the decreasing rate is. So ω is more sensitive to β when β approaches zero.

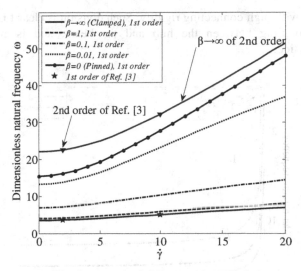

Fig. 3. ω vs. $\dot{\gamma}$ of different values of β

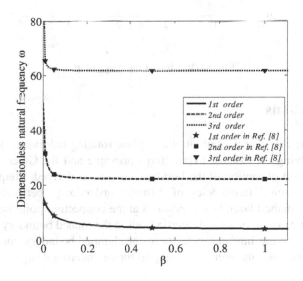

Fig. 4. ω vs. β of different orders

Based on the definition of the clamped and the pinned boundary conditions, it can be regarded that the connecting rigidity between the hub and the beam is infinity in the case of the clamped-free boundary condition, and zero in the case of the pinned-free boundary condition. As a result, the range of β can be divided into three regions, as shown in Fig. 5. Region I is the region with very low connecting rigidity, region II is the highly rigid-flexible coupling region with intermediate connecting rigidity, and region

III is the region with high connecting rigidity. Such division is clear to understand the rigid-flexible coupling between the hub and the beam, and is meaningful for engineering practice.

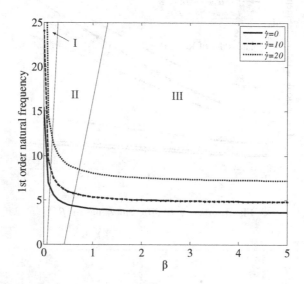

Fig. 5. Division of the range of β

4   Conclusions

In this paper, the effect of the hub on the in-plane rotating hub-beam system is investigated. By applying the extended Hamilton's principle and the Galerkin method, the governing equation of motion of the hub-beam system is derived. Frequency analysis shows that the natural frequencies of different orders vary between those of the clamped and the pinned boundary conditions at the connecting point between the hub and the beam. Zero value of the hub inertia leads to the pinned boundary condition, and the infinity value of the hub inertia leads to the clamped boundary condition. Due to this flexibility, the phenomenon is meaningful for engineering design.

Acknowledgements. This work was supported by the National Natural Science Foundation of China under grant 51375312, for which the authors are grateful.

References

1. Kane, T.R., Ryan, R.R., Banerjee, A.K.: Dynamics of a cantilever beam attached to a moving base. Journal of Guidance, Control, and Dynamics **10**(2), 139–151 (1987)
2. Seo, S., Yoo, H.H.: Dynamic analysis of flexible beams undergoing overall motion employing linear strain measures. Aiaa J. **40**(2), 319–326 (2002)

3. Yoo, H.H., Shin, S.H.: Vibration analysis of rotating cantilever beams. Journal of Sound and Vibration **212**(5), 807–808 (1998)
4. Yoo, H.H., Seo, S., Huh, K.: The effect of a concentrated mass on the modal characteristics of a rotating cantilever beam. Proceedings of the Institution of Mechanical Engineers, Part C: Journal of Mechanical Engineering Science **216**(2), 151–164 (2002). doi:10.1243/0954406021525098
5. Liu, J.Y., Hong, J.Z.: Dynamics of three-dimensional beams undergoing large overall motion. European Journal of Mechanics, A/Solids **23**(6), 1051–1068 (2004). doi:10.1016/j.euromechsol.2004.08.003
6. Liu, J.Y., Hong, J.Z.: Geometric stiffening effect on rigid-flexible coupling dynamics of an elastic beam. Journal of Sound and Vibration **278**(4–5), 1147–1162 (2004). doi:10.1016/j.jsv.2003.10.014
7. Low, K.H., Lau, M.W.S.: Experimental investigation of the boundary condition of slewing beams using a high-speed camera system. Mechanism and Machine Theory **30**(4), 629–643 (1995)
8. Low, K.H.: A note on the effect of hub inertia and payload on the vibration of a flexible slewing link. Journal of Sound and Vibration **204**(5), 823–828 (1997)
9. Bellezza, F., Lanari, L., Ulivi, G.: Exact modeling of the flexible slewing link. In: Proceedings of the 1990 IEEE International Conference on Robotics and Automation, pp. 734-739. Publ. by IEEE, Cincinnati
10. Morris, K.A., Taylor, K.J.: A Variational Calculus Approach to the Modelling of Flexible Manipulators. SIAM Review **38**(2), 294–305 (1996)
11. Zhu, T.L.: The vibrations of pre-twisted rotating Timoshenko beams by the Rayleigh-Ritz method. Computational Mechanics **47**(4), 395–408 (2011). doi:10.1007/s00466-010-0550-9
12. Blevins, R.D. (ed.): Formulas for natural frequency and mode shape. Krieger, Malabar, Fla (1979)
13. Yoo, H.H., Cho, J.E., Chung, J.: Modal analysis and shape optimization of rotating cantilever beams. Journal of Sound and Vibration **290**(1–2), 223–241 (2006). doi:10.1016/j.jsv.2005.03.014
14. Hamdan, M.N., Al-Bedoor, B.O.: Non-linear free vibrations of a rotating flexible arm. Journal of Sound and Vibration **242**(5), 839–853 (2001). doi:10.1006/jsvi.2000.3387
15. Lin, S.C., Hsiao, K.M.: Vibration analysis of a rotating Timoshenko beam. Journal of Sound and Vibration **240**(2), 303–322 (2001). doi:10.1006/jsvi.2000.3234

Research on Attitude Adjustment Control for Large Angle Maneuver of Rigid-Flexible Coupling Spacecraft

Xiao Yan[1], Ye Dong[1(✉)], Yang Zhengxian[2], and Sun Zhaowei[1]

[1] Research Center of Satellite Technology, Harbin Institute of Technology,
Harbin 150001, People's Republic of China
yed@hit.edu.cn
[2] The Second Department of the Second Academy, China Aerospace
Science and Industry Corporation, Beijing 100854, People's Republic of China

Abstract. A rigid - flexible coupling dynamic model is established for full reflection of the coupling between deformation of flexible body and large angle maneuver. Considering uncertainties of the flexible spacecraft a traditional PD controller and an adaptive attitude adjustment controller are designed to solve the attitude adjustment problem of the rigid-flexible coupling spacecraft based on Lyapunov method. These two controllers are compared in simulation analyses; results show that compared with the traditional PD controller, the adaptive controller can effectively suppress the influence of disturbance torques, with higher precision and better robustness, and be able to fulfill the important demand of attitude adjustment for the rigid-flexible coupling spacecraft.

Keywords: Rigid-flexible coupling · Attitude control · Large angle maneuver · Adaptive control

1 Introduction

Mechanic systems such as spacecraft and robot consist of more and more flexible structures as they are developed to a lighter, faster and more precise direction. Most previous studies are based on the zero order approximation of the rigid-flexible coupling model while ignoring the coupling term of the deformation of flexible body and large angle maneuver [1]. So the zero order approximation model cannot be able to predict dynamic behaviors of the rigid-flexible coupling spacecraft accurately. The reliability of this approximation is quite limited even if results of the simulations are not divergent.

On the other hand, the flexible spacecraft is an under-actuated system [2], which means that the design and analysis of it are very hard and complicated. Because when we use the torque from the central rigid body to accomplish a large angle maneuver, there will be serious vibration appeared on the flexible cantilever beam at the same time. Even though some actuators like piezoelectric ceramics are added to the flexible body for active vibration suppression, challenges of the system for its nonlinearity, strong coupling and infinite dimensional distributed parameters are still there [3-5].

© Springer International Publishing Switzerland 2015
H. Liu et al. (Eds.): ICIRA 2015, Part I, LNAI 9244, pp. 24–33, 2015.
DOI: 10.1007/978-3-319-22879-2_3

In recent years, some scholars have introduced the first order approximation model to the rigid-flexible coupling spacecraft which is proved to be more accurate than the zero order approximation models by simulations. Meanwhile, control problems for the first order approximation model of the rigid-flexible spacecraft draw much attention: Ref. [6] gives a feedback linearization control method, but accurate model parameters and state variables are needed in advance. For parameter uncertainties, a variable structure control algorithm is proposed in [7], while the parameter uncertainty region must be known when a high order filter is incorporated. A high gain feedback observer which can compensate the nonlinear dynamic model is discussed in [8], but its high sensitivity to the noise of measurements makes it too complicated to use.

Adaptive control is mainly used in the condition when parameters of the controlled structure are mainly uncertain, because adaptive control has weak dependence of the structure of the controlled system, certain robustness to the variable parameters and fault-tolerant ability [9-11]. With all the abilities mentioned above, new approaches based on the adaptive control theorem are studied to solve control problems of uncertainty of parameters, nonlinearity and unknown systems. Recently, some of the research results are already used in manipulator control and spacecraft control. An adaptive controller is presented in [12] to solve the control problem of uniaxial maneuvering of flexible spacecraft, in which the influence of the orbital movement to the attitude is considered. An adaptive output feedback controller is designed in [13] for the control of a uniaxial flexible spacecraft, where parameters of the model are mainly unknown.

This paper investigates attitude adjustment control problem of spacecraft based on rigid-flexible dynamic model. An adaptive adjustment attitude controller is proposed based on Lyapunov method, and then a traditional PD controller is designed as a contrast controller. Simulations are set to compare the two controllers and demonstrate the superiority and effectiveness of the adaptive attitude adjustment controller to the PD controller.

2   Problem Definition

The model of rigid-flexible coupling spacecraft, shown in Fig.1, consists of a central rigid body of which the radius is a, a uniform cantilever as the flexible appendage whose length is l, and a tip mass weights m_t.

The central rigid body rotates around the rigid point O_N, while the flexible cantilever beam is fixed to the central rigid body and the tip mass is fixed to the end of the flexible cantilever. The moment of inertia of the central rigid body is J_h, and the control torque and the disturbance torque acted on the central rigid body is T_c and T_d respectively. E, A and P is the elastic modulus, sectional area and volume density of the flexible cantilever beam respectively. $\underline{e}^n = \begin{pmatrix} N_1 & N_2 \end{pmatrix}^T$ in Fig.1 is the inertial frame, and $\underline{e}^b = \begin{pmatrix} B_1 & B_2 \end{pmatrix}^T$ is the floating coordinate system fixed to the medial axis of the flexible cantilever beam.

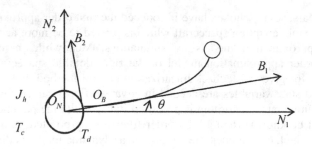

Fig. 1. Model of flexible spacecraft

Given that the deformation and the strain are both small, with small axial tension and some high order nonlinear terms neglected, the dynamic model of the rigid-flexible coupling spacecraft is given as follows

$$\left(J_{\mathrm{mb}} - p^{\mathrm{T}} G p \right) \ddot{\theta} - U M_{\mathrm{pp}}^{-1} C_{\mathrm{f}} \dot{p} - U M_{\mathrm{pp}}^{-1} \left(K_{\mathrm{f}} + \dot{\theta}^2 G \right) p - 2\dot{\theta} p^{\mathrm{T}} G \dot{p} = T_{\mathrm{c}} + T_{\mathrm{d}} \quad (1)$$

$$\ddot{\eta} + M_{\mathrm{pp}}^{-1} C_{\mathrm{f}} \dot{p} + M_{\mathrm{pp}}^{-1} \left(K_{\mathrm{f}} + \dot{\theta}^2 G \right) p = 0 \quad (2)$$

In which:

$$J_{\mathrm{mb}} = J_{\mathrm{h}} + J_{\mathrm{f}} + J_{\mathrm{t}} - U M_{\mathrm{pp}}^{-1} U^{\mathrm{T}} > 0 \quad (3)$$

$$G^{\mathrm{T}} = G = D - M_{\mathrm{pp}} > 0 \quad (4)$$

$$\eta = \dot{p} + M_{\mathrm{pp}}^{-1} U^{\mathrm{T}} \dot{\theta} \quad (5)$$

Where J_{f}, J_{t} is the moment of inertia of the cantilever beam and the tip mass, respectively; U is the matrix of the rigid-flexible coupling coefficients; the symmetric positive definite matrix $M_{\mathrm{pp}} = M_{\mathrm{pp}}^{\mathrm{T}} > 0$, $K_{\mathrm{f}} = K_{\mathrm{f}}^{\mathrm{T}} > 0$ are the mass matrix and the stiffness matrix of the flexible cantilever beam, respectively; the symmetric positive definite matrix $D = D^{\mathrm{T}} > 0$ is the dynamic stiffness matrix of the rotating flexible cantilever beam; C_{f} is the structural damping of the flexible cantilever beam.

The attitude adjustment controller is designed in the condition that the desired attitude of the flexible spacecraft is constant. Therefor the system error dynamic equations are as below:

$$\left(J_{\mathrm{mb}} - p^{\mathrm{T}} G p \right) \ddot{\theta}_{\mathrm{e}} - U M_{\mathrm{pp}}^{-1} C_{\mathrm{f}} \dot{p} - U M_{\mathrm{pp}}^{-1} \left(K_{\mathrm{f}} + \dot{\theta}_{\mathrm{e}}^2 G \right) p - 2\dot{\theta}_{\mathrm{e}} p^{\mathrm{T}} G \dot{p} = T_{\mathrm{c}} + T_{\mathrm{d}} \quad (6)$$

$$\ddot{\eta} + M_{\mathrm{pp}}^{-1} C_{\mathrm{f}} \dot{p} + M_{\mathrm{pp}}^{-1} \left(K_{\mathrm{f}} + \dot{\theta}_{\mathrm{e}}^2 G \right) p = 0 \quad (7)$$

$$\eta = \dot{p} + M_{pp}^{-1} U^T \dot{\theta}_e \tag{8}$$

In which $\theta_e = \theta - \theta_d$ is the attitude error angle, and θ_d is the desired constant attitude angle with $\dot{\theta}_d = \ddot{\theta}_d = 0$. p is the displacement coordinate matrix of the finite element nodes.

Aiming at the desired constant attitude angle θ_d, the control objective is to design an attitude adjustment controller T_c such that the closed-loop systems (6) (7) (8) are asymptotically stable, namely $\lim\limits_{t\to\infty} \theta_e = 0, \lim\limits_{t\to\infty} \dot{\theta}_e = 0, \lim\limits_{t\to\infty} p = 0$ and $\lim\limits_{t\to\infty} \dot{p} = 0$.

3   Controller Design

In this section, two different controllers based on the Lyapunov function are developed to solve the control problem stated in Section 2. The traditional PD controller will be designed fist, and then extended to a more complicated controller based on adaptive control theorem.

Theorem 1. For the systems (6) (7) (8) of the flexible spacecraft in the absence of disturbance torque, there exists the controller (9)

$$T_c = -K_p \theta_e - K_d \dot{\theta}_e \tag{9}$$

Such that the closed-loop system is asymptotically stable, in which $K_p > 0$, $K_d > 0$.

Proof. The candidate lyapunpv function is defined as below

$$V = \frac{1}{2}\left(J_{mb} - p^T G p\right)\dot{\theta}_e^2 + \frac{1}{2}\eta^T M_{pp}\eta + \frac{1}{2}p^T K_f p + \frac{1}{2}K_p \theta_e^2 \tag{10}$$

The inequality $J_{mb} - p^T G p > 0$ holds when the node variables of the beam are bounded; and $V \geq 0$ when the deformation of the cantilever beam is small.

Computing the first-order derivative of V yields

$$\dot{V} = \dot{\theta}_e\left(J_{mb} - p^T G p\right)\ddot{\theta}_e - p^T G\dot{p}\dot{\theta}_e^2 + \eta^T M_{pp}\dot{\eta} + p^T K_f \dot{p} + K_p \theta_e \dot{\theta}_e \tag{11}$$

Substituting (6) (7) (8) yields

$$\dot{V} = \dot{\theta}_e \left(UM_{pp}^{-1} \left(C_f \dot{p} + K_f p + \dot{\theta}_e^2 G \right) + 2\dot{\theta}_e p^T G \dot{p} + T_c \right) - p^T G \dot{p} \dot{\theta}_e^2$$
$$+ \left(\dot{p} + M_{pp}^{-1} U^T \dot{\theta}_e \right)^T M_{pp} \left(-M_{pp}^{-1} C_f \dot{p} - M_{pp}^{-1} \left(K_f + \dot{\theta}_e^2 G \right) p \right) \qquad (12)$$
$$+ p^T K_f \dot{p} + K_p \theta_e \dot{\theta}_e$$

Reorganizing (12) yields

$$\dot{V} = \dot{\theta}_e UM_{pp}^{-1} \left(C_f \dot{p} + K_f p + \dot{\theta}_e^2 Gp \right) + 2\dot{\theta}_e^2 p^T Gp + \dot{\theta}_e T_c$$
$$- \dot{\theta}_e UM_{pp}^{-1} \left(C_f \dot{p} + K_f p + \dot{\theta}_e^2 Gp \right) + \dot{p}^T \left(-C_f \dot{p} - K_f p - \dot{\theta}_e^2 Gp \right) \qquad (13)$$
$$- \dot{\theta}_e^2 \dot{p}^T Gp + \dot{p}^T K_f p + K_p \theta_e \dot{\theta}_e$$
$$= \dot{\theta}_e \left(T_c + K_p \theta_e \right) - \dot{p}^T C_f \dot{p}$$

Substituting the controller (9) yields

$$\dot{V} = -K_d \dot{\theta}_e^2 - \dot{p}^T C_f \dot{p} \le 0 \qquad (14)$$

According to LaSalle's invariance principle, the system will be convergent to the largest invariant set [14]. $\dot{\theta}_e = 0$, $\dot{p} = 0$ as $\dot{V} = 0$, so according to the dynamic equations (6) (7) (8) and (9) ,we can conclude that if $\dot{\theta}_e \equiv 0, \dot{p} \equiv 0$,then $\theta_e = 0$, $\dot{\theta}_e = 0$,hence the sets when $\theta_e = 0, \dot{\theta}_e = 0, p = 0, \dot{p} = 0$ are the largest invariant sets of the system , so that the system is asymptotically stable, namely

$$\lim_{t \to \infty} \theta_e = 0 \quad \lim_{t \to \infty} \dot{\theta}_e = 0 \quad \lim_{t \to \infty} p = 0 \quad \lim_{t \to \infty} \dot{p} = 0 \qquad (15)$$

The proof is completed.

The structure of the traditional PD controller is simple and easy to implement. In addition it has certain robustness to the model uncertainty[15,16]. When the disturbance is neglected, the convergence of the theorem.1 can be proved. However, when the disturbance is considered, the disturbance suppression is only valid in a local area theoretically [17,18].

Theorem 2. For the systems of flexible spacecraft (6) (7) (8) in the presence of the disturbance torque T_d , suppose that the disturbance torque satisfies $\left| T_d \right| < \alpha_{dmax}$ and $\dot{\theta}_e \to 0 \Rightarrow T_d \to 0$. Define the estimating error $\tilde{\alpha}_{dmax} = \alpha_{dmax} - \hat{\alpha}_{dmax}$ as the estimation of the maximum disturbance $\hat{\alpha}_{dmax}$. Then there exists the controller (16)

$$T_c = -K_p \theta_e - K_d \dot{\theta}_e - \hat{\alpha}_{dmax} \operatorname{sgn} \left(\dot{\theta}_e \right) \qquad (16)$$

$$\dot{\alpha}_{\text{dmax}} = \alpha_{\text{d}} \left| \dot{\theta}_{\text{e}} \right| \tag{17}$$

Such that the closed-loop system is asymptotically stable, where $K_{\text{p}} > 0$, $K_{\text{d}} > 0$ and $\alpha_{\text{d}} > 0$, and $\text{sgn}(\cdot)$ is a sigh function.

Proof: Selecting the candidate Lyapunov function similar to the Theorem 1

$$V = \frac{1}{2}\left(J_{\text{mb}} - \boldsymbol{p}^{\text{T}} \boldsymbol{G} \boldsymbol{p} \right) \dot{\theta}_{\text{e}}^2 + \frac{1}{2}\boldsymbol{\eta}^{\text{T}} \boldsymbol{M}_{\text{pp}} \boldsymbol{\eta} + \frac{1}{2}\boldsymbol{p}^{\text{T}} \boldsymbol{K}_{\text{f}} \boldsymbol{p} + \frac{1}{2} K_{\text{p}} \theta_{\text{e}}^2 + \frac{1}{2\alpha_{\text{d}}} \tilde{\alpha}_{\text{dmax}}^2 \geq 0 \tag{18}$$

Computing the first- order derivative of the candidate Lyapunov function yields

$$\dot{V} = \dot{\theta}_{\text{e}}\left(T_{\text{c}} + T_{\text{d}} + K_{\text{p}}\theta_{\text{e}} \right) - \dot{\boldsymbol{p}}^{\text{T}} \boldsymbol{C}_{\text{f}} \dot{\boldsymbol{p}} + \tilde{\alpha}_{\text{dmax}} \dot{\hat{\alpha}}_{\text{dmax}} / \alpha_{\text{d}} \tag{19}$$

Substituting the controllers (16) (17) yields

$$\begin{aligned}
\dot{V} &= \dot{\theta}_{\text{e}}\left(-K_{\text{d}}\dot{\theta}_{\text{e}} - \hat{\alpha}_{\text{dmax}} \, \text{sgn}\left(\dot{\theta}_{\text{e}}\right) + T_{\text{d}} \right) - \dot{\boldsymbol{p}}^{\text{T}} \boldsymbol{C}_{\text{f}} \dot{\boldsymbol{p}} - \tilde{\alpha}_{\text{dmax}} \dot{\hat{\alpha}}_{\text{dmax}} / \alpha_{\text{d}} \\
&= -K_{\text{d}}\dot{\theta}_{\text{e}}^2 - \dot{\boldsymbol{p}}^{\text{T}} \boldsymbol{C}_{\text{f}} \dot{\boldsymbol{p}} - \hat{\alpha}_{\text{dmax}}\left|\dot{\theta}_{\text{e}}\right| + T_{\text{d}}\dot{\theta}_{\text{e}} - \left(\alpha_{\text{dmax}} - \hat{\alpha}_{\text{dmax}} \right)\left|\dot{\theta}_{\text{e}}\right| \\
&= -K_{\text{d}}\dot{\theta}_{\text{e}}^2 - \dot{\boldsymbol{p}}^{\text{T}} \boldsymbol{C}_{\text{f}} \dot{\boldsymbol{p}} + T_{\text{d}}\dot{\theta}_{\text{e}} - \alpha_{\text{dmax}}\left|\dot{\theta}_{\text{e}}\right| \\
&\leq -K_{\text{d}}\dot{\theta}_{\text{e}}^2 - \dot{\boldsymbol{p}}^{\text{T}} \boldsymbol{C}_{\text{f}} \dot{\boldsymbol{p}} \leq 0
\end{aligned} \tag{20}$$

Where $\dot{V}_2$ is negative semi-definite, then V_2 is bounded, such that $\dot{\theta}_{\text{e}}$ and $\dot{\boldsymbol{p}}$ are bounded, that is $\left(\dot{\theta}_{\text{e}}, \dot{\boldsymbol{p}} \right) \in \mathcal{L}_{\infty}$. According to the dynamic equations (6) (7) (8), $\ddot{\theta}_{\text{e}}$ and $\ddot{\boldsymbol{p}}$ are both bounded with $\left(\ddot{\theta}_{\text{e}}, \ddot{\boldsymbol{p}} \right) \in \mathcal{L}_{\infty}$. Compute the integrations of both sides of the inequality (19) from 0 to ∞, then $\dot{\theta}_{\text{e}} \in \mathcal{L}_1$, $\dot{\boldsymbol{p}} \in \mathcal{L}_1^n$, such that $\lim\limits_{t \to \infty} \dot{\theta}_{\text{e}} = 0$ and $\lim\limits_{t \to \infty} \dot{\boldsymbol{p}} = 0$. When $\dot{\theta}_{\text{e}} \equiv 0$ and $\dot{\boldsymbol{p}} \equiv \boldsymbol{0}$, there is $T_{\text{d}} = 0$, therefore $\lim\limits_{t \to \infty} \theta_{\text{e}} = 0$ and $\lim\limits_{t \to \infty} \boldsymbol{p} = 0$. The proof is completed.

Because of the sign function in the controller (16), there is chattering phenomenon, which is difficult to suppress in the application [19,20]. In order to achieve the vibration suppression, the controller can be adjusted as below

$$T_{\text{c}} = -K_{\text{p}}\theta_{\text{e}} - K_{\text{d}}\dot{\theta}_{\text{e}} - \hat{\alpha}_{\text{dmax}} \overline{f}\left(\dot{\theta}_{\text{e}} \right) \tag{21}$$

$$\dot{\hat{\alpha}}_{\text{dmax}} = \alpha_{\text{d}} \left| \dot{\theta}_{\text{e}} \right| \tag{22}$$

$$\overline{f}\left(\dot{\theta}_e\right) = \tanh\left(\frac{\dot{\theta}_e}{\varepsilon}\right) \tag{23}$$

In which ε is a small positive constant and $\tanh(\cdot)$ is a hyperbolic tangent function which is defined as $\tanh(u) = \dfrac{e^u - e^{-u}}{e^u + e^{-u}}$.

4   Simulation and Results

In this section, different simulation cases are presented to validate and compare the traditional PD controller and the adaptive attitude adjustment controller introduced above.

The initial condition is set as $\theta(0) = 0\,\text{rad}$, $\dot{\theta}(0) = 0\,\text{rad}/s$, $p(0) = 0\text{m}$, $\dot{p}(0) = 0\,\text{m/s}$. The desired attitude of the two controllers is $\theta_d = 0.5\,\text{rad}$. Suppose that the disturbance torque is $T_d = 0.4\dot{\theta}\sin(0.1t)\text{N}\cdot\text{m}$. The parameters of the controller (9) are $K_p = 30$, $K_d = 70$. The parameters of the controller (16) are $K_p = 30$, $K_d = 70$, $\alpha_d = 20$ and $\varepsilon = 0.1$, with the initial estimate $\hat{\alpha}_{\text{dmax}}(0) = 0.01$. In addition, the magnitude of the control torque is bounded as $|T_c| \leq 5\text{N}\cdot\text{m}$. The simulation results are shown in Figs. 2-5, in which Case 1 is the result of the PD controller and Case 2 is the result of the adaptive attitude adjustment controller.

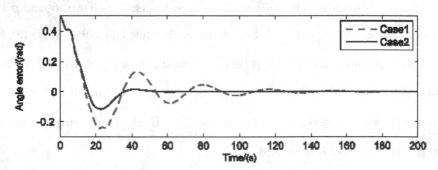

Fig. 2. Response of attitude angle error

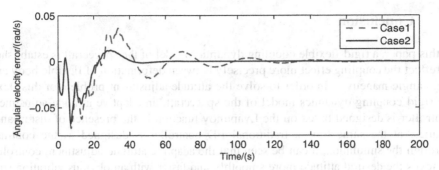

Fig. 3. Response of attitude velocity error

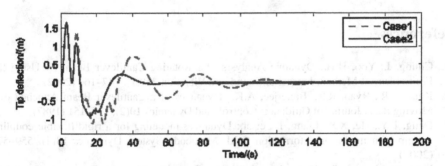

Fig. 4. Tip deflection of the beam

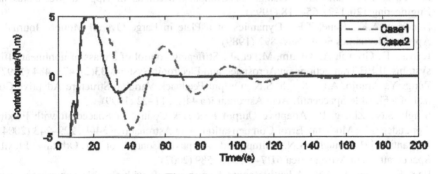

Fig. 5. Control torque

From Figs. 2-5, it can be seen that comparing to the PD controller(9) , the adaptive attitude adjustment controller (16) can converge to the desired attitude in a better way, and it has a more obvious vibration suppression effect, in the presence of limited control torque, environment disturbance and initial attitude error. This is because the vibration of the flexible cantilever beam is based on the state of the system, so the adaptive controller can suppress this perturbation faster and achieve a stable state more smoothly.

5   Conclusion

In this paper, a rigid-flexible coupling dynamics model of the spacecraft is established to reflect the coupling effect more precisely between deformation of flexible body and large angle maneuver. In order to solve the attitude adjustment problem of this flexible-rigid coupling dynamics model of the spacecraft, an adaptive attitude adjustment controller is designed based on the Lyapunov function in the presence of disturbance torques, at the same time a traditional PD controller is designed as the contrast. Through the simulation, it can be seen that the adaptive attitude adjustment controller achieves the desired attitude more smoothly and faster with an obvious vibration suppression effect compared to the PD controller.

References

1. Chung, J., Yoo, H.H.: Dynamic Analysis of a Rotating Cantilever Beam by Using the Finite Element Method. Journal of Sound and Vibration 249(1), 147–164 (2002)
2. Kane, T.R., Ryan, R.R., Banerjee, A.K.: Dynamics of a Cantilever Beam Attached to a Moving Base. Journal of Guidance, Control, and Dynamics 10(2), 139–151 (1987)
3. Deng, F.Y., He, X.S., Liang, L., et al.: Dynamics modeling for a rigid-flexible coupling system with nonlinear deformation field. Multibody System Dynamics 18(4), 559–578 (2007)
4. Vu-Quoc, L., Li, S.: Dynamics of Sliding Geometrically-Exact Beams: Large Angle Maneuver and Parametric Resonance. Computer Methods in Applied Mechanics and Engineering 120(1–2), 65–118 (1995)
5. Banerjee, A.K., Kane, T.R.: Dynamics of a Plate in Large Overall Motion. Journal of Applied Mechanics 56(4), 887–892 (1989)
6. Karray, F., Grewal, A., Glaum, M., et al.: Stiffening control of a class of nonlinear affine systems. IEEE Transactions on Aerospace and Electronic Systems 33(2), 473–484 (1997)
7. Zeng, Y., Araujo, A.D., Singh, S.N.: Output Feedback Variable Structure Adaptive Control of a Flexible Spacecraft. Acta Astronautica 44(1), 11–22 (1999)
8. Singh, S.N., Zhang, R.: Adaptive Output Feedback Control of Spacecraft with Flexible Appendages by Modeling Error Compensation. Acta Astronautica 54(4), 229–243 (2004)
9. Maganti, G.B., Singh, S.N.: Simplified Adaptive Control of an Orbiting Flexible Spacecraft. Acta Astronautica 61(7–8), 575–589 (2007)
10. Lee, K.W., Singh, S.N.: Adaptive control of spacecraft with flexible appendages using nussbaum gain. In: AIAA Guidance, Navigation and Control Conference and Exhibit, pp. 1288–1299. AIAA Press, South Carolina (2007)
11. Khorrami, F., Jain, S., Tzes, A.: Experimental Results on Adaptive Nonlinear Control and Input Preshaping for Multi-link Flexible Manipulators. Automatica 31(1), 83–97 (1995)
12. Bošković, J.D., Li, S.M., Mehra, R.K.: Robust Adaptive Variable Structure Control of Spacecraft under Control Input Saturation. Journal of Guidance, Control, and Dynamics 24(1), 14–22 (2001)
13. Hu, Q., Shi, P., Gao, H.: Adaptive Variable Structure and Commanding Shaped Vibration Control of Flexible Spacecraft. Journal of Guidance, Control, and Dynamics 30(3), 804–815 (2007)

14. Cheng, L., Hou, Z.G., Tan, M.: Adaptive Neural Network Tracking Control for Manipulators with Uncertain Kinematics, Dynamics and Actuator Model. Automatica **45**(10), 2312–2318 (2009)
15. Hongli, D., Zidong, W., Steven, X.D., Huijun, G.: Finite-horizon reliable control with randomly occurring uncertainties and nonlinearities subject to output quantization. Automatica **52**, 355–362 (2015)
16. Hongli, D., Zidong, W., Steven, X.D., Huijun, G.: Finite-horizon estimation of randomly occurring faults for a class of nonlinear time-varying systems. Automatica **50**(12), 3182–3189 (2014)
17. Leeghim, H., Choi, Y., Bang, H.: Adaptive Attitude Control of Spacecraft Using Neural Networks. Acta Astronautica **64**(7–8), 778–786 (2009)
18. Dong, C.Y., Xu, L.J., Chen, Y., et al.: Networked Flexible Spacecraft Attitude Maneuver Based on Adaptive Fuzzy Sliding Mode Control. Acta Astronautica **65**(11–12), 1561–1570 (2009)
19. Calise, A., Yang, B.J., Craig, J.I.: Augmenting Adaptive Approach to Control of Flexible Systems. Journal of Guidance, Control, and Dynamics **27**(3), 387–396 (2004)
20. Khorrami, F., Jain, S., Tzes, A.: Experimental Results on Adaptive Nonlinear Control and Input Preshaping for Multi-link Flexible Manipulators. Automatica **31**(1), 83–97 (1995)

Artificial Bee Colony Algorithm Based on K-means Clustering for Droplet Property Optimization

Liling Sun[1,2], Jingtao Hu[1], Maowei He[1,2], and Hanning Chen[3(✉)]

[1] Department of Information Service and Intelligent Control,
Shenyang Institute of Automation, Chinese Academy of Sciences,
Shenyang 110016, China
{Sunliling,Hujingtao,Hemaowei}@sia.cn
[2] University of Chinese Academy of Sciences, Beijing 100040, China
[3] School of Computer Science and Software, Tianjin Polytechnic University,
Tianjin 300387, China
perfect_chn@hotmail.com

Abstract. The major challenge in printable electronics fabrication is to effectively and accurately control a drop-on-demand (DoD) inkjet printhead for high printing quality. In this paper, a prediction model based on Lumped Element Modeling (LEM) is proposed to search the parameters of driving waveform for obtaining the desired droplet properties. Although the evolution algorithms are helpful to solve this problem, the classical evolution algorithms may get trapped into local optimal due to the inefficiency of local search. To overcome it, we present an improved artificial bee colony algorithm based on K-means clustering (KCABC), which enhances the population diversity by dynamically clustering and increases the convergence rates by the modification of information communication in the employed bees' phase. Combined with KCABC, the prediction model is applied to optimize the droplet volume and velocity of nano-silver ink for high printing quality. Experimental results demonstrate the proposed prediction model with KCABC plays a good performance in terms of efficiency and accuracy of searching the appropriate combination of waveform parameters for printable electronics fabrication.

Keywords: Piezoelectric inkjet system · Lumped element modeling · Artificial bee colony algorithm · Swarm intelligent algorithm

1 Introduction

In recent years, Drop-on-Demand (DoD) inkjet printing has received considerable attention in mechanical engineering, life sciences, and electronics industry [1]. This revolutionary technology can offer a high-resolution, high-speed, solution-thrift and high-compatibility way of depositing picoliter droplets with diverse physical and chemical properties on various printing surfaces without contact. Nowadays the developments of DoD printing are moving towards higher productivity and quality,

© Springer International Publishing Switzerland 2015
H. Liu et al. (Eds.): ICIRA 2015, Part I, LNAI 9244, pp. 34–44, 2015.
DOI: 10.1007/978-3-319-22879-2_4

requiring adjustable small droplet sizes fired at high jetting frequencies. Generally, the print quality delivered by an inkjet printhead depends on the properties of the jetted droplet, i.e., the droplet velocity, the jetting direction, and the droplet volume. However, in order to meet the challenging requirements of printable electronics fabrication, the conductive ink droplet properties have to be tightly controlled for higher inkjet performance [2].

To implement this issue, the theory field and engineering application field follow two different paths for high printing quality. For theory researchers, the analytical or numerical techniques are utilized to model the DoD piezoelectric inkjet printhead. Numerical modeling of droplet formation process has been focus on the liquid filament evolution [3]. However, numerical models are rather complex and therefore computationally expensive. In the last decade, analytical modeling approach is gradually employed to describe the ink channel dynamics, although their accuracy is lower than numerical models [4]. Based on several assumptions and simplifications, analytical models provide a simple and timesaving way to model-based control droplet generator with a sufficient accuracy. In this category, the lumped element modeling (LEM) approach introduces an equivalent electric circuit to describe the dynamics of the ink channel [5]. The LEM results explain the driving force difference between different operating conditions. Unlike the theory researchers' enthusiasm towards the inkjet mechanism, practice engineering applications need to find an effective means to solve the problem of residual vibrations. A common method is to search an appropriate combination of the parameters of driving waveform for the used ink with specific physical properties. Generally, these combinations of the parameters are obtained by exhaustive experiments and a wise guess. However, the search process is extremely unbearable.

In order to efficiently and accurately search the appropriate waveform parameters with the model-based approach, this work constructs a prediction model combining the lumped element modeling method with the swarm-intelligence optimization technique. In order to overcome the drawbacks of the classical evolution algorithms such as trapping into local optimal, we present an improved artificial bee colony based on K-means clustering (KCABC), which adopts dynamically clustering technology to avoid prematurity and modifies the information communication in the employed bees' phase to enhance the local search ability. Then, this work applies KCABC for automatically searching the appropriate waveform parameters. The experimental results reveal that the proposed prediction model combining the DLEM with the KCABC algorithm can efficiently and accurately predict the optimal combinations of high-frequency driving waveform parameters for high printing quality.

The rest of this paper is organized as follows. An introduction of the prediction system for droplet property based on lumped element modeling and swarm intelligent algorithm is given in Section 2. The proposed KCABC algorithm is detailed in Section 3. Some prediction results are demonstrated in Section 4. Finally, concluding remarks are collected in Section 5.

2   Droplet Property Optimization Problem

2.1   Original Lumped Element Model

A classical LEM is given to simulate droplet formation process of PZT printhead by *Gallas*, as shown in Fig. 1. The equivalent circuit model is constructed by the energy storage elements and ideal dissipative terminal. In an electro-acoustic system, pressure and voltage are independent variables, while current and volumetric flow rate are dependent variables. Model structure shows that the energy converts from electrical energy to mechanical energy, then to fluidic/acoustic energy and finally to kinetic energy. The droplet generator structure can be characterized by equivalent acoustic mass (representing stored kinetic energy) and acoustic compliance (representing stored potential energy), as represented in Fig. 1 (a) and (b). Furthermore, the piezoceramic model is constructed based on the electro-fluidic/acoustic theory [6]. The neck model is built on the velocity profile function [7]. The nozzle model is constructed according to the end correction for an open tube theory [8]. In Fig. 1, excitation voltage V_{ac} is applied to piezoelectric ceramic to create mechanical deformation. Coupling coefficient N represents a conversion from mechanical domain to acoustic domain. C_{eb} is the blocked electrical capacitance of the piezoelectric material. In the acoustic domain, C_{aD} and C_{aC} represent the acoustic compliance of piezo-ceramic and channel. R_{aD}, R_{aN}, and R_{aO} are the acoustic resistance due to structural damping, neck tapering, and fluid flowing out of nozzle, respectively. M_{aD}, M_{aN}, and M_{aRad} represent the acoustic mass of piezo-ceramic, neck and nozzle in proper order, respectively. The calculation formulas of LEM model parameters are provided in Table. 1.

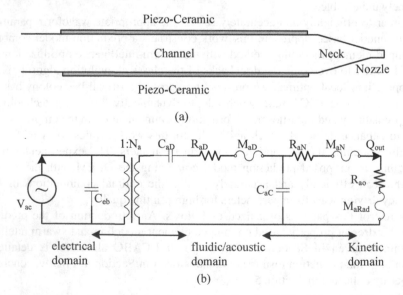

Fig. 1. Schematic overview of lumped-element modeling for PZT printhead. (a) Droplet generator structure (b) Equivalent circuit model

Table 1. Lumped elements modeling parameter estimation

LEM		Description
Neck	R_{aN}	$R_{aN} = (2 + \dfrac{a_0}{r})\dfrac{\sqrt{2\mu\rho_0\omega}}{\pi r^2}$ where u is viscosity, w is wave frequency, a_0 / r is gradients ratio
	M_{aN}	$M_{aN} = \dfrac{\rho_0(t + \Delta t)}{\pi r^2}$ where $\Delta t = 0.85r(1 - 0.7\dfrac{r}{a_0}) + 0.85r$
Cavity	C_{aC}	$C_{aC} = V_0 / \rho_0 c_0^2$ where V_0 is volume of the cavity,
Piezoelectric	C_{aD}	$C_{aD} = \dfrac{\pi r^6(1 - v^2)}{16Eh^3}$ where E is the elastic modulus, v is Poisson's ratio, h is the thickness
	M_{aD}	$M_{aD} = 4\rho_0 L / 3\pi a_0^2$
	R_{aD}	$R_{aD} = 2\xi\sqrt{(M_{aD} + M_{aRad})/C_{aD}}$ where ξ is the experimentally determined damping factor [37]
Nozzle	R_{aO}	$R_{aO} = \dfrac{\rho_0 c_0}{\pi r^2}[1 - \dfrac{2J_1(2kr)}{2kr}]$ where J_1 is the Bessel function of the first kind and the $k = w / c_0$
	M_{aRad}	$M_{aRad} = 8\rho_0 / 3\pi^2 a_0$
	C_{eB}	$C_{eD} = C_{eF}(1 - \kappa^2)$ where κ^2 is electroacoustic coupling factor

2.2 Driving Waveform

As the excitation and the control subject of the system, the driving waveform including a pair of negative and positive trapezoid waveforms can effectively restrain residual vibrations for high-frequency printing, as shown in Fig. 2. When the rising edge of the positive trapezoid waveform is applied to droplet generator, negative pressure waves are generated and then propagate inside the channel. The waveform design issue is focused on determining the optimal value for dwell time. The theoretical optimal value is calculated as $T_{dwell} = L / C$ (where L is the length of channel and C is the sonic speed of the used ink) and the negative pressure can amplify to maximum. To damp the residual oscillations after jetting a droplet, an additional negative trapezoid waveform is applied after the positive one. The falling edge of the second negative waveform generates larger pressure in channel to accelerate the speed of droplet than conventional single trapezoidal waveform. The rising edge, which sets the voltage back to 0, works to restrain the meniscus vibration [9].

2.3   Prediction Model with LEM and Intelligent Optimization Algorithm

For industry applications in printable electronics fabrication, the printhead must work at a certain status to meet many restrictive conditions. Choosing the appropriate combinations of the driving waveform parameters for the used conductive inks becomes the key problem. However, an exhausting manual search process depending on a wise guess inevitably wastes a lot of time because many adjustable parameters exist in the driving waveform. Therefore, the computer-aided methods are urgent needed for searching these appropriate combinations efficiently and robustly.

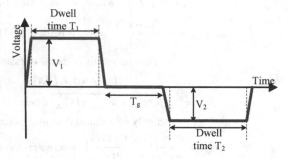

Fig. 2. Applied driving waveform of printhead

Here, we combine the proposed lumped element model with the swarm-intelligent optimization technology to search the appropriate combinations of the driving waveform parameters. A schematic diagram of the approach is illustrated in Fig. 3.

The driving waveform is parameterized in a vector with several dimensions, which is regarded as the input of lumped element model. LEM can predict the information of the droplet status, such as droplet volume and velocity. Through optimizing procedure with intelligent optimization algorithm, the optimal pulse parameters are obtained by minimizing the error between the desired droplet volume/velocity and simulated droplet volume/velocity. In such optimization process, the restraint relationships among the parameters of the driving waveform and the fitness function for evaluating droplet status must be established.

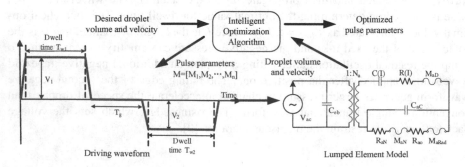

Fig. 3. A schematic diagram of predication system

3 Artificial Bee Colony Algorithm Based on K-means Clustering

This section presents the detailed description of the proposed artificial bee colony algorithm based on K-means clustering (KCABC). The proposed algorithm makes improvements on two aspects: 1) modifying the method of information communication in employed bees' phase; 2) partitioning the population using k-means clustering. The number of the clusters will be chosen from the predefined set G. After each specific iterations, when the defined criteria reaches, parts of individuals in clusters will be removed and an equal number of new individuals are re-generated randomly. Then the number of clusters will change into a new value of the set G and the population will be re-clustered basing on k-means clustering.

3.1 Modified Employed Bees' Phase of ABC

To increase the exploitation of ABC algorithm and fasten the convergence rate, the employed bees' phase is modified. The new food source (candidate solution) generates using the following way:

$$\begin{cases} v_{ij} = x_{ij} + \phi \left(x_{ij} - x_{kj} \right) & \text{rand} \geq C_r \\ v_{ij} = x_{ij} + \varphi \left(x_{best,j} - x_{ij} \right) & \text{rand} < C_r \end{cases} \tag{1}$$

$$Cr = \frac{iter}{iter_{max}} \tag{2}$$

where $x_{best,j}$ is the position of the global best solution, and the term $x_{best,j}$ can drive the new candidate solution towards the global best solution, iter is the current iteration, itermax is the maximum iteration.

The parameter Cr in Eq. (2) plays an important role in balancing the exploration and exploitation of the candidate solution search. If $Cr = 0$, Eq. (2) is identical to position update equation of the canonical artificial bee colony algorithm. With the increase of Cr, the probability of the candidate solution learning to the best solution increases correspondingly. In this way, in the beginning of optimization process, the proposed algorithm operates as the canonical ABC algorithm, which can well keep the population diversity; in the end of optimization process, the modified ABC algorithm has a considerable improvement on both convergence rate and local search.

3.2 Cluster Setting

The K-means clustering method is employed to partition the population into subpopulations. The basic concepts of K-means clustering are presented firstly, and then we will give a detailed description of its application in our proposed algorithm.

3.2.1   Basic Parameters of K-means

The cluster centers are substituted for center positions of food sources and the formula of computing the centers is shown in Eq. (3). If the ith cluster contains n_i members and the members are denoted as $x_1{}^i, x_2{}^i, \ldots, x_{ni}{}^i$, then the center ($cluster_i^{center}$) is determined as

$$cluster_i^{center} = \frac{\sum_{i=1}^{n_i} x_i}{n_i} \qquad (3)$$

The radius (R) of a cluster is defined as the mean distance (Euclidean) of the members from the center of the cluster. Thus, R can be written as Eq. (4):

$$R_i = \frac{\sum_{p=1}^{n_i} \left\| x_p - cluster_i^{center} \right\|}{n_i} \qquad (4)$$

3.2.2   Clusters in the Proposed Algorithm KCABC

In the proposed algorithm, the stochastically generated population is partitioned into n subpopulations based on the widely adopted k-means cluster method [11]. The number of clusters is determined by the predefined set $G = \{g_1, g_2, \ldots, g_m\}$, where $g_1 > g_2 > \ldots > g_m$. Every cluster operates as the modified ABC introduced in the above section. During optimization, it may happen that two or more clusters come close to each other or get overlapped to a high degree. Then, they will practically search the same domain of the functional landscape. To avoid this scenario, the distances between each two clusters are calculated as following equation:

$$Dis_{cluster} = \left\| cluster_i^{center} - Nei\_cluster_i^{center} \right\| \qquad (5)$$

where $Dis_{cluster}$ is the distance between one cluster and its neighbor, $Nei\_cluster_i^{center}$ is the center of the ith cluster's neighbor. $cluster_i^{center}$ is the center of the ith cluster.

If the distance between one cluster and its neighbors is smaller than the specific distance DIS_m, one of the clusters will be removed and its non-domination solutions are store.

$$DIS_m = 0.2 * \min \left(R_i, R_{i\_neighbor} \right) \qquad (6)$$

where R_i is the radius of $cluster_i$ and $R_{i\_neighor}$ is the radius of the neighbors of $cluster_i$.

During optimization, information communication is a not inconsiderable aspect. For keeping the good ability of exploration, in the proposed algorithm, there is no information communication among g_i clusters in specific iterations. Therefore, in order to exchange information among individuals, the whole population is re-partitioned into g_{i+1} clusters based on k-means clustering after each TI iterations, where g_i and g_{i+1} are orderly chosen from the predefined set $G = \{g_1, g_2, \ldots, g_m\}$. That is, the individuals in a cluster may be distributed into different new clusters when the number of the clusters is changing.

Table 2. Pseudocode of KCABC

//Step 1: Initialization
Generalize a population of *NP* individuals in the search region randomly;
Set the numbers of clusters *G*; Create the external archive *EA*; initial the current
iteration *iter* = 1;
//Step 2: Loop
while stopping criteria are not satisfied **do**
    Set *TI*
    Decide the number g_i of clusters according to *G*
    Partition the whole population based on K-means clustering
    For all cluster
        Compute $Dis_{cluster}$ according to Eq. (6)
            Decide whether some clusters need to be removed according to DIS_m
        Select x_{up} individuals based on non-domination.
    End for
    For *iter*=1:*TI*
        Update the individuals' position in each cluster according to Eq. (1)
        iter = *iter*+1;
    End for
    Calculate the number of individuals in each cluster needing to be regenerated
    Cenerate a certain number of the new individuals in every cluster;
 End while

---

4   Prediction Results

4.1   General Prediction Procedure

The general procedure of droplet volume/velocity prediction based on LEM is as
follows:

- Step 1: According to droplet generator parameters and properties of fluid,
  calculating most of the parameters in LEM.
- Step 2: According to experimental measurements of droplet volume driven by
  standard waveform, determining the rest of the adjustment coefficients and
  damping coefficient.
- Step 3: Based on the constructed LEM, predicating the values of droplet volume
  and velocity.
- Step 4: According to the need of applications and restrictive conditions, finding out
  several groups of the appropriate parameters of driving waveform from predictive
  data lists of volume and velocity.
- Step 5: After making a subtle adjustment in practice, deciding the most appropriate
  parameters of driving waveform.

4.2   Predication Based on Intelligence Optimization Algorithm

For applications, the equispaced-sampling method of adjusting less than two parameters does not always meet the restrictive conditions. Furthermore, many adjustable parameters exist in high-frequency driving waveform. The exhausting search process for high dimensions with above method is unaccepted. Here, we utilize the proposed KCABC algorithm to accelerate the search process, as introduced in Section 2.3.

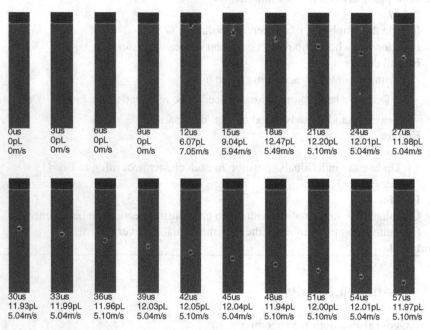

Fig. 4. A sequence of pictures of droplet falling from nozzle

In this experiment, the rising or falling time of the trapezoidal pulse is fixed as $0.3us$. There are five adjustable parameters in high-frequency driving waveform, including dwell time T_{w1}, T_{w2}, voltage magnitude V_1, V_2 and gap time T_g. Thus, the high-frequency driving waveform is defined as a vector with five dimensions $M = [T_1, V_1, T_g, T_2, V_2]^T$. Here, the restraint relationships between these parameters are given as follows:

$$T_{w1} + T_g + T_{w2} + 4*0.3us \leq T_s - 20us \qquad (7)$$

$$V_1 \geq V_2 \qquad (8)$$

$$V_1 + V_2 \leq 28V \qquad (9)$$

$$T_{w1}, T_{w2}, T_g \geq 0.3us \quad and \quad V_1, V_2 \geq 5V \qquad (10)$$

where T_s is the working period of printhead.

For the equation (7), the sum time of waveform must be less than the working period and also must leave enough time (20us) for system response. Equations (8) and (9) are set up according to project experience. Equations (8) and (9) avoid that so many satellite droplets emerge after jetting the main droplet. Equation (10) indicates the minimum of each adjust parameter.

The fitness function is defined as:

$$fitness = \min[w(1)*(V - V_{t\,\arg et})^2 + w(2)*(S - S_{t\,\arg et})^2]$$ (11)

where V is volume, S is velocity, $w(1)$ and $w(2)$ are the user-defined weighting.

The restraint conditions include that the droplet volume is smaller than $12pL$ and the droplet velocity is larger than 5m/s. The weighting is set as $w = [0.2, 0.8]$. The optimization process is carried out with 60 population size and 100 iterations. The optimal parameter vector $M = [T_1, V_1, T_g, T_2, V_2]^T$ is obtained as $M = [4.6, 13.5, 3.2, 9.7, 7.2]^T$. From the droplet dynamics shown in Fig. 4, the optimized results are satisfactory.

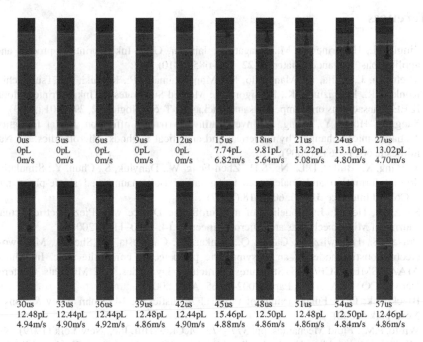

0us	3us	6us	9us	12us	15us	18us	21us	24us	27us
0pL	0pL	0pL	0pL	0pL	7.74pL	9.81pL	13.22pL	13.10pL	13.02pL
0m/s	0m/s	0m/s	0m/s	0m/s	6.82m/s	5.64m/s	5.08m/s	4.80m/s	4.70m/s

30us	33us	36us	39us	42us	45us	48us	51us	54us	57us
12.48pL	12.44pL	12.44pL	12.48pL	12.44pL	15.46pL	12.50pL	12.48pL	12.50pL	12.46pL
4.94m/s	4.90m/s	4.92m/s	4.86m/s	4.90m/s	4.88m/s	4.86m/s	4.86m/s	4.84m/s	4.86m/s

Fig. 5. A sequence of pictures of droplet falling from nozzle

Additionally, as a comparison, the genetic algorithm (GA) is chosen to search the appropriate combination of the parameters in the high-frequency driving waveform. The GA algorithm is a mature method for solving search problem; interested readers may consult References [12] for more information.

For optimization based on GA, the restraint condition, fitness function and the weighting are used as the same as KCABC algorithm. The population size and iterations are also chosen as 60 and 100, respectively. The optimal parameter vector $M = [T_1, V_1, T_g, T_2, V_2]^T$ is $M = [5.8, 14.0, 3.7, 9,9, 5.9]^T$. The dynamic effect of droplet formation driven by the proposed pulse parameters are shown in Fig. 5. Compared with Fig. 4, the optimization accuracy obtained by GA is lower than the results by KCABC. Moreover, there are obvious satellite droplets after the main droplet.

5   Conclusion

In this work, the lumped element modeling is employed to predict the droplet volume/velocity. Through combining with an improved artificial bee colony algorithm based on K-means Clustering, an optimal prediction model is constructed for searching the appropriate waveform parameters. Unlike the exhausting manual search process, the proposed prediction model can effectively accelerate the search process for high dimensions with sufficient accuracy.

References

1. Singh, M., Haverinen, H.M., Dhagat, P., Jabbour, G.E.: Inkjet printing: process and its applications. Advanced Materials 22, 673–685 (2010)
2. Miettinen, J., Kaija, K., Mantysalo, M., Mansikkamaki, P., Kuchiki, M., Tsubouchi, M., Ronkka, R., Hashizume, K., Kamigori, A.: Molded Substrates for Inkjet Printed Modules. IEEE Transactions on Components and Packaging Technologies 32, 293–301 (2009)
3. Sang, L., Hong, Y., Wang, F.: Investigatin of viscosity effect on droplet formation in T-shaped micro-channels by numerical and analytical methods. Microfluidics and Nanofluidics 6, 6621–6635 (2009)
4. Xiuqing, X., Butler, D.L., Ng, S.H., Zhen-Feng, W., Danyluk, S., Chun, Y.: Simulation of droplet formation and coalescence using lattice Boltzmann-based single-phase model. J. Colloid Interface 311(2), 609–618 (2007)
5. Seitz, H., Heinzl, J.: Modeling of a Microfluidic Device with Piezoelectric Actuators. Journal of Micromechanics and Microengineering 14, 1140–1147 (2004)
6. Prasad, S., Horowitz, S., Gallas, Q., Sankar, B., Cattafesta, L., Sheplak, M.: Two-port electroacoustic model of an axisymmetric piezoelectric composite plate. In: The 43rd AIAA/ASME/ASCE/AHS Structures, Structural Dynamics, and Materials Conference, Denver, CO, USA, AIAA Paper 2002-1365, April 2002
7. Blackstock, D.T.: Fundamentals of Physical Acoustics, p. 145. John Wiley & Sons Inc., New York (2000)
8. White, F.M.: Fluid Mechanics, pp. 377–379. McGraw-Hill, Inc., New York (1979)
9. Minolta, K.: Inkjet Head Application Note-KM1024 series. Konica Minolta IJ Technologies, Inc.
10. Karaboga, D., Basturk, B.: On the performance of Artificial Bee Colony (ABC). Applied Soft Computing 8(1), 687–697 (2008)
11. Li, W.-X., Zhou, Q., Zhu, Y., Pan, F.: An improved MOPSO with a crowding distance based external archive maintenance strategy. In: Tan, Y., Shi, Y., Ji, Z. (eds.) ICSI 2012, Part I. LNCS, vol. 7331, pp. 74–82. Springer, Heidelberg (2012)
12. Goldberg, D.E.: Genetic Algorithms in Search, Optimization and Machine Learning. Addison-Wesley, Reading (1989)

SMCSPO Based Force Estimation for Jetting Rate Control of 3D Printer Nozzle to Build a House

Keum-Gang Cha[1], Min-Cheol Lee[1(✉)], and Hee-Je Kim[2]

[1] Department of Mechanical Engineering, Pusan National University,
Jangjeon 2-dong, Geumjeong-gu, Busan, Korea
chagmagng@naver.com, mclee@pusan.ac.kr
[2] Department of Electrical Engineering, Pusan National University,
Jangjeon 2-dong, Geumjeong-gu, Busan, Korea
heeje@pusan.ac.kr

Abstract. To use a 3D print for building, a method for jetting some materials should be different from the conventional method of an office type 3D print because the material usually used as mixed concrete with viscosity. This research discuss about SMCSPO based force estimation for a jetting rate control of the nozzle for simulation of the 3D building print. In this study, the concrete material is replaced with silicon in a pseudo 3D building print. When building an outer wall for simulation, it is important to jet a silicon with constant velocity and force. In order to jet the silicon with constant velocity and force, the reaction force occured from silicon should be estimated. This paper address that the jetting force is estimated by sliding perturbation observer and the actuator is controlled by sliding mode control algorithm.

Keywords: 3D print for building · Sliding perturbation observer · Sliding mode control

1 Introduction

Recently, the 3D print has been a hot issue for the third industrial revolution. A variety of manufacturing facilities produces many types of 3D print. Many 3D print companies are interested in building by 3D print and doing research in this area widely. The disaster authorities cannot predict the natural disasters happened suddenly such as earthquake, so they cannot provide an emergency house for the disaster refugees. In this case, 3D print for building the houses can be used for constructing house for refugees quickly.

The building is constructed by a lot of procedures from the past. Exterior cement siding, piping, electrical construction and the rest are existed in many procedures. Among the procedures, the work that spends the most time to build are the exterior cement siding. Therefore, the concept of 3D print is applied to building construction n for reducing the working time and cost.

© Springer International Publishing Switzerland 2015
H. Liu et al. (Eds.): ICIRA 2015, Part I, LNAI 9244, pp. 45–55, 2015.
DOI: 10.1007/978-3-319-22879-2_5

A method of jetting materials from the 3D printer nozzle is important to build houses. The jetting method by a nozzle of an office 3D print is different from the jetting method for building because material of 3D print is not plastic but concrete. In this paper, silicon is used to simulate real 3D print for building.

It is important to jet silicon from nozzle with constant velocity and force for constructing building model. In order to jet the silicon with constant force and velocity, estimating the silicon jetting force from nozzle is necessary. The nozzle actuator is controlled by sliding mode control(SMC) and force is estimated by sliding perturbation observer(SPO).

In section 2, a dynamic model equation of the nozzle for 3D building printer simulation is addressed. In section 3, SMC and SPO are introduced respectively. Sliding mode control with sliding perturbation observer(SMCSPO) is robust controller with good performance [1]. It is proved that SPO can estimate uncertainties such as disturbance, nonlinear term and reaction force. In section 4, the result analysis is used to verify the reaction force estimation. Finally, in section 5, the review with overall contents is contained.

2   The Design of 3D Print Jetting Nozzle System

Fig. 1 show a 3D printer simulation nozzle system for building a small model house.

The silicon is located at load cell position. Motor is attached and rotates in down side of nozzle system. The pusher to jet silicon though the nozzle has a straight movement by linear guide and ball screw. For obtaining the dynamics equation of nozzle system, following equation is essentially discussed.

The inertia of system can be obtained by ball screw equation. We assume that damper and spring term is not contained in this system. So only inertia is existed in modeling equation and then inertia can be obtained by Eq.(1). By inertia equation, modeling equation of experiment system can be defined as Eq.(3).

$$J = 4(GD_i + \frac{WP^2}{\pi}) \tag{1}$$

$$GD_i = \frac{1}{8}MR^2 \tag{2}$$

$$\tau = J\ddot{\theta} \tag{3}$$

where M is mass of ball screw, R is radial of ball screw, P is pitch of ball screw, W is weight of pusher.

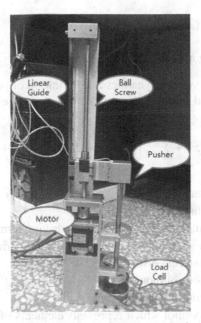

Fig. 1. Experiment system for 3D print for building

Fig. 2. 3D print for building system

3   System Robust Control with SMCSPO

3.1   System Modeling and the Perturbation Concept

This section presents the system dynamics and definition of perturbation. The governing equation for general second dynamics with n-degree-of-freedom defined as Eq.(4).

$$\ddot{x}_j = f_i(x) + \Delta f_i(x) + \sum_{i=1}^{n} \left[\left(b_{ij}(x) + \Delta b_{ij}(x) \right) u_i \right] + d_i(t) \tag{4}$$

where
$x \quad \equiv [x_1, \cdots, x_n]^T$: state vector,
$x_j \quad \equiv [x_j, \dot{x}_j]^T$: state variable,
$\Delta f_j(x)$: uncertainties of nonlinear driving terms,
$\Delta b_{ji}(x)$: uncertainties of the control gain matrix,
d_j : external disturbance,
u_j : control input,
f_j, b_{ji} : continuous functions of states,
"i" : symbol which represents elements of control gain matrix effected by control input.

Perturbation is defined as the combination of all the uncertainties of Eq. (5).

$$\psi_j(x, t) = \Delta f_j(x) + \sum_{i=1}^{n} [\Delta b_{ji}(x) u_i] + d_j(t) \tag{5}$$

It is assumed that the perturbations are upper bounded by a known functions of the state as Eq. (5).

$$\Gamma_j(x, t) = F_j(x) + \sum_{i=1}^{n} |\Phi_{ji}(x) u_i| + |D_j(t)| \geq |\psi_j(x, t)| \tag{6}$$

where, $F_j > |f_j(x)|$, $\Phi_{ji} > |\Delta b_{ji}(x)|$, $D_j > |d_j|$ represent the expected upper bounds of the uncertainties, respectively.

3.2   Sliding Perturbation Observer

The new control variable which is used in order to decouple the control of Eq.(4).

$$f_j(\hat{x}) + \sum_{i=1}^{n} b_{ji}(\hat{x}) u_i = \alpha_{3j} \bar{u}_j \tag{7}$$

where $\hat{x}$ is the estimated state vector, α_{3j} is an arbitrary positive number and $\bar{u}_j$ is the new control variable[5]. Throughout the text, "~", refers to estimation errors whereas "^" symbolizes the estimated quantity. The state representation of the simplified dynamics is given by

$$\dot{x}_{1j} = x_{2j} \tag{8}$$

$$\dot{x}_{2j} = \alpha_{3j}\bar{u}_j + \psi_j \tag{9}$$

$$y_j = x_{1j} \tag{10}$$

where "j" is the number of joint number.

Let x_{3j} be a new state variable defined as

$$x_{3j} = \alpha_{3j}x_{2j} - \psi_j/\alpha_{3j} \tag{11}$$

It is desirable to observe the variable x_{3j} and calculate ψ_j using Eq. (11) instead of estimating the perturbation directly. Consequently, it is necessary to estimate x_{2j} in order to obtain the estimated perturbation $\hat{\psi}_j$. This structure can be achieved by writing the observer equation as

$$\dot{\hat{x}}_{1j} = \hat{x}_{2j} - k_{1j}sat(\tilde{x}_{1j}) - \alpha_{1j}\tilde{x}_{1j} \tag{12}$$

$$\dot{\hat{x}}_{2j} = \alpha_{3j}\bar{u}_j - k_{2j}sat(\tilde{x}_{1j}) - \alpha_{2j}\tilde{x}_{1j} + \hat{\psi}_j \tag{13}$$

$$\dot{\hat{x}}_{3j} = \alpha_{3j}{}^2(-\hat{x}_{3j} + \alpha_{3j}\hat{x}_{2j} + \bar{u}_j) \tag{14}$$

where $\hat{\psi}_j$ is derived as

$$\hat{\psi}_j = \alpha_{3j}(-\hat{x}_{3j} + \alpha_{3j}\hat{x}_{2j}) \tag{15}$$

$k_1, k_2, \alpha_1, \alpha_2$ are the positive numbers, $\tilde{x}_{1j} = \hat{x}_{1j} - x_{1j}$ is the estimation error of the measurable state, and $sat(\tilde{x}_{1j})$ is the saturation function for the existence of sliding mode[6].

3.3   Sliding Mode Control with Sliding Perturbation Observer

This section presents the integration of SMC control law and SPO observer scheme. For the system of Eq.(8), Eq.(9), Eq.(10), we define the estimated sliding function as Eq.(16).

$$\hat{s}_j = \dot{\hat{e}}_j + c_{j1}\hat{e}_j \tag{16}$$

where $c_{j1} > 0$, $\hat{e}_j = \hat{x}_{1j} - x_{1dj}$ is the estimated position tracking error and $[x_{1dj} \quad \dot{x}_{1dj}]$ is the desired motion for the "j-th" degree of freedom. The actual sliding function is

$$s_j = \dot{e}_j + c_{j1}e_j \tag{17}$$

where $e_j = x_{1j} - x_{1dj}$ is the actual position tracking error.

The estimation error of the sliding function is defined as $\tilde{s}_j = \hat{s}_j - s_j$. Using Eq.(16) and Eq.(17)

$$\tilde{s}_j = \dot{\tilde{x}}_{1j} + c_{1j}\tilde{x}_{1j} \tag{18}$$

The control $\bar{u}_j$ is selected to enforce $\dot{\hat{s}}_j\hat{s}_j < 0$ out side a prescribed manifold. A desired $\hat{s}_j$-dynamics is selected as

$$\dot{\hat{s}}_j = -K_j sat(\hat{s}_j) \tag{19}$$

where

$$sat(\hat{s}_j) = \begin{cases} \hat{s}_j/|\hat{s}_j|, & if \ |\hat{s}_j| \geq \epsilon_{0j} \\ \hat{s}_j/\epsilon_{0j}, & if \ |\hat{s}_j| \leq \epsilon_{0j} \end{cases} \tag{20}$$

is used due to its desirable anti-chatter properties [7]. In this equation, ϵ_{0j} stands for boundary layer of the SMC controller, as opposed to the ϵ_{0j} in SPO.

Using the results of previous sections it is possible to compute $\dot{\hat{s}}_j$ as

$$\dot{\hat{s}}_j = \alpha_3\bar{u}_j - \left[\frac{k_{2j}}{\epsilon_{0j}} + c_{j1}\left(\frac{k_{1j}}{\epsilon_{0j}}\right) - \left(\frac{k_{1j}}{\epsilon_{0j}}\right)^2\right]\tilde{x}_{1j} - \left(\frac{k_{1j}}{\epsilon_{0j}}\right)\tilde{x}_{2j} \\ - \ddot{x}_{1dj} + c_{j1}(\hat{x}_{2j} - \dot{x}_{1dj}) + \hat{\psi}_j \tag{21}$$

In order to enforce Eq.(19) when $\tilde{x}_{2j} = 0$, a control law is selected as

$$\bar{u}_j = \frac{1}{\alpha_{3j}}\{-K_j sat(\hat{s}_j) + \left[\frac{k_{2j}}{\epsilon_{0j}} + c_{j1}\left(\frac{k_{1j}}{\epsilon_{0j}}\right) - \left(\frac{k_{1j}}{\epsilon_{0j}}\right)^2\right]\tilde{x}_{1j} \\ + \ddot{x}_{1dj} - c_{j1}(\hat{x}_{2j} - \dot{x}_{1dj}) - \hat{\psi}_j\} \tag{22}$$

The resulting $\hat{s}_j$-dynamics including the effects of $\tilde{x}_{2j}$, becomes Eq. (23).

$$\dot{\hat{s}}_j = -K_j sat(\hat{s}_j) - \left(\frac{k_{1j}}{\epsilon_{0j}}\right)\tilde{x}_{2j} \tag{23}$$

From the sliding conditions [4], the state estimation error is bounded by $|\tilde{x}_{2j}| \leq k_{1j}$. Therefore, in order to satisfy $\hat{s}_j \dot{\hat{s}}_j < 0$ outside the manifold $|\hat{s}_j| \leq \epsilon_{0j}$, the robust control gains must be chosen such that

$$K_j \geq k^2_{1j}/\epsilon_{0j} \tag{24}$$

After the approaching phase, selection Eq.(24) assures $|\hat{s}_j| \leq \epsilon_{0j}$. Using $\tilde{s}_j = \hat{s}_j - s_j$, Eq.(12), Eq.(13), and Eq.(14), the actual $\hat{s}_j$-dynamics within the boundary layer $|\hat{s}_j| \leq \epsilon_{0j}$ becomes Eq.(25)

$$\dot{\hat{s}}_j + \frac{K_j}{\epsilon_{0j}} s_j = \left[\frac{k_{2j}}{\epsilon_{0j}} - \left(\frac{k_{1j}}{\epsilon_{0j}} - \frac{K_j}{\epsilon_{0j}}\right)\left(c_{j1} - \frac{k_{1j}}{\epsilon_{0j}}\right)\right] \tilde{x}_{1j}$$
$$- \left(c_{j1} + \frac{K_j}{\epsilon_{0j}}\right)\tilde{x}_{2j} - \hat{\psi}_j \tag{25}$$

If each variables are upper bounded as $|\hat{s}_j| \leq \epsilon_{0j}$, $|\tilde{x}_{1j}| \leq \epsilon_{0j}$, control performance can be improved with sliding mode control, sliding perturbation observer operating well. With this assumption, Eq.(25) can be defined as observer error dynamics Eq.(26) vector form for selection of gain.

$$\begin{bmatrix} \dot{\tilde{x}}_{1j} \\ \dot{\tilde{x}}_{3j} \\ \dot{\tilde{x}}_{3j} \\ \dot{\hat{s}}_j \end{bmatrix} = \begin{bmatrix} -\dfrac{k_{1j}}{\epsilon_{0j}} & 1 & 0 & 0 \\ -\dfrac{k_{2j}}{\epsilon_{0j}} & \alpha_{3j}^2 & -u_{3j} & 0 \\ 0 & \alpha_{3j}^3 & -\alpha_{3j}^2 & 0 \\ -\dfrac{k_{2j}}{\epsilon_{0j}} + \left(c_{j1} - \dfrac{k_{1j}}{\epsilon_{0j}}\right)^2 & -(2c_{j1} + \alpha_{3j}^2) & \alpha_{3j} & c_{j1} \end{bmatrix} \begin{bmatrix} \tilde{x}_{1j} \\ \tilde{x}_{2j} \\ \tilde{x}_{3j} \\ s_j \end{bmatrix} + \begin{bmatrix} 0 \\ 0 \\ 1 \\ 0 \end{bmatrix}\begin{matrix} \hat{\psi}_j \\ \\ \alpha_{3j} \end{matrix} \tag{26}$$

Since s_j and e_j are in same frequency bandwidth, $c_{1j} = K_j/\epsilon_{0j}$ can apply. In addition, if α_{3j} is selected larger enough than time derivative of perturbation, we can let system vector form have parameter of safe parameter for securing performance. In order to select parameter of safe parameter, characteristic equation can be obtained as Eq.(27) from system vector form.

$$[\lambda + c_{j1}][\lambda^3 + (k_{1j}/\epsilon_{0j})\lambda^2 + (k_{2j}/\epsilon_{0j})\lambda + \alpha^2_{3j}(k_{2j}/\epsilon_{0j})] = 0 \tag{27}$$

From Eq.(27), system is stable when eigenvalue of vector form is negative integer. Therefore polynomial can be defined as Eq.(28).

$$p(\lambda_d) = (\lambda + \lambda_d)^4 : \text{p is polynomial}, \lambda_d > 0 \tag{28}$$

System is stable when eigenvalue of observer error dynamics and system vector form have negative value. Therefore, following relationship can be obtained from Eq.(27) and Eq.(28).

$$k_{1j}/\epsilon_{0j} = 3\lambda_d, \quad k_{2j}/k_{1j} = \lambda_d, \quad \alpha_{3j} = \sqrt{\lambda_d/3}$$
$$c_{j1} = K_j/\epsilon_{0j} = \lambda_d \tag{29}$$

In fact, selection of eigenvalue is related with sample time, time constant with more reason. In this paper, eigenvalue is selected based on slope of sliding hyperplane.

4   Experiment Result Analysis and Discussion

The experiment is done with following condition. First, silicon is attached in load cell and pusher is a straight movement with constant velocity by PDI controller and SMCSPO controller respectively. Second, silicon is replaced with load cell. Jetting force estimation is done for 20mm and 40mm final pusher displacement respectively with SMCSPO

4.1   Position Tracking by SMCSPO

Fig. 3 and Fig. 4 show experiment result of position tracking by SMCSPO and PDI controller respectively. Reference is constant velocity. Experiment result of position tracking by SMCSPO controller is compared with that of PDI controller. Each position tracking is compliance with reference. But difference between SMCSPO and PDI is that result of SMCSPO controller is more precisely following reference than result of PDI.

Table 1. shows parameters of SMCSPO and PDI controller

Controller	Parameter values
PDI	P=30, D=0.5, I=0.001
SMCSPO	$K_j = 300,$ $k_{1j} = 30,$ $k_{2j} = 300,$ $c_{j1} = 10$ $\alpha_{3j} = 1.8257,$ $\epsilon_{0j} = 1$

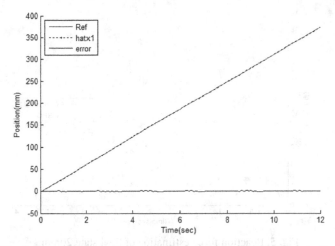

Fig. 3. Experiment result of position tracking by SMCSPO controller

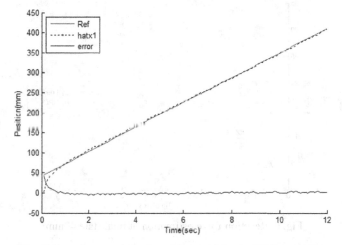

Fig. 4. Experiment result of position tracking by PDI controller

4.2   Reaction Force Estimation by SPO

Fig. 5 and Fig. 6 show force estimation of final state 20mm and 40mm respectively. Each estimated result follows well the measured value of estimation by load cell. In the transient period the estimated forces are oscillated with small magnitude caused by friction, system error, and ball screw design error. In steady state period the error of reaction force estimation is observed with small magnitude. This error comes from modeling error, friction with more reason. The pusher is contacted with the load cell at 10mm. So the final force difference between 20mm and 40mm means that how deep the load cell is compressed. If SPO with low pass filter is operated in this system, this error can be reduced.

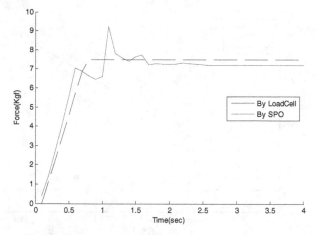

Fig. 5. Reaction force estimation of final state 20mm

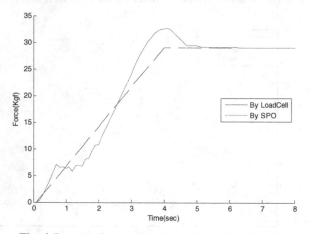

Fig. 6. Reaction force estimation of final state 40mm

5   Conclusion

This paper showed that SMCSPO jetting force estimation performance of 3D print nozzle is good. In position tracking condition, the PDI and SMCSPO is compared with each other results. It is confirmed that the position control accuracy of SMCSPO controller is better than that of PDI controller. In jetting force estimation condition, the force estimated by SPO is well following the result measured by load cell. Therefore, this nozzle is proper system and controller to apply simulation.

Acknowledgment. This research was supported by the MOTIE (Ministry of Trade, Industry & Energy), Korea, under the Industry Convergence Liaison Robotics Creative Graduates Education Program supervised by the KIAT (N0001126).

This work was supported by the National Research Foundation of Korea(NRF) grant funded by the Korea Government(MSIP) (No.NRF-2012M2B2B 1055503).

References

1. Yoon, S.M., Lee, M.C.: An identification of the single rod hydraulic cylinder using signal compression method and applying SMCSPO. In: URAI 2014, pp. 438–443 (2014)
2. Song, Y.E., Lee, M.C., Kim, C.Y.: Sliding mode control with sliding perturbation observer for surgical robots. In: ISIE 2009, pp. 2153-2158 (2009)
3. Yoon, S.M., Lee, M.-C., Kim, C.Y.: Sliding perturbation observer based reaction force estimation method in surgical robot instrument. In: Lee, J., Lee, M.C., Liu, H., Ryu, J.-H. (eds.) ICIRA 2013, Part I. LNCS, vol. 8102, pp. 227–236. Springer, Heidelberg (2013)
4. Elmali, H., Olgac, N.: Sliding Mode Control With Perturbation Estimation (SMCSPE). International Journal of Control **56**, 923–941 (1992)
5. Terra, M.J., Elmali, H., Olgac, N.: Sliding Mode Control With Perturbation Observer. Journal of Dynamics System, Measurement, and Control **119**, 657–665 (1997)
6. Slotine, J.J., Hedrick, J.K., Misawa, E.A.: On Sliding Observers for Non-Linear Systems. ASME Journal of Dynamic System, Measurement and Control **109**, 245–252 (1987)
7. Stewart, D.: A platform with six degree of freedom. In: Proc, of the Institute of Mechanical Engineering, vol. 180, pp. 317–386 (1966)
8. Lee, M.C., Aoshima, N.: Real time multi-input sliding mode control of a robot manipulator based on DSP. In: Proc. of SICE, pp. 1223–1228 (1993)
9. You, K.S., Lee, M.C., You, W.S.: Sliding Mode Controller with Sliding Perturbation Observer Based on Gain Optimization using Genetic Algorithm. KSME Int. J, **18**(4), 630–639 (2004)
10. Stopp, S., et al.: A new method for printer calibration and contour accuracy manufacturing with 3D-print technology. Rapid Prototyping Journal **14**(3), 167–172 (2008)

Sufficient Conditions for Input-to-State Stability of Spacecraft Rendezvous Problems via Their Exact Discrete-Time Model

Kun Li$^{(\boxtimes)}$, Haibo Ji, and Shaomin He

Department of Automation, University of Science and Technology of China,
Hefei, Anhui 230027, People's Republic of China
jihb@ustc.edu.cn,{zkdlk,hshaomin}@mail.ustc.edu.cn

Abstract. This paper investigates the sampled-data control problem for spacecraft rendezvous with target spacecraft on an arbitrary elliptical orbit. The exact discrete-time dynamic model and its Euler approximation are established based on the Lawden equations. With bounded external disturbances attached on the discrete-time model, input-to-state stability (ISS) analysis is introduced to design a sampled-data controller, which can achieve the rendezvous mission. Finally, simulation results show that the proposed control laws are effective.

Keywords: Spacecraft rendezvous · Sampled-data control · Input-to-state stability

1 Introduction

Recent decades have witnessed a lot of research interest in spacecraft rendezvous and docking technology, which is essential for present and future space missions such as re-supplying of orbital platform and stations, repairing of spacecraft in orbit and so on [1,2]. And the goal of autonomous rendezvous is to design a controller with which the chaser spacecraft can hold at a fixed relative position near to the target spacecraft.

Over the course of the past decades, many control approaches have been proposed to deal with the spacecraft rendezvous problems [3–6]. A study on an adaptive output feedback control law for spacecraft rendezvous and docking problems was done in [3]. In [4], the problem of robust H_∞ control for a class of spacecraft rendezvous systems was investigated, which contained parametric uncertainties, external disturbances and input constraints. And in [5], a nonlinear trajectory control algorithm of rendezvous with maneuvering target spacecraft was developed. [6] utilized genetic algorithms to find a close approximation to spacecraft rendezvous problems.

K. Li—This work was supported by the National Natural Science Foundation of China under Grant 61273090 and the National Basic Research Program of China under Grant 973-10001.

© Springer International Publishing Switzerland 2015
H. Liu et al. (Eds.): ICIRA 2015, Part I, LNAI 9244, pp. 56–65, 2015.
DOI: 10.1007/978-3-319-22879-2_6

The control approaches mentioned above have shown adequate reliability on continuous systems in spacecraft rendezvous problems. However, in actual rendezvous systems, the control thrusts applied to the chaser spacecraft are almost in discrete forms, that is to say, previous continuous control laws may not work in actual systems. Naturally, it is vital to design discrete-time control laws to deal with the problem.

So far, many research studies have been carried out on sampled-data control for nonlinear systems. [7] has given the sufficient conditions for stabilization of sampled-data nonlinear systems via discrete-time approximations. In [8], a unified framework for design of stabilizing controllers for sampled-data differential inclusions via their approximate discrete-time models was presented considering both fixed and fast sampling. And in [9], backstepping control for sampled-data nonlinear systems was discussed. Futhermore, some basic conclusions on input-to-state stabilization for sampled-data nonlinear systems were presented in [10] and [11]. In this paper, control laws based on input-to-state stability for the exact discrete-time model are proposed. It takes good performance on spacecraft position control with bounded external disturbances.

The remainder of this paper is organized as follows. Section 2 introduces the linear dynamic model based on the Lawden equations. Then, after presenting some basic concepts, the control laws are designed in Section 3, stability analysis is also obtained. Section 4 shows the simulation results and analysis to validate the proposed approach. Section 5 summarizes the conclusions.

2   Problem Statement

2.1   Dynamic Model

The dynamic model is established in this subsection. Figure 1 shows the target-orbital coordinate system $X - Y - Z$, where the origin is fixed at the center of mass of the target, the x axis points in the direction of motion of the target, the

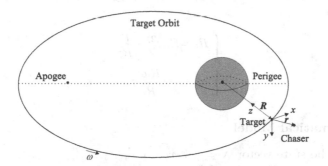

Fig. 1. Spacecraft rendezvous system

z axis points to the center of the earth, and the y axis is normal to the orbital plane, opposite the angular momentum vector.

The chaser motion relative to the target spacecraft in the coordinate system shown in Figure 1 can be captured by [12,13]:

$$
\begin{cases}
\ddot{x} = 2\omega\dot{z} + \dot{\omega}z + \omega^2 x - \dfrac{\mu x}{\|\mathbf{R}_t + \mathbf{r}_{tc}\|^3} + u_x \\[2mm]
\ddot{y} = -\dfrac{\mu y}{\|\mathbf{R}_t + \mathbf{r}_{tc}\|^3} + u_y \\[2mm]
\ddot{z} = \omega^2 z - 2\omega\dot{x} - \dot{\omega}x - \mu\left(\dfrac{z - R_t}{\|\mathbf{R}_t + \mathbf{r}_{tc}\|^3} + \dfrac{1}{R_t{}^2}\right) + u_z
\end{cases}
\tag{1}
$$

where $\mathbf{R}_t$ represents the relative displacement of target from the center of the Earth; $\mathbf{r}_{tc} = [x, y, z]^T$ refers to the relative displacement of chaser from target; ω represents the orbital rate of the rotating coordinate system; μ is the gravitational constant; $\|\mathbf{r}_{tc}\| = r_{tc}$; $\|\mathbf{R}_t\| = R_t$; $\mathbf{u} = [u_x, u_y, u_z]^T$ stands for the control acceleration vector in the target-orbital coordinate frame applied to chaser.

In this paper, the distance between the target and the chaser is much smaller than the distance between the target and the center of the Earth, that is to say:

$$
R_t \gg r_{tc}
\tag{2}
$$

Then, the nonlinear system in Equation (1) can be linearized at the origin as follows [12]:

$$
\begin{cases}
\ddot{x} = 2\omega\dot{z} + \dot{\omega}z + \omega^2 x - \dfrac{\mu x}{R_t{}^3} + u_x \\[2mm]
\ddot{y} = -\dfrac{\mu y}{R_t{}^3} + u_y \\[2mm]
\ddot{z} = \omega^2 z - 2\omega\dot{x} - \dot{\omega}x + 2\dfrac{\mu z}{R_t{}^3} + u_z
\end{cases}
\tag{3}
$$

which are known as the Lawden equations.

The additional equations that describe the evolution of the target's orbit are as follows [14]:

$$
\begin{cases}
\ddot{R}_t = R_t\omega^2 - \dfrac{\mu}{R_t^2} \\[2mm]
\dot{\omega} = -2\dfrac{\dot{R}_t\omega}{R_t}
\end{cases}
\tag{4}
$$

2.2   Mathmatical Model

By defining the state vector $X = [x_1, x_2, x_3, x_4, x_5, x_6]^T = [x, y, z, \dot{x}, \dot{y}, \dot{z}]^T$, the state-space equations of the model can be expressed as:

$$\begin{cases} \dot{x}_1 = x_4 \\ \dot{x}_2 = x_5 \\ \dot{x}_3 = x_6 \\ \dot{x}_4 = 2\omega x_6 + \dot{\omega} x_3 + \omega^2 x_1 - \dfrac{\mu x_1}{R_t^{\;3}} + u_x \\ \dot{x}_5 = -\dfrac{\mu x_2}{R_t^{\;3}} + u_y \\ \dot{x}_6 = \omega^2 x_3 - 2\omega x_4 - \dot{\omega} x_1 + 2\dfrac{\mu x_3}{R_t^{\;3}} + u_z \end{cases} \tag{5}$$

In general, because of some bounded external disturbances attached on the original model, it can be rewritten as follows:

$$\begin{cases} \dot{x}_1 = x_4 + d_1 \\ \dot{x}_2 = x_5 + d_2 \\ \dot{x}_3 = x_6 + d_3 \\ \dot{x}_4 = 2\omega x_6 + \dot{\omega} x_3 + \omega^2 x_1 - \dfrac{\mu x_1}{R_t^{\;3}} + u_x + d_4 \\ \dot{x}_5 = -\dfrac{\mu x_2}{R_t^{\;3}} + u_y + d_5 \\ \dot{x}_6 = \omega^2 x_3 - 2\omega x_4 - \dot{\omega} x_1 + 2\dfrac{\mu x_3}{R_t^{\;3}} + u_z + d_6 \end{cases} \tag{6}$$

Definition 1. *[9] (One-step strong consistency) The exact plant model F_T^e and its Euler approximation F_T^{Euler} corresponding to the original system $\dot{x}(t) = f(x(t), u(t), w(t))$ are denoted as*

$$x[k+1] = F_T^e(x[k], u[k], w[k]) \tag{7}$$

$$x[k+1] = F_T^{Euler}(x[k], u[k], w[k]) \tag{8}$$

where $u[k] := u(kT) = u(t), \forall t \in [kT, (k+1)T], k \in N$ and $T > 0$, which will be a piecewise constant signal. And $w : R \geq 0 \rightarrow R^n$ is a given function with the notation: $w[k] := w(t), t \in [kT, (k+1)T], k \in N$ and $T > 0$.

The family F_T^{Euler} is said to be one-step strongly consistent with F_T^e if given any strictly positive real numbers (Δ_1, M, Δ_2), there exists a function $\rho \in K_\infty$ and $T^* > 0$ such that, for all $T \in (0, T^*)$, all $x \in R^{n_x}, u \in R^m, w \in L_\infty$ with $\| x \| \leq \Delta_1, \| u \| \leq M, \| w \|_\infty \leq \Delta_2$, we have $\| F_T^e - F_T^{Euler} \| \leq T\rho(T)$.

Definition 2. *The family of system $x[k+1] = F_T(x[k], w[k])$ is semiglobally practically input-to-state stable if there exist functions $\alpha_1, \alpha_2, \alpha_3 \in K_\infty$ and $\gamma \in K$, and for any strictly positive real numbers $(\Delta_1, \Delta_2, \delta_1, \delta_2)$ there exist strictly positive real numbers T^* and L such that for all $T \in (0, T^*)$ there exists a function $V_T : R^{n_x} \rightarrow R \geq 0$ such that, for all $x \in R^{n_x}$ with $\| x \| \leq \Delta_1$ and all $w \in L_\infty$ with $\| w \|_\infty \leq \Delta_2$, the following holds:*

$$\alpha_1(\| x \|) \leq V_T(x) \leq \alpha_2(\| x \|) \tag{9}$$

$$\frac{1}{T}[V_T(F_T(x,w)) - V_T(x)] \leq -\alpha_3(\| x \|) + \gamma(\| w \|_\infty) + \delta_1 \qquad (10)$$

and, moreover, for all x_1, x_2, z with $\| (x_1^T, z^T)^T \|, \| (x_2^T, z^T)^T \|$ and all $T \in (0, T^*)$, we have $|V_T(x_1, z) - V_T(x_2, z)| \leq L \| x_1 - x_2 \|$.

Note that the above definition was firstly proposed by Nesic in [10], which was called Lyapunov semiglobally practically input-to-state stability (Lyapunov-SP-ISS).

3   Control Laws Design

Corresponding to the original plant model (6), Its Euler approximation can be denoted as:

$$\begin{cases}
x_1[k+1] = x_1[k] + T(x_4[k] + d_1[k]) \\
x_2[k+1] = x_2[k] + T(x_5[k] + d_2[k]) \\
x_3[k+1] = x_3[k] + T(x_6[k] + d_3[k]) \\
x_4[k+1] = x_4[k] + T(2\omega x_6[k] + \dot{\omega} x_3[k] + \omega^2 x_1[k] - \frac{\mu}{R_t^3} x_1[k] + u_x[k] + d_4[k]) \\
x_5[k+1] = x_5[k] + T(-\frac{\mu}{R_t^3} x_2[k] + u_y[k] + d_5[k]) \\
x_6[k+1] = x_6[k] + T(\omega^2 x_3[k] - 2\omega x_4[k] - \dot{\omega} x_1[k] + 2\frac{\mu}{R_t^3} x_3[k] + u_z[k] + d_6[k])
\end{cases}$$
$$(11)$$

With the Euler approximation of the original system (6), its exact sampled model can be written as

$$\begin{cases}
x_1[k+1] = x_1 + T(x_4 + d_1) + f_{x_1}^e - f_{x_1}^{Eu} \\
x_2[k+1] = x_2 + T(x_5 + d_2) + f_{x_2}^e - f_{x_2}^{Eu} \\
x_3[k+1] = x_3 + T(x_6 + d_3) + f_{x_3}^e - f_{x_3}^{Eu} \\
x_4[k+1] = x_4 + T(2\omega x_6 + \dot{\omega} x_3 + \omega^2 x_1 - \frac{\mu}{R_t^3} x_1 + u_x + d_4) + f_{x_4}^e - f_{x_4}^{Eu} \\
x_5[k+1] = x_5 + T(-\frac{\mu}{R_t^3} x_2 + u_y + d_5) + f_{x_5}^e - f_{x_5}^{Eu} \\
x_6[k+1] = x_6 + T(\omega^2 x_3 - 2\omega x_4 - \dot{\omega} x_1 + 2\frac{\mu}{R_t^3} x_3 + u_z + d_6) + f_{x_6}^e - f_{x_6}^{Eu}
\end{cases}$$
$$(12)$$

According to (4), the bound of $\dot{\omega}$ can be expressed as [14]

$$|\dot{\omega}| = \left| -2\frac{\dot{R}_t \omega}{R_t} \right| \leq 2\frac{\mu e}{R_p^3} = \xi \qquad (13)$$

where e is the eccentricity of target orbit, R_p represents the perigee radius of the target orbit which is much longer than the radius of earth R_E, hence, $\xi = 2\frac{\mu e}{R_p^3} < 2\frac{\mu e}{R_E^3} < \frac{1}{2}$ can be obtained, the upper bound ξ is assumped to be known.

Theorem 1. *Consider the exact closed-loop discrete-time model (12) of the spacecraft rendezvous system with target spacecraft on an arbitrary elliptical orbit. For any given strictly positive real numbers (Δ_1, δ), there exists $\parallel X \parallel \leq \Delta_1$, and the external disturbances attached on the model are bounded. Take the following control laws*

$$u = \begin{bmatrix} u_x \\ u_y \\ u_z \end{bmatrix} = \begin{bmatrix} -2x_4 - c(x_1 + x_4) - \phi_1(x_1, \cdots, x_6) \\ -2x_5 - c(x_2 + x_5) - \phi_2(x_1, \cdots, x_6) \\ -2x_6 - c(x_3 + x_6) - \phi_3(x_1, \cdots, x_6) \end{bmatrix} \qquad (14)$$

where $c > \frac{\xi - \xi^2}{1 - 2\xi} > 0, \phi_1 = 2\omega x_6 + \omega^2 x_1 - \frac{\mu}{R_t^3} x_1, \phi_2 = -\frac{\mu}{R_t^3} x_2, \phi_3 = \omega^2 x_3 - 2\omega x_4 + 2\frac{\mu}{R_t^3} x_3$, the exact closed-loop discrete-time model (12) is semiglobally practically input-to-state stable.

Proof. Consider the Lyapunov function

$$V_T = \frac{1}{2} \sum_{i=1}^{3} x_i^2 + \frac{1}{2} \sum_{j=1}^{3} (x_j + x_{(j+3)})^2 \qquad (15)$$

Owing to one-step strong consistence between the exact discrete-time model and its Euler approximation, there exist some K_∞ functions $\rho_{x_1}, \cdots, \rho_{x_6}$ that satisfy the following inequalities

$$\parallel f_{x_i}^e - f_{x_i}^{Eu} \parallel < T\rho_{x_i}(T) \qquad i = 1, 2, \cdots, 6 \qquad (16)$$

It can be obtained that for any given positive numbers $\frac{\delta}{60\Delta_1}, \frac{\delta}{60\Delta_2}$, there exist positive numbers δ_{x_i} to make $\rho_{x_i} < \frac{\delta}{60\Delta_1}, \rho_{x_i} < \frac{\delta}{60\Delta_2}$ establish with $T < \delta_{x_i}$. It is because that $\rho_{x_i}, i = 1, 2, \cdots, 6$ are K_∞ functions and they are right-continuous at the origin.

Take $T^* = min\{\frac{1}{4}, \delta_{x_1}, \delta_{x_2}, \cdots, \delta_{x_6}, \frac{\delta}{30\Delta_1^2}, \frac{60\Delta_1^2}{\delta}, \frac{\delta\Delta}{2\Delta_1 + 2\Delta_2}, \frac{\delta}{90\Delta_2^2}, \frac{\delta}{120\Delta_1\Delta_2(1+2\xi)}, \frac{\delta}{240\Delta_1\Delta_2}\}$, when $T \in (0, T^*)$, we have

$$\frac{1}{T}\left(V_T(x[k+1]) - V_T(x[k])\right)$$

$$= \frac{1}{2T} \sum_{i=1}^{3} \left(x_i^2[k+1] - x_i^2[k] + (x_i[k+1] + x_{i+3}[k+1])^2 - (x_i[k] + x_{(i+3)}[k])^2\right)$$

$$= \frac{1}{2T} \sum_{i=1}^{3} (x_i^2[k+1] - x_i^2[k]) + \frac{1}{T} \sum_{i=1}^{3} \left((x_i[k+1] + x_{(i+3)}[k+1]) - (x_i[k]\right.$$

$$+ x_{(i+3)}[k]))(x_i[k] + x_{(i+3)}[k]) + \frac{1}{2T} \sum_{i=1}^{3} \left((x_i[k+1] + x_{(i+3)}[k+1])\right.$$

$$\left. - (x_i[k] + x_{(i+3)}[k])\right)^2 \qquad (17)$$

Substituting (12) into (17) yields

$$\frac{1}{T}\big(V_T(x[k+1]) - V_T(x[k])\big)$$

$$= \frac{T}{2}\sum_{i=1}^{3} d_i^2 + \sum_{i=1}^{3} x_i x_{(i+3)} + \sum_{i=1}^{3}\Big(\frac{T}{2}x_{(i+3)}^2 + Tx_{(i+3)}d_i + \frac{1}{2T}(f_{x_i}^e - f_{x_i}^{Eu})^2$$

$$+ x_i d_i + \frac{1}{T}x_i(f_{x_i}^e - f_{x_i}^{Eu}) + (x_{(i+3)} + d_i)(f_{x_i}^e - f_{x_i}^{Eu})\Big) - \sum_{i=1}^{3} x_i x_{(i+3)}$$

$$- \sum_{i=1}^{3}\big(x_{(i+3)}^2 + c(x_i + x_{(i+3)})^2\big) + \frac{1}{T}\sum_{i=1}^{3}\big((f_{x_i}^e - f_{x_i}^{Eu}) + (f_{x_{(i+3)}}^e - f_{x_{(i+3)}}^{Eu})$$

$$+ T(d_i + d_{(i+3)})\big)(x_i + x_{(i+3)}) + \dot{\omega}x_3(x_1 + x_4) - \dot{\omega}x_1(x_3 + x_6) \qquad (18)$$

$$+ \frac{1}{2T}\sum_{i=1}^{3}\big(-Tx_{(i+3)} - cT(x_i + x_{(i+3)}) + (f_{x_i}^e - f_{x_i}^{Eu}) + (f_{x_{(i+3)}}^e - f_{x_{(i+3)}}^{Eu})$$

$$+ T\dot{\omega}(x_3 - x_1)\big)^2 + \frac{1}{2T}\sum_{i=1}^{3} T^2(d_i + d_{(i+3)})^2 + \sum_{i=1}^{3}\Big[\big(-Tx_{(i+3)} - cT(x_i$$

$$+ x_{(i+3)})\big) + (f_{x_i}^e - f_{x_i}^{Eu}) + (f_{x_{(i+3)}}^e - f_{x_{(i+3)}}^{Eu}) + T\dot{\omega}(x_3 - x_1)\Big](d_i + d_{(i+3)})$$

Combining (13), (16) and (18), the following inequality can be obtained

$$\frac{1}{T}\big(V_T(x[k+1]) - V_T(x[k])\big)$$

$$\leq \frac{T}{2}\sum_{i=1}^{3} d_i^2 + \sum_{i=1}^{3} x_i x_{(i+3)} + \sum_{i=1}^{3}\Big(\frac{T}{2}\Delta_1^2 + T\Delta_1\Delta_2 + \frac{T}{2}\rho_{x_i}^2 + \Delta_1\Delta_2 + \Delta_1\rho_{x_i}(T)$$

$$+ T(\Delta_1 + \Delta_2)\rho_{x_i}(T)\Big) - \sum_{i=1}^{3} x_i x_{(i+3)} - \sum_{i=1}^{3}\big(x_{(i+3)}^2 + c(x_i + x_{(i+3)})^2\big)$$

$$+ 2\sum_{i=1}^{3}\big(\rho_{x_i}(T) + \rho_{(i+3)}(T) + 2\Delta_2\big)\Delta_1 + \frac{1}{2}(\xi x_1^2 + \xi x_3^2 + \xi x_4^2 + \xi x_6^2) \qquad (19)$$

$$+ \frac{T}{2}\sum_{i=1}^{3} d_{(i+3)}^2 + \frac{1}{2T}\sum_{i=1}^{3}\big((T\Delta_1 + 2cT\Delta_1) + T\rho_{x_i}(T) + T\rho_{x_{(i+3)}}(T) + 2T\xi\Delta_1\big)^2$$

$$+ \frac{3T}{2}\sum_{i=1}^{3}\Delta_2^2 + 2\sum_{i=1}^{3}\big(T\Delta_1 + 2c\Delta_1 + T\rho_{x_i}(T) + T\rho_{x_{(i+3)}}(T) + 2T\xi\Delta_1\big)\Delta_2$$

$$\leq -\sum_{i=1}^{3}\big(x_{(i+3)}^2 + c(x_i + x_{(i+3)})^2\big) + \frac{T}{2}\sum_{i=1}^{6} d_i^2 + \frac{1}{2}(\xi x_1^2 + \xi x_3^2 + \xi x_4^2 + \xi x_6^2) + \delta$$

Based on $0 < \xi < \frac{1}{2}$, and $c > \frac{\xi - \xi^2}{1 - 2\xi}$, we have

$$x_4^2 + c(x_1 + x_4)^2 - \frac{1}{2}\xi x_1^2 - \frac{1}{2}\xi x_4^2 \geq \frac{1}{2}\xi x_1^2 + \frac{1}{2}\xi x_4^2 \qquad (20)$$

$$x_5^2 + c(x_2 + x_5)^2 \geq x_5^2 + c(x_2 + x_5)^2 - \frac{1}{2}\xi x_2^2 - \frac{1}{2}\xi x_5^2 \geq \frac{1}{2}\xi x_2^2 + \frac{1}{2}\xi x_5^2 \quad (21)$$

$$x_6^2 + c(x_3 + x_6)^2 - \frac{1}{2}\xi x_3^2 - \frac{1}{2}\xi x_6^2 \geq \frac{1}{2}\xi x_3^2 + \frac{1}{2}\xi x_6^2 \quad (22)$$

According to the definition of L_∞ norm, the following inequality can be obtained, where M is a positive number.

$$\| d(t) \|_{L_\infty} \leq M, t \geq 0 \quad (23)$$

Choose $\alpha_1(r) = \frac{1}{8}r^2, \alpha_2(r) = 2r^2, \alpha_3(r) = \frac{1}{2}\xi r^2$ and $\gamma(r) = \frac{T}{2}r$, (17) can be rewritten and additional condition can be verified as follows:

$$V_T(x[k+1]) - V_T(x[k]) \leq -T\alpha_3(\| x \|) + \gamma(\| d(t) \|_{L_\infty}) + \delta \quad (24)$$

$$\alpha_1(\| x \|) \leq V_T(\| x \|) \leq \alpha_2(\| x \|) \quad (25)$$

Above all, by definition 2, (25) and (24) imply that the exact closed-loop discrete-time system with bounded external disturbances (12) is semiglobally practically input-to-state stable.

4 Simulation Results

In this section, we provide numerical simulations to validate the effectiveness of the proposed control algorithms for a particular rendezvous and docking mission. The orbital parameters are as follows: the semi-major axis $a = 4.44 \times 10^7 m$, the eccentricity $e = 0.6$, the gravitational constant $\mu = 3.98 \times 10^{14} m^3/s^2$, and the perigee radius $R_p = 4.4178 \times 10^7 m$. For simulation purposes, we set control coeffcient $c = 1$ and the upper bound $\xi = 0.0005$.

Moreover, the initial values of states are chosen as $x_1(0) = 100$, $x_2(0) = 55$, $x_3(0) = -70$, $x_4(0) = -4$, $x_5(0) = -3$, $x_6(0) = 2.5$, representing that the distance between the target spacecraft and the chaser spacecraft in all three directions are respectively $100m$, $55m$, $-70m$, and the relative velocity in all three directions are respectively $-4m/s$, $-3m/s$, $2.5m/s$. Besides, the external disturbances attached on the model are chosen as $d_1(t) = 0.8\sin t$, $d_2(t) = 0.2\cos t$, $d_3(t) = 0.4 \times (0.6\sin t + 0.9\cos t)$, $d_4(t) = 0.5 \times (0.2\sin t + 0.8\cos t)$, $d_5(t) = 0.7\sin t$, $d_6(t) = 0.1\cos t$.

Fig.2-Fig.4 show the state variables and the control input with sampling time $T = 0.5s$, and Fig.5-Fig.7 show the state variables and the control input with sampling time $T = 0.7s$. It can be seen all these state variables converge to a bounded region near the origin finally. It means that the controller (14) can achieve a good performance on the relative position and the relative velocity stabilization with bounded external disturbances.

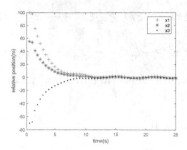

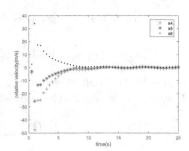

Fig. 2. Relative position with controller (14) (T=0.5s)

Fig. 3. Relative velocity with controller (14) (T=0.5s)

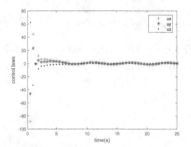

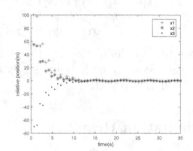

Fig. 4. Control input with controller (14) (T=0.5s)

Fig. 5. Relative position with controller (14) (T=0.7s)

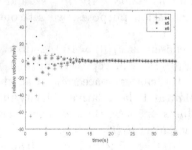

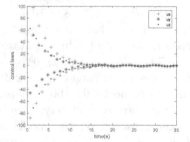

Fig. 6. Relative velocity with controller (14) (T=0.7s)

Fig. 7. Control input with controller (14) (T=0.7s)

5   Conclusions

In this paper, for a class of spacecraft rendezvous problems with target spacecraft on an arbitrary elliptical orbit, considering bounded external disturbances attached on the discrete-time dynamic model, we propose sampled-data control laws based on ISS method to achieve the whole system stabilization. It is obvious that relative position and relative velocity are semiglobally practically input-to-state stable with sampling time satisfying some certain conditions. Finally, numerical simulation results validate the effectiveness of the proposed control laws.

References

1. Zhang, D.W., Song, S.M., Pei, R.: Safe guidance for autonomous rendezvous and docking with a noncooperative target, In: Proc. of the AIAA Guidance, Navigation, and Control Conference (2010)
2. Fehse, W.: Automated Rendezvous and Docking of Spacecraft, vol. 16. Cambridge University Press (2003)
3. Singla, P., Subbarao, K., Junkins, J.L.: Adaptive Output Feedback Control for Spacecraft Rendezvous and Docking Under Measurement Uncertainty. Journal of Guidance, Control, and Dynamics 29(4), 892–902 (2006)
4. Gao, H., Yang, X., Shi, P.: Multi-Objective Robust Control of Spacecraft Rendezvous. IEEE Transactions on Control Systems Technology 17(4), 794–802 (2009)
5. Zhiqiang, Z.: Trajectory control of rendezvous with maneuver target spacecraft. In: AIAA/AAS Astrodynamics Specialist Conference (2012)
6. Kim, Y.H., David, B.: Spencer, Optimal Spacecraft Rendezvous Using Genetic Algorithms. Journal of Spacecraft and Rockets 39(6), 859–865 (2002)
7. Nesic, D., Teel, A.R., Kokotovi, P.V.: Sufficient Conditions For Stabilization of Sampled-data Nonlinear Systems via Discrete-time Approximations. Systems and Control Letters 38(4), 259 270 (1999)
8. Nesic, D., Teel, A.R.: A Framework for Stabilization of Nonlinear Sampled-data Systems Based on Their Approximate Discrete-time Models. IEEE Transactions on Automatic Control 49(7), 1103–1122 (2004)
9. Nesic, D., Teel, A.R.: Stabilization of Sampled-data Nonlinear Systems via Backstepping on Their Euler Approximate Model. Automatica 42(10), 1801–1808 (2006)
10. Nesic, D., Laila, D.S.: A Note on Input-to-state Stabilization for Nonlinear Sampled-data Systems. IEEE Transactions on Automatic Control 47(7), 1153–1158 (2002)
11. Liu, X., Marquez, H.J., Lin, Y.: Input-to-state Stabilization for Nonlinear Dual-rate Sampled-data Systems via Approximate Discrete-time Model. Automatica 44(12), 3157–3161 (2008)
12. Yamanaka, K., Ankersen, F.: New State Transition Matrix for Relative Motion on An Arbitrary Elliptical Orbit. Journal of guidance, control, and dynamics 25(1), 60–66 (2002)
13. Zhou, B., Lin, Z., Duan, G.-R.: Lyapunov Differential Equation Approach to Elliptical Orbital Rendezvous with Constrained Controls. Journal of Guidance, Control, and Dynamics 34(2), 345–358 (2011)
14. Curtis, H.: Orbital mechanics for engineering students. Butterworth-Heinemann (2013)

Marine Vehicles and Oceanic Engineering

Maximum Power Tracking Control for Current Power System Based on Fuzzy-PID Controller

Zhaoyong Mao[1(✉)], Weichao Huang[1], Chenguang Yang[2,3], Rongxin Cui[1], and Sanjay Sharma[4]

[1] School of Marine Science and Technology,
Northwestern Polytechnical University, Xi'an 710072, China
maozhaoyong@nwpu.edu.cn
[2] Center for Robotics and Neural Systems, University of Plymouth, Plymouth PL4 8AA, UK
[3] College of Automation Science and Engineering,
South China University of Technology, Guangzhou, China
[4] School of Marine Science and Engineering, University of Plymouth, Plymouth PL4 8AA, UK

Abstract. In order to solve the problem of maximum energy capture in the dynamic changing current, the maximum power point tracking (MPPT) control strategy of an extensible vertical-axis blade current power generation system was developed. Consider the current speed with the characteristics of randomness and uncertainties. Affixed set of PID parameters can hardly achieve desired generator speed. Therefore, this paper presents a Fuzzy-PID control, which combines the characteristics of traditional PID and fuzzy control. The simulation results show that the Fuzzy-PID control is able to guarantee desired dynamic characteristics, such as fast response, robustness and stability, and is able to effectively track the maximum power point.

Keywords: Maximum power tracking · Extensible vertical-axis blade · Fuzzy-PID control

1 Introduction

With the high speed development of economy, global problems gradually change seriously, such as air pollution, resource shortage and ecological deterioration. In order to solve the energy crisis and air pollution effectively, development of new energy and the use the renewable energy to replace the non-renewable energy have become the trend [1]. The global ocean area is vast, which potential energy is enormous. The ocean current energy mainly includes the kinetic energy of water flow, which has many advantages in comparison to other renewable energy. The ocean current energy has strong regularity and stability, which are predictable. The ocean current generation device lies below sea level, which does not occupy the land area, would not affect the surrounding landscape and would not pollute the environment. Therefore, the development and utilization of the ocean current energy is connected by more and more people [2].

© Springer International Publishing Switzerland 2015
H. Liu et al. (Eds.): ICIRA 2015, Part I, LNAI 9244, pp. 69–80, 2015.
DOI: 10.1007/978-3-319-22879-2_7

The basic composition of ocean current power generation device and the wind power generation device is similar, it is also known as the "underwater windmill". The principle of Ocean current power generation is that the kinetic energy of flowing water into mechanical energy and then the mechanical energy into electrical energy. Consider that the undersea environment is more complex, sea velocity uncertainty, randomness and other characteristics, a fixed set of PID parameters can hardly satisfy specified control performance to guarantee de-sired ocean current generator speed. Therefore, in this paper we propose a Fuzzy-PID controller to investigate the MPPT control strategy based on a scalable vertical axis impeller ocean current power generation system, which constantly adjust the PID parameters online based on the change of ocean current velocity [3, 4]. By analyzing the principle and character of the ocean current power generation system, we established simulation mathematic models in the MATLAB / Simulink environment. Simulation results show that the method can make the generator work at maximum power, and improve the dynamic process.

2   The Power Generation System

2.1   Power Generation System

Depending on the relative direction between the turbine rotational axis and the flow direction Ocean current power generation system water turbines can be classified into two groups: horizontal axis water turbine (HAWT) and vertical axis water turbine (VAWT) [5, 6]. Compared to the HAWT, the VAWT generation sys-tem can efficient capture any direction of ocean current energy. Generally the VAWT generation system in the structure is mainly composed of the impeller, gearbox, generator, power conditioning devices and so on (See Fig. 1).

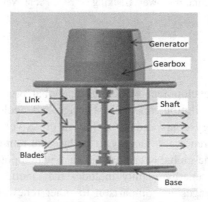

Fig. 1. VAWT generation system

Consider that a rotation period of VAWT can be divided into the advancing half cycle and the returning half cycle. The blades generate positive torque in the advancing half cycle, and generate negative torque in the returning half cycle. In order to reduce the negative torque in the returning half cycle, this paper presents a novel vertical axis water turbine with retractable arc-type blades (See Fig. 2)[7]. In the advancing half cycle blades are pushed out, generating positive torque to drive the rotor, while in the returning half cycle, blades are pulled to the surface of the drum, reducing the negative torque.

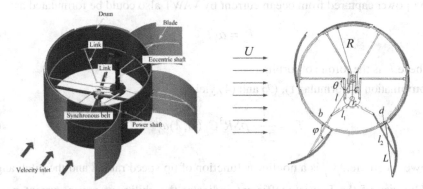

Fig. 2. The structure and working schematic plot of extensible vertical axis turbine

U is the velocity of the incident flow in fig. 2. The following table shows the main geometric parameters of turbine.

Table 1. Main geometric parameters of turbine

Parameters	values
Airflow shroud radius	$R=0.5\text{m}$
Eccentric distance	$C=0.225\text{m}$
Eccentric disc radius	$r=0.05\text{m}$
Link length	$l_1 = 0.225\,m, l_2 = 0.45\,m$
Blade chord length	$L=0.5\text{m}$

2.2 Power System Kinematics

Known by the Betz theory, the limited actual power captured from ocean current by VAWT could be formulated as [8]:

$$P_a = \frac{1}{2}\rho S C_p(\beta, \lambda) v^3 \qquad (1)$$

$$\lambda = \frac{\omega_r R}{v} \qquad (2)$$

$$S = 2RH \qquad (3)$$

Where ρ is the seawater density, S is the area swept by the blades, and v is the seawater speed, C_p is called the "power coefficient," and β represents the blade pitch angle. λ is the tip-speed ratio, R is the equivalent radius of the turbine and ω_r is the rotational speed. H is the height of the turbine.

The power captured from ocean current by VAWT also could be formulated as:

$$P_a = \omega_r T_a \tag{4}$$

Where T_a is the torque of turbine.

Combination of formula (1), (2) and (4) yields:

$$T_a = \frac{1}{2\lambda^3}\rho S R^3 C_p(\beta,\lambda)\omega_r^2 \tag{5}$$

Power coefficient C_p is a nonlinear function of tip speed ratio λ and the pitch angle β. The size of the Power coefficient indicates the ability of ocean current power generation system to absorb energy currents. When β is constant, C_p can only achieve the maximum in determining λ. λ known as the best tip speed ratio λ_{opt}, C_p known as the best Power coefficient $C_{p\max}$. The reference torque of the turbine is:

$$T_{opt} = k_{opt}\omega_r^2 \tag{6}$$

Where k_{opt} is constant.

Thus, when the system enters a stable state, the actual output torque of the turbine is equal to the output reference torque for any seawater speed and rotational speed.

According VAWT model, when the blade pitch angle is 0, the torque coefficient k_{opt} can be expressed as:

$$k_{opt} = \frac{1}{2\lambda_{opt}^3}\rho S R^3 C_{p\max}(\lambda) \tag{7}$$

In order to study the motion characteristics of ocean current power generation system, it's necessary to obtain the function of Power coefficient C_p and tip speed ratio λ. The relationship of C_p and λ shown in table 2 by analyzing based on the turbine model established in paper [7].

In order to achieve maximum power tracking, according to the data in Table 2 after several least-squares fitting, the function between and can be ex-pressed as:

$$Cp = 4.9167\lambda^5 - 5.2029\lambda^4 - 0.7712\lambda^3 + 0.4228\lambda^2 + 0.8053\lambda \tag{8}$$

Fig. 3 shows the curve of tip speed ratio and power coefficient. When the best power coefficient $C_{p\max}$ is 0.2595, the best tip speed ratio λ_{opt} is 0.4.

Table 2. Utilization coefficient under different tip speed ratio

λ	C_P
0.2	0.165503937
0.3	0.227308438
0.4	0.259535922
0.5	0.239151179
0.6	0.177328974

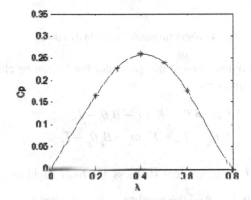

Fig. 3. The curve of tip speed ratio and power coefficient

3   The Control Strategy

3.1   A Simplified Model of the System

The second-order model of the VAWT generation system is shown in Fig. 4.

Where T_a is the output torque of the turbine, ω_r is the rotational speed of turbine, T_{ls} is the anti-torque of low-speed shaft, K_r is the out damping of the turbine, B_{ls} is the equivalent stiffness of low-speed shaft, K_{ls} is the equivalent damping of low-speed shaft. T_{hs} is the torque of high-speed shaft, J_g is the moment of inertia of the PMSG, K_g is the out damping of the PMSG, and T_{em} is the electromagnetic torque of PMSG.

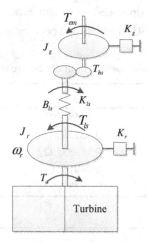

Fig. 4. Water turbine drive train dynamics

Combined the rotor dynamics and the generator inertia can be characterized by the following differential equations:

$$\begin{cases} J_r \dot{\omega}_r = T_a - K_r \omega_r - B_r \theta_r - T_{ls} \\ J_g \dot{\omega}_g = T_{hs} - K_g \omega_g - B_g \theta_g - T_g \end{cases} \tag{9}$$

Where $B_r, \theta_r, B_g, \theta_g$ for equivalent stiffness, the rotor angle of the turbine and the generator rotor, $T_g = n_g T_{em}$. And the gearbox ratio is defined as:

$$n_g = \frac{\omega_g}{\omega_r} = \frac{T_{ls}}{T_{hs}} \tag{10}$$

Thus, the equation 9 also can be expressed as:.

$$J_t \dot{\omega}_r = T_a - K_t \omega_r - B_t \theta_r - T_g \tag{11}$$

Where

$$\begin{cases} J_t = J_r + n_g^2 J_g \\ K_t = K_r + n_g^2 K_g \\ B_t = B_r + n_g^2 B_g \end{cases} \tag{12}$$

For simplicity, let us ignore the external stiffness B_t which is very low. Consequently, we can use the following simplified equation for control purposes.

$$J_t \dot{\omega}_r = T_a - K_t \omega_r - T_g \qquad (13)$$

For the control of PMSG we use zero d-axis current (ZDC) control. The principle of ZDC control is the control of PMSG d-axis stator current is 0, so that by the linear electromagnetic torque control to adjust the motor speed [9, 10].

In the dq coordinate, the electromagnetic torque equation of PMSG is expressed as:

$$T_{em} = \frac{3p}{2} \left[\lambda_r i_{qs} - (L_d - L_q) i_{ds} i_{qs} \right] \qquad (14)$$

where L_d, L_q is the d-axis inductance and q-axis inductance λ_r is the rotor flux, i_{ds}, i_{qs} is the dq axis stator current, P is the number of pole-pairs.

When the d-axis current of PMSG is equal to 0, the electromagnetic torque equation is expressed as:

$$T_{em} = \frac{3p}{2} \lambda_r i_{qs} = \frac{3p}{2} \lambda_r i_s \qquad (15)$$

As can be seen from the above equation, the electromagnetic torque is proportional to q-axis current when the generator rotor flux is a constant.

3.2   The Design of MPPT Control

This paper analyzes the model of the VAWT generation system, and proposed a new maximum power point tracking control strategy (optimal torque control). This method is based on the actual speed of the turbine to obtain the corresponding reference torque as electromagnetic reference torque, and then adjust the speed of the turbine at the optimum, that realization ocean current power generation MPPT control [11].

Fig. 5 is a control strategy designed in this paper, no need to measure sea velocity signal and does not require closed speed-loop control, only two current-loop control. First calculate the electromagnetic torque T_e^* according to the optimum power curve of the turbine and the actual speed of the turbine w_m, and then calculates the generator q-axis current reference i_q^* based on equation 10. Second, measure the generator three-phase stator currents i_{as}, i_{bs}, i_{cs} and position angle of the rotor magnetic, and convert the three-phase stator currents to the dq-axis stator current i_{ds}, i_{qs} by abc / dq coordinate transformation. In order to achieve the ZDC control, setting the d-axis current reference i_d^* is 0, compare the dq-axis stator current i_{ds}, i_{qs} with the reference value i_d^*, i_q^*, then get reference voltage of dq-axis v_d^*, v_q^* by two Fuzzy-PID controller, and then get the reference voltage V_α^*, V_β^* by abc/ $\alpha\beta$ coordinate transformation. Finally, control the active power of generator through SVPWM, and achieve maximum power tracking control.

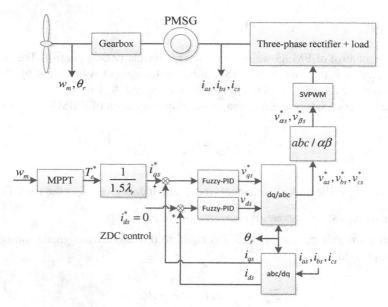

Fig. 5. Control strategy

4 Fuzzy-PID Controller

The conventional PID controller is a class of the most widely used basic controller with simple algorithm yet high reliability. However, the actual process of power generation system with nonlinear, time-varying, non-stable, and the con-trolled object load varied, complex confounding factors. To obtain satisfactory control results, we need to constantly tune the PID parameters online, so this paper proposes a Fuzzy-PID control algorithm to achieve maximum power tracking control [12, 13].

Fuzzy-PID control combines fuzzy control and traditional PID control. It is generally difficult for Fuzzy controller to completely eliminate the steady-state error, but the integrator of PID controller is able to eliminate the steady-state error perfectly. Therefore, the fuzzy control and PID regulator can be combined to increase the steady state control performance [14]. Fuzzy-PID control principle is shows in Fig. 6.

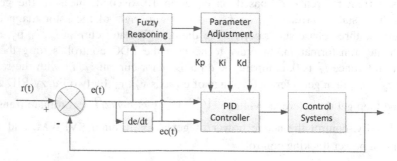

Fig. 6. The fuzzy PID controller

In this paper, two-dimensional fuzzy controller is adopted. First, let us define the fuzzy controller input, which are the absolute value of system deviation |e| and the derivate of error |de/dt|:{-6,-4,-2,0,2,4,6}, fuzzy subsets respectively{NB, NM, NS, ZO, PS, PM, PB}. According to the characteristics of the power generation system, we choose membership function is triangular membership functions (See Fig. 7).

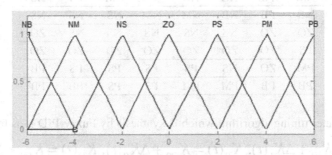

Fig. 7. The membership function

According to the principles of adaptive PID, after repeated operations and training, we can obtain the fuzzy control rule as follows [15, 16]:

Table 3. Fuzzy rule of kp

e	ec						
	NB	NM	NS	ZO	PS	PM	PB
NB	PB	PB	PM	PM	PS	PS	ZO
NM	PB	PM	PM	PS	PS	ZO	NS
NS	PM	PM	PS	PS	ZO	ZO	NS
ZO	PM	PS	PS	ZO	NS	NS	NM
PS	PS	PS	ZO	NS	NS	NM	NM
PM	PS	ZO	NS	NS	NM	NM	NB
PB	ZO	NS	NS	NM	NM	NB	NB

Table 4. Fuzzy rule of ki

e	ec						
	NB	NM	NS	ZO	PS	PM	PB
NB	NB	NB	NM	NM	NS	NS	ZO
NM	NB	NM	NM	NS	NS	ZO	PS
NS	NM	NM	NS	NS	ZO/	ZO	PS
ZO	NM	NS	NS	ZO	PS	PS	PM
PS	NS	NS	ZO	PS	PS	PM	PM
PM	NS	ZO	PS	PS	PM	PM	PB
PB	ZO	PS	PS	PM	PM	PB	PB

Table 5. Fuzzy rule of kd

e	ec						
	NB	NM	NS	ZO	PS	PM	PB
NB	PS	PS	NB	NB	NB	NB	PS
NM	PS	NB	NB	NM	NM	PS	NS
NS	PB	NB	NM	NM	NS	NS	NS
ZO	ZO	NS	NS	NS	NS	NS	ZO
PS	ZO	ZO	ZO	ZO	ZO	ZO	ZO
PM	ZO	NS	PS	PS	PS	PS	PB
PB	PB	PM	PM	PS	PS	PB	PB

PID parameter tuning algorithm which is synthesis by Fuzzy-PID is as follows:

$$K_I(t) = K_{I0} + \Delta K_I(t), \ K_P(t) = K_{P0} + \Delta K_P(t), \ K_D(t) = K_{D0} + \Delta K_D(t)$$

5 Simulation Results

This paper first established the maximum power point tracking control model for the current power generation system, and then validated the design scheme based on MATLAB / Simulink simulation software. In the simulation process, because the density of seawater is particularly large, the output torque of the blades is relatively large, increased a gearbox to prevent damage to the motor. By calculating the out torque of the turbine in the sea velocity 0.5m / s to 2m / s, select gearbox growth ratio of 1:20. In order to research the grid after, select the DC bus voltage Vdc = 380V. The parameters of PMSG are set as follows: Stator resistance R = 2.875Ω, stator inductance $Ld = Lq$ = 8.5mH, pole pairs P = 6, permanent magnet flux linkage 0.175Wb, moment of inertia $J = 8.0 \times 10^{-3}$kg.m^2. PWM wave period T = 0.0002S. PID parameters of the traditional PID controller are set offline as Kp= 10, Ki = 3 and Kd = 0.05. Fig. 8 is the seawater velocity signal in simulation. The detailed simulation results are shown in Fig. 9-11.

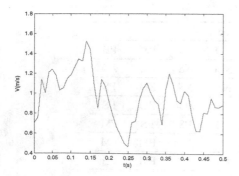

Fig. 8. Curve of the velocity

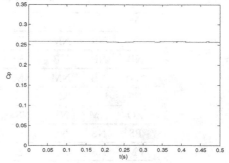

Fig. 9. Curve of Cp

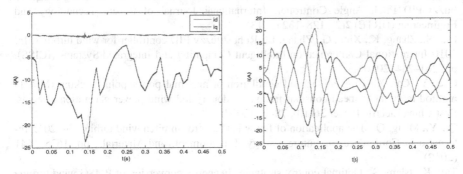

Fig. 10. Curves of power   **Fig. 11.** Curve of three-phase current

As shown in Fig. 9, the power coefficient always is 0.2596 with the constantly changing of seawater speed. Fig. 10 shows that the simulation has a good zero d-axis current control of PMSG with Fuzzy-PID control. Fig. 11 shows that the three-phase current constantly changing with d-axis current, but the overall trend showed a sine wave, which meet the running rules of PMSG. The above analysis shows that the fuzzy PID control with excellent control effect and more rapid response speed, high regulation accuracy, good steady-state performance.

6   Conclusions

This paper has investigated the characteristics of ocean current power generation system and has developed a Fuzzy-PID control strategy for the stator current control of ocean current power generator system. Maximum power point tracking has been achieved with the optimum torque control. The simulation results have clearly shown that: for the system parameter uncertainty, randomness and other characteristics, Fuzzy-PID control strategy has good dynamic characteristics, fast response, good stability, can effectively track the maximum power point. These provide practical guidance to the next step to design and improve physical prototypes.

Acknowledgement. This work was supported by the National Natural Science Foundation of China (NSFC) under grants 51179159, 61472325, 61473120, the NSFC-RS joint research project under grants 51311130137 in China and IE121414 in the UK, the Guangdong Provincial Natural Science Foundation of China 2014A030313266, and the Fundamental Research Funds for the Central Universities under Grant 2015ZM065.

References

1. Yu, H., Gao, Y., Zhang, H.: Fuzzy Self-adaptive PID control of the variable speed constant frequency variable-pitch wind turbine system. In: 2014 IEEE International Conference on System Science and Engineering (ICSSE), pp. 124–127 (2014)

2. Goyal, S., Gaur, M., Bhandari, S.: Power Regulation of a Wind Turbine Using Adaptive Fuzzy-PID Pitch Angle Controller. International Journal of Recent Technology and Engineering (IJRTE) 2(2), 128–132 (2013)
3. Liu, S., Zhang, K., Xing, G., Zhang, L.: Robust fuzzy PID controller for wind turbines. In: 2010 International Conference on Intelligent Computing and Integrated Systems (ICISS), pp. 261–264 (2010)
4. Cao, Y., Cai, X.: Comparison and simulation of maximal power point tracking (MPPT) method in variable speed constant frequency doubly-fed wind power generation systems. East China Electric Power 36(10), 78–82 (2008)
5. Qi, Y., Meng, Q.: The application of fuzzy PID control in pitch wind turbine. In: 2012 International Conference on Future Energy, Environment, and Materials, pp. 1635–1641 (2012)
6. Tan, K., Islam, S.: Optimal control strategies in energy conversion of PMSG wind turbine system without mechanical sensors. IEEE Trans. Energy Convers. 19(2), 392–399 (2004)
7. Wenlong, T., Baowei, S., Zhaoyong, M., Hao, D.: Design of a novel vertical axis water turbine with retractable arc-type blades. Marine Technology Society Journal 47(4), 94–100 (2013)
8. Li, D.-X., Li, B., Liu, L.-N.: The simulation of VSCF wind power system based on fuzzy-PID controller. In: 2011 International Conference on Computer Distributed Control and Intelligent Environmental Monitoring, pp. 1896–1899 (2011)
9. Yang, M., Wang, X.: Fuzzy PID controller using adaptive weighted PSO for permanent magnet synchronous motor drives. In: 2009 Second International Conference on Intelligent Computation Technology and Automation, pp. 736–739 (2009)
10. Liao, M., Dong, L., Jin, L., Wang, S.: Study on rotational speed feedback torque control for wind turbine generator system. In: 2009 International Conference on Energy and Environment Technology, pp. 853–856 (2009)
11. Fu, M.-L., Wu, X.-H., Lei, T.: Research of three-phase high power factor rectifier based on space-vector pulse width modulation 36(10), 1527–1529+1541 (2012)
12. Jeng, Y.-F., Zhang, L., Xu, B.-J., Wen, K.-L.: Fuzzy PID controler in permanent magnetic. In: 2012 IEEE International Conference on Intelligent Control, Automatic Detection and High-End Equipment (ICADE), pp. 176–180 (2012)
13. Elwer, A.S., Wahsh, S.A., Khalil, M.O., Nur-Eldeenl, A.M.: Intelligent fuzzy controller using particle swarm optimization for control of permanent magnet synchronous motor for electric vehicle. In: Proceedings of Industrial Electronics Society (ICEON 2003), vol. 2, pp. 1762–1766 (2003)
14. Arulmozhiyal, R., Kandiban, R.: Design of fuzzy PID controller for brushless DC motor. In: International Conference on Computer Communication and Informatics, pp. 10–12 (2012)
15. Deng, W., Wang, W.-Q., Liu, L., Liu, M.: Power Control Study of Wind Power Generation Based on Fuzzy-PID. Journal of Electric Power 28(1), 40–43 (2013)
16. Errami, Y., Hilal, M., Benchagra, M., Maaroufi, M., Ouassaid, M.: Nonlinear Control of MPPT and Grid Connected for Wind Power Generation Systems Based on the PMSG. Multimedia Computing and System 10(1109), 1055–1060 (2012)

Optimization of Composite Cylindrical Shell Subjected to Hydrostatic Pressure

Guang Pan, Jiangfeng Lu[✉], Kechun Shen, and Jiujiu Ke

School of Marine Science and Technology, Northwestern Polytechnical University,
Xi'an 710072, Shaanxi, People's Republic of China
winf.23@gmail.com

Abstract. Composite is used to substitute aluminum alloy as material of the underwater vehicle in this study. Nonlinear buckling behaviors of composite cylindrical underwater shell are studied using ANSYS software. Carbon/Epoxy is selected as material of the vehicle through comparative analysis. Optimization of ply sequence, thickness and angle of composite cylindrical shell subjected to hydrostatic pressure is investigated. Both buckling and material damage is considered in the optimization, the results show that buckling is the major destroy form. The vehicle's buckling pressure is greatly increased after optimization. And the vehicle's working depth has a 32% increase without changing its shape and structure.

Keywords: Underwater vehicle · Composite material · Finite element · Optimization

1 Introduction

Cylindrical shell has high pressure resistance and good hydrodynamic characteristics, so most automatic underwater vehicles (AUV) are cylindrical. And composite material has high strength: weight ratio, good corrosion resistance, good sound absorption performance, good manufacturability, long service life etc.[1]. These good properties make composite a suitable material for underwater pressure vessels.

R.K.H.Ng [2] studies E-glass/Epoxy shallow water composite pressure vessel, finds that its strain and stress safety factor are 17.17 and 11.95, but buckling safety factor is only 1.25. That means buckling damage occurs before stress damage, which is the main failure mode. Therefore, it is very meaningful to study the buckling behavior of composite cylindrical shell subjected to external hydrostatic pressure.

Gyeong-Chan Lee [3] uses a micro-genetic algorithm to optimize underwater composite sandwich cylinders, the result shows that both buckling and material failure should be considered.

Tanguy Messager [4] uses a genetic algorithm to optimize ply angle and sequence of thin underwater composite sandwich cylindrical vessels, the result shows that specific $[90_{N1}\psi_1\phi_{N2}\psi_290_{N3}]$ patterns can induce significant increase of buckling pressure.

© Springer International Publishing Switzerland 2015
H. Liu et al. (Eds.): ICIRA 2015, Part I, LNAI 9244, pp. 81–90, 2015.
DOI: 10.1007/978-3-319-22879-2_8

The author has studied buckling behaviors of underwater composite cylindrical hull by nonlinear method, finds that buckling pressure changes through varying ply thickness, angle and sequence. In other words, the buckling pressure of underwater cylindrical shell can maximize through optimization of ply thickness, angle and sequence.

Nonlinear buckling numerical analysis method is adopted to study the buckling behaviors of AUV shell, which greatly improves the accuracy. By contrast, proper material Carbon/Epoxy is selected as AUV shell's material. And the method of distribution optimization is used to design composite layers to achieve maximum of AUV detection depth.

2   Description of the Problem

An aluminum automatic underwater vehicle (AUV) named V8 has been made for underwater detection several years ago, the max detection depth is 10 m. Due to working requires, the vehicle's working scope needs to be elevated to 16 m. V8's detection depth is increased by substituting aluminum to composite material in this study, maintaining its shape and wall thickness the same. Then, the AUV's working scope will be increased without redesigning, saving a lot of time and money.

The AUV's cross-sectional view is showed in Fig. 1. And its dimensions are as follows: a length of 1850 mm, an outer diameter of 200 mm, a wall thickness of 3 mm, a rib number of 7, a rib space of 211 mm, a rib thickness of 3 mm, and a rib width of 5 mm.

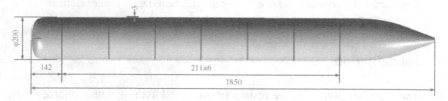

Fig. 1. The cross-sectional view of the vehicle

3   Finite Element Analysis Verification

Seong-Hwa Hur [5] uses a nonlinear finite element software, ACOSwin, to study the postbuckling behavior of composite cylindrical underwater shell. The deviation between simulation and experiment is about 15%.

The concept of defects factor (*DF*) is introduced when considering model's geometric imperfections [6], and finite element software, ANSYS, is used to study nonlinear buckling behaviors of composite cylindrical shell subjected to hydrostatic pressure in this paper. The deviation between simulation and experiment is about 5%.

In order to apply the proper geometry imperfections, *DF* needs to be estimated reliably. *DF* is equal to the ratio of manufacturing errors (*MT*) with maximum displacement (*MD*) of first-order linear eigenvalue buckling, as shown in Equation 1.

$$DF = \frac{MT}{MD} = \frac{0.17}{1.035} = 0.164 \tag{1}$$

The finite element model and boundary conditions are showed in Fig. 2. The model's dimension is 316 mm (diameter) ×600 mm (length). Left end of the model is unclosed, and right end is closed by 13 mm thick carbon steel. Left end is applied fixed constraint, right end is only allowed axial displacement, circumferential direction and right end are exert uniform pressure load.

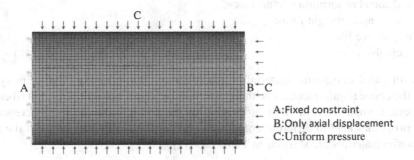

Fig. 2. The finite element model and boundary conditions

The displacement of node 1078 is max of the model. And pressure vs. displacement curve of node 1078 is showed in Fig. 3. As can be seen from the figure, the curve reaches the peak at 0.57MPa, then change of pressure causes a sharp change of displacement, which shows that buckling behavior of cylindrical shell has occurred. Experimental buckling pressure [5] is 0.55MPa, and finite element buckling pressure in this paper is 0.57MPa, relative error is 5%, agreed well with the experimental results.

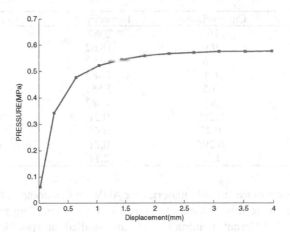

Fig. 3. Pressure vs. displacement curve of node 1078

4   Material Selection

The selected material must be able to withstand the harsh marine environment, because the optimization purpose is realization of the AUV's bigger detection depth. Under the deep-sea environment, requires of material are as follows [7]:

- High resistance to water pressure;
- Good corrosion resistance;
- Good sound absorption performance;
- High strength: weight ratio;
- Long service life;
- Acceptable price.

Resin-based composite materials Carbon/Epoxy, Boron/Epoxy, Glass/Epoxy can meet the above requirements. The three materials' matrixes are all Epoxy, but fibers are different. Comparisons of fiber properties are showed in Table 1 [8]. Differences of fiber properties lead to different composites' properties, elastic properties of the three composite materials are showed in Table 2 [9-11].

Table 1. Comparisons of fiber properties

Fiber	Tensile Modulus ,GPa	Tensile Strength ,GPa	Density ,g/cm^3	Price ,$/kg
Carbon	207-345	2.41-6.89	1.75-1.9	44-220
Boron	400	5.03-6.89	2.3-2.6	220-550
Glass	69-86	3.03-4.61	2.48-2.6	11-88

Table 2. Elastic properties of composite materials

Material	Carbon/Epoxy	Boron/Epoxy	Glass/Epoxy
E_{11}/GPa	162	206.8	45.6
E_{22}/GPa	9.6	18.62	16.2
E_{33}/GPa	9.6	18.62	16.2
G_{12}/GPa	6.1	4.482	5.83
G_{23}/GPa	3.5	2.551	5.78
G_{13}/GPa	6.1	4.482	5.83
v_{12}	0.298	0.21	0.278
v_{23}	0.47	0.45	0.4
v_{13}	0.298	0.21	0.278
$\rho/g/cm^3$	1.32	2.11	1.85

In order to choose a more suitable material for AUV shell, a comparative analysis of the three composite materials is done. As shown in Fig. 4, buckling pressure vs. ply thickness curves of different materials shell are studied at specific $[90_5/60/60_6]_s$ stacking sequence.

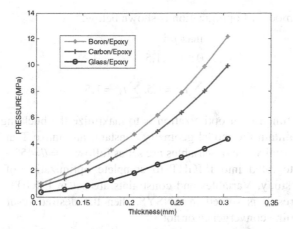

Fig. 4. Buckling pressure vs. ply thickness curves of different materials shell

As can be seen from Fig. 4, the buckling pressure of Boron/Epoxy shell is the largest, but the price is the highest, and the density (2.11g/cm3) is the largest, so the material is only applicable to improve the buckling pressure; the buckling pressure of Glass/Epoxy shell is the smallest, but the price is the lowest, and the density (1.85g/cm3) is smaller, the material is only applicable to pursue cost; the buckling pressure of Carbon/Epoxy shell is the larger, and the price is lager, but the density (1.32g/cm3) is the smallest, the material is applicable to pursue comprehensive performance. Based on the above analysis, Carbon/Epoxy is chosen as AUV shell's material in this study.

5   Optimization

The method of distribution optimization is used to design composite layers in this paper. Firstly, ply angle and thickness are optimized, the numbers of layers can be got through total thickness divided by single thickness. Then according to the number of layers, ply sequence is optimized by Genetic algorithm (GA) [12-14].

Taking into account the case of the same design parameters, such as rib geometric parameter, rib arrangement, shell thickness, destroy is most likely to occur in the middle of the AUV shell [15]. Thus only buckling behavior of the shell's middle section is studied.

Optimization of Ply Angle and Thickness

$[0/\pm\phi/90]_s$ and $[0/\pm\theta/90]_s$ are set as stacking sequence of casing and ribs respectively. In order to facilitate the expression of ply thickness, some rules are set as follows: t_{ij} represents ply thickness, $i=1,2$ represent casing and rib respectively, $j=1,2,3,4$ represent $0°$, $\phi°\ \&\ \theta°$, $-\phi°\ \&\ -\theta°$, $90°$ respectively. Optimization variables include ply angles ϕ, θ and ply thickness t_{ij}.

Mathematical model of optimization is shown below:

$$\text{max } pre.$$

$$0 \le t_{ij} \le 1.5$$

$$\sum_{j=1}^{4} t_{1j} = 1.5, \sum_{j=1}^{4} t_{2j} = 1.5 \tag{2}$$

The objective function of optimization is to maximize the buckling pressure, constraints are to maintain the model geometry constant, and make Tsai-Wu coefficient less than 1. The initial value of variables are set as follows: $\varphi=\theta=45°$, $t_{ij}=0.375$ mm.

ANSYS is integrated into ISIGHT to complete optimization of ply angle and thickness in this study. Variables and constraints are set in ISIGHT, finite model is established and result is solved in ANSYS, then the desired result is returned to ISIGHT to determine convergence or not.

Optimization result of ply angle and thickness is shown in Table 3. Due to restrictions on the laying process, ply angle is rounded to 45°. Single-layer material thickness of Carbon/Epoxy is 0.1 mm, and ply layers are integer's layer [16]. Thus the thickness of each layer value needs to be rounded after optimization, the rounded value is an integer multiple of 0.1. Round result is also shown in Table 3.

Table 3. Optimization result of ply angle and thickness

variable	ϕ	θ	t_{11}	t_{12}	t_{13}	t_{14}	t_{21}	t_{22}	t_{23}	t_{24}
Opti.	46	45	0.862	0.095	0.094	0.447	1.032	0.092	0.092	0.283
Round	45	45	0.9	0.1	0.1	0.4	1.0	0.1	0.1	0.3

Buckling pressure comparison chart between before and after optimization is shown in Fig. 5. The initial buckling pressure is 1.4456 MPa, buckling pressure becomes 1.6512 MPa which is 14.22% larger than the former after optimization.

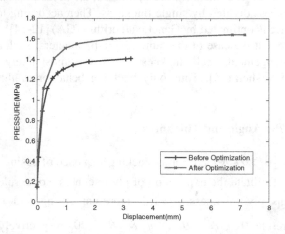

Fig. 5. Buckling pressure comparison chart between before and after optimization

Tsai-Wu criterion is used in this paper, because the theory and experiments are well consistent with each other [17-19]. Tsai-Wu failure factor graph of the vehicle is showed in Fig. 6. As shown in Fig. 6, the maximum Tsai-Wu failure factor is 0.92, less than 1, thus the vehicle's material is safe.

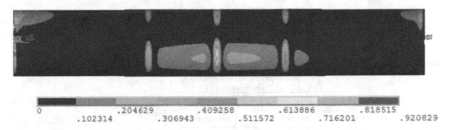

0		.204629		.409258		.613886		.818515	
	.102314		.306943		.511572		.716201		.920829

Fig. 6. Tsai-Wu failure factor graph of the vehicle

After optimization, stacking sequence of casing is $\left[0_9 / \pm 45 / 90_4\right]_s$, and stacking sequence of ribs is $\left[0_{10} / \pm 45 / 90_3\right]_s$.

Optimization of Ply Sequence

Variables of stacking sequence are all discrete variables, and GA's integer encoding strategy is good at optimization of discrete variables, thus GA is selected as optimization algorithm.

The principle of GA can be described by biological evolution: parameters of problems need to be solved are encoded to form a gene by binary coded or decimal code; then a chromosome is consisted by several genes; initial population is consisted by several chromosomes; new chromosomes and population are consisted after selection, crossover, mutation; optimization is over after several times iteration until optimal chromosomes are got.

After ply angle optimization, $0°, 45°, -45°, 90°$ are chosen as laying angle, they are all discrete. According to integer encoding strategy, inters 1,2,3,4 represent $0°, 45°, -45°, 90°$ respectively. Thus $\left[1_9 / 2 / 3 / 4_4\right]_s$ represents stacking sequence of casing, and $\left[1_{10} / 2 / 3 / 4_3\right]_s$ represents stacking sequence of ribs.

As layers of the vehicle are symmetrical, shell needs to define the stacking sequence of 15 layers, and ribs need to define the stacking sequence of 15 layers too. A total of 30 layers need to be defined, therefore gene number of chromosome is 30. The chromosome is called parent chromosome, and it's expressed as: $[1_{10}2343/1_92344]$.

Flow chart of ply sequence optimization is shown in Fig. 7. To expand on the procedure flowchart, each step is explained here.

Step 1: The chromosome is decoded into ply angle using Matlab.
Step 2: ANSYS is called to complete modeling and solving by Matlab.
Step 3: The result is returned to Matlab to calculate the fitness value by ANSYS.
Step 4: Selection, crossover and mutation is conducted in Matlab.
Step 5: The above iteration is conducted until convergence.

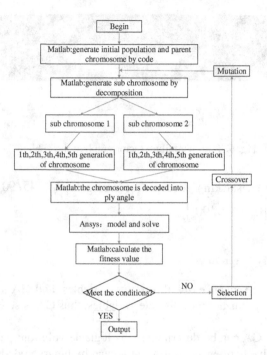

Fig. 7. Flow chart of ply sequence optimization

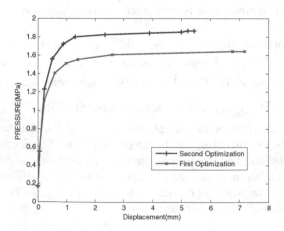

Fig. 8. Comparison chart of two times of optimization results

After 106 iteration, the result has been convergence. The ply sequence of casting and ribs are [±45/0/90/0₂/90/0₂/90/0₃/90/0] and [±45/0₂/90/0₃/90/0₃/90/0₂] respectively through optimization. Comparison chart of two times of optimization results is shown in Fig. 8. As shown in Fig. 8, the buckling pressure is 1.8651MPa which is bigger 13% than last optimization result.

Tsai-Wu failure factor graph of the vehicle after two times of optimization is shown in Fig. 9. As shown in Fig. 9, the maximal Tsai-Wu failure factor is 0.97 which is smaller than 1, thus the vehicle's material is safe.

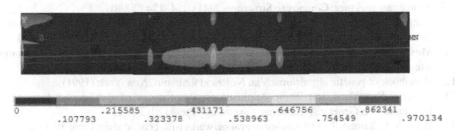

0	.215585	.431171	.646756	.862341
.107793	.323378	.538963	.754549	.970134

Fig. 9. Tsai-Wu failure factor graph after two times of optimization

6   Conclusion

Nonlinear number analysis method is used to study the bucking behaviors of the vehicle, well consistency with experiments is verified through an example. Buckling behaviors of three kinds of materials cylinder underwater shell is contrasted, including Carbon/Epoxy, Boron/Epoxy and Glass/Epoxy, the result shows that tensile modulus of fibers has great impact on the pressure capacity of cylindrical shell, and Carbon/Epoxy is an idea material for underwater vehicle.

The optimization of composite cylindrical shell subjected to hydrostatic pressure is investigated. Ply angle, thickness and sequence are studied using the method of distribution optimization. Buckling pressure is increased from 1.4456 to 1.8651 after two times of optimization. The buckling pressure has been greatly improved without changing shape and structure of the vehicle, saving a lot of time and energy.

References

1. Elsayed, F., Qi, H., Tong, L.L., et al.: Optimal Design Analysis of Composite Submersible Pressure Hull. Applied Mechanics and Materials **578**, 89–96 (2014)
2. Ng, R.K.H., Yousefpour, A., Uyema, M., et al.: Design, analysis, manufacture, and test of shallow water pressure vessels using e-glass/epoxy woven composite material for a semi-autonomous underwater vehicle. Journal of Composite Materials **36**(21), 2443–2478 (2002)
3. Lee, G.C., Kweon, J.H., Choi, J.H.: Optimization of composite sandwich cylinders for underwater vehicle application. Composite Structures **96**, 691–697 (2013)

4. Messager, T., Pyrz, M., Gineste, B., et al.: Optimal laminations of thin underwater composite cylindrical vessels. Composite Structures **58**(4), 529–537 (2002)
5. Hur, S.H., Son, H.J., Kweon, J.H., et al.: Postbuckling of composite cylinders under external hydrostatic pressure. Composite Structures **86**(1), 114–124 (2008)
6. Amabili, M., Karagiozis, K., Paidoussis, M.P.: Effect of Geometric Imperfections on Nonlinear Stability of Cylindrical Shells Conveying Fluid (2011)
7. Ross, C.T.F.: A conceptual design of an underwater vehicle. Ocean Engineering **33**(16), 2087–2104 (2006)
8. Kaw, A.K.: Mechanics of composite materials. CRC press (2005)
9. Hur, S.H., Son, H.J., Kweon, J.H., et al.: Postbuckling of composite cylinders under external hydrostatic pressure. Composite Structures **86**(1), 114–124 (2008)
10. Tanov, R., Tabiei, A.: A simple correction to the first-order shear deformation shell finite element formulations. Finite Elements in Analysis and Design **35**(2), 189–197 (2000)
11. Messager, T., Pyrz, M., Gineste, B., et al.: Optimal laminations of thin underwater composite cylindrical vessels. Composite Structures **58**(4), 529–537 (2002)
12. Handbook of genetic algorithms. Van Nostrand Reinhold, New York (1991)
13. Genetic Algorithms and Their Applications: Proceedings of the Second International Conference on Genetic Algorithms. Psychology Press (2013)
14. Harp, S.A., Samad, T.: Optimizing neural networks with genetic algorithms. In: Proceedings of the 54th American Power Conference, Chicago, vol. 2 (2013)
15. Carvelli, V., Panzeri, N., Poggi, C.: Buckling strength of GFRP under-water vehicles. Composites Part B: Engineering **32**(2), 89–101 (2001)
16. Barbero, E.J.: Introduction to composite materials design. CRC press (2010)
17. Tsai, S.W., Wu, E.M.: A general theory of strength for anisotropic materials. Journal of Composite Materials **5**(1), 58–80 (1971)
18. Tsai, S.W.: Theory of composites design. Think composites, Dayton (1992)
19. Groenwold, A.A., Haftka, R.T.: Optimization with non-homogeneous failure criteria like Tsai–Wu for composite laminates. Structural and Multidisciplinary Optimization **32**(3), 183–190 (2006)

Application of Federated and Fuzzy Adaptive Filter on the Velocity and Angular Rate Matching in the Transfer Alignment

Lijun Song[(✉)], Zongxing Duan, Dengfeng Chen, Jiwu Sun, and Shengjun Xu

Electronic Information and Control Engineering College,
Xi'an University of Architecture and Technology, Xi'an 710055, China
Songlijun9071@sina.com

Abstract. The federated and fuzzy adaptive filter is composed of the federated filter and the fuzzy adaptive filter which has been improved. To adjust the structure coefficient and algorithm for the new innovation of the matching filter, the federated and fuzzy adaptive filter adjust the weight of locally estimated state on the velocity matching and the angular rate matching in the transfer alignment by the difference between the real innovation variance and the theory innovation variance. And it guarantees the stability of matching filter and the accuracy of transfer alignment. The result of simulation shows that the federated and fuzzy adaptive filter is better to estimate the initial attitude misalignment angle of inertial navigation system (INS) when the system dynamic model and noise statistics characteristics of inertial navigation system are unclear.

Keywords: The federated and fuzzy adaptive filter · Velocity and angular rate matching · Transfer alignment · The fuzzy adaptive filter · The federated filter

1 Introduction

The coordinate of INS is uncertain in the beginning, the INS must have been given the initial parameters to establish the coordinate of INS before the INS start, and it is called initial alignment. The precision of INS is determined by the precision of initial alignment. The INS of vector used as reference of alignment in the initial alignment of INS on moving base because the movement environment of moving base is complex, and the slave inertial navigation systems (SINS) is finished initial alignment on dynamic matching of the output date of the main inertial navigation system (MINS) and SINS, it is called Transfer Alignment (TA) [1,2].

2 The Federated and Fuzzy Adaptive Filter

The state noise variance and observation noise variance are given in advance, the calculation of gain matrix and variance matrix toward the steady state, so the Kalman

This thesis is supported by Project supported by the National Natural Science Foundation of China, project Number: 51178373.

H. Liu et al. (Eds.): ICIRA 2015, Part I, LNAI 9244, pp. 91–100, 2015.
DOI: 10.1007/978-3-319-22879-2_9

filter has lower capability for the change noise of maneuvering target. The precision of filter will decreased when the characteristics of practical noise has a little change. It is the basic idea of federated and fuzzy adaptive filter that model parameters or system characteristics are estimate and corrected by utilization of observation data and the real-time filter have been improved [3,4].

2.1   The Federated Filter

The federal filter is composed of several subfilters and a main filter, where the sub filter can independently update the time and measurement, and the result will send to the main filter. And the main filter will feed back to the sub filter after data-fusion is analyzed. It is the initial value of the next treatment cycle [5,6].

The fusion algorithm of the traditional federated filter is:

$$\hat{X}_g = P_g \left(\sum_{i=1}^{m} P_{ci}^{-1} \hat{X}_{ci} \right) \tag{1}$$

$$P_g = \left(\sum_{i=1}^{m} P_{ci}^{-1} \right)^{-1}$$

Where $\hat{X}_g$ is global state estimation after fusion, $\hat{X}_{ci}$ is the local optimum estimation of sub filter on the system of public state, P_{ci} is variance matrix of error of the local optimum estimation, and the local estimation is not related with each other.

2.2   The Fuzzy Adaptive Filter

In the actual engineering application, It is center of the fuzzy adaptive filter that the observation data which come from system used as the information of optimal filter when the system model have the unknown parameters.

It is assume that the system equation and measurement equation of linear discrete system can be written as:

$$\begin{cases} X_k = \Phi_{k,k-1} X_{k-1} + W_{k-1} \\ Z_k = H_k X_k + V_k \end{cases} \tag{2}$$

Where X_k is estimated state, Z_k is the measurement of system, $\phi_{k,k-1}$ is transfer matrix from t_{k-1} to t_k. W_{k-1} is noise sequence of system incentive. H_k is the measurement matrix, V_k is noise sequence of measurement and $\mathrm{E}\left[V_k V_k^{\mathrm{T}}\right] = R$.

In order to avoid the divergence of system used to system noise and measurement noise, it is Supposed that the variance matrix of system noise and measurement noise are exponential function, kalman filter equation can modified .

System noise: $Q_k = Q\alpha^{-(2k+1)}$

Measurement noise: $R_k = R\alpha^{-(2k+1)}$

Matrix of weighted filtering: $P_k^{\alpha-} = P_k \alpha^{-2k}$

Matrix of state prediction: $\hat{X}_{k/k-1} = \Phi_{k,k-1}\hat{X}_{k-1}$

Covariance matrix of prediction error: $P_{k/k-1}^{\alpha-} = \alpha^2 \Phi_{k,k-1} P_k^{\alpha-} \Phi_{k,k-1}^{T} + Q_k$

Matrix of filter gain: $K_k = P_{k/k-1}^{\alpha-} H_k^{T}\left(H_k P_{k/k-1}^{\alpha-} H_k^{T} + R_k\right)^{-1}$

Covariance matrix of estimation error: $P_k^{\alpha-} = (I - K_k H_k) P_{k/k-1}^{\alpha-}$

Matrix of state estimation: $\hat{X}_k = \hat{X}_{k/k-1} + K_k(Z_k - H_k \hat{X}_{k/k-1})$

Innovation of measurement (residual error) : $r_k = Z_k - H_k \hat{X}_k$

Covariance matrix of innovation: $P_k^r = H_k P_{k/k-}^{\alpha-} H_k^{T} + R_k$

3   The Federated and Fuzzy Adaptive Filter

In information fusion of transfer alignment, it is the federated and fuzzy adaptive filter that the fuzzy adaptive filter used in the sub filter and the global estimate is accepted by the weight of fuzzy adaptive filter of in the main filter.

3.1   The Design of Fuzzy Adaptive Filter

In the period, it is assumed that statistical number is N, and the mean value of innovation at K time is

$$\bar{r}_k = \begin{bmatrix} \bar{r}_{k1} & \bar{r}_{k2} & \bar{r}_{k3} \end{bmatrix}^{T} = \frac{1}{N} \sum_{j=k-N+1}^{k} r_j$$

Where $r_j = \begin{bmatrix} r_{j1} & r_{j2} & r_{j3} \end{bmatrix}^{T} = Z_j - H_j \hat{X}_j$

Estimation of innovation covariance at K time is

$$\hat{P}_k^r = \begin{bmatrix} \hat{P}_{k1}^r & \hat{P}_{k2}^r & \hat{P}_{k3}^r \end{bmatrix}^{T}$$

Where $\hat{P}_{ki}^r = \frac{1}{N-1} \sum_{j=k-N+1}^{k} \left(r_{ji} - \bar{r}_{ki}\right)^2$, $i = 1, 2, 3$

Theoretical value of innovation covariance at K time is

$$P_k^r = H_k P_{k/k-1}^{\alpha-} H_k^{\mathrm{T}} + R_k$$

The diagonal elements of theoretical value of innovation covariance at K time is

$$P_{ki}^r = P_k^r(i,i) \quad , \quad i = 1,2,3$$

So the ratio of estimation variance of innovation and theoretical variance of innovation at K time is

$$\beta_{ki} = \hat{P}_{ki}^r / P_{ki}^r \quad , \quad i = 1,2,3$$

In the process of filtering, three T-S fuzzy logic system adjusted by using adaptivev, the mean value of innovation $\bar{r}_{ki}$ and the ratio of estimation variance of innovation and theoretical variance of innovation β_{ki} used as inputs, and parameter α used output.

3.2   The Weight of Fuzzy Adaptive Filter

There are some difficulties in the process of transfer alignment when you want to design the system model and noise model. In order to avoid the noise and error of model, the main filter used the fuzzy adaptive filter of weight to fuse the information which comes from the sub filter. It is weight assignment of local estimation in the global state that comes from the parameter of sub filter [7,8].

In the weight of fuzzy adaptive filter, the difference of theoretical variance and practical variance of innovation DOM_k and variance matrix of measurement noise R used as input, the weight of subfilter's local estimation used as output.

The theoretical variance of innovation is

$$\hat{P}_k^r = \begin{bmatrix} \hat{P}_{k1}^r & \hat{P}_{k2}^r & \hat{P}_{k3}^r \end{bmatrix}^{\mathrm{T}} \tag{3}$$

Where $\hat{P}_{ki}^r = \dfrac{1}{N-1} \displaystyle\sum_{j=k-N+1}^{k} \left(r_{ji} - \bar{r}_{ki} \right)^2 \quad , \quad i = 1,2,3$

The practical variance of innovation is

$$P_k^r = H_k [\boldsymbol{\Phi}_{k,k-1} P_k \boldsymbol{\Phi}_{k,k-1}^{\mathrm{T}} + Q_k] H_k^T + R_k$$

Then

$$DOM_k = P_k^r - \hat{P}_k^r$$

T-S fuzzy logic system shows as figure 1:

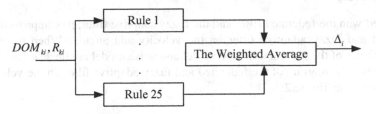

Fig. 1. The structure of T-S fuzzy logic system

There are 5 grades of the input parameters: Z、XS、S、B、XB, there are 5 grades of the output weights: Z=0、S=0.25、M=0.5、B=0.75、L=0.75 and the domain is [0,1].the fuzzy rules show as tables 1:

Table 1. The fuzzy rules

DOM_{ki} / R_{ki}	Z	XS	S	B	XB
Z	L	L	B	M	M
XS	L	B	B	M	S
S	B	B	M	S	S
B	M	M	S	S	Z
XB	M	S	S	Z	Z

After the weight Δ_i of the locally estimated state $\hat{X}_i$, it is global state estimtion that completed by center of gravity method

$$\hat{X}(k) = \frac{\sum_{j=1}^{n} X_j(k)\Delta_j(k)}{\sum_{j=1}^{n} \Delta_j(k)} \quad (4)$$

Where n is the number of subfilter.

3.3   Structure Design of Federated and Fuzzy Adaptive Filter on the Velocity and Angular

Combined with the federated filter and the fuzzy adaptive filter, it is improved that the federated and fuzzy adaptive filter on the velocity and angular When the statistics characteristics of the system dynamic model and noise model are unclear [9-11].

The principle diagram of the federated and fuzzy adaptive filter on the velocity and angular shows as figure 2:

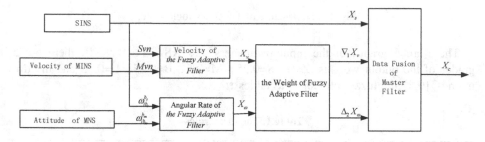

Fig. 2. The principle diagram of the federated and fuzzy adaptive filter on the velocity and angular

The principle of the federated and fuzzy adaptive filter on the velocity and angular show as follows:

（1）The output date of MINS used as the common reference system, the velocity and angular of SINS regarded as independent subsystems. In order to solve the problem, the statistics characteristics of the system dynamic model and noise model are unclear, the fuzzy adaptive filter used in the two subsystems.

（2）The output date of two subsystems used the weight of fuzzy adaptive filter to distribute the information, the common reference system used the federal filter to obtain global sub-optimal estimation.

4   Application of Federated and Fuzzy Adaptive Filter on the Velocity and Angular Rate Matching in the Transfer Alignment

4.1   The Velocity Matching

It is supposed that $X_v = \begin{bmatrix} \boldsymbol{\varphi}^{n\mathrm{T}} & \delta V_e^{n\mathrm{T}} & \boldsymbol{\varepsilon}^{b_s\mathrm{T}} & \boldsymbol{\nabla}^{b_s\mathrm{T}} \end{bmatrix}^T$ is system state of the velocity matching. Where $\boldsymbol{\varphi}^n = \begin{bmatrix} \varphi_x & \varphi_y & \varphi_z \end{bmatrix}^{\mathrm{T}}$ is misalignment angle of SINS. $\delta V_e^n = \begin{bmatrix} \delta V_{ex}^n & \delta V_{ey}^n & \delta V_{ez}^n \end{bmatrix}^{\mathrm{T}}$ is velocity error of SINA. $\boldsymbol{\varepsilon}^{b_s} = \begin{bmatrix} \varepsilon_x^{b_s} & \varepsilon_y^{b_s} & \varepsilon_z^{b_s} \end{bmatrix}^{\mathrm{T}}$ is gyro's drift of SINS.

$\nabla^{b_s} = \begin{bmatrix} \nabla^{b_s}_x & \nabla^{b_s}_y & \nabla^{b_s}_z \end{bmatrix}^{\mathrm{T}}$ is accelerometer's constant error of SINS, state equation of the velocity matching is

$$\dot{X}_v = \begin{bmatrix} -(\omega^n_{in}\times) & 0_{3\times3} & -C^n_{b_s} & 0_{3\times3} \\ (f^n\times) & -[(2\omega^n_{ie}+\omega^n_{en})\times] & 0_{3\times3} & C^n_{b_s} \\ 0_{3\times3} & 0_{3\times3} & 0_{3\times3} & 0_{3\times3} \\ 0_{3\times3} & 0_{3\times3} & 0_{3\times3} & 0_{3\times3} \end{bmatrix} X_v + \begin{bmatrix} -C^n_{b_s}\varepsilon^{b_s}_b \\ C^n_{b_s}\nabla^{b_s}_b \\ 0_{3\times1} \\ 0_{3\times1} \end{bmatrix} \tag{5}$$

Measurement equation of the velocity matching is

$$Z_V = \begin{bmatrix} 0_{3\times3} & I_{3\times3} & 0_{3\times3} & 0_{3\times3} \end{bmatrix} X_v + V_V$$

Where V_V is noise sequence of measurement.

4.2   The Angular Rate Matching

It is supposed that $X_\omega = \begin{bmatrix} \varphi^{nT} & \mu^{b_I T} & \lambda^{b_m T}_f & \omega^{b_m T}_f & \varepsilon^{b_s T}_b \end{bmatrix}^{\mathrm{T}}$ is system state of the angular rate matching. Where $\varphi^n = \begin{bmatrix} \varphi_x & \varphi_y & \varphi_z \end{bmatrix}^{\mathrm{T}}$ is misalignment angle of SINS. $\mu^{b_I} = \begin{bmatrix} \mu^{b_I}_x & \mu^{b_I}_y & \mu^{b_I}_z \end{bmatrix}^{\mathrm{T}}$ is error of missile body's installation angle, $\lambda_f = \begin{bmatrix} \lambda_{fx} & \lambda_{fy} & \lambda_{fz} \end{bmatrix}^{\mathrm{T}}$ is flexure deformation angle of wing. $\omega_f = \begin{bmatrix} \omega_{fx} & \omega_{fy} & \omega_{fz} \end{bmatrix}^{\mathrm{T}}$ is flexure deformation angular rate of wing, $\varepsilon^{b_s} = \begin{bmatrix} \varepsilon^{b_s}_x & \varepsilon^{b_s}_y & \varepsilon^{b_s}_z \end{bmatrix}^{\mathrm{T}}$ is gyro drift of SINS . State equation of the angular rate matching is

$$\dot{X}_\omega = \begin{bmatrix} -(\omega^n_{in}\times) & 0_{3\times3} & 0_{3\times3} & 0_{3\times3} & -C^n_{b_s} \\ 0_{3\times3} & 0_{3\times3} & 0_{3\times3} & 0_{3\times3} & 0_{3\times3} \\ 0_{3\times3} & 0_{3\times3} & 0_{3\times3} & I_{3\times3} & 0_{3\times3} \\ 0_{3\times3} & 0_{3\times3} & -[\beta^2] & -[\beta] & 0_{3\times3} \\ 0_{3\times3} & 0_{3\times3} & 0_{3\times3} & 0_{3\times3} & 0_{3\times3} \end{bmatrix} X_\omega + \begin{bmatrix} -C^n_{b_s}\varepsilon^{b_s}_b \\ 0_{3\times1} \\ 0_{3\times1} \\ \eta \\ 0_{3\times1} \end{bmatrix} \tag{6}$$

Measurement equation of the angular rate matching is

$$Z_\omega = \begin{bmatrix} \omega^{b_m}_{ib_m}\times(C^n_{b_m})^{-1} & 0_{3\times3} & \omega^{b_m}_{ib_m}\times & I_{3\times3} & I_{3\times3} \end{bmatrix} X_\omega + V_\omega$$

Where $\varepsilon_w^{b_s}$ is Gaussian white noise of gyro, $\eta = \begin{bmatrix} \eta_x & \eta_y & \eta_z \end{bmatrix}^T$ is Gaussian white

noise sequence of second-order, which $\eta_i \sim N(0, Q_i)$, $Q_i = 4\beta_i^3 \sigma_\eta^2$, σ_η^2 is the va-

riance of flexure deformation angle; $[\boldsymbol{\beta}] = \mathrm{diag}(\beta_x, \beta_y, \beta_z)$, $[\boldsymbol{\beta}^2] = \mathrm{diag}(\beta_x^2, \beta_y^2, \beta_z^2)$

4.3   The Algorithms of Data Processing System

1）the Fuzzy Adaptive Filter of the subsystems

It is supposed that the system equation and measurement equation of the velocity
matching and the angular rate matching can be written as equation (1)[12-14], and
equation of the fuzzy adaptive filter is

$$
\begin{aligned}
\hat{X}_{k/k-1} &= \boldsymbol{\Phi}_{k,k-1} \hat{X}_{k-1} \\
P_{k/k-1}^{\alpha-} &= \alpha^2 \boldsymbol{\Phi}_{k,k-1} P_k^{\alpha-} \boldsymbol{\Phi}_{k,k-1}^{\mathrm{T}} + Q_k \\
K_k &= P_{k/k-1}^{\alpha-} H_k^{\mathrm{T}} \left(H_k P_{k/k-1}^{\alpha-} H_k^{\mathrm{T}} + R_k \right)^{-1} \\
P_k^{\alpha-} &= (I - K_k H_k) P_{k/k-1}^{\alpha-} \\
\hat{X}_k &= \hat{X}_{k/k-1} + K_k (Z_k - H_k \hat{X}_{k/k-1})
\end{aligned}
\tag{7}
$$

Where $Q_k = Q \alpha^{-(2k+1)}$ is the system noise, $R_k = R \alpha^{-(2k+1)}$ is the measurement noise,
and $P_k^{\alpha-} = P_k \alpha^{-2k}$ is matrix of weighted filtering.

2)   The algorithm of Federated and fuzzy adaptive filter

It is supposed that $\hat{X}_{cV}$ is local optimum estimation of the velocity matching and

P_{cV} is variance of estimation error; $\hat{X}_{c\theta}$ is local optimum estimation of the angular

rate matching and $P_{c\theta}$ is variance of estimation error.

Then the algorithm of federated and fuzzy adaptive filter is

$$
\begin{cases}
\hat{X}_g = P_g^{-1} (P_{CV}^{-1} \Delta_1 \hat{X}_{CV} + P_{C\theta}^{-1} \Delta_2 \hat{X}_{C\theta}) \\
P_g = (P_{CV}^{-1} + P_{C\theta}^{-1})^{-1}
\end{cases}
\tag{8}
$$

Where P_g is covariance matrix of error, and $\Delta_1 + \Delta_2 = 1$.

5   Simulation Analysis of Federated and Fuzzy Adaptive Filter on the Velocity and Angular Rate Matching in the Transfer Alignment

It is supposed that unknown measurement noise is included measurement matrix and the wing flexure of subfilter is modeled.

The condition for simulation is as follows:

Error parameters of SINS: Constant drift of gyro is $1°/h$, random walk of gyro is $0.1°/\sqrt{h}$, constant offset error of accelerometer is $5\times10^{-4}g$, and standard deviation of accelerometer is $5\times10^{-5}g\cdot\sqrt{s}$.

Installing-error angle of missile body: $\mu=\begin{bmatrix}0.1° & 0.1° & 0.1°\end{bmatrix}^{T}$

Misalignment initial angle of SINS: $\varphi(0)=\begin{bmatrix}0.1° & 0.1° & 0.5°\end{bmatrix}^{T}$

Velocity initial error of SINS: $\delta V_e^n(0)=\begin{bmatrix}3m/s & 3m/s & 3m/s\end{bmatrix}^{T}$

There are a series of wing motion by aircraft, the angle of shake wing is $30°$. The motion of wing which is simulation trajectory of aircraft based on the federated and fuzzy adaptive filter on the velocity and angular rate matching is adopted by the paper. it has the same conditions as kalman filter. The simulation period of kalman filter is 20ms. where solid line is result of the federated and fuzzy adaptive filter, and dashed line is result of kalman filter.

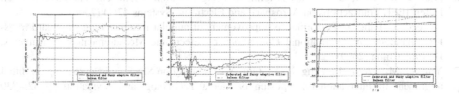

Fig. 3. Estimate error of misalignment angle

Analysis of simulation results:

1）It is the main difference between the federated and fuzzy adaptive filter and kalman filter that the velocity matching and the angular rate matching used the fuzzy adaptive filter. In order to the global optimization, the data of the velocity matching and the angular rate matching fused in the main filter. From the Figure 3, it is showed that the convergence of misalignment angle of the federated and fuzzy adaptive filter has been in $10'$ after 10s.

2) When the system used the federated and fuzzy adaptive filter, it has considered the installing-error angle and the wing flexure in the angular rate matching, and the system carried on the estimate and compensate. The simulation results show that the federated and fuzzy adaptive filter can better estimate the misalignment angle of SINS

and the federated and fuzzy adaptive filter more accurate and faster than Kalman filter when the statistics characteristics of the system dynamic model and noise model are unclear.

References

1. Groves, P.D.: Optimising the Transfer Alignment of Weapon INS. Journal of Navigation **56**(2), 323–335 (2003)
2. Dong, L., Chen, S.: F Study on Transfer Alignment with Dynamic Flexure Deformation. Aero Weaponry **1**(5), 28–30 (2015)
3. Gu, D.Q., Qin, Y.Y., Zheng, J.B.: Fuzzy adaptive Kalman filter for marine SINS initial alignment. In: 6th International Symposium on Test and Measurement, Dalian (2005)
4. Cadcna, C., Neira, J.: SIAM in O with the combined Kalmam information filter. Robotics and Autonomous Systems **58**(11), 1207–1219 (2010)
5. Du, H.Y., Huo, Y., Zhuo, Y.: SLAM algorithm based on fuzzy adaptive Kalman filter. J. Huazhong Univ. of Sci. & Tech. (Natural Science Edition) **40**(1), 58–62 (2012)
6. Escamilla-Ambrosio, P.J., Mort, N.: A hybrid Kalman filter-fuzzy logic adaptive multisensor data fusion architectures. In: Proceedings of 42nd IEEE Conference on Decision and Control, Hawaii (2003)
7. Zhou, R.-H.: The Application of Fuzzy Control Theory in Attitude Control of Launch Vehicle. National University of Defense Technology (2003)
8. He, S.Z., et al.: Fuzzy self-tuning of PID controllers. Fuzzy Sets and Systems **56**(1), 37–46 (1993)
9. Liu, H.-G., Chen, G., Zhou, C.: Analysis of angular velocity matching transfer alignment for vessel. Journal of Chinese Inertial Technology **21**(5), 565–569 (2013)
10. Ahn, H.-S., Won, C.-H.: Fast Alignment using Rotation Vector and Adaptive Kalman Filter. IEEE Transactions on Aerospace and Electronic Systems **42**(1) (2006)
11. Majeed, S., Fang, J.: Performance improvement of angular rate matching shipboard transfer alignment. In: International Conference on Electronic Measurement & Instruments, Beijing, China (2009)
12. Yan, G.-M., Weng, J., Yang, P.-X., Qin, Y.-Y.: Study on SINS rapid gyrocompass initial alignment. In: International Symposium on Inertial Technology and Navigation, pp. 323-330 (2010-10)
13. Shortelle, K.J., Graham, W.R.: F-16 flight tests of a rapid transfer alignment procedure. In: IEEE RLANS 1998, pp. 379–3869 (1998)
14. Niu, E.-Z., Ren, J.-X., Tan, J.: Design and analysis of simulation track in the test of strap-down inertial navigation system. Computer Simulation **26**(8), 18–21 (2010)

A Suitable Decoding Circuit Design of Underwater Acoustic Data Transmission System

Gou Yanni[✉] and Wang Qi

College of Marine Science and Technology,
Northwestern Polytechnical University, Xi'an, China
googogu@nwpu.edu.cn

Abstract. Because of the complexity of underwater acoustic data transmission channel for the design of the signal transmission system has brought great challenges, including the stand or fall of decoding circuit design seriously affects the data transmission rate and bit error rate. Manchester decoding because of its strong anti-interference ability is widely used in the field of industrial control network.In combination with the characteristics of acoustic data transmission on detecting the underwater targets of detection system. This paper designed a circuit which is applicable to the underwater acoustic data transmission system based on the Manchester encoding and decoding method and realized the function of encoding and decoding on CPLD by using the VHDL language. The experiment results given by the MAX+PLUS II numerical simulation under certain rate and laboratory test shows that the stability and reliability of data transmission are guaranteed by using strong anti-interference and clock recovery ability of mature Manchester encoding and decoding technology, the effectiveness and feasibility of decoding circuit design is proved.

Keywords: Acoustic detection · Underwater acoustic data transmission · CPLD · Manchester encoding and decoding

1 Introduction

Underwater acoustic data transmission channel is a very complicated time, space, frequency and random, multipath transmission channel. There is high environmental noise, narrow bandwidth, applicable low carrier frequency, long time-delay, large transmission attenuation, and many other adverse factors, is by far the most difficult of wireless communication channel. How to find an effective channel encoding and decoding method enables the underwater acoustic signal to undistorted reached the requirements of certain transmission rate is one of the subjects has been struggling to beg. The main task of the underwater acoustic detection system is to locate and track targets effectively in the time domain through continuous measurement and estimation on target motion parameters, such as position, speed and course by using underwater acoustic sensors. The data transmission section of detection system convert the collected

© Springer International Publishing Switzerland 2015
H. Liu et al. (Eds.): ICIRA 2015, Part I, LNAI 9244, pp. 101–110, 2015.
DOI: 10.1007/978-3-319-22879-2_10

multiple underwater acoustic signal into baseband digital signal which is easy to be transported, and then transfer these digital signal to water-surface data receiving device through the transmission cable with high speed for terminal processing [1]. Decoding circuit suitable for underwater acoustic data transmission system is an important part of the relationship to the underwater acoustic signal transmission error rate control. Manchester encoding uses the transition in the middle of the timing window to determine the binary value for that bit period. The top waveform moves to a lower position so it is interpreted as a binary zero. The second waveform moves to a higher position and is interpreted as a binary one. From above we can see the type of Manchester code is often used in high-speed transmission of baseband data for its simple features without DC component but rich of clock information [2]. Therefore it has been widely used in industrial control networking such as computerized logging and wireless monitoring [3–5]. Such as Manchester II codec was designed and implemented in 1553B bus interface based on FPGA [6–8]. In [9] it describs a method for encoding and decoding digital data signals using a modified form of the standard Manchester code for the application in galvanically isolated data transmission in smart power electronics. A chip-encoded version of the Manchester-II decoding algorithm that can be used for long-distance transmission for well logging [10] is introduced. A new Manchester-II encoding/decoding system is used for nuclear logging by a 7000m armoring cable with a bit error rate of better than 10-10. A Robust, Low-Complexity and Ultra-Low Power Manchester Decoder for Wireless Sensor Nodes [11] is discussed. The implementation of Manchester Encoder and Decoder modules that is suitable for RF communication systems [12]. It uses Manchester encoding in order to interchange data among field devices on board. Literature [13] focused on the independent design & development of Manchester encoding module, Manchester decoding module and parallel/serial and serial/parallel conversion module by hardware description language (HDL) with FPGA as main control unit and Manchester code and low-voltage differential signaling as theoretical basis. These research direction above are not involved in Manchester decoding circuit in the application of underwater data transmission system, implement to carry out the relevant field of study is very necessary. Based on the features of underwater acoustic data transmission system and the strong anti-interference ability of Manchester code , baseband at high speed data transmission system used in Manchester code method, we design a applicability of Manchester decoding circuit.In this circuit, the conversion between paralle-serial signal and achieve function of Manchester encoding and decoding will be completed in the internal CPLD. The advantages of the circuit include good versatility of encoding and decoding ability, low cost and high coding rate. Section 2 describes composition of underwater acoustic data transmission system and functions of CPLD in the part of encoding and decoding. Section 3 introduces the design principles of Manchester decoding circuit. Detailed numerical simulation through MAX + PLUS by VHDL language and circuit test validation are carried out and the results are presented in Section 4. Finally the paper is concluded in Section 5.

2   Acoustic Data Transmission System

2.1   System Structure

A typical structure of underwater data transmission system is shown in Figure 1. Both receiver and transmitter part of the transmission channel using high-speed signal processors DSP of TI Company as the central control unit. CPLD in MAX7000 series of Altera Corporation is selected as signal encoding and decoding module to implement Manchester encoding and decoding. In order to communicate with PC and increase the flexibility of signal processing, some peripheral expansion circuit are configured to the DSP, includes corresponding program memory, clock, power supply and so on.

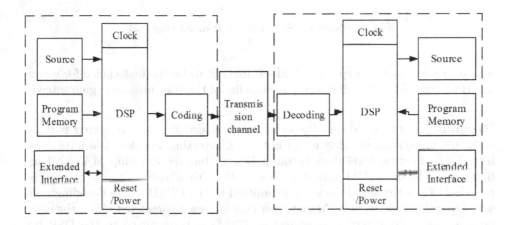

Fig. 1. An underwater Data Transmission System Structure

Transmitting Part. Hardware frameworks of underwater data collecting and transmitting part of transmission system are shown in Figure 2. The underwater part is required to collect multiple hydrophone signals simultaneously and synchronously. In order to implement simultaneous sampling in multi-channel, the circuit chooses ADC manufactured by MAXIM company with 12-bit and 8-channel simultaneous sampling and the highest conversion rate is 2Msps. After the conversion ADC will send a signal to notice DSP read the conversion result.

This part works as follows: Multi-channel hydrophone signals and auxiliary sensor signals are arranged in a certain order(Frame Format) in the internal DSP and transmitted to the CPLD in parallel after compressed, and then the parallel-serial conversion and Manchester encoding are completed in internal CPLD so that the data is suitable for transmission in the channel. Finally, the encoded data will be transmitted in serial by using RS-485 interface. The RS-485 interface has the advantages of good anti-noise performance, farther transmission distance

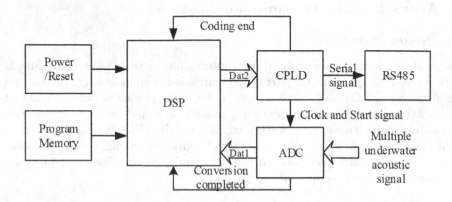

Fig. 2. Hardware Structure of Transmitting Part

and well multi-station capacity of which make it to be the first choice for serial interface, therefore the reliability and stability of the transmission is guaranteed.

Receiving Part. Hardware frameworks of underwater data receiving part of transmission system are shown in Figure 3. Converting 485 electrical level data into TTL electrical level data through RS-485 bus receiver chip MAX3291 at first. That is to say, Manchester code data with a special frame format is received. Then the serial data is transmitted to the CPLD where decoding and serial/parallel conversion of Manchester code data are completed. After the conversion the parallel signal converted by CPLD is transmitted to the DSP by 16-bit data lines and restored in the internal DSP, and then some necessary processing is implemented as follows.

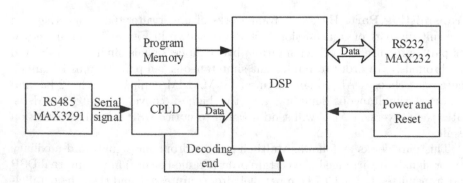

Fig. 3. Hardware Structure of Receiving Part

2.2   CPLD

CPLD is the most important module in the transmission system, and also is the key part of encoding and decoding circuit. In the transmitting and receiving parts of the system, EPM7128AE in MAX7000 series of Altera Corporation is selected as CPLD. The available gates are 2,500, and with 128 macro-cells inside. The most I/O pins are 100 and the clock frequency is up to 192.3MHz. Voltage of power supply is 3.3V [14]. The functions of encoding and decoding are completed in two CPLD respectively in this paper. The internal program diagram of CPLD in encoding part is shown in Figure 4. As can be seen from this figure, parallel signal is latched firstly, and then be converted into serial data by shift register according to the encoding clock. Finally, the encoded serial NRZ data through by Manchester encoding module as the encoding parts output.

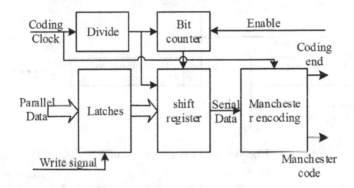

Fig. 4. Function Diagram of CPLD in Encoding Part

The internal program diagram of CPLD in decoding part is shown in Figure 5. First, Manchester code which input is decoded into NRZ code by Manchester encoding module and extract a synchronous clock from that code element. Then the serial data will be converted into parallel data by shift register to latch and output at last.

3   Manchester Encoding and Decoding Circuit Design

The design concept of encoding and decoding circuit is to formulate the frame data format according to the data interface protocol. It can be proceed from two parts including data transmission module encoding and receiving module decoding of which each frame of data encoding and decoding are implemented in the way of Manchester.

3.1   Encoder Design

Sync tip and Manchester code element sequentially output during the encoding as shown in Figure 6 which is the frame format of Manchester encoded data. Sync

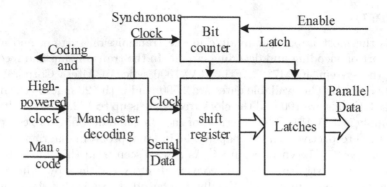

Fig. 5. The Internal Diagram of CPLD in Decoding Part

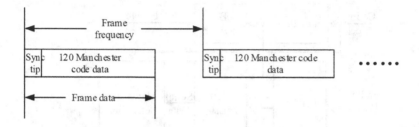

Fig. 6. Encoded Frame Format

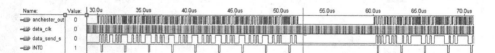

Fig. 7. Simulation Figure of Encoded Frame Format

tip is Barker code is encoded form of Manchester code and 120-bit valid data are followed. In this figure, the frame frequency is the frequency of occurrence from a frame of data.

Figure 7 shows a simulation of encoded frame format. The 'anchester_out' refers to uplinking Manchester data. The 'data_send_sis' actual serial acoustic data before coding and 'data_clk' represents the corresponding serial data clock.

Take the second harmonic generation of data transmission rate as clock frequency in encoder circuit according to the characteristics of Manchester code. When the number of clocks is odd, the Manchester code equals to NRZ code and on the other hand the Manchester code equals to the negated NRZ code. This method of encoding is simple and it can solve the burr problem generated by conventional XOR encoding.

3.2   Decoder Design

The beginning time of a decoding cycle is the key point of whether decoding is implemented correctly during the processing. A decoding cycle is started when the decoder detects the sync tip as represents in below figure. Diagram of Manchester decoding is shown in Figure 8, by consisting of four sections as sync tip detection, synchronous clock extraction, decider and counter. The sync tip detection and decider are designed by VHDL language, and the other parts are designed by using schematics.

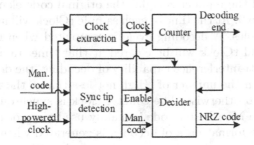

Fig. 8. Diagram of Manchester Decoding

Two functions of sync tip detection circuit are detecting and removing sync tip. This part of circuit detects the received data continuously, and the enable signal will be transmitted to the other three sections to control to start work when the sync header is catched. At the same time, standard Manchester code of which sync tip is removed is transmitted to decider for Manchester decoding. Synchronous clock is extracted from the Manchester code through synchronous clock extractor. A hopping will occur in the middle of each code element from the characteristics of Manchester code that is picked up by clock extraction circuit to recover the bit-synchronous clock. Synchronous clock extraction circuit is shown in Figure 9. From the figure we can seen, DataIn is the input Manchester code. CLK16xis the high-power clock when decoding. DataIn2is the Manchester code signal after twice-delayed that XOR with once-delayed signal to get the hopping information. CLR is the hopping information extracts from the Manchester code which is regarded as a clear signal of counter, while synchronous clock can be recovered from the input Manchester code by using high-power clock. CLKOUT is the output of the synchronous clock which can be used on other parts of the circuit.

Decider is the core part of Manchester decoding in which the sixteen times frequency transmission rate is used as decoding clock. This clock is used sampled after receiving the Manchester code signal. In order to decode a NRZ code correctly sixth samples are needed, that is to say, one decoding cycle is equal to sixteen clock cycles. tHigh and tClock are set as two counters during decoding. tHigh is used to record the high-level numbers of the front eight samples in every

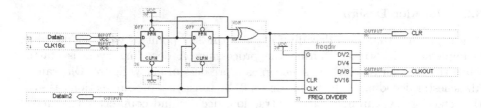

Fig. 9. Synchronous Clock Extraction Circuit

sixteenth samples. If the number is eight, the original code element is1. On the contrary it is equal to 0. At the time of each sample tClock will add one to record the numbers of samples. A decoding cycle is completed when number is up to 16 and the tHigh and tClock will be cleared at that time. In actual design, so as to improve the anti-interference capability of decoding, the decision threshold is set to 8-2=6. When the number of tHigh not less than 6 the original signal is considered to be1or 0 otherwise. The counter clock is synchronous clock which is recovered from the Manchester code is mainly used to record the bit number of decoding. A frame format data of decoded is generated when the bit number up to 16-bit, and then the decoding complete signal can be used to inform the external processing unit for reading the decoded signal.

4   Simulation and Results

4.1   Encoding Simulation

In the experiment, VHDL language based on MAX + PLUS II is used in the numerical simulation, and the data encoding rate is set to 2Mbit/s. The graphic result is shown in Figure 10, where D [0..5] is parallel signal in transmission part. DSP_CLK is provided by an external 40MHz clock signal which produce 4MHz second harmonic generation clock signal after frequency division in the CPLD and MAN_OUT is the ultimate Manchester code.

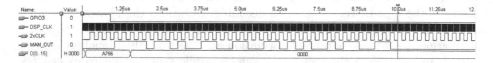

Fig. 10. Simulation Diagram of Manchester Encoding

As can be seen from Figure 10 the high level of Manchester code in first four clock cycles is Sync tip of encoded signal. Due to the LSB data first and MSB data behind in program, when input parallel signal is 0xA756, the NRZ code after parallel/serial conversion is 0110,1010,1110,0101 and corresponding Manchester code is 01 10 10 01,10 01 10 01,10 10 10 01,01 10 01 10 (as shown as MAN_OUT in Figure 10).

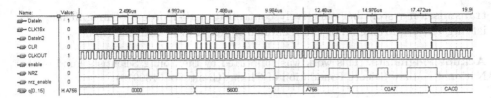

Fig. 11. Simulation Chart of Manchester Decoding

4.2 Decoding Simulation

The decoding clock used in the simulation is 16MHz and the result is shown in Figure 11, where DataIn is 4Mbit/s the input Manchester code of which the rate is two times the original data rate. DataIn2 is generated by the input signal delayed by one clock cycle. CLK16x is 16MHz high power clock signal provided by external part. CLR is hopping information extracted from the Manchester code. CLKOUT is the recovered clock signal. Enable is the enable signal generated by the Synchronous detection circuit. NRZ is the output NRZ code through decider and q [0 .. 15] is the decoded data.

From Figure 11,as can be seen the decoded data is equal to 0xA756 and the output Manchester code is decoded correctly.

4.3 Circuit Testing

The circuit has been successfully used on a certain type of underwater acoustic detection system equipment, in order to verify the transmission system transmission rate and bit error rate, circuit was carried out by laboratory testing. Experiments using 350m long standard transmission cables, bit-rate Manchester code is 7.5megabits per second, the sampling rate is set to 15kHz, a frame of data is 416 bit, monitoring time continuous 30min. During this time altogether receives the total number of data (16 bits) is 721501025 including 113 total error data. After calculation, the effective data rate is 6.24 Mbps, bit error rate is 1.57×10^{-7}.

5 Conclusion

Manchester encoding and decoding circuit design is implemented through Alteras CPLD by using VHDL language in a typical underwater acoustic data transmission system. In order to improve the flexibility of software design the VHDL language and schematic mode are mixed use to design of encoding and decoding devices. Encoding and decoding numerical simulations based on MAX + PLUS II are implemented under the clock of 2MHz. Experimental results show that the design idea of circuit by using Manchester encoding and decoding technology in the transmission system is correct and the circuit can be of engineering realization and valid. And through the circuit experiment test to verify the data transmission system ber effectively controlled in 10-7 orders of magnitude, satisfying the requirements of the technical index of data transmission.

Acknowledgments. This work was supported by the Young Scientists Fund of the National Natural Science Foundation of China (Grant No. 5130919).

References

1. Wang, L.J., Ling, Q., Yuan, Y.Y.: U.S. sonar equipment and technology. National Defense Industry Press, Beijing (2011)
2. Changxin, F., Lina, C.: Principles of Communication, 7th edn. Nation Defense Industry Press, Beijing (2012)
3. Copeland, B.J.: IEEE J. The Manchester Computer: A Revised History Part 1: The Memory Annals of the History of Computing **33**(1) (2011)
4. Copeland, B.J.: IEEE J. The Manchester Computer: A Revised History Part 2: The Baby Computer the Memory Annals of the History of Computing **33**(1) (2011)
5. Jingzhuo, S., Yingxi, X., Jing, S.: ICIT J. Manchester encoder and decoder based on CPLD Industrial Technology (2008)
6. Li, J., Chai, M.: Design of 1553B avionics bus interface chip based on FPGA. In: IEEE International Conference on Electronics, Communications and Control (ICECC), Ningbo, People's Republic of China, pp. 3642–3645 (2011)
7. Jemti, J.: Design of manchester II bi-phase encoder for MIL-STD-1553 protocol. In: IEEE International Multi Conference on Automation, Computing, Control, Communication and Compressed Sensing (iMac4s), Kottayam, India, pp. 240–245 (2013)
8. Wang, C.-Y., Zheng, Y.-L.: Design and implementation of manchester codec based on FPGA. In: International Conference on Communications, Electronics and Automation Engineering, Xian, People's Republic of China, pp. 681–687(2013)
9. Niedermeier, M.T., Wenger, M.M., Filimon, R.: Galvanically isolated differential data transmission using capacitive coupling and a modified Manchester algorithm for smart power converters. In: IEEE 23rd International Symposium on Industrial Electronics (ISIE), Istanbul, Turkey, pp. 2596–2600 (2014)
10. Cao, P., Song, K.-Z., Yang, J.-F.: A Hardware Decoding Algorithm for Long-distance Transmission for Well Logging. Journal of Environmental and Engineering Geophysics **17**(1), 39–47 (2012)
11. Pang, C.Y., Gopalakrishnan, P.K.A.: Robust, low-complexity and ultra-low power manchester decoder for wireless sensor nodes. In: 12th International Symposium on Integrated Circuits, Singapore, pp. 396–399 (2009)
12. John, F., Jagadeesh Kumar, P.: Custom processor implementation of a self clocked coding scheme for low power RF communication systems. In: Das, V.V., Stephen, J., Chaba, Y. (eds.) CNC 2011. CCIS, vol. 142, pp. 643–648. Springer, Heidelberg (2011)
13. Yang, R., Deng, M., Zhang, Q.: Realization of high-speed data transmission in seismic data acquisition. In: International Conference on Information Science, Automation and Material System, Zhengzhou, People's Republic of China, pp. 560–564(2011)
14. Data Sheet of MAX7000 series, Altera Corporation (2010)

Approximation-Based Control of a Marine Surface Vessel with Full-State Constraints

Zhao Yin[1], Wei He[1($\boxtimes$)], Weiliang Ge[1], Chenguang Yang[2,3], and Sanjay Sharma[3,4]

[1] School of Automation Engineering and Center for Robotics,
University of Electronic Science and Technology of China, Chengdu 611731, China
hewei.ac@gmail.com
[2] College of Automation Science and Engineering,
South China University of Technology, Guangzhou 510640, China
[3] Centre for Robotics and Neural Systems, Plymouth University,
Plymouth PL4 8AA, UK
[4] School of Marine Science and Engineering, Plymouth University,
Plymouth PL4 8AA, UK

Abstract. In this paper, a trajectory tracking control law is proposed for a class of marine surface vessels in the presence of constraints and uncertainties. A barrier Lyapunov function (BLF) is employed to prevent states from violating the constraints. Neural networks are used to approximate the system uncertainties in our design, and the control law is designed by using the Moore-penrose inverse. Under the proposed control, the multiple state constraints are never violated, the signals of the closed loop system are semiglobally uniformly bounded (SGUB), and the asymptotic tracking is achieved. Finally, the performance of the proposed control is illustrated via the simulations.

Keywords: Neural networks · State constraints · Marine surface vessel · Barrier lyapunov function · Adaptive control

1 Introduction

In recent years, the marine surface vessels have been broadly applied in ocean engineering. The control design of marine surface vessels has gained an increasing research attention due to the development of ocean engineering techniques [1–3].

This is a challenging problem for nonlinear control design to ensure stability of the marine surface vessel in the harsh ocean environment. On the one hand, the vessel is usually unstable without the closed-loop control and the vessel dynamics contain unknown parametric and external disturbance. Thus, it is

This work was supported by the National Natural Science Foundation of China under Grants 61203057, 61473120, the National Basic Research Program of China (973 Program) under Grant 2014CB744206, the Fundamental Research Funds for the Central Universities under Grant 2015ZM065, and the NSFC-RS joint research project under grants 51311130137 in China and IE121414 in the UK.

H. Liu et al. (Eds.): ICIRA 2015, Part I, LNAI 9244, pp. 111–125, 2015.
DOI: 10.1007/978-3-319-22879-2_11

necessary to design robust controllers for the unknown parametric and external disturbances. Neglecting this problem may lead to performance degradation or even destabilization. On the other hand, state constraint is also a problem need be solved in the tracking of a marine surface vessel. In practical systems, constraints on inputs and states are ubiquitous and always manifest themselves as performance requirement and safety specifications [4,5]. Violation of the state constraint may cause the performance degradation, hazards, system damage or environment pollution. Therefore, it is important to handle these two problems in the control design.

To tackle the problem of the system with uncertainties, numerous control approaches have been proposed for marine surface vessels to track the desired trajectories. For example, in [6], a sliding-mode control law is presented. In [7], the authors consider the problem of tracking a desired trajectory for fully actuated ocean vessels, in the presence of uncertainties and unknown disturbances. Under constant disturbances, the author design a ship trajectory tracking control law to the nonlinear ship surface movement mathematical model including the Coriolis and centripetal matrix and nonlinear damp term in [8]. In [9], the authors through the backstepping technique, a discontinuous feedback control law has been proposed for the underactuated surface vessels. The authors present the problems of accurate identification and learning control of ocean surface vessel with uncertain dynamical environments in [10]. In this paper, we use the NN techniques to approximate the unknown parameters to deal with the first problem in the control design.

In [11], the authors investigate optimized adaptive control and trajectory generation for a class of wheeled inverted pendulum (WIP) models of vehicle systems, and handle the problem of internal and external uncertainties. However, they do not consider the constraints of the position and velocity which is not practical in the actual life. To avoid the violation of constraints, many methods for dealing with constrained nonlinear systems have been proposed such as constrained model predictive control [12], reference governor [13], command governor [14], coordinate transformation [15] and extremum seeking control [16]. Among them, the barrier Lyapunov function (BLF) is a kind of control Lyapunov functions which has been developed to guarantee the constraints are not vilation [17,18]. Inspired by the BLF's property, BLF based methods have been used in constrained nonlinear systems in Brunovsky form [19] and output feedback form [20]. In [21], the authors propose robust adaptive control strategies for Remotely Operated Vehicles (ROVs) with velocity constraints in the presence of uncertainties and disturbances. Nonetheless, all of the aforementioned achievements on BLFs require the system dynamic to be at least partially known, which have been limited in the theoretical scope. In this paper, we investigate the tracking control of state constrained vessel systems with unknown system dynamics, which are approximated by the neural networks.

Recently, the NN control of the nonlinear systems with uncertainties and constraints has been proposed in [22–27]. Apart from this, the authors provide a framework for synchronised tracking control of a general class of high-order single-input-single-output (SISO) systems with unknown dynamics in [17]. Frank presents a multilayer neural network-based controller for a class of single-input single-output continuous-time nonlinear system in [28]. It need to construct a barrier Lyapunov function that increases to infinity whenever its arguments closes to some values. Therefore, keeping the BLF bounded in the closed-loop system, it is to ensure that the constraints are never violated. In [29], the authors address the problem of control design for strict-feedback systems with constraints on the partial states. In addition, the authors present a control for state-constrained nonlinear systems in strict feedback form to achieve output tracking [30]. In this paper, the full-states are constrained and the system dynamics are a multiply degree-of-freedom marine surface vessel with multiple-input-multiple-output which is a more difficult problem. The primary contributions of the paper include:

The rest of this paper is organized as follows. Section 2 includes the preliminaries and the dynamics of a 3 degree-of-freedom marine surface vessel with multiple-input-multiple-output. In Section 3, we design an adaptive neural networks control by employing a BLF and neural networks. In Section 4, simulations are carried out to illustrate the feasibility of the proposed control. The last section concludes our paper.

2   Problem Formulation

2.1   Useful Technical Lemmas and Definitions

Definition 1. *[31] A barrier Lyapunov function is a scalar function $V(x)$, defined with respect to the system $\dot{x} = f(x)$ on an open region $\mathbb{D}$ containing the origin, that is continuous, positive definite, has continuous first-order partial derivatives at every point of $\mathbb{D}$, has the property $V(x) \to \infty$ as x approaches the boundary of $\mathbb{D}$, and satisfies $V(x(t)) \leq b, \forall t \geq 0$ along the solution of $\dot{x} = f(x)$ for $x(0) \in \mathbb{D}$ and some positive constant b.*

The following lemma formalizes the result for general forms of barrier function and is used in the control design and analysis for strict feedback system to ensure that state constraints are not violated.

Lemma 1. *[32] Rayleigh-Ritz theorem: Let $A \in \mathbb{R}^{n \times n}$ be a real, symmetric, positive-definite matrix; therefore, all the eigenvalues of A are real and positive. Let $\lambda_{\min}$ and $\lambda_{\max}$ denote the minimum and maximum eigenvalues of A, respectively; then for $\forall x \in \mathbb{R}^n$, we have*

$$\lambda_{\min}||x||^2 \leq x^T A x \leq \lambda_{\max}||x||^2$$

where $|| \cdot ||$ denotes the standard Euclidean norm.

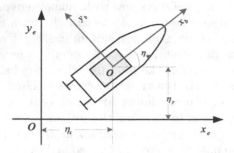

Fig. 1. The diagram of the marine surface vessel system

2.2 Problem Formulation

The motions and state variables of the single point mooring systems are defined and measured with respect to two important reference frames: earth-fixed frame, body-fixed frame. Fig. 1 shows the earth-fixed frame is denoted as (x_e, y_e) with its origin located at the connection of the mooring line and the mooring terminal. The body-fixed frame, denoted as (x_b, y_b), is fixed to the vessel body, that the origin coincides with the center of gravity of the moored vessel. The dynamics of a 3 degree-of-freedom (DOF) marine surface vessel with multiple-input-multiple-output (MIMO) [7] are described as follow

$$\dot{\eta} = J(\eta)v$$
$$M\dot{v} + C(v) + D(v) + g(\eta) = \tau \tag{1}$$

where the output $\eta = [\eta_x, \eta_y, \eta_\psi] \in \mathbb{R}^3$ represents the Earth-frame positions and heading, respectively, $\tau \in \mathbb{R}^3$ is the control input, $v = [v_x, v_y, v_\psi] \in \mathbb{R}^3$ denotes the velocities of vessel in the vessel-frame system. $M \in \mathbb{R}^{3 \times 3}$ is a symmetric positive definite inertia matrix, $C(v)v \in \mathbb{R}^3$ is the Centripetal and Coriolis torques, and $D(v)$ is the damping matrix, $g(\eta)$ represents the restoring forces caused by force of gravity, ocean currents and floatage, $J(\eta)$ is the transformation matrix which is assumed to be nonsingular, and it is defined as

$$J(\eta) = \begin{bmatrix} \cos\eta_\psi & -\sin\eta_\psi & 0 \\ \sin\eta_\psi & \cos\eta_\psi & 0 \\ 0 & 0 & 1 \end{bmatrix}$$

Let $x_1 = \eta, x_2 = v$, then the vessel system can be described as

$$\dot{x}_1 = J(x_1)x_2$$
$$\dot{x}_2 = M^{-1}[\tau - C(x_2)x_2 - D(x_2)x_2 - g(x_1)] = a \tag{2}$$

The control objective is to track a desired trajectory of the earth-frame positions $x_d(t) = [x_{d1}(t), x_{d2}(t), x_{d3}(t)]^T$, and desired trajectory of the velocities

$x_{2d} = [x_{2d1}(t), x_{2d2}(t), x_{2d3}(t)]^T$. While ensuring that all signals are bounded and that the full-state constraints are not violated, i.e., $|x_1| \leq k_{c1}, |x_2| \leq k_{c2}, \forall t \geq 0$, where $k_{c1} = [k_{c11}, k_{c12}, k_{c13}]^T$, $k_{c2} = [k_{c21}, k_{c22}, k_{c23}]$ are positive constant vectors.

3   Control Design

3.1   Model Based Control

In case that the parameters $M, C(v), D(v)$ and $g(\eta)$ are known, we denote $z_1 = [z_{11}, z_{12}, z_{13}]^T = x_1 - x_d$, and $z_2 = [z_{21}, z_{22}, z_{23}]^T = x_2 - \alpha$. Choosing the asymmetric barrier Lyapunov function as

$$V_1 = \frac{1}{2} \sum_{i=1}^{3} \log \frac{k_{ai}^2}{k_{ai}^2 - z_{1i}^2} \tag{3}$$

where $k_a = k_{c1} - \mathbf{X_0} = [k_{a1}, k_{a2}, k_{a3}]^T$, then differentiating of V_1 with respect to time we have

$$\dot{V}_1 = \sum_{i=1}^{3} \frac{z_{1i}\dot{z}_{1i}}{k_{ai}^2 - z_{1i}^2} \tag{4}$$

Differentiating of z_1 with respect to time, we have

$$\dot{z}_{1i} = J_i(x_1)(z_2 + \alpha) - \dot{x}_{di} \tag{5}$$

where $J_i(x_1)$ is the ith line of $J(x_1)$. We propose α as

$$\alpha = J^T(\dot{x}_d - A) \tag{6}$$

where

$$A = \begin{bmatrix} (k_{a1}^2 - z_{11}^2)k_{11}z_{11} \\ (k_{a2}^2 - z_{12}^2)k_{12}z_{12} \\ (k_{a3}^2 - z_{13}^2)k_{13}z_{13} \end{bmatrix} \tag{7}$$

k_{1i}, $i = 1, 2, 3$ are positive constants. Substituting (5), (6) and (7) into (4) we can obtain

$$\dot{V}_1 = -\sum_{i=1}^{3} k_{1i}z_{1i}^2 + \sum_{i=1}^{3} \frac{z_{1i}J_iz_2}{k_{ai}^2 - z_{1i}^2} \tag{8}$$

Then we consider a barrier Lyapunov function candidate as

$$V_2 = V_1 + \frac{1}{2} \sum_{i=1}^{3} \log \frac{k_{bi}^2}{k_{bi}^2 - z_{2i}^2} + \frac{1}{2}z_2^T M z_2 \tag{9}$$

where $k_b = k_{c2} - \mathbf{Y_0} = [k_{b1}, k_{b2}, k_{b3}]^T$, then differentiating (9) with respect to time leads to

$$\dot{V}_2 = \dot{V}_1 + \sum_{i=1}^{3} \frac{z_{2i}\dot{z}_{2i}}{k_{bi}^2 - z_{2i}^2} + z_2^T M \dot{z}_2 \tag{10}$$

$$= -\sum_{i=1}^{3} k_{1i}z_{1i}^2 + \sum_{i=1}^{3} \left(\frac{z_{1i}J_i z_2}{k_{ai}^2 - z_{1i}^2} + \frac{z_{2i}\dot{z}_{2i}}{k_{bi}^2 - z_{2i}^2} \right) + z_2^T M \dot{z}_2$$

Differentiating z_2 with respect to time, we have

$$\dot{z}_2 = M^{-1}[\tau - C(x_2)x_2 - D(x_2)x_2 - g(x_1)] - \dot{\alpha} \tag{11}$$

According to the Moore-Penrose inverse, we can obtain

$$z_2^T (z_2^T)^+ = \begin{cases} 0, & z_2 = [0,0,0]^T \\ 1, & \text{Otherwise} \end{cases} \tag{12}$$

When $z_2 = [0,0,0]^T$, $\dot{V}_2 = -\sum_{i=1}^{3} k_{1i}z_{1i}^2 \le 0$. Then asymptotic stability of the system can still be drawn by the Barbalat's lemma [33]. Otherwise in case of $z_2 \ne [0,0,0]^T$, we designed the model-based control as

$$\tau = C(x_2)x_2 + D(x_2)x_2 + g(x_1) + M\dot{\alpha}$$

$$- (z_2^T)^+ \sum_{i=1}^{3} \frac{z_{1i}J_i^T}{k_{ai}^2 - z_{1i}^2} - (z_2^T)^+ \sum_{i=1}^{3} \frac{z_{2i}(a_i - \dot{\alpha}_i)}{k_{bi}^2 - z_{2i}^2} - K_2 z_2 \tag{13}$$

where k_{2i}, $i = 1, 2, 3$ are positive constants. Then substituting (11) and (13) into (10), we can get

$$\dot{V}_2 = -\sum_{i=1}^{3} k_{1i}z_{1i}^2 - z_2^T K_2 z_2 < 0 \tag{14}$$

According to Lemma 1, we know the signal z_1 remains in the interval $-k_a \le z_1 \le k_a, \forall t > 0$, similarly, the signal z_2 remains in the interval $-k_b \le z_2 \le k_b, \forall t > 0$.

3.2   Adaptive Neural Network Control with Full-State Feedback

The parameters of the marine vessel system $M, C(x_2), D(x_2), g(x_1)$ may be unknown in practise, and in this case the control law above can be implementable. To handle this problem we use a approximator based on neural networks to approximate the unknown parameters. In the following we will design an adaptive neural network control.

The adaptive law is proposed as follows

$$\dot{\hat{W}}_i = \Gamma_i[S_i(Z_i)z_{2,i} - \sigma_i|z_{2i}|\hat{W}_i], i = 1, 2, 3 \tag{15}$$

where $\hat{W} = [\hat{W}_1, \hat{W}_2, \hat{W}_3]^T$ are the weights of the neural networks, $S(Z) = [S(Z)_1, S(Z)_2, S(Z)_3]$ is the basis functions, and $Z = [x_1^T, x_2^T, \alpha^T, \dot{\alpha}^T]$ are the inputs of the neural networks, and $\Gamma_i = \Gamma_i^T > 0$ $(i = 1, 2, 3)$ is the constant gain matrix, $\sigma_i > 0, i = 1, 2, 3$ are small constants. The neural network $\hat{W}^T S(Z)$ is used to approximate $W^{*T} S(Z)$.

$$W^{*T} S(Z) = -(C(x_2)x_2 + D(x_2)x_2 + g(x_1) + M\dot{\alpha}) - \epsilon(Z) \qquad (16)$$

where $\tilde{W}_i = \hat{W}_i - W_i^*$ and $\tilde{W}_i$, $\hat{W}_i$, W_i^* are the NN weight errors, estimate and actual value respectively.

Lemma 2. *[34] For adaptive law (15), there exits a compact set*

$$\Omega_{\omega 1} = \left\{ \hat{W}_i \big| \|\hat{W}_i\| \leq \frac{s_i}{\sigma_i} \right\}$$

where $\|S_i(Z)\| \leq s_i$ with $\phi_i > 0$, such that $\hat{W}_i(t) \in \Omega_{\omega 1}$, $\forall t \geq 0$ provided that $\hat{W}_i(0) \in \Omega_{\omega 1}$.

Proof: Let $V_{\omega 1} = \frac{1}{2} \hat{W}_i^T \Gamma_i^{-1} \hat{W}_i$, its time derivative is

$$\dot{V}_{\omega 1} = \hat{W}_i^T (S_i(Z) z_{2,i} - \sigma_i |z_{2i}| \hat{W}_i) \leq -|z_{2i}| \|\hat{W}_i\| (\sigma_i \|\hat{W}_i\| - s_i)$$

$\dot{V}_{\omega 1}$ will become negative as long as $\|\hat{W}_i\| > \frac{s_i}{\sigma_i}$. Therefore, $\hat{W}_i \in \Omega_{\omega 1}$ if $\hat{W}_i(0) \in \Omega_{\omega 1}$ for $t \geq 0$. ∎

Then, we propose the following control as

$$\tau = -(z_2^T)^+ \sum_{i=1}^{3} \left[\frac{z_{2i}(a_i - \alpha_i)}{k_{bi}^2 - z_{2i}^2} + \frac{k_{1i} z_{1i}^2}{k_{ai}^2 - z_{1i}^2} + \frac{k_{2i} z_{2i}^2}{k_{bi}^2 - z_{2i}^2} \right]$$

$$- \sum_{i=1}^{3} \frac{z_{1i}(J_i)^T}{k_{ai}^2 - z_{1i}^2} - \hat{W}^T S(Z) - K_3 z_2 \qquad (17)$$

where K_2 is the control gain, Γ_i is the constant gain matrix, and $\sigma_i > 0$, $(i = 1, 2, 3)$ are small positive constants.

In the following part, we are ready to present the stability theorem of the closed-loop system.

Theorem 1. *Consider the marine surface vessel dynamics (1), with the state feedback control law (17) together with adaption law (15), for initial conditions satisfy $z_1(0) \in \Omega_0 := \{z_1 \in \mathbb{R}^3 : -k_a < z_1 < k_a\}$, and $z_2(0) \in \Omega_0 := \{z_2 \in \mathbb{R}^3 : -k_b < z_2 < k_b\}$, i.e., the initial conditions are bounded. The signals of the closed loop system are semiglobally uniformly bounded (SGUB). The multiple full-state constraints are never violated, i.e., $|x_1| < k_{c1}$, $|x_2| < k_{c2}$, $\forall t > 0$, and the closed-loop error signals z_1 and z_2 will remain within the compact sets*

Ω_{z1}, Ω_{z2}, respectively, defined by

$$\Omega_{z1} := \{z_1 \in \mathbb{R}^3| \, \|z_{1i}\| \leq \sqrt{k_{ai}^2(1 - e^{-D})}, i = 1, 2, 3\} \tag{18}$$

$$\Omega_{z2} := \{z_2 \in \mathbb{R}^3| \, \|z_{2i}\| \leq \sqrt{\frac{D}{\lambda_{\min}(M)}}, i = 1, 2, 3\}$$

$$\bigcap \{z_{2i} \in \mathbb{R}^3| \, \|z_2\| \leq \sqrt{k_{bi}^2(1 - e^{-D})}, i = 1, 2, 3\} \tag{19}$$

where $D = 2(V_3(0) + C/\rho)$, ρ and C are two positive constants.

Proof: Consider the following Lyapunov candidate function

$$V_3 = V_2 + \frac{1}{2}\sum_{i=1}^{3} \tilde{W}_i^T \Gamma_i^{-1} \tilde{W}_i \tag{20}$$

Differentiating V_3 with respect to time yields, and taking $\dot{z}_2 = \dot{x}_2 - \dot{\alpha} = a - \dot{\alpha}$ into (20), we have

$$\dot{V}_3 = -\sum_{i=1}^{3} k_{1i} z_{1i}^2 + \sum_{i=1}^{3} \frac{z_{1i} J_i z_2}{k_{ai}^2 - z_{1i}^2} + \sum_{i=1}^{3} \frac{z_{1i}(a_i - \dot{\alpha}_i)}{k_{bi}^2 - z_{2i}^2}$$

$$+ z_2^T[\tau - C(x_2)x_2 - D(x_2)x_2 - g(x_1) - M\dot{\alpha}] + \sum_{i=1}^{3} \tilde{W}_i^T \Gamma_i^{-1} \dot{\tilde{W}}_i \tag{21}$$

Substituting (15), (16) and (17) into (21), we can obtain

$$\dot{V}_3 = \sum_{i=1}^{3} \frac{z_{2i}(a_i - \dot{\alpha}_i) - z_2^T(z_2^T)^+ z_{2i}(a_i - \dot{\alpha}_i)}{k_{bi}^2 - z_{2i}^2} - \sum_{i=1}^{3} \frac{z_2^T(z_2^T)^+ k_{1i} z_{1i}^2}{k_{ai}^2 - z_{1i}^2}$$

$$- \sum_{i=1}^{3} \frac{z_2^T(z_2^T)^+ k_{2i} z_{2i}^2}{k_{bi}^2 - z_{2i}^2} + \sum_{i=1}^{3} \tilde{W}_i^T S_i(Z) z_{2,i} - \sum_{i=1}^{3} \tilde{W}_i^T |z_{2i}| \hat{W}_i - \sum_{i=1}^{3} k_{1i} z_{1i}^2$$

$$+ z_2^T(W^{*T}S(Z) - \hat{W}^T S(Z) + \epsilon(Z)) - z_2^T K_3 z_2 \tag{22}$$

When $z_2 = [0, 0, 0]^T$, $\dot{V}_2 = -\sum_{i=1}^{3} k_{1i} z_{1i}^2$, according to the Barbalat's lemma, we can still be drawn the asymptotic stability of the system. Otherwise, in case of $z_2 \neq [0, 0, 0]^T$, we have

$$\dot{V}_3 \leq -z_2^T(K_3 - I)z_2 - \sum_{i=1}^{3} \frac{k_{1i} z_{1i}^2}{k_{ai}^2 - z_{1i}^2} - \sum_{i=1}^{3} \frac{k_{2i} z_{2i}^2}{k_{bi}^2 - z_{2i}^2}$$

$$+ \sum_{i=1}^{3} \frac{\sigma_i^2}{4}(\|W_i^*\|^4 + \|\tilde{W}_i\|^4 - 2\|W_i^*\|^2 \|\tilde{W}_i\|^2) + \frac{1}{2}\|\bar{\epsilon}(Z)\|^2 \tag{23}$$

From lemma 2, we can obtain

$$\|\tilde{W}_i\| = \|\hat{W}_i - W_i^*\| \le \frac{s_i}{\sigma_i} + \|W_i^*\| \tag{24}$$

Therefore, we have

$$\dot{V}_3 \le -\rho V + C \tag{25}$$

where

$$\rho = \min(\min(2k_{1i}), \min(2k_{2i}), \frac{2\lambda_{\min}(K_3 - I)}{\lambda_{\max}(M)}, \min(\frac{\sigma_i^2 \|W_i^*\|^2}{\lambda_{\max}(\Gamma_i^{-1})}))$$

$$C = \frac{1}{2}\|\bar{\epsilon}(Z)\|^2 + \sum_{i=1}^{3} \frac{\sigma_i^2}{4}(\|W_i^*\|^4 + \|\tilde{W}_i\|^4) \tag{26}$$

where $\lambda_{\min}(\bullet)$ and $\lambda_{\max}(\bullet)$ denote the minimum and maximum eigenvalues of matrix $\bullet$, where $\lambda(A)$ are real, respectively. To ensure $\rho > 0$, the control gain K_3 is chosen to satisfy the following conditions:

$$\lambda_{\min}(K_3 - I) > 0 \tag{27}$$

According to Lemma 1, $z_1(t)$ remains in the open set $z_1 \in (-k_a, k_a), \forall t \in [0, +\infty)$, provided that $z_1(0) \in (-k_a, k_a)$. As we know $x_1(t) - z_1(t) + x_d(t)$, $-\mathbf{X}_0 \le x_d(t) \le \mathbf{X}_0$, $k_a = k_{c1} - \mathbf{X}_0$, and we known $z_2(t)$ remains in the open set $z_2 \in (-k_b, k_b), \forall t \in [0, +\infty)$, provided that $z_2(0) \in (-k_b, k_b)$, and $x_2(t) = z_2(t) + \alpha(t)$, $-\mathbf{Y}_0 \le \alpha(t) \le \mathbf{Y}_0$, $k_b = k_{c2} - \mathbf{Y}_0$ similarly. Hence, the output constraints are not violated, i.e., $|x_1| \le k_{c1}$, $|x_2| \le k_{c2}, \forall t \ge 0$.
Multiplying (25) by $e^{\rho t}$, we can obtain

$$\frac{d}{dt}(V_2 e^{\rho t}) \le C e^{\rho t} \tag{28}$$

Integrating the above inequality, we obtain

$$V_3 \le V_3(0) + \frac{C}{\rho} \tag{29}$$

For z_1, we have

$$\frac{1}{2}\log\frac{k_{ai}^2}{k_{ai}^2 z_{1i}^2} \le V_3(0) + \frac{C}{\rho}$$

$$\|z_{1i}\| \le \sqrt{k_{ai}^2(1 - e^{2(V_3(0) + \frac{C}{\rho})})} \tag{30}$$

■

Therefore, we can conclude that the signals z_1 and z_2 are semiglobally uniformly bounded.

4   Simulation

Consider the vessel with model of Cybership II, which is a 1:70 scale supply vessel replica built in a marine control laboratory in the Norwegian University of Science and Technology [7]. In this section, we choose the desired trajectories as follows:

$$\begin{cases} x_{1xd}(t) = \sin 0.5t \\ x_{1yd}(t) = 2\cos 1.1t \\ x_{1\psi d}(t) = 3\sin t \end{cases} \tag{31}$$

We have proposed two cases for the simulation studies. Firstly, we give the mode-based control (13). Subsequently, the adaptive neural network control (17) with the state feedback is evaluated.

For the model based control, to guarantee the state constraints $|x_1| < k_{c1} = [1.17, 2.14, 3.15]^T$, we can choose the constraints of z_1 will be $k_a = k_{c1} - X_0 = [0.17, 0.14, 0.15]^T$. In the same reason, we can proposed $|x_2|$ and z_2 as above. The tracking performance of the closed-loop system for the vessel are given in Figs. 2 and 4. From the two figures, we can obtain that all the x_1 and x_2 can successfully track the desired trajectory. According the Figs. 3 and 5 we can state the system

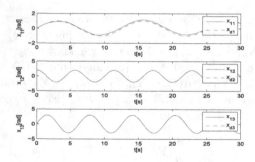

Fig. 2. Comparison between x_1 and x_d (dashed line-x_d, solid line-x_1).

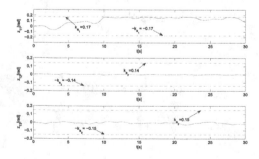

Fig. 3. Tracking error z_1.

errors are converging to a small value close to zero. The corresponding control inputs are given in Fig. 6.

For adaptive neural network control with the state feedback. The control objectives are to make the state of the system x_1 and x_2 track the ideal trajectory x_{1d} and x_{2d}, then guaranteeing the state constraints $|x_1| < k_{c1} = [1.17, 2.14, 3.15]^T$. The simulation results are shown in Figs. 7-11. The Fig. 7 shows that the state x_1 can successful track the desired trajectory. From Fig. 8.

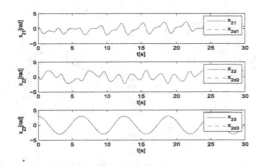

Fig. 4. Comparison between x_2 and x_{2d} (dashed line-x_{2d}, solid line-x_2).

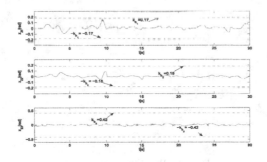

Fig. 5. Tracking error z_2.

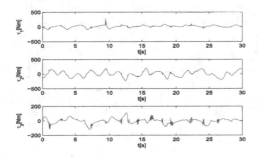

Fig. 6. Control input τ.

we can know the tracking error is converging to a small value that close to zero. In Fig. 9, we can see that the x_2 tracks the desired trajectory with a high accuracy. The tracking error z_2 is proposed in Fig. 10, and the control input τ is shown in Fig. 11.

We have got the exciting conclusion that the proposed control is able to track a desired trajectory with an excellent performance.

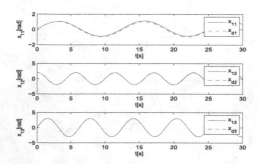

Fig. 7. Comparison between x_1 and x_d (dashed line-x_d, solid line-x_1).

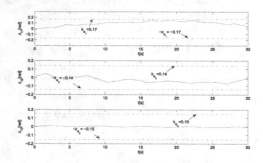

Fig. 8. Tracking error z_1.

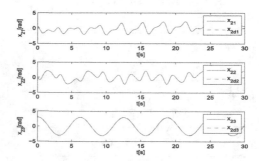

Fig. 9. Comparison between x_2 and x_{2d} (dashed line-x_{2d}, solid line-x_2).

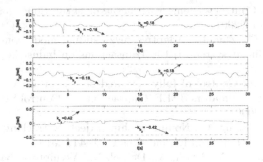

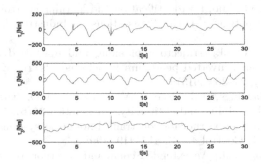

Fig. 10. Tracking error z_2.

Fig. 11. Control input τ

5   Conclusion

In this paper, we consider the control design for a marine surface vessel with full-state constraints and unknown parameters using the barrier Lyapunov function and neural networks. We have proven that under the proposed control laws, the signals of the closed loop system are semiglobally uniformly bounded (SGUB), the asymptotic tracking is achieved, and the multiple state constraints are never violated. Simulation results illustrate the effectiveness of the proposed design techniques.

References

1. Ge, S.S., Zhao, Z., He, W.: Localization of drag anchor in mooring systems via magnetic induction and acoustic wireless communication network. IEEE Journal of Oceanic Engineering **39**(3), 515–525 (2014)
2. Yang, C., Li, Z., Cui, R., Xu, B.: Neural network-based motion control of an underactuated wheeled inverted pendulum model **25**(11), 2004–2016 (2014)
3. Cui, R., Guo, J., Mao, Z.: Adaptive backstepping control of wheeled inverted pendulums models. Nonlinear Dynamics **79**(1), 501–511 (2015)
4. Liu, D., Michel, A.N.: Dynamical systems with saturation nonlinearities: analysis and design. Springer (1994

5. Hu, T., Lin, Z.: Control systems with actuator saturation: analysis and design. Brikhuser, Bostion (2001)
6. Chen, Z., Shan, C., Zhu, H.: Adaptive fuzzy sliding mode control algorithm for a non-affine nonlinear system. IEEE Transactions on Industrial Informatics **3**(4), 302–311 (2007)
7. Tee, K.P., Ge, S.S.: Control of fully actuated ocean surface vessels using a class of feedforward approximators. IEEE Transactions on Control Systems Technology **14**(4), 750–756 (2006)
8. Du, J., Wang, L., Jiang, C.: Nonlinear ship trajectory tracking control based on backstepping algorithm. Ship Engineering **1**, 41–44 (2010)
9. Ghommam, J., Mnif, F., Benali, A., Derbel, N.: Asymptotic backstepping stabilization of an underactuated surface vessel. IEEE Transactions on Control Systems Technology **14**(6), 1150–1157 (2006)
10. Dai, S.-L., Wang, C., Luo, F.: Identification and learning control of ocean surface ship using neural networks. IEEE Transactions on Industrial Informatics **8**(34), 801–810 (2012)
11. Yang, C., Li, Z., Li, J.: Trajectory planning and optimized adaptive control for a class of wheeled inverted pendulum vehicle models. IEEE Transactions on Cybernetics **43**(1), 24–36 (2013)
12. Mayne, D.Q., Rawlings, J.B., Rao, C.V., Scokaert, P.O.: Constrained model predictive control: Stability and optimality. Automatica **36**(6), 789–814 (2000)
13. Alberto, B.: Reference governor for constrained nonlinear systems. IEEE Transactions on Automatic Control **43**(3), 415–419 (1998)
14. Gilbert, E.G., Ong, C.J.: An extended command governor for constrained linear systems with disturbances. In: Proceedings of the 48th IEEE Conference on Decision and Control, pp. 6929–6934 (2009)
15. Do, K.: Control of nonlinear systems with output tracking error constraints and its application to magnetic bearings. International Journal of Control **83**(6), 1199–1216 (2010)
16. DeHaan, D., Guay, M.: Extremum-seeking control of state-constrained nonlinear systems. Automatica **41**(9), 1567–1574 (2005)
17. Cui, R., Ren, B., Ge, S.S.: Synchronised tracking control of multi-agent system with high order dynamics. IET Control Theory & Applications **6**(5), 603–614 (2012)
18. He, W., Chen, Y., Yin, Z.: Adaptive neural network control of an uncertain robot with full-state constraints (2015)
19. Ngo, K.B., Mahony, R., Jiang, Z.P.: Integrator backstepping using barrier functions for systems with multiple state constraints. In: Proceedings of the 44th IEEE Conference on Decision and Control, pp. 8306–8312 (2005)
20. Ren, B., Ge, S.S., Tee, K.P., Lee, T.H.: Adaptive neural control for output feedback nonlinear systems using a barrier lyapunov function. IEEE Transactions on Neural Networks **21**(8), 1339–1345 (2010)
21. Li, Z., Yang, C., Ding, N., Bogdan, S., Ge, T.: Robust adaptive motion control for remotely operated vehicles with velocity constraints. International Journal of Control, Automation, and System **10**(2), 421–429 (2012)
22. Chen, M., Ge, S.S., Ren, B.: Adaptive tracking control of uncertain MIMO nonlinear systems with input constraints. Automatica **47**(3), 452–465 (2011)
23. How, B.V.E., Ge, S.S., Choo, Y.S.: Dynamic Load Positioning for Subsea Installation via Adaptive Neural Control. IEEE Journal of Oceanic Engineering **35**(2), 366–375 (2010)

24. Yang, C., Ganesh, G., Haddadin, S., Parusel, S., Albu-Schaeffer, A., Burdet, E.: Human-like adaptation of force and impedance in stable and unstable interactions. IEEE Transactions on Robotics **27**(5), 918–930 (2011)
25. Yang, C., Ge, S.S., Lee, T.H.: Output feedback adaptive control of a class of nonlinear discrete-time systems with unknown control directions. Automatica **45**(1), 270–276 (2009)
26. Yang, C., Ge, S.S., Xiang, C., Chai, T., Lee, T.H.: Output feedback NN control for two classes of discrete-time systems with unknown control directions in a unified approach. IEEE Transactions on Neural Networks **19**(11), 1873–1886 (2008)
27. He, W., Ge, S.S., Li, Y., Chew, E., Ng, Y.S.: Neural network control of a rehabilitation robot by state and output feedback. Journal of Intelligent & Robotic Systems, 1–17 (2014)
28. Yeşildirek, A., Lewis, F.L.: Feedback linearization using neural networks. Automatica **31**(11), 1659–1664 (1995)
29. Tee, K.P., Ren, B., Ge, S.S.: Control of nonlinear systems with time-varying output constraints. Automatica **47**(11), 2511–2516 (2011)
30. Tee, K.P., Ge, S.S.: Control of nonlinear systems with full state constraint using a barrier lyapunov function. In: 2009 IEEE 48th Conference on Decision and Control, pp. 8618–8623 (2009)
31. He, W., Ge, S.S., How, B.V.E., Choo, Y.S.: Dynamics and Control of Mechanical Systems in Offshore Engineering. Springer (2014)
32. He, W., Ge, S.S., How, B.V.E., Choo, Y.S., Hong, K.-S.: Robust adaptive boundary control of a flexible marine riser with vessel dynamics. Automatica **47**(4), 722–732 (2011)
33. Slotine, J., Li, W.: Applied Nonlinear Control. Prentice Hall, Englewood Cliffs (1991)
34. Meng, W., Yang, Q., Pan, D., Zheng, H., Wang, G., Sun, Y.: NN-based asymptotic tracking control for a class of strict-feedback uncertain nonlinear systems with output constraints. In: 2012 IEEE 51st Annual Conference on Decision and Control, pp. 5410–5415 (2012)

Optimization of Parameters for an USV Autopilot

Andy S.K. Annamalai[1] and Chenguang Yang[2,3(✉)]

[1] Moray College, University of Highlands and Islands, Inverness, UK
[2] Center for Robotics and Neural Systems, Plymouth University, Plymouth, UK
cyang@ieee.org
[3] Key Lab of Autonomous System and Network Control and College of Automation Science
and Engineering, South China University of Technology, Guangzhou, China

Abstract. The purpose of this research is to provide an insight into the effect of different parameters on the autopilot design. Various independent parameters of three autopilots namely, proportional integral derivative, linear quadratic regulator and model predictive controller are analyzed and evaluated to obtain optimum performance. Further these optimal parameters are employed in a controller design which is integrated with a Kalman filter and an interval Kalman filter based navigation system and a line of sight based guidance system. Overall performance of the autopilots with the optimum parameters are presented in a tabular form to enable easier comparison and to serve as a benchmark to tune autopilot parameters of an uninhabited surface vehicles.

Keywords: Proportional integral derivative · Linear quadratic regulator · Model predictive control · Kalman filtering · Navigation, guidance and control · Uninhabited surface vehicle

1 Introduction

In order to meet the testing mission demands being imposed by the various sectors, autopilots have been designed based on, for example, fuzzy (Park et al 2005), gain scheduling (Alves et al 2006), H infinity (Elkaim and Kelbley 2006), sliding mode (Ashrafiuon et al 2008) and neural network (Qiaomei et al 2011) techniques, all of which have met with varying degrees of success.

The intention of this paper is to focus on obtaining the optimal parameters for the proportional integral derivative (PID), linear quadratic regulator (LQR) and model predictive controller (MPC) autopilots of an uninhabited surface vehicle (USV). The next section will explain the test platform used in this study.

2 Materials and Methods: The Test Platform

2.1 Hardware

A comprehensive account of the USV hardware have already been published in Sutton et al (2011), only an outline will be presented here for the sake of completeness.

© Springer International Publishing Switzerland 2015
H. Liu et al. (Eds.): ICIRA 2015, Part I, LNAI 9244, pp. 126–135, 2015.
DOI: 10.1007/978-3-319-22879-2_12

The Springer USV was designed by Sutton's group at Plymouth University as a medium water plane twin hull vessel which is versatile in terms of mission profile and payload. It is approximately 4m long and 2.3m wide with a displacement of 0.6 tonnes. Each hull is divided into three watertight compartments. The navigation, guidance and control (NGC) system is carried in watertight Pelican cases and secured in a bay area between the crossbeams. This facilitates the quick substitution of systems on shore or at sea. The batteries which are used to provide the power for the propulsion system and onboard electronics are carried within the hulls, accessed by a watertight hatch. In order to prevent any catastrophe resulting from a water leakage, leak sensors are utilized within the motor housing. If a breach is detected the onboard computer immediately issues a warning to the user and/or takes appropriate action in order to minimize damage to the onboard electronics (Sutton et al 2011, Sharma et al 2012b). A mast has also been installed to carry the global positioning system (GPS) and wireless antennas. The wireless antenna is used as a means of communication between the vessel and its user and is intended to be utilized for remote monitoring purpose, intervention in the case of erratic behavior and to alter the mission parameters. The Springer is shown in Figure 1.

Fig. 1. The Springer uninhabited surface vehicle [adapted from the Springer project website at http://www.tech.plymouth.ac.uk/sme/springerusv/2011/Springer.html]

The Springer propulsion system consists of two propellers powered by a set of 24V 74lbs (334N) Minn Kota Riptide transom mounted saltwater trolling motors. As will be seen in the next section 3, steering of the vessel is based on differential propeller revolution rates.

In Springer, the integrated sensor suite combines a GPS, three different types of compasses, speed log and depth sensor. All of these sensors are interfaced to a personal computer (PC) via a NI-PCI 8430/8 (RS232) serial connector. Since the GPS, depth and speed sensors are not used in this study, their characteristics will not be expanded upon any further.

However, TCM2, HMR3000 and KVH-C100 are the three different types of electronic compass installed in the Springer. All of the compasses can output NMEA 0183 standard sentences with special sentence head and checksum. As all of these compasses are very sensitive, they were mounted as far as possible from any source of magnetic field and from ferrous metal objects. In addition, each compass was individually housed in a small waterproof case to provided further isolation and insulation.

3   Experiments for Modelling the Yaw Dynamics

Hydrodynamic modelling was initially explored to obtain the model of the USV. However, this approach is usually very expensive, time consuming and requires the use of specialist equipment in the form of a tank testing facility. Nevertheless, the approach does produce detailed models based upon hydrodynamic derivatives. In addition, costs can also rise further if vehicle configurations change and thus, the tank testing and modelling procedure have to be repeated. Since the hiring and running costs for such a facility were deemed to be prohibitive, it was considered more appropriate to model the vehicle dynamics using Black box system identification (SI) techniques.

Thus SI techniques have been applied to obtain the Springer model. For this, several trials were carried out where the vessel was driven for some calculated maneuvers and data recorded at Roadford Reservoir, Devon, UK.

The vehicle has a differential steering mechanism and thus requires two inputs to adjust its course. This can be simply modelled as a two input, single output system in the form depicted in Figure 2.

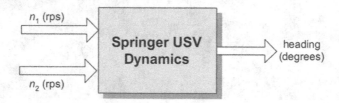

Fig. 2. Block diagram representation of a two-input USV (Sharama et al, 2012a)

where n1 and n2 being the two propeller thrusts in revolutions per minute (rpm). Clearly, straight line maneuvers require both the thrusters running at the same speed whereas the differential thrust is zero in this case. In order to linearize the model at an operating point, it is assumed that the vehicle is running at a constant speed of 3 knots. This corresponds to both thrusters running at 900 rpm. To clarify this further, let nc and nd represents the common mode and differential mode thruster velocities defined to be (Sharma et al, 2012a)

$$n_c = \frac{n_1 + n_2}{2} \tag{1}$$

$$n_d = \frac{n_1 - n_2}{2} \tag{2}$$

In order to maintain the velocity of the vessel, nc must remain constant at all times. The differential mode input, however, oscillates about zero depending on the direction of the maneuver. For data acquisition, several inputs including a pseudo random binary sequence (prbs) was applied to the thrusters and the heading response was recorded. Figures 3a and 3b depict two data sets obtained from those trials. The input shown is the differential rpm, nd, which cause the vehicle to maneuver as required. The acquired data was processed and down sampled to 1 Hz since this frequency was deemed to be adequate for controller design.

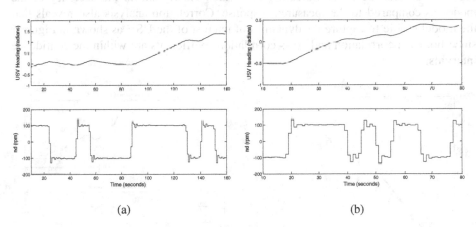

(a)   (b)

Fig. 3. Experimental data sets from trials conducted by Sutton's group at Roadford Reservoir in Devon, UK

SI was then applied to the acquired data set and a dynamic model of the vehicle is obtained in the following form.

$$y(z) = G_1(z)u_1 + G_2(z)u_2 \tag{3}$$

where G1 and G2 denotes the discrete transfer functions from inputs u1 and u2 respectively and y being the output of the system. In this case, only nd has been manipulated and therefore act as the sole input to the system. This alters both n1 and n2 whereas nc is maintained to conserve the operating regime. Two models of second and fourth order were identified from the data, however, subsequent simulation study revealed that there was no significant advantage of using a more complex fourth order model. Hence, the second order model shown in (4) and (5) in state space form is selected for further analysis and controller design.

$$x(k+1) = \mathbf{A}\,x(k) + \mathbf{B}\,u(k)$$
$$y(k) = \mathbf{C}\,x(k)$$
(4)

where

$$\mathbf{A} = \begin{bmatrix} 1.002 & 0 \\ 0 & 0.9945 \end{bmatrix}, \quad \mathbf{B} = \begin{bmatrix} 6.354 \\ -4.699 \end{bmatrix} \times 10^{-6},$$
$$\mathbf{C} = \begin{bmatrix} 34.13 & 15.11 \end{bmatrix}$$
(5)

with a sampling time of 1s, where u(k) represents the differential thrust input in rpm and y(k) the heading angle in radians.

To validate the model, cross validation test and residual analysis are carried out. Figure 4 depicts the cross validation test which clearly shows the accuracy of the model as compared to the measured response. Correlation analysis also reveals that the model is able to capture the dynamics of interest of the USV as shown in Figure 5 since both autocorrelation and cross correlation coefficients are within the confidence intervals.

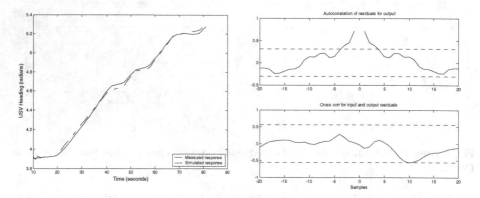

Fig. 4. Actual measurements and simulated output of the USV model

Fig. 5. Cross correlation and autocorrelation of the residuals

4   Results and Discussion of Significant Results

One the model of the USV was obtained, then in order to compare the three types of autopilot, each one must be tuned to offer the best possible performance. In the case of PID controllers, there is a widespread literature on tuning rules based on the transient step-response of a plant, and in all cases, it would be straightforward to tune the parameters based on optimizing some performance measure of the step-response, such as minimization of the integral of the absolute error (ITAE). However, since the system under consideration contains non-linear elements (actuator saturation limits) and performance is based not only on tracking the reference heading angle but overall on how well the vessel follows the desired trajectory, based not just on the deviation from the ideal trajectory but also upon other criteria such an energy used, it was decided to tune each controller based on the evaluation of a cost function over simulations of the actual NGC system for the mission plan under consideration.

In addition to the random disturbance affecting the yaw dynamics of the vessel, a disturbance consisting of an imposed variation in the vessel speed was applied in order to recreate a more realistic scenario in which surface currents exist, transporting the vessel as if on a conveyor belt. The magnitude of this variation, though not constant, was nominally around 25% of the constant forward speed of the vessel (Figure 6) with a constant direction given by the positive direction of the y-coordinate axis. Furthermore, this disturbance solely affects the position of the vessel without affecting its heading.

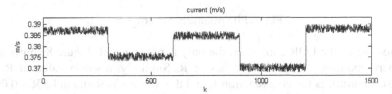

Fig. 6. Disturbance signal

Under this scenario, the mission plan was simulated using the NGC with the three types of controllers for different values of their respective parameters. The simulations were carried out using an ordinary Kalman filter (KF) as state estimator for the yaw-dynamics model. The performance indices measured were: number of waypoints reached (out of 7), total distance travelled, average deviation and average controller energy. For each set of values of the parameters, the simulation was repeated ten times and the graphs shown reflect the mean values obtained.

Figure 7 shows the values obtained of each measure for each triplet (Kp,Ti,Td), which are color based. The best parameter values for each measure are indicated on the graph by an asterisk. Because, as can be expected, the optimum parameters are different for each measure of performance, in order to obtain a single performance value a single cost function that takes each of these into account can be constructed. The cost function chosen is simply a weighted average of the four performance measures:

$$J = -\rho_1 \frac{wp\ reached}{7} + \rho_2 \frac{distance}{1500} + \rho_3 \frac{av.deviation}{30} + \rho_4 \frac{av.energy}{5} \qquad (6)$$

Each measure is divided by a representative value in order to normalize it. The values of the weights were chosen as $\rho_1 = 100$, $\rho_2 = 75$, $\rho_3 = 75$, $\rho_4 = 50$, whereby the fraction of waypoints reached is given the most importance, and the controller energy the least. The values of PID parameters that minimize J are found to be Kp = 130, Ti = 13, Td = 0.

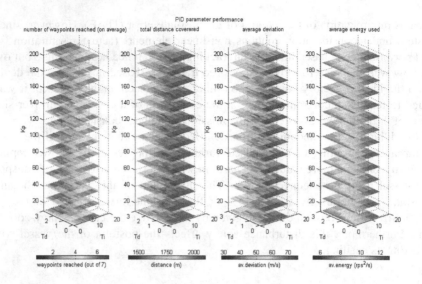

Fig. 7. PID parameter performance

In the case of the LQR controller, the only parameter is R. Figure 8 shows the average performances for different values of R. Similarly, a single value of R is obtained that minimizes the cost function J, and this value was found to be R = 0.0049.

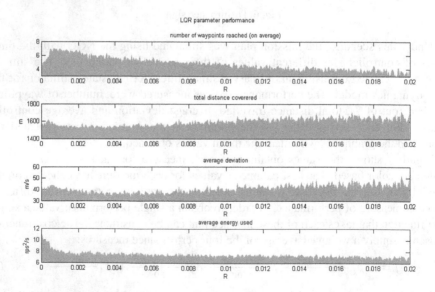

Fig. 8. LQR parameter performance.

For the MPC controller, the four performance measures are shown as a function of the prediction horizon Hp and the control horizon Hc (Figure 9). The optimum values that minimize J were found to be Hp = 10, Hc = 2.

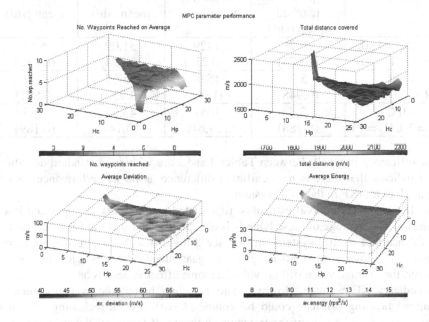

Fig. 9. MPC parameter performance.

Table 1 summarizes the results for the optimal parameter values of each controller.

Table 1. Average performance measures for NGC systems using KF estimation.

Case	No. of way-points reached mean (std)	Total dis-tance (m) mean (std)	$\overline{r_d}$ (m/s) mean (std)	$\overline{CE_u}$ (rps²/s) mean (std)
PID ($K_p = 130$, $T_i = 13, T_d = 0$)	6.5 (0.641)	1409 (41.219)	28.78 (3.751)	5.41 (0.615)
LQR (R = 0.0049)	6.1 (0.939)	1521 (40.157)	36.78 (3.278)	6.72 (0.901)
MPC ($H_p = 10, H_c = 2$)	5.5 (0.768)	1616 (70.61)	39.65 (10.144)	8.70 (0.189)

Using the optimal parameters found for each of the controllers, the following table shows the results obtained by simulating the mission plan using the NGC with an interval Klaman filter (IKF) rather than a standard KF to estimate the heading angle. As before, ten simulations were carried out for each case and the table presents the mean and standard deviation for each one.

Table 2. Average performance measures for NGC systems using IKF estimation.

Case	No. of way-points reached mean (std)	Total distance (m) mean (std)	$\overline{r_d}$ (m/s) mean (std)	$\overline{CE_u}$ (rps^2/s) mean (std)
PID ($K_p = 130$, $T_i = 13$, $T_d = 0$)	6.7 (0.674)	1429 (40.258)	31.09 (3.824)	5.52 (0.652)
LQR ($R = 0.0049$)	5.7 (0.948)	1526 (41.905)	35.57 (3.730)	7.21 (0.891)
MPC ($H_p = 10$, $H_c = 2$)	5.8 (0.788)	1651 (69.64)	43.91 (10.144)	8.65 (0.189)

The difference in results between Tables 1 and 2 are within the standard deviations and so indicate that there is no significant difference as far as performance goes in using a KF or an IKF in the NGC system.

The IKF provides interval estimates, represented on the plots as upper and lower boundaries. The average of these bounds is also plotted and it is the value fed back to the guidance and control systems in practice. Notice that the width of the intervals increases over time, and the advantage of the guaranteed estimates is lost in practice. However, the average value still provides the same effectiveness as an estimate as that of an ordinary KF estimate. However, should tighter bounds be available, then the guaranteed heading estimate could be combined with a dead-reckoning system to provide guaranteed bounds of the trajectory estimate. If this is sufficiently narrow, it could be very useful to feedback such information to the guidance system which could generate a trajectory that is guaranteed being clear of submerged sandbanks and shoals.

5   Conclusions

In this study a NGC system is built based on LQR and MPC autopilots with KF / IKF feedback, and benchmarked against a standard PID control algorithm. With regard to the autopilots, the LQR and MPC, and particular the latter, require that the nonlinearities of the physical limitations (actuator limits) be incorporated into the optimization algorithms in order to achieve better performance. This is evidenced by the results obtained when such limitations were partially lifted. The feasibility of the integrated NGC system was explored with various permutations and combinations of the parameters involved in the controller design. The optimum parameters for each of the controllers were obtained. Furthermore, finally the performance of these optimum controllers were presented in a tabular form. The same set of optimum parameters of the controller design was utilized and the performance of the overall NGC system, incorporating IKF to provide navigation estimates was analyzed and presented in the tabular form. Autopilots based on MPC, LQR and PID in conjunction with an LOS guidance system, achieved reasonably good results (the worst case still achieving over 5 waypoints reached successfully on average out of the 7 waypoints, even though the vessel's position was distorted by the effect of adding 25% of its nominal speed along a constant direction to simulate a current along that direction).

Acknowledgement. The authors would like to thank Professor R. Sutton of Plymouth University and EPSRC project EP/I012923/1, for the opportunity to conduct research on Springer USV.

References

1. Alves, J., Oliveira, P., Oliveira, R., Pascoal, A., Rufino, M., Sebastiao, L., Silvestre, C.: Vehicle and mission control of the Delfim autonomous surface craft. In: Proc of 14th Mediterranean Conference on Control Automation, Ancona, Italy, pp. 1–6, June 2006
2. Ashrafiuon, H., Muske, K.R., McNinch, L.C., Soltan, R.A.: Sliding-mode tracking control of surface vessels. IEEE Trans. on Industrial Electronics **55**(11), 4004–4012 (2008)
3. Elkaim, G.H., Kelbley, R.: Measurement based H infinity controller synthesis for an autonomous surface vehicle. In: Proc of 19th International Technical Meeting of the Satellite Division of the Institute of Navigation, Fort Worth, USA, pp. 1973–1982, September 2006
4. Park, S., Kim, J., Lee, W., Jang, C.: A study on the fuzzy controller for an unmanned surface vessel designed for sea probes. In: Proc of International Conference on Control, Automation and Systems, Kintex, Korea, pp. 1–4, June 2005
5. Qiaomei, S., Guang R., Jin, Y., Xiaowei, Q.: Autopilot design for unmanned surface vehicle tracking control. In: Proc of 3rd International Conference on Measuring Technology and Mechatronics Automation, vol. 1, Shangshai, China, pp. 610–613, January 2011
6. Sutton, R., Sharma, S., Xu, T.: Adaptive navigation systems for an unmanned surface vehicle. Proc. IMarEST - Part A: Journal of Marine Engineering and Technology **10**(3), 3–20 (2011)
7. Sharma, S., Sutton, R.: Modelling the yaw dynamics of an uninhabited surface vehicle for navigation and control system design. Proc IMarEST - Part A: Journal of Marine Engineering and Technology **11**(3), 9–20, September 2012a
8. Sharma, S., Naeem, W., Sutton, R.: An Autopilot Based on a Local Control Network Design for an Unmanned Surface Vehicle. The Journal of Navigation **65**, 1–21 (2012b). The Royal Institute of Navigation, doi:10.1017/S0373463311000701

Research on 3-D Motion Simulation of Mooring Buoy System Under the Effect of Wave

Qiaogao Huang[✉] and Guang Pan

School of Marine Science and Technology,
Northwestern Polytechnical University, Xi'an, People's Republic of China
{huangqiaogao,panguang601}@163.com

Abstract. A mooring buoy system usually contains a tether and a buoy. The mathematical model of mooring system should be developed to research the three-dimensional motion of the mooring buoy system under the effect of wave force. Based on the Newton's Second Law, a motion governing equation of tether is established. Moreover a discrete motion equation of tether is deduced through the lumped mass method. According to the momentum and angular momentum theories of rigid body, a motion governing equation of buoy is developed. The tether model and the buoy model constitute the three-dimensional motion mathematical model of the mooring buoy system together with the added corresponding boundary condition. The three-dimensional motion of a single-point mooring buoy system is simulated respectively under two level sea condition and four level sea condition. The results show that the single-point mooring buoy system regularly sways near the initial position under the effect of wave force and achieves a dynamic steady state after a period of time.

Keywords: Marine monitoring technology · Single-point mooring buoy system · Wave force · Three-dimensional motion simulation

1   Introduction

As a new-type marine monitoring tool, the mooring buoy system can monitor ocean environment of certain points at specific times continuously and automatically. It has been extensively used for various oceanic applications, geographical measurements and military tasks etc [1-4]. The motion of tether and buoy interact under the effect of wave force, which is a coupled and complex physical system. Due to large water depth, it is almost impossible to measure the movement response of the mooring buoy system under laboratory conditions. So it is extremely necessary to develop a numerical method of the movement response of the mooring buoy system under the effect of wave force [5,6].

The mathematical model of mooring buoy system under the effect of wave force is established. The three-dimensional movement of a single-point mooring buoy system is simulated and its movement rule is accurately predicted, which can provide a scientific and reasonable reference for design and application of mooring buoy system.

© Springer International Publishing Switzerland 2015
H. Liu et al. (Eds.): ICIRA 2015, Part I, LNAI 9244, pp. 136–147, 2015.
DOI: 10.1007/978-3-319-22879-2_13

2   Motion Mathematical Model of Mooring Buoy System

2.1   Coordinates [7]

In order to study movement rule of mooring buoy system under the effect of wave force, the overall coordinate system $O_0x_0y_0z_0$, the tether coordinate system $Obtn$ and the buoy coordinate system $Oxyz$ are established. Fig. 1 shows the definition of coordinate systems.

The overall coordinate system $O_0x_0y_0z_0$ is a spatial Cartesian coordinate, its origin is on the horizontal plane $z_0=0$, (i, j, k) are the unit vectors of the overall coordinate system's three directions, where k is vertically upward. The all mathematical models in this paper are established based on the overall coordinate system. The tether coordinate system $Obtn$ is attached to the tether, the tangential direction of the tether is assumed as the t-axis (positive—the increase direction of the cable length s). Euler angle (γ, ϕ) is the attitude angle of tether micro unit relative to the overall coordinate system. The buoy coordinate system $Oxyz$ moves together with the buoy, the centroid of the buoy is treated as the origin, The coordinate line starting from the origin to the head direction is assumed as the x-axis, the one pointing to the right is the y-axis, and the one which is perpendicular to the xoy-plane and satisfies the right hand rule is the z-axis. Reference AUV maneuverability definition, the attitude angles of the buoy are pitch angle θ, the yaw angle ψ and roll angle φ.

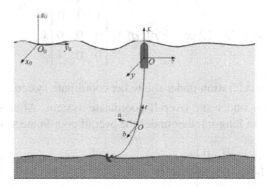

Fig. 1. Definition of coordinates

2.2   The Motion Model of Tether

(1) Governing equation

Assumed that the tether is a slender cylindrical body, without considering the bending moment and torque. For the tether micro unit ds, motion governing equation according to Newton's second law is

$$M\ddot{x} = \frac{\partial T}{\partial s} + \sum F \tag{1}$$

Where M is mass matrix of the tether; $\ddot{x}$ is acceleration of the tether; T is tension of the tether, only internal restoring force; F represents various external forces acting on the tether, typically includes gravity, buoyancy and wave load of the external environment. The each force is relative to the unstretched tether.

(2) Added mass of tether
Due to a large slenderness, the inertial hydrodynamic force of the tangential direction is small compared with the transverse direction, so it is negligible. The tether has only added mass in the transverse direction [8]. Under the tether coordinate system, the added mass of per unit length tether is

$$\rho k_a \sigma \begin{bmatrix} 1 & 0 & 0 \\ 0 & 0 & 0 \\ 0 & 0 & 1 \end{bmatrix} \tag{2}$$

Where k_a is added mass factor, $k_a = 1.0$ for a circular-section tether; ρ is fluid density; σ is cross-sectional area of tether. Under the overall coordinate system, the added mass matrix is the function of attitude angles and can be obtained by the coordinate transformation

$$F = M_a \cdot [A]\ddot{x}_{btn} = \rho k_a \sigma [A] \cdot \begin{bmatrix} 1 & 0 & 0 \\ 0 & 0 & 0 \\ 0 & 0 & 1 \end{bmatrix} \ddot{x}_{btn} \tag{3}$$

Where $\ddot{x}_{btn}$ is acceleration under the tether coordinate system; $[A]\ddot{x}_{btn}$ is corresponding acceleration under the overall coordinate system. After simplification, the added mass of per unit length tether under the overall coordinate system is

$$
\begin{aligned}
M_a &= \rho k_a \sigma [A] \cdot \begin{bmatrix} 1 & 0 & 0 \\ 0 & 0 & 0 \\ 0 & 0 & 1 \end{bmatrix} \cdot [A]^{-1} \\
&= \rho k_a \sigma \begin{bmatrix} 1 - \sin^2 \gamma \cos^2 \phi & -\cos \gamma \sin \gamma \cos^2 \phi & -\sin \gamma \cos \phi \sin \phi \\ -\cos \gamma \sin \gamma \cos^2 \phi & 1 - \cos^2 \gamma \cos^2 \phi & -\cos \gamma \cos \phi \sin \phi \\ -\sin \gamma \cos \phi \sin \phi & -\cos \gamma \cos \phi \sin \phi & \cos^2 \phi \end{bmatrix}
\end{aligned} \tag{4}
$$

(3) Mathematical model of lumped mass method [9]
The tether is discretized into N segments from the upper mooring point to the end fixing point, N+1 nodes. The upper mooring point is the first node, i=1, corresponding

to the tether length is $s_1 = 0$; the end fixing point is the last node, i=N+1, corresponding to the tether length is $s_{N+1} = S$. The tether length of each node satisfies the following basic relations:

$$0 = s_1 < s_2 < \cdots < s_N < s_{N+1} = S \tag{5}$$

Where S is the total length of the tether. Each node can distribute with uniform spacing or nonuniform spacing.

For lumped mass method, the relationship of stress – strain is generally used as continuity condition:

$$T = f(\varepsilon) \tag{6}$$

Referring to equation (1), the motion governing equation of the i-th node can be expressed as:

$$M_i \ddot{x}_i = F_i \tag{7}$$

Where $M_i = m_i I + M_{ai}$ is mass matrix, including inertial mass $m_i = (\overline{m}_{i-1/2} l_{i-1/2} + \overline{m}_{i+1/2} l_{i+1/2})/2$ and added mass $M_{ai} = (M_{a(i-1/2)} + M_{a(i+1/2)})/2$, I is 3×3 unit matrix, $\overline{m}_i$ is mass for unit length tether, l is length between nodes, the subscript i+1/2 represents physical quantity between node i and i +1; F_i is all external forces acting on node i, including tension ΔT_i, buoyancy B_i, gravity G_i and fluid drag D_i, namely

$$F_i = \Delta T_i + B_i + G_i + D_i \tag{8}$$

The calculation method of ΔT_i, B_i, G_i, and D_i can be referred to the literature [10].

2.3   The Motion Model of Buoy

(1) Governing equation
The motion governing equations of buoy is established based on the momentum and angular momentum theories.

$$\begin{cases} \dfrac{dQ_t}{dt} + \omega \times Q_t = F \\[2mm] \dfrac{dK_t}{dt} + \omega \times K_t + v_O \times Q_t = M \end{cases} \tag{9}$$

Where F is all external forces acting on the buoy; M is all external torque acting on the buoy. The all external forces and torque on the buoy are taken into equation (9), six degrees of freedom motion governing equation of buoy can be obtained

$$A_{m\lambda} \begin{bmatrix} \dot{v}_x \\ \dot{v}_y \\ \dot{v}_z \\ \dot{\omega}_x \\ \dot{\omega}_y \\ \dot{\omega}_z \end{bmatrix} + \frac{dA_{m\lambda}}{dt} \begin{bmatrix} v_x \\ v_y \\ v_z \\ \omega_x \\ \omega_y \\ \omega_z \end{bmatrix} = -A_{vw} \left\{ A_{m\lambda} \begin{bmatrix} v_x \\ v_y \\ v_z \\ \omega_x \\ \omega_y \\ \omega_z \end{bmatrix} \right\} + A_{FM} \tag{10}$$

Where $A_{m\lambda}$ is inertia matrix; A_{vw} is speed matrix; A_{FM} is external force matrix and external torque matrix.

(2) Momentum and angular momentum of buoy

The momentum of rigid body is the product of mass and centroid speed, so the momentum of buoy is

$$Q_t = mv_c \tag{11}$$

Where m is mass of buoy, v_c is centroid speed of buoy:

$$v_c = v_O + \omega \times r_c \tag{12}$$

Where v_O is buoyant centre speed of buoy under the overall coordinate system, ω is rotational angular speed; r_c is buoyant centre vector relative to centroid of buoy.

The angular momentum of buoy relative to buoyant centre is

$$\begin{aligned} K_t &= \int_m r \times v \, dm = \int_m r \times (v_O + \omega \times r) dm \\ &= \int_m \left[(x^2 + y^2 + z^2)\omega - (x\omega_x + y\omega_y + z\omega_z) \right] dm \end{aligned} \tag{13}$$

Where r is vector of any buoy's point relative to buoyant centre; v is speed of any buoy's point.

The momentum and angular momentum of buoy is expressed as:

$$\begin{bmatrix} Q_{tx} \\ Q_{ty} \\ Q_{tz} \end{bmatrix} = \begin{bmatrix} m & 0 & 0 & 0 & mz_c & -my_c \\ 0 & m & 0 & -mz_c & 0 & mx_c \\ 0 & 0 & m & my_c & -mx_c & 0 \end{bmatrix} \tag{14}$$

$$\begin{bmatrix} K_{tx} \\ K_{ty} \\ K_{tz} \end{bmatrix} = \begin{bmatrix} 0 & -mx_c & my_c & J_{xx} & 0 & 0 \\ mz_c & 0 & -mx_c & 0 & J_{yy} & 0 \\ -my_c & mx_c & 0 & 0 & 0 & J_{zz} \end{bmatrix} \tag{15}$$

(3) Wave force acting on buoy

The force acting on buoy includes ideal fluid inertia force, fluid viscosity position force and fluid viscous damping force, buoyancy, gravity and wave force. Here the calculation method of wave force is illustrated detailedly, and the remaining forces can be calculated according to the literature [11].

Based on potential flow theory, fluid is inviscous and irrotational, the tension on the fluid free surface is neglected. $\phi(p,t)$ represents total fluid velocity potential at t moment and p point, including two parts of steady and unsteady. The steady part is fluid velocity potential ϕ_S after the structure reaches a steady state in still water. The unsteady part has incident potential ϕ_I and perturbation potential ϕ_B, ϕ_B including diffraction potential and radiation potential.

$$\phi(p,t) = -ux + \phi_I(p,t) + \phi_B(p,t) + \phi_S(p,t) \tag{16}$$

Where $\phi_B(p,t) = \sum_{k=1}^{6} \phi_k(p,t) + \phi_7(p,t)$, $\phi_k(p,t)$ is radiation potential of buoy; $\phi_7(p,t)$ is diffraction potential. Under the overall coordinate system, the boundary conditions of diffraction potential and radiation potential are

$$\nabla^2 \phi_j(p,t) = 0, z \le 0, j = 1,2,3,4,5,6,7 \tag{17}$$

$$\left[\left(\frac{\partial}{\partial t} - u\frac{\partial}{\partial x} \right)^2 + g\frac{\partial}{\partial z} \right] \phi_j(p,t) = 0, z = 0, j = 1,2,3,4,5,6,7 \tag{18}$$

$$\frac{\partial \phi_j}{\partial n} = \dot{\eta}_j n_j + u\eta_j m_j, at\ s(t), j = 1,2,3,4,5,6 \tag{19}$$

$$\frac{\partial \phi_7}{\partial n} = -\frac{\partial \phi_0}{\partial n}, at\ s(t), (m_1,m_2,m_3) = -\frac{1}{u}(\boldsymbol{n} \cdot \nabla)\boldsymbol{w}, (m_4,m_5,m_6) = -\frac{1}{u}(\boldsymbol{n} \cdot \nabla)(\boldsymbol{r} \times \boldsymbol{w}) \tag{20}$$

In addition, the far-field condition and bottom condition are considered. $\eta_j (j = 1,2,3,4,5,6)$ is displacement; w is induced velocity; $s(t)$ is buoy's surface. Consider second-order Stokes wave is incident wave, the incident potential is

$$\phi_I = \left(\frac{Ag\cosh k(z+H)}{\omega\ \sinh kH} \sin\theta + \frac{3A^2\omega\cosh 2k(z+H)}{8\sinh^4 kH} \sin 2\theta \right) \sin(kx - \omega x + \varepsilon) \tag{21}$$

Where $\omega^2 = gk\tanh(kH)$, H is water depth; A is wave amplitude; ω is incident wave frequency; θ is incident angle; k is wave number.

The wave force acting on buoy is as follows:

$$F(t) = -\rho\frac{d}{dt}\iint_{s_0}(\phi_I + \phi_B)\boldsymbol{n}ds - \rho\iint_{s_0}\phi_B\frac{\partial}{\partial n}(\nabla\phi_I) - \nabla\phi_I\frac{\partial\phi_B}{\partial n}ds \tag{22}$$

The wave moment acting on buoy is as follows:

$$M(t) = -\rho \frac{d}{dt} \iint_{s_0} (\phi_I + \phi_B) \mathbf{r} \times \mathbf{n} ds - \rho \iint_{s_0} \phi_B \frac{\partial}{\partial n} (\mathbf{r} \times \nabla \phi_I) - \frac{\partial \phi_B}{\partial n} (\mathbf{r} \times \nabla \phi_I) ds \qquad (23)$$

2.4   The Motion Governing Equations and Numerical Solution Method

Combine tether's motion governing equation and buoy's motion governing equation, additional the boundary conditions and the speed differential definition $v = dx / dt$, a complete movement mathematical model of the mooring buoy system is formed, as follows:

$$\begin{cases} \dfrac{d\ddot{x}_i}{dt} = M_i^{-1} \cdot F_i \\[2mm] \dfrac{dx_i}{dt} = \dot{x}_i \end{cases} \qquad (i = 1, 2, \cdots, N)$$

$$A_{m\lambda} \begin{bmatrix} \dot{v}_x \\ \dot{v}_y \\ \dot{v}_z \\ \dot{\omega}_x \\ \dot{\omega}_y \\ \dot{\omega}_z \end{bmatrix} + \frac{dA_{m\lambda}}{dt} \begin{bmatrix} v_x \\ v_y \\ v_z \\ \omega_x \\ \omega_y \\ \omega_z \end{bmatrix} = -A_{vw} \left\{ A_{m\lambda} \begin{bmatrix} v_x \\ v_y \\ v_z \\ \omega_x \\ \omega_y \\ \omega_z \end{bmatrix} \right\} + A_{FM}$$

$$\dot{\theta} = \omega_y \sin \varphi + \omega_z \cos \varphi$$

$$\dot{\psi} = \omega_y \sec \theta \cos \varphi - \omega_z \sec \theta \sin \varphi$$

$$\dot{\varphi} = \omega_x - \omega_y \tan \theta \cos \varphi + \omega_z \tan \theta \sin \varphi$$

$$\dot{x}_O = v_x \cos \theta \cos \psi + v_y (\sin \psi \sin \varphi - \sin \theta \cos \psi \cos \varphi) + v_z (\sin \psi \cos \varphi + \sin \theta \cos \psi \sin \varphi)$$

$$\dot{y}_O = v_x \sin \theta + v_y \cos \theta \cos \varphi - v_z \cos \theta \sin \varphi$$

$$\dot{z}_O = -v_x \cos \theta \sin \psi + v_y (\cos \psi \sin \varphi + \sin \theta \sin \psi \cos \varphi) + v_z (\cos \psi \cos \varphi - \sin \theta \sin \psi \sin \varphi)$$

The equations are solved by Fourth-order Runge-Kutta method, the integral formula from t_n to $t_{n+1} = t_n + \Delta t$ is

$$\begin{cases} y_{n+1} = y_n + (k_1 + 2k_2 + 2k_3 + k_4) \\ k_1 = \Delta t f(t_n, y_n) \\ k_2 = \Delta t f(t_n + \Delta t / 2, y_n + k_1) \\ k_3 = \Delta t f(t_n + \Delta t / 2, y_n + k_2 / 2) \\ k_4 = \Delta t f(t_n + \Delta t / 2, y_n + k_3) \end{cases}$$

To ensure accuracy, the node number N should be large enough to accurately describe the tether's geometry and dynamic characteristics; otherwise it will produce large model errors.

3   3-D Motion Simulation of Mooring Buoy System

In this paper, the three-dimensional motion of a single-point mooring buoy system is simulated respectively under two level sea condition (wave height 0.6m, average period 2.7s) and four level sea condition (wave height 2.2m, average period 5.3s). The end of tether is fixed on the structure or seafloor in a 40m depth. The following table shows the physical parameters of the tether.

Table 1. Physical parameters of the tether

length/m	38
fixed depth/m	40
diameter/m	0.032
mass of per unit length tether/kg	0.87
Young's modulus/N/m^2	5.5×10^9
tangential damping coefficient	0.015
normal damping coefficient	1.1

The simulation results are shown from different angles under two level sea condition, including the shape of tether, the track of buoy and so on. Fig. 2 and Fig. 3 respectively show the shape of tether on $x_0 O_0 z_0$ and $y_0 O_0 z_0$ within 600s. As can be seen from the figure, the tether is steady within a certain range, within ± 0.3m, and concentrated in the intermediate section. Overall, the buoy can reach a dynamic steady state from the static state and maintain a good posture.

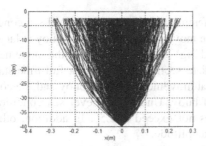
Fig. 2. Shape of tether on $x_0 O_0 z_0$

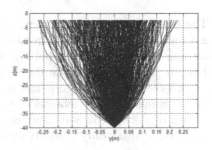
Fig. 3. Shape of tether on $y_0 O_0 z_0$

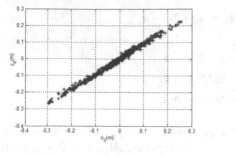

Fig. 4. Track of buoy on $x_0 O_0 z_0$

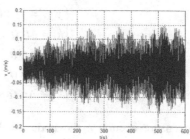

Fig. 5. Buoy speed v_x

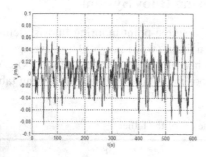

Fig. 6. Buoy speed v_y

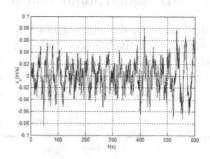

Fig. 7. Buoy speed v_z

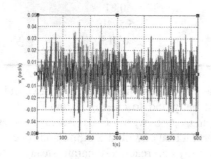

Fig. 8. Buoy rotational angular speed ω_y

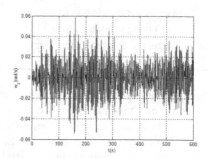

Fig. 9. Buoy rotational angular speed ω_z

The simulation results are shown from different angles under four level sea condition. Compared to two level sea condition, the movement amplitude of buoy is larger under four level sea condition, but Fig. 12 and Fig. 13 can be seen that the movement of buoy is still concentrated in the intermediate section. Fig. 14 is a full-time-state track on $x_0 O_0 y_0$. The scattered points are initial movement track of buoy and the concentrated points are later movement track of buoy. Under four level sea condition, the range of pitch angle and yaw angle does not exceed ± 2.5°. In addition, other parameters are also steady in relatively small amplitude.

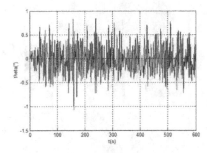

Fig. 10. Buoy pitch angle θ

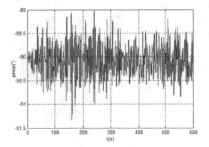

Fig. 11. Buoy yaw angle ψ

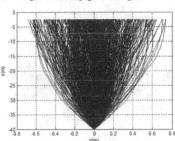

Fig. 12. Shape of tether on $x_0 O_0 z_0$

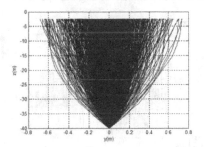

Fig. 13. Shape of tether on $y_0 O_0 z_0$

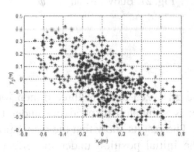

Fig. 14. Track of buoy on $x_0 O_0 z_0$

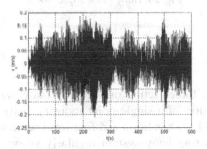

Fig. 15. Buoy speed v_x

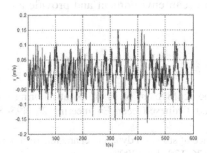

Fig. 16. Buoy speed v_y

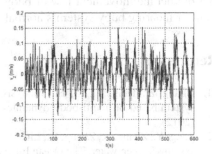

Fig. 17. Buoy speed v_z

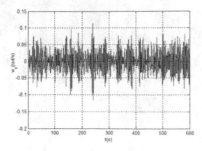

Fig. 18. Buoy rotational angular speed ω_y

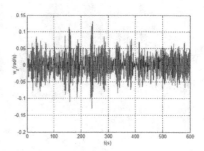

Fig. 19. Buoy rotational angular speed ω_z

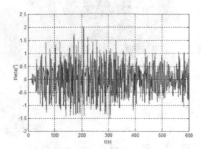

Fig. 20. Buoy pitch angle θ

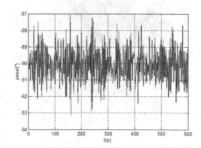

Fig. 21. Buoy yaw angle ψ

4   Conclusions

The forces acting on mooring buoy system are analyzed and the three-dimensional motion of a single-point mooring buoy system is simulated respectively under two level sea condition and four level sea condition. The results show that the single-point mooring buoy system regularly sways near the initial position under the effect of wave force and achieves the dynamic steady state after a period of time. The research of the mooring buoy system's movement response under wave force is significant. It can predict the movement law of buoy under ocean environment and provide a criterion for mooring buoy system's normal work.

References

1. Chang, Z., Tang, Y., Li, H.: Analysis for the deployment of single-point mooring buoy system based on multi-body dynamics method. Chinese Ocean Engineering **26**, 495–506 (2012)
2. Tang, Y., Zhang, S., Zhang, R., Liu, H.: Advance of study on dynamic characters of mooring systems in deep water. The Ocean Engineering **26**, 120–126 (2008)

3. Jiaming, W., Jiawei, Y., Ning, L.: Comments on the research of hydrodynamic and control performances of a multi-parameter profile sampling towed system. The Ocean Engineering **22**, 111–120 (2004)
4. Xiaofeng Kuang, K., Wang, Y., Miao, Q.: Theory prediction of wave forces for a deep-submerged body. Journal of Ship Mechanics **10**, 28–35 (2006)
5. Song, B., Du, X., Hu, H.: Effect of wave on AUV hovering near sea surface. Journal of Northwestern Polytechnical University **25**, 482–486 (2007)
6. Pan, G., Liu, Y., Du, X.: Computational analysis and prediction of wave forces on portable underwater vehicle. Acta Armamentarii **35**, 371–378 (2014)
7. Guang, P.: Torpedo Mechanics. Shaanxi Normal University General Publishing House CO., LTD., Xi'an (2013)
8. Pan, G., Yang, Z., Du, X.: Longitudinal motion research on UUV underwater recovery with a deployable tether. Journal of Northwestern Polytechnical University **29**, 245–250 (2011)
9. Chai, Y.T., Varyani, K.S., Barltrop, N.D.P.: Three dimensional lump-mass formulation of a catenary riser with bending, torsion and irregular seabed interaction effect. Ocean Engineering **29**, 1503–1525 (2002)
10. Pan, G., Yang, Z., Du, X.: Research on longitudinal motion simulation of mooring system under the effect of wave. Acta Armamentarii **34**, 1431–1436 (2013)
11. Dai, Y., Duan, W.: Potential low theory of ship motions in waves. National Defense Industry Press, Beijing (2008)

The Frequency Characteristic Analysis of Ocean Rocking Energy Conversion System

Zhaoyong Mao$^{(\boxtimes)}$, Na Tian, and Zhijun Shen

School of Marine Science and Technology,
Northwestern Polytechnical University, Xi'an 710072, China
maozhaoyong@nwpu.edu.cn

Abstract. In order to solve the energy problem of the underwater mooring platforms, a horizontal pendulum-type ocean rocking energy conversion system scheme is proposed. Based on the multi-body dynamics software LMS, the motion characteristic simulation model is established, and the influence parameter on motion characteristic is analyzed, such as pendulum mass, pendulum length, excitation frequency and amplitude, and system damping coefficient. The simulation results show that output power is increasing with the pendulum mass, pendulum length, excitation frequency, and excitation amplitude, which has a practical guiding significance to the design of the mode of the current power generation system.

Keywords: Ocean energy · Horizontal pendulum · Kinetic energy power · Motion characteristics

1 Introduction

Underwater Mooring Platform (UMP), self-explanatorily, works underwater, and it is chained to the seabed directly with an anchor. Many studies have been done on its applications in both military and civilian area. In order to increase the working time of the UMP, not only need we consider low-energy-consumption design, but also solve energy support problem which is the bottleneck problem of UMP to work sustainably underwater.

Numerous achievements in the study of energy utilization of ocean environments have been made, hundreds of different design schemes of ocean energy power generator devices have been designed all over the world in recent years [1–11]. The Ocean Rocking Energy Conversion System (ORECS) collects the ocean energy by the energy collecting mechanism and then converts the ocean energy to the electricity. At present, the energy harvester is mainly the single eccentric pendulum-type structure [12-14]. According to the relative position between the eccentric pendulum axis of the eccentric pendulum and the flow direction, ORECS can be categorized into two types, namely the vertical pendulum-type and horizontal pendulum-type. The pendulum axis of the horizontal-pendulum ORECS is vertical to the flow direction, thus it can capture the energy excitation from multiple directions, which makes it efficient for energy extraction.

© Springer International Publishing Switzerland 2015
H. Liu et al. (Eds.): ICIRA 2015, Part I, LNAI 9244, pp. 148–155, 2015.
DOI: 10.1007/978-3-319-22879-2_14

Considering the underwater mooring platform is fixed through the chain floating in the water, which is at rocking motion state under the influence of waves, currents and other movements. Therefore, the project groups propose a horizontal-pendulum ORECS. This system takes advantage of the fact that the UMP suspends in the water and it will always be in the influence of ocean power, to a certain extent, it will solve the problem of mooring platform energy demand under water. In the course of the study, according to the horizontal-pendulum kinetic energy conversion system we proposed, we establish the simulation model for the horizontal-pendulum ORECS based the multi-body dynamics analysis software LMS Virtual.lab for the characteristics analysis of the system and analyze several parameters which affect the generating power. It will provide a theoretical basis for the engineering application of the subsequent rocking kinetic energy conversion system.

2   Kinematic Modeling of the Ocean Rocking Energy Conversion System

The main structure of proposed horizontal-pendulum ORECS consists of the pendulum, swinging rod, and generators as shown in Fig.1. When it is motivated by the external disturbance, the pendulum will oscillate accordingly, which results in the rotation of the axis through the swinging rod. Then this rotation is passed to the power generator to generate electricity.

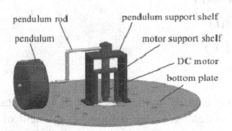

Fig. 1. Schematic of horizontal-pendulum ocean rocking kinetic energy conversion system

For the horizontal-pendulum ORECS we proposed in this work, the pendulum bob is the carrier of wave energy absorption, whose movement affects the efficiency of wave energy absorption directly. The force distribution schematic diagram is described in Fig. 2. The model of the force distribution can be simplified as a single pendulum damping system under the external excitation.

According to the law of rigid body rotating around a fixed axle, when the rigid body is rotating around a fixed axle, the algebra sum of all the torques ΣM exerted on the rigid body by the external forces equals the product of the moment of inertia of rigid body to the axle ΣJ and the gained angular acceleration.

$$\sum M = (\sum J)\beta \tag{1}$$

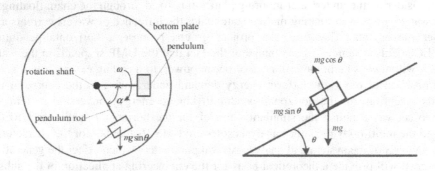

Fig. 2. Force diagram of rocking kinetic energy conversion system

Analyzing the force distribution of the ORECS, we deduce that external torque ΣM is the function of the swing angle $\alpha(t)$ and tilt angle $\theta(t)$. Therefore, the kinetic model of the power generator system can be described as [15]

$$
\begin{aligned}
&\sum M = J\beta(t) \\
&\sum M = T_t(t) - T_d(t) \\
&T_t(t) = mgL\sin\theta(t)\cos\alpha(t) - \mu mg\cos\theta(t) \\
&T_d(t) = C\omega(t) \\
&\beta(t) = \ddot{\alpha}(t) \\
&\omega(t) = \dot{\alpha}(t) \\
&J = mL^2
\end{aligned}
\tag{2}
$$

where T_t is the torque exerted on the generator, T_d is the total damping moment of the power generating system, C is the damping coefficient of the system, m is the mass of the pendulum, g is the gravity acceleration, μ is the rolling friction coefficient between the pendulum and the body plate, L is the length of the pendulum rod, J is the rotational inertia of the pendulum, ω is the angular velocity of the pendulum.

When the ORECS is under the ocean current influence, the tilt angle $\theta(t)$ is a time-variant function of external excitation amplitude A and its frequency f is given by

$$
\theta(t) = f(A, f, t)
\tag{3}
$$

Deriving from (2) and (3), we obtain the natural frequency of kinetic energy generation system as

$$
f_0(t) = \frac{\omega(t)}{2\pi}
\tag{4}
$$

The power of kinetic energy collection as

$$P = T_d(t)\omega(t) = C\omega(t)^2 = \frac{Cf_0(t)^2}{4\pi^2} \tag{5}$$

We can find that the power of the kinetic energy collecting of the electricity gene-rating system is depending on multiple parameters such as the mass of pendulum M, the length of pendulum rod L, the external excitation amplitude A , and the damping coefficient C, etc.

3   Ocean Rocking Energy Conversion System Frequency Characteristic Analyzing

In Fig. 3, we present the relationship between the power of kinetic energy collection and the external excitation frequency (from 0.1Hz to 1.0 Hz) under the situation that the length of pendulum is 0.15m, the external excitation amplitude is 6^0 , and the damping coefficient are 0.1 and 0.5.

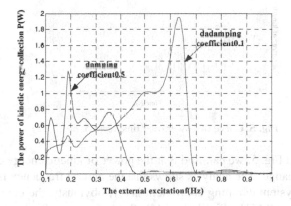

Fig. 3. Curve of power versus excitation frequency

From Fig. 3, we can find that the relation between the power of kinetic energy col-lection and the external excitation frequency is non-line under the same situation. In addition, Fig. 3 shows the maximum power of kinetic energy collection reaches its peak when the damping coefficient is 0.1 and the external excitation frequency is approximately 0.65Hz. Similarly, it reaches its peak when the damping coefficient is 0.5 and the external excitation frequency is approximately 0.2Hz.

In order to analyze the relationship between the natural frequency and the excita-tion frequency while the power reaches the maximum value, Fig. 4 and Fig. 5 present the curves of the power of kinetic energy collection along with the time when the damping coefficient are 0.1 and 0.5.

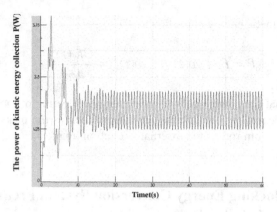

Fig. 4. Curve of power over time (c=0.1, f=0.65 Hz)

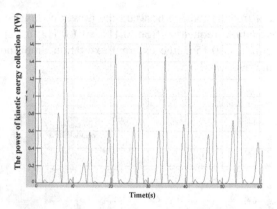

Fig. 5. Curve of power over time (c=0.5, f=0.2 Hz)

From Fig. 4 and Fig. 5, we can find that the power of kinetic energy collection with time is periodic change. The analysis shows that the natural frequency is about 0.6Hz-0.7Hz when the system damping coefficient is 0.1, obviously, the external excitation frequency is close to 0.65Hz. Similarly, while the damping coefficient is 0.5, the natural frequency of the system is about 0.15Hz-0.25Hz, and the outside excitation frequency is close to 0.2Hz.

4   Experimental Test Analysis

Our project team designed a prototype of ORECS (the parameter specifications are given by Table 1) and conducted a series of measurement experiments on a six-degree-freedom shaking platform (shown in Fig. 6). In the experiment, we connected the output port of the power generator to a resistance of the load port directly. Then we used an oscilloscope to monitor the output power of the shaking power generator by adjusting the resistance. In this way, we can gain the trend curve of the output power.

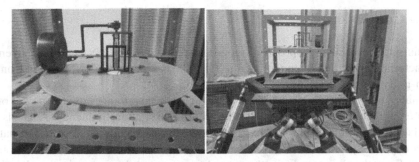

Fig. 6. Six-degree-freedom shaking platform

Table 1. Parameter Specifications

coeffecients	value
Length of the pendulum rod	0.15 m
Mass of the pendulum	10 kg
Type of electricity generator	Maxon DC22S

In the experiment, we assumed that the ORECS underwent steady state sinusoidal oscillation. Under the experimental test and simulation analysis condition, Fig. 7 shows the contrast curves that the power of kinetic energy collection changes along with the external excitation frequency (from 0.1Hz to 1.0 Hz) at the wave amplitude as 6^0 and the frequency as 0.25Hz.

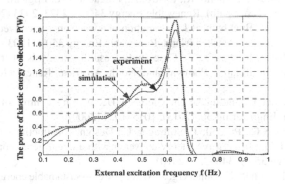

Fig. 7. Contrast curves of the test and simulation analysis

Conclusion that the trend of both curves are almost at the same pace, they reach the maximum output power when the external excitation frequency is about 0.65Hz.

5   Conclusions

In this paper, we have proposed a horizontal-pendulum ORECS, and the movement characteristics simulation model of the ORECS has been established based on the multi-body dynamics software LMS Virtual.lab.

Analyzing from the perspective of frequency, we have found how the power of kinetic energy collection changed with the external excitation frequency. Analysis and simulation show that, when the external excitation frequency is close to the natural frequency, the system will be in resonance state and will have the maximum output power. From the above analysis, the natural frequency of the system is directly related to the damping coefficients.

The results of theoretical analysis and experiment test are consistent with performance test of our prototype of the ORECS. It has shown that the validness of the theoretical model we proposed in this paper, and can provide a basis for the following research of shaking the kinetic energy power system control design and engineering application.

Acknowledgement. This work was supported by the National Natural Science Foundation of China (NSFC) under grants 51179159.

References

1. You, Y., Li, W., Liu, W., et al.: Development status and perspective of marine enrgy conversion systems. Automation of Electric Power Systems 34(14), 25–37 (2010)
2. Iglesias, G., López, M., Carballo, R., Castro, A., Fraguela, J.A., Frigaard, P.: Wave energy potential in Galicia (NW Spain). Renewable Energy 34(11), 2323–2333 (2009)
3. Beatty, S.J., Wild, P., Buckham, B.J.: Integration of a wave energy converter into the electricity supply of a remote Alaskan island. Renewable Energy 35(6), 1203–1213 (2010)
4. Dalton, G.J., Alcorn, R., Lewis, T.: Case study feasibility analysis of the Pelamis wave energy convertor in Ireland. Portugal and North America. Renewable Energy 35(2), 443–455 (2010)
5. Babarit, A.: Michel Guglielmi, Alain Clément: Declutching control of a wave energy converter. Ocean Engineering 36(12–13), 1015–1024 (2009)
6. Elwood, D., Yim, S.C., Prudell, J., Stillinger, C.: Design, construction, and ocean testing of a taut-moored dual-body wave energy converter with a linear generator power take-off. Renewable Energy 35(2), 348–354 (2010)
7. Langhamer, O., Haikonen, K., Sundberg, J.: Wave power—Sustainable energy or environmentally costly? A review with special emphasis on linear wave energy converters. Renewable and Sustainable Energy Reviews 14(4), 1329–1335 (2010)
8. Falcao, A.: Wave energy utilization: A review of the technologies. Renewable and Sustainable Energy Reviews 14(3), 899–918 (2010)
9. Dunnett, D., Wallace, J.S.: Electricity generation from wave power in Canada. Renewable Energy 34(1), 179–195 (2009)
10. Serena, A.: Marta Molinas, Ignacio Cobo: Design of a direct drive wave energy conversion system for the seaquest concept. Energy Procedia 20, 271–280 (2012)

11. Song, B., Ding, W., Mao, Z.: Conversion System of Ocean Buoys Based on Wave Energy. Journal of Mechanical Engineering **48**(12), 139–143 (2012)
12. Mitcheson, P.D., Toh, T., Wong, K.H., Burrow, S.G., Holmes, A.S.: Tuning the resonant frequency and damping of an energy harvester using power electronics. IEEE Trans. Circ. & Syst. II **58**(12), 792–796 (2011)
13. Mao, Z., Song, B., Zheng, K.: Design of generate electricity device for UUV based on ocean energy. Measurement & Control Technology **31**(6), 127–129 (2012)
14. Deng, Q.: Research and Development of a Novel Pendulum Wave Energy Converter. Dalian University of Technology (2013)
15. Wu, B.: University physics. Higher education, Beijing (2004)

A Stable Platform to Compensate Motion of Ship Based on Stewart Mechanism

Zhongqiang Zheng[1,2], Xiaopeng Zhang[3], Jun Zhang[1], and Zongyu Chang[1,2(✉)]

[1] Engineering College, Ocean University of China,
Qingdao 266100, People's Republic of China
zongyuchang@qq.com
[2] Key Lab of Ocean Engineering of Shandong Province,
Laoshan 266100, People's Republic of China
[3] LongKou Port, Longkou, Shandong, People's Republic of China

Abstract. Severe vibration and swing of ship or offshore structure in hostile sea environment have great effects on instruments, equipment and crews aboard. The paper brings out a stable platform based on a 6 SPS Stewart mechanism. It can compensate the motion of vessel by adjusting the length of 6 support cylinders. Kinematic analysis is carried out for this mechanism, and the motion of cylinder for different motion of ship is calculated. The workspace of motion compensation is given out by applying numerical method. This mechanism can provide a solution for stable platform for high precision instruments or equipment on ship or vessel.

Keywords: Stewart mechanism · Offshore platform · Kinematic analysis · Work space · Transformation matrix

1 Introduction

Vessels or floating oil platforms in hostile sea are suffering from severe load of wind, wave, current etc. Motions and vibrations of vessel caused by severe sea load have significant effects on instrument aboard. Swaying and vibration of vessel are one of the most serious threats to analyzing instruments on board [1]. It can not only lead to measurement errors, but also damage the sophisticated instruments and crews on platform. Series swing also can cause the flip of crews and overturn of equipment on vessel or offshore platform.

Gyroscope-stabilized platform is a general solution for military vessels. But it can only keep the stability of orientation or compensate the rotation motion, and can't compensate the other translation motions as heave, sway and surge. Actually, effects of the heave motion can't be neglected for sailing ship or floating platform in serious weather condition [2]. Generally, a completed stabilized platform is essential for oceanographics, offshore industry and naval national defense that can compensate all the motion in 6 dimensions especially heave, pitch and roll.

Stewart platform is a 6-DOF spatial mechanism which can generation motion in all six dimension. It consists of two rigid bodies (referred to as the fixed and moving

© Springer International Publishing Switzerland 2015
H. Liu et al. (Eds.): ICIRA 2015, Part I, LNAI 9244, pp. 156–164, 2015.
DOI: 10.1007/978-3-319-22879-2_15

platform) connected through six extensible legs, each spherical joints at both ends or with spherical joint at one end and with universal joint at the other. In particular, the kinematic structure with spherical joints at both ends of each leg will be referred to as 6-SPS (spherical-prismatic-spherical) Stewart platform as Fig. 1. Similarly, the structure with universal joint at the base and spherical joint at the top (platform-end) of each leg will be referred to as 6-UPS (universal-prismatic-spherical) Stewart platform.

It was invented by V. E. Gough and D. Stewart in 1950's individually[2]. Soon after the mechanism is carried out, the possible utilization of the mechanism to absorb motion for stabilizing a platform was first suggested by Meier for conducting operation on a ship in privately correspondence to Stewart[3] in 1965. But it is not been caused more attention. But until 1995 the control strategy of the mechanism for stabilization is studied by O.E. Hatip and M.K. Ozgoren(1995)[4]. Based on the measurement of position and orientation of vessel platform, in 2007, Delft University has built Stewart platform as assess corridor to wind turbine tower[5,6].

This mechanism can implement complex motions in 6 degree of freedom by controlling the length of cylinders of six legs and is widely used in flight simulator and other simulator as ship, car and so on. It has the advantages of high stiffness, high payload capacity and easy to install actuators. Nowadays it also is served as virtual axis machine, astronautics docking interface mechanism. Research works are carried out based on machine tool, docking interface, simulator and so on. The mechanism principles are introduced systematically including kinematic, work space and all kinds of mechanism in [6-11]

A motion-compensation platform in this paper is given based on Stewart mechanism and measurements of position and orientation. It can counteract all the motion as orientation motion of roll, pitch and yaw, and translation motion as surge, sway and heave as Fig. 2. The inverse and forward dynamics of motion compensation platform is analyzed. The workspace is calculated by numerical method. And the compensation range in heave-roll-pitch is obtained.

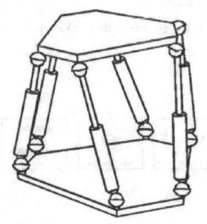

Fig. 1. Structure of Stewart platform

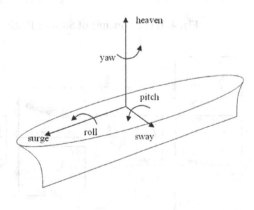

Fig. 2. Motion of vessel in six dimensions

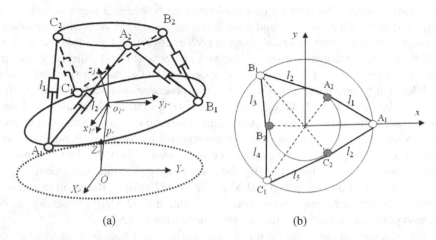

(a)   (b)

Fig. 3. The scheme of motion compensation

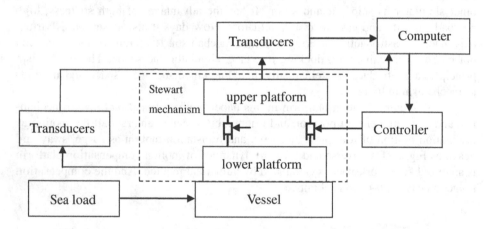

Fig. 4. Basic Principle of Stewart Platform to Compensate Motions of Vessel

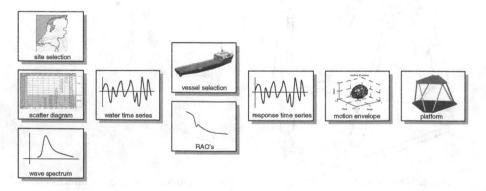

Fig. 5. Scheme of motion compensation based on Stewart mechanism

2   Structure and Kinematics of Motion-Compensation Stewart Platform

Generally, Stewart Mechanism's upper platform is moving with the varied length of cylinders and lower platform is used as fixed base. However, motion-compensation platform has a lower platform installed on the vessel, and upper platform keeping stable based on Stewart mechanism as shown in Fig. 3.

Not lost generously, the Stewart platform is considered a 6SPS double-triangle mechanism shown in Fig. 3(a). The absolute frame O-XYZ is located in center of low platform. Other body frame o_1-$x_1y_1z_1$ is also located in center of low platform and moves with the low platform. The distribution of joint points on the platform is shown in Fig. 3(b).

We supposed an initial situation of the platform, which two platforms have horizon orientation and the length of cylinder is medium position. In this situation, coordinates of three joints on the lower platform of mechanism are

$$p_{1a} = \begin{bmatrix} r_1 & 0 & 0 \end{bmatrix}^T \tag{1a}$$

$$p_{1b} = \begin{bmatrix} -r_1 \cos\frac{\pi}{3} & r_1 \sin\frac{\pi}{3} & 0 \end{bmatrix} \tag{1b}$$

$$p_{1c} = \begin{bmatrix} -r_1 \cos\frac{\pi}{3} & -r_1 \sin\frac{\pi}{3} & 0 \end{bmatrix} \tag{1c}$$

The coordinates of three joints on the upper platform will keep constant if the stabliziton.

$$P_{2a} = \begin{bmatrix} r_2 \cos\frac{\pi}{3} & r_2 \sin\frac{\pi}{3} & h_{12} \end{bmatrix} \tag{2a}$$

$$P_{2b} = \begin{bmatrix} -r_2 & 0 & h_{12} \end{bmatrix} \tag{2b}$$

$$P_{2c} = \begin{bmatrix} r_2 \cos\frac{\pi}{3} & -r_2 \sin\frac{\pi}{3} & h_{12} \end{bmatrix} \tag{2c}$$

Before kinematics analysis, the two space vectors are established, Cartesian space of platform $[x, y, z, \alpha, \beta, \gamma]^T$, where $[x, y, z]$ represent the position of origin of movable platform; $[\alpha, \beta, \gamma]$ presents the orientation of platform: angles relative to x, y, z axis.

If the platform only rotation with the axis, the rotation matrix can be obtained as Equation (1)

$$R_{x,\theta_x} = \begin{bmatrix} 1 & 0 & 0 \\ 0 & \cos\theta_x & -\sin\theta_x \\ 0 & \sin\theta_x & \cos\theta_x \end{bmatrix}$$

$$R_{y,\theta_y} = \begin{bmatrix} \cos\theta_y & 0 & -\sin\theta_y \\ 0 & 1 & 0 \\ \sin\theta_y & 0 & \cos\theta_y \end{bmatrix}$$

$$R_{z,\theta_z} = \begin{bmatrix} \cos\theta_z & -\sin\theta_z & 0 \\ \sin\theta_z & \cos\theta_z & 0 \\ 0 & 0 & 1 \end{bmatrix}$$

$$R = R_{z,\theta_z} R_{y,\theta_y} R_{x,\theta_x}$$
$$= \begin{bmatrix} \cos\theta_z\cos\theta_y & \cos\theta_z\sin\theta_y\sin\theta_x - \sin\theta_z\cos\theta_x & \cos\theta_z\sin\theta_y\cos\theta_x + \sin\theta_z\sin\theta_x \\ \sin\theta_z\sin\theta_y & \sin\theta_z\sin\theta_y\sin\theta_x + \cos\theta_z\cos\theta_x & \sin\theta_z\sin\theta_y\cos\theta_x - \cos\theta_z\sin\theta_x \\ -\sin\theta_y & \cos\theta_y\sin\theta_x & \cos\theta_y\cos\theta_x \end{bmatrix}$$

where θ_x, θ_y, θ_z represent the rotation angle about x, y and z axis; R_{x,θ_x}, R_{y,θ_y}, R_{z,θ_z} represent the rotation matrixes for angle of roll, pitch and yaw.

Considering the translation motion, the transformation matrix is

$$T = \begin{bmatrix} c\theta_z c\theta_y & c\theta_z s\theta_y s\theta_x - s\theta_z c\theta_x & c\theta_z s\theta_y c\theta_x + s\theta_z s\theta_x & p_x \\ s\theta_z s\theta_y & s\theta_z s\theta_y s\theta_x + c\theta_z c\theta_x & s\theta_z s\theta_y c\theta_x - c\theta_z s\theta_x & p_y \\ -s\theta_y & c\theta_y s\theta_x & c\theta_y c\theta_x & p_z \\ & 0 & & 1 \end{bmatrix}$$

Joint space [11, 12, 13, 14, 15, 16]T includes displacement of six linear actuators of six cylinder.

Then if we can obtain the information of low platform mounted on moving platform as six degree of freedom,

The absolute coordinates of joint points on low platform can be obtained by transformation representation

$$P_{li} = T \cdot p_{li}$$

$$l_i = |P_{1a}P_2|(p_{2ax} - p_{1ax})^2 + (p_{2ay} - p_{1ay})^2 + (p_{2az} - p_{1az})^2 = l_{a1a2}^2$$

Then the length of cylinders can be obtained by below.

$$l_1^2 = (p_{2ax} - p_{1ax})^2 + (p_{2ay} - p_{1ay})^2 + (p_{2az} - p_{1az})^2 = l_{a1a2}{}^2 \tag{6a}$$

$$l_2^2 = (p_{2ax} - p_{1bx})^2 + (p_{2ay} - p_{1by})^2 + (p_{2az} - p_{1bz})^2 = l_{b1a2}{}^2 \tag{6b}$$

$$l_3^2 = (p_{2bx} - p_{1bx})^2 + (p_{2by} - p_{1by})^2 + (p_{2bz} - p_{1bz})^2 = l_{b1b2}{}^2 \tag{6c}$$

$$l_4^2 = (p_{2bx} - p_{1cx})^2 + (p_{2by} - p_{1cy})^2 + (p_{2bz} - p_{1cz})^2 = l_{c1a2}{}^2 \tag{6d}$$

$$l_5^2 = (p_{2cx} - p_{1cx})^2 + (p_{2cy} - p_{1cy})^2 + (p_{2cz} - p_{1cz})^2 = l_{c1c2}{}^2 \tag{6e}$$

$$l_6^2 = (p_{2cx} - p_{1ax})^2 + (p_{2cy} - p_{1ay})^2 + (p_{2cz} - p_{1az})^2 = l_{a1c2}{}^2 \tag{6f}$$

3   Compensation Analysis and Motion Envelop of Stewart Mechanism

Because of structure and dimension of Stewart Mechanism, the motion compensation condition is limited. The motion compensation space is closely related to the properties of the stable platform and it is the base to design and control the mechanism. According to workspace analysis method [6], the factors affecting the compensation workspace include:

(1) the limit of stroke length of each hydraulic cylinder between up and below platform;
(2) limitation of rotation of spherical ball joint and U-joint;
(3) interfaces between the links.

The compensation space is different from workspace of stewart mechanism. It is the compensation motion in different dimensions by adjusting the length of hydraulic cylinders when keep the upper platform stationary.

Generally, heave, pitch and roll are the most important components in the 6 dof motions for mooring vessel or floating production storge operation structure (FPSO). The pitch angle θ_x Roll angle θ_y is roller angle and pitch angle the height h_{12} is the heave displacement. The factor (1) and (2) above are the key ones to determine the boundary of compensation space.

The space search method is used to calculate the points on the boundary of motion compensation. The motion compensation spaces in solid mode and in frame mode are shown in Fig. 8 and Fig. 9 respectively.

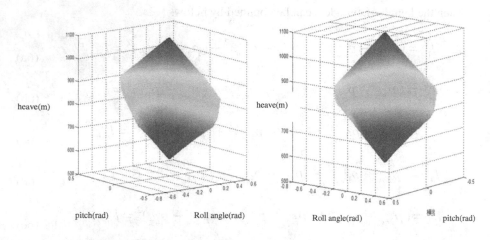

Fig. 6. The motion compensation space in solid mode

From the figures, we can find that the compensation space is near elliptical sphere in three dimensions. When the heave motion compensated reach almost (almost 400 or 600), the ability to compensate the motion of swing disappears. At the same, when the heave motion compensated is zero, the angle of roll can reach most.

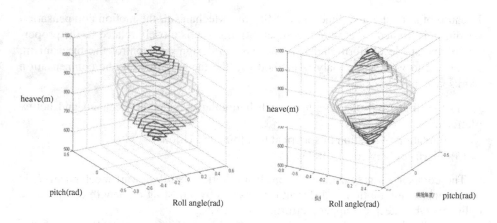

Fig. 7. Motion compensation space in frame mode

In order to understand the three parameter relationship between three parameters, the relationship between two parameters are obtained if one is fixed.The compensation range of roller angle and pitch angle is shown in Fig. 10(a) when heave motioan is 700mm, 800mm and 900mm. 800 is the intermedia position. From this fig, the roll-pitch motion compensation range is larger than the h=700 and 900. In addtion, the roll-pitch motion compensation, roll-heave motion compensation and the pitch heave motion compensation is shown in Fig. (b) and Fig. (c).

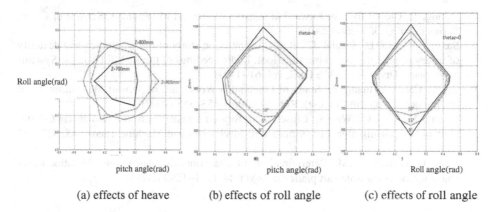

| (a) effects of heave | (b) effects of roll angle | (c) effects of roll angle |

Fig. 8. Effects of parameters on compensation range

4 Conclusion

This paper studies a stable platform of stewart mechanism for motion compensation of vessel or offshore structure. By adjusting the length of six cylinders, the upper platform can keep constant in position and orientation while the lower platform is moving with the swing and motion of vessel. The kinematic analysis of this mechanism is carried out and the displacement motions of cylinders are given for different vessel motion. The workspace of are also given when the limit of adjust length of cylinder are given. This work is useful to obtain a stable platform in three translation motion and three rotation motion for aboard instrument and weapon.

Acknowledgement. We are grateful for the support of NSFC (No. 5716484) and support of Key Ocean Engineering Laboratory of Shandong Province.

References

1. Jinping, Z.: Thinking and Experience of development of Oceanographic Observation Technology (In Chinese). Ocean Press, Beijing (2006)
2. Dasguptaa, B., Mruthyunjayab, T.S.: The Stewart platform manipulator: a review. Mechanism and Machine Theory **35**, 15–40 (2000)
3. Hatip, O.E., Ozgoren, M.K.: Utilization of a stewart platform mechanism as a stabilizator. In: Proceedings of 9th World Congress Theory of Machin and Mechanism, pp. 1393–1396 (1995)
4. Salzmann, D.J.C.: Ampelmann: Development of the access System for Offshore Wind Turbines. PhD dissertation of technology University of Delft (2010)
5. Taempel, J.V., Salzmann, D.C., Koch, J.M.L., et al.: Future applications of ampelmann systems in offshore wind (2010)
6. Huang, Z., Kong, L., Fang, Y.: Mechanism Theory and Control of Parallel Robot (In Chinese). China Machine Press, Beijing (1997)

7. Huang, T.: Closed form of forward kinematics of Stewart Manipulator. Science of China, Series E. **86**(3), 324–330 (2001)
8. Saxena, V., Liu, D., Daniel, C., et al.: A simulation study of the workspace and dexterity of a stewart platform based machine tool. In: Proceedings of the ASME Dynamic Systems and Control Division, ASME DSC, vol. 61 (1997)
9. Li, K., Cao, Y., Huang, Z.: Study of Orientation Workspace of Stewart Mechanism based on unit quaternion. (in Chinese). Robotics **30**(4), 353–358 (2007)
10. Huang, X., Liao, Q., Wei, S.: Closed-form forward kinematics for a symmetricla 6-6 stewart platform using algebraic elimination. Mechanism and Machine Theory **45**, 327–334 (2010)
11. Jiang, Q., Gosselin, C.M.: Determination of the maximal singularity-free orientation workspace for the Gough-Stewart platform. MMT **44**, 1281–1293 (2009)

Drives and Actuators' Modeling

Motion Simulation and Statics Analysis of the Stator and Rotor of Low Speed High Torque Water Hydraulic Motor

Zhiqiang Wang[1(✉)], Huang Yao[2], Weijun Bao[1], and Dianrong Gao[3]

[1] School of Mechanical Engineering, Hangzhou Dianzi University,
Hangzhou, People's Republic of China
wangzq@hdu.edu.cn
[2] School of Mechanical Engineering, Zhejiang University, Hangzhou,
People's Republic of China
huangyao36@sina.com
[3] School of Mechanical Engineering, Yanshan University, Qinhuangdao,
People's Republic of China
gaodr@ysu.edu.cn

Abstract. In order to verify the rationality of the design of low speed high torque hydraulic motor structure, on the one hand, motion simulation for the motor, the change of radial velocity, acceleration and displacement of roller and the law of impulse force with time were researched under different rotational speed. On the other hand, the static simulation of the motor was completed by FEA (finite element analysis) software. And the overall and local equivalent stress, deformation and the contact stress between the stator and rotor position were obtained. Meanwhile, the maximum contact stress of roller contact with inner surface of the stator was calculated by using Hertz theory. By analyzing the two aspects simulation results and the theoretical calculation results, the validity of the analysis solution and the rationality of the structure design of the motor were verified.

Keywords: Water hydraulic motor · Stator · Rotor · Motion simulation · Statics analysis

1   Introduction

Low speed high torque hydraulic motor water in fresh water or seawater as the working medium, and the water pressure can be converted into mechanical energy to realize continuous rotary movement of the water hydraulic actuators. So it plays an important role in water hydraulic system. In order to research the performance of water hydraulic component, international scholars make lots of research about this. The effect of port plate structure on the flow characteristic of water axial piston hydraulic pump was researched by Tang and Zhao [1]. The finite element analysis was performed on connecting rod of low speed high torque hydraulic motors with the help of ANSYS by Li and Shi [2]. Wang et al. [3] had analyzed the stress, strain and

© Springer International Publishing Switzerland 2015
H. Liu et al. (Eds.): ICIRA 2015, Part I, LNAI 9244, pp. 167–180, 2015.
DOI: 10.1007/978-3-319-22879-2_16

contact state of actuators by using the FEA software with the static structural analysis module. The results show that the anti-deformation and intensity performance of the actuators can meet requirements of production. Xing et al. [4] researched the motion characteristic and oil allocation characteristic of the cam curve curve radial piston motor by Pro/E platform.

2   The Structure Characteristics and Principle of Motor

As shown in figure 1 is the conception of valve plate flow distribution of piston structure of roller type low speed high torque hydraulic motor. The motor is composed of front end cover, stator, replaceable wear plate, after the end cover, rotor, plunger and roller, bearing, seal ring, spring and damper. The flow distribution of hydraulic motor makes the rubbing wear of port plate pair larger, replaceable wear plate can be a good solution to solve this problem, increasing the working life of the hydraulic motor. As can be seen from the chart 1, hydraulic motor adopts flow-distributing pair fitting with the rotor, reducing the supporting shaft and rotor inner hole this pair of friction pair, making the motor structure simpler, but also reducing the collocation of dual pair material problem. If the geometry size does not meet the requirements , which are likely to make the contact form of stator roller pair from line contact become a point contact, this is greatly increased the matching pair of failure probability, reducing the working life of the hydraulic motor.

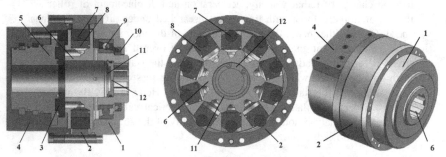

1-front end housing 2-stator 3-replaceable wear-resistant plate 4-rear housing 5-port plate 6-rotor 7-roller 8-piston 9-bearings 10-seal ring 11-catch spring 12-sheeting

Fig. 1. The structure of low speed high torque hydraulic motor

2.1   The Principle of Works

From the Fig. 1, it can also be seen that the pistons (roller and plunger) are arranged radially on piston bores of rotor with an equal distance apart. When pressurized water enters into the rotor, it is distributed into each chamber of the piston bottom at the inlet region, forcing the pistons to extend out. The roller bear against the inner surfaces of the stator. The inner surface of the stator then produces a normal reaction force, which can be decomposed into two branch forces: a radial branch force and a tangent branch force. The tangent branch force develops a torque at output shaft through the pistons, make the rotor rotate the output shaft, and form a certain speed and torque on the output shaft.

2.2   Design Parameters

The main parameters of water hydraulic motor as shown in table 1.

Table 1. Parameters of water hydraulic motor

Pressure of pump p_s/MPa	Speed of motor n/(r/min)	Plunger stroke s/mm	Diameter of plunger d/mm	The smallest diameter ρ_0/mm	Diameter of roll r_g/mm
10	10~200	6	20	47	8

3   The Motion Simulation of Hydraulic Motor Assembly Body

Through the simulation of motor assembly body movement, the radial velocity, acceleration and displacement of roller were got. When the speed is 50 °/s, the change of displacement, velocity and acceleration of roller with time are shown in figure 2 to 4. The impulse and reaction force between the rolling element and stator and the change of the relative speed curve as shown in figure 5 to 6. At the same time, in order to verify the correctness of the simulation, these results of the motion simulation were compared with the design parameters.

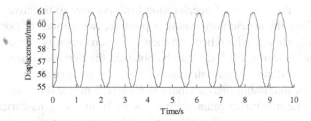

Fig. 2. The curve of the radial displacement of the roller

Fig. 2 shows the change of the radial displacement of the center of ball to the cylinder body with the rotor rotation time. From the figure, it can be seen that the change rule and the design of water hydraulic motor parameters are corresponding.

The change rule of the rolling body radial velocity is shown in figure 3. From the figure, it can be seen that when the speed change from negative to positive, the velocity is zero in the process. This is due to the existence of zero velocity section. Although the velocity has a constant velocity section when maximum speed reduced, due to the constant speed section of the interval is small, and speed is opposite bigger. So the process of constant speed is not displayed.

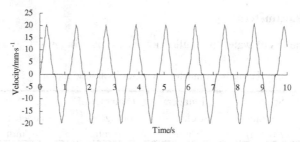

Fig. 3. The curve of the radial roller velocity

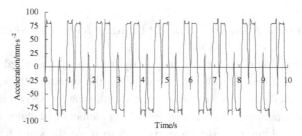

Fig. 4. The curve of roller radial acceleration

The change rule of the rolling body radial acceleration is as shown in figure 4. From the figure, it can be seen that the rising segment acceleration value basically remain unchanged and in the process of acceleration from positive to negative is the constant phase curve. When the acceleration value is positive, two periods of positive acceleration will suddenly change which because there are a zero velocity curve segments.

When the rotor speed is 50 °/s, it can be obtained the maximal displacement of motor rolling body is 61 mm, the minimum is 55 mm by numerical calculation. The motion simulation of maximum displacement is 60.98022 mm and the minimum value is 54.71665 mm. The maximum radial velocity is 18.74 mm/s by numerical calculation, and the simulated maximum is 19.769 mm/s. The maximum acceleration of acceleration section is 89.5 mm/s^2 by numerical calculation and the acceleration value of deceleration section is -89.5 mm/s^2. The simulated acceleration value of the acceleration section is 89.60481 mm/s^2 and the value of deceleration section is -89.6587 mm/s^2. It can be seen that the results of calculation and simulation result are very close, so these data can provide design bases for the hydraulic motor.

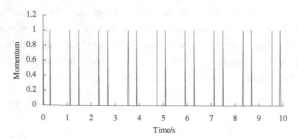

Fig. 5. The curve of the momentum of roller and stator

The momentum change rule between roller and the stator is shown in figure 5. From the figure, it can be seen that when the ordinate is 1, that is the location of the mutation momentum and the position of the momentum change dramatically. These could provide the basis for the design of the stator. In addition, it can be found that the impact position mutation occurred in acceleration mutation takes place by comparison of Fig. 4 and Fig. 5. The locations are zero speed segment and accelerating segment joint and the joint of uniform velocity section and deceleration section. So the stator two positions should strengthen.

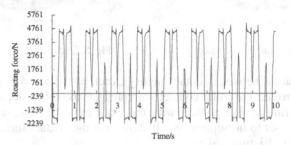

Fig. 6. The curve of roller and the normal force of stator

The change of the normal force of the roller with the time is shown in figure 6. From the figure, it can be seen that the change trend of normal force with time between roller and the stator during the scroll body exercise. When the normal force is negative, showing that rolling element is in accelerating section. When the normal force is positive, showing that rolling body section is in deceleration section. And the normal force of accelerating section is a lot smaller than the normal force of the deceleration section. In addition, the friction of the simulation model between the piston and cylinder is ignored and there is no fluid pressure in the bottom of the plunge. According to the change law of acceleration plunger group, the change law of the normal force between rolling element and the stator is got. Due to the motor assembly is in ideal conditions, the simulation will have some error.

4   Statics Analysis of Model of the Motor

As being known from the structure of the internal curve radial piston motor, there are five plungers connected high pressure water and five plungers connecting backwater flow channel in ten plungers at the same time. When the plunger is in backwater flow channel, the acting force between the plunger and the stator is extremely small. The force can be ignored relative to high pressure water plunger. So, in order to simplify the calculation, the simulation model is only the high pressure water plunger model.

4.1   Setting Simulation Analysis Model and Boundary Conditions

First of all, the geometric model of hydraulic motor is established by 3d software of Solidworks. Then it is imported to FEA software. And the key components of hydraulic motor are simulated by structural analysis modules of software. All materials in the model are set to structural steel. The motor structure as shown in figure 7 (rotor has hidden), and the motor meshing diagram is shown in figure 8.

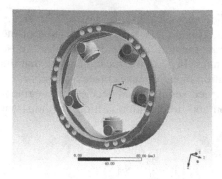

Fig. 7. The structure of motor   **Fig. 8.** The grid model of the motor

About the relation between contact Settings, each piston has 4 contacts. They are the contact with roller; the roller contact with plunger; the roller side contact with the stator; and the plunger side contact with the stator cylinder hole, respectively. The contact relation is set to no separation. It shows that the contact surface could have a small amount of no friction sliding.

Grids are generated by manual meshing, grid node number is 129940, and grid cell number is 64558.The grid model is shown in figure 10.

For the settings of the model load and constraint, each of the bottom of plunger is applied 10MPa and the rotor is applied 194N·m. For the constraint which could be divided into three parts. On the two side of the stator is set to the fixed constraint. The cylinder of stator is set to the column constraints. And the shaft of stator is set to the frictionless constraint. The constraint and load model is shown in figure 9.

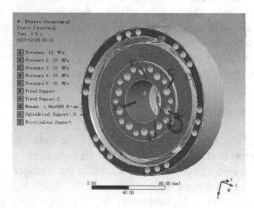

Fig. 9. The diagram of putting constraints and load model

4.2   Key Components of the Results of Simulation Analysis

When the motor is running, global and local equivalent stress and contact deformation diagrams are shown in figure 10 to 12.

For the statics analysis of the model by FEA software, selecting different physical parameters will have a lot of kinds of results. Here, the model is analyzed by the equivalent stress and total deformation.

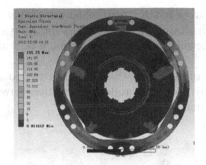
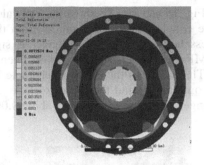

Fig. 10. Equivalent stress of the motor model   **Fig. 11.** Deformation of the motor model

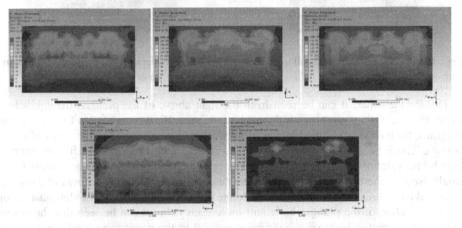

Fig. 12. Equivalent stress of the five plungers

Fig. 10 and Fig. 11 are the equivalent stress and total deformation of whole figure, and Fig. 12 is the equivalent effective stress of five plungers. It can seen form the three charts, the maximum deformation and stress have occurred in the position of the plunger 8. That is to say, the stress and deformation of here have bigger influence on the structure performance of the motor.

When the motor is running, the equivalent stress and deformation between roller and stator inner surface are shown in figure 13 and figure 14.

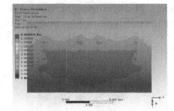

Fig. 13. Equivalent stress of roller contact with inner surface of the stator   **Fig. 14.** The overall deformation of roller contact with inner surface of the stator

Fig. 15 and Fig. 16 show the equivalent stress and deformation between inner surface contact roller and plunger. From the figures, it can be seen that the equivalent stress between roller and plunger is significantly smaller, the maximum is around 50MPa, and the distribution is very even.

This is due to the structure characteristic of the roller which makes it loaded area become larger, and then lead to stress evenly. However, roller deformation between plunger and relatively is large, about 0.0058 mm.

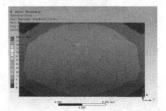   

Fig. 15. Equivalent stress of roller contact with surface of the plunger   **Fig. 16.** The overall deformation of roller contact with surface of the plunger

Hidden the roller, it can be seen from the right above of the plunger, the equivalent stress and strain diagram shown in figure 17 and figure 18. As can be seen from the stress diagram, the inner wall of the plunger of stress should be greater than the stress of roller bottom. This is because the roller in contact with the stator inner surface after the oppression by the reaction, and make its force to transfer the results of the plunger. Stress has a small rise near the damping hole, this is because the lower structure stiffness of damping hole make the stress concentration at the edge of the damping hole. The deformation of the deformation of the plunger body still more evenly, but it can be seen that the deformation of the piston body on the direction parallel to the paper from top to bottom has obvious transition change. The deformation of the upper part of the plunger is larger than the lower. And the problem will be carried out in detail in the rear.

The stress and deformation at the bottom of plunger are shown in figure 19 and 20. From the figures, it can be seen clearly that the stress distribution and deformation of the bottom of the plunger. The equivalent stress at the bottom of plunger is distributed into rings and most areas of stress are under 10MPa, which is consistent with the inflow water pressure. There is a ring of high stress area on the outside of the middle of radius and the value is about 20MPa. In the overall deformation diagram can be seen that the deformation of piston underside more than the top surface deformation. The maximum is 0.007 mm, the minimum is around 0.0055 mm, and the transition change trend of the deformation is aligned with the top surface.

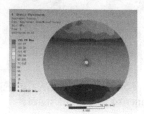   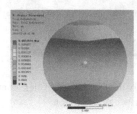

Fig. 17. Equivalent stress of the plunger   **Fig. 18.** The overall deformation of the plunger

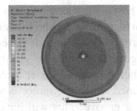 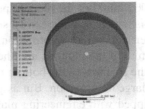

Fig. 19. Equivalent stress of the bottom of plunger

Fig. 20. The deformation of the bottom of plunger

Looking from the main direction of the plunger, the stress and deformation of the plunger outside are shown in figure 21 and 22. It can be seen that the piston ring groove space affected by the stress is larger, so the design of plunger should consider this aspect. And when the water hydraulic motor is working, the piston ring groove location is equipped with antifriction, anti-abrasive and corrosion resistant engineering plastic ring which will weaken the stress of the piston ring groove. And it can be also seen that the stress value of groove side is small at the cylindrical roller nest slot and the tress value of nest groove bottom is bigger. This is because for cylindrical form of bearing surface, the effective bearing area is the actual bearing surfaces in perpendicular to the plane of the reaction arca. From Fig. 22, it can be seen that the distribution of deformation of plunger and the position of largest deformation is in ring groove bottom ring surface.

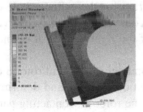

Fig. 21. Equivalent stress of the plunger lateral

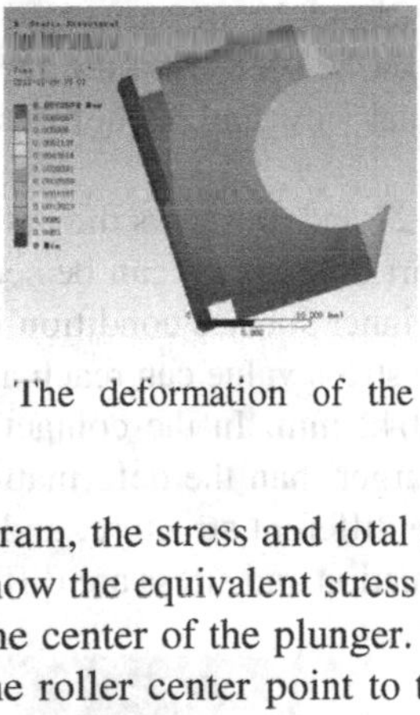

Fig. 22. The deformation of the plunger lateral

For hell phenomenon in overall deformation diagram, the stress and total deformation are analyzed by sections. Fig. 23 and Fig. 24 show the equivalent stress and total deformation in the distribution of cross section in the center of the plunger. It can be seen that the force of the stator of roller through the roller center point to the lower right. If the friction and the influence of inertia force are ignored, this force in the vertical direction of component is used to offset the piston bottom water fluid pressure. And horizontal component force will provide motor rotation torque, and the component is bound to make roll plunger. The result of the hell phenomenon is the top right corner and the lower left corner of the plunger extruded cylinder hole, this can be seen in the stress and deformation figure.

For the rotor, cylinder hole right side bears certain stress in equivalent stress figure, and it is about 5 or 6MPa. And it can be seen that the right side of deformation of the cylinder hole is considerably larger than the left in the deformation diagram, it is about 0.002 mm. For the plunger parts, it caused by hell phenomenon shown in stress diagram is not clear, but it can be seen from the deformation figure and the left corner of plunger had a great deformation, it has reached around 0.007 mm. It is because of the structure of the plunger, and the side of the plunger has the annular groove which reduces the stiffness of the bottom of the plunger. When it bears a greater force, it will create a large deformation. But as previously mentioned, when the motor is working, it is equipped with engineering plastic ring in annular groove. This will have greatly improved for the large deformation.

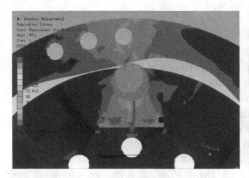

Fig. 23. Equivalent stress of the plunger center section   **Fig. 24.** The deformation of the plunger center section

Figure 25and 26 are stress and deformation diagrams of the cylinder hole of plunger 8. It can be seen from the figure, the result is consistent with the observed from the sectional view and the deformation and stress of the right surface of cylinder hole is larger.

Figure 27 and 28 shows the pattern of the stress and deformation of rotor 8 contacting the part of stator. It can be seen that the distribution of stress and deformation of the stator inner surface condition is consistent with the rotor 8. Contacting part of the maximum stress value can reach around 55MPa, and the deformation also can achieve about 0.0042 mm. In the contact pair, the equivalent stress and deformation of the stator is larger than the deformation and stress of the rotor. It is because of different parts have different structure, and this will result in some differences of stiffness. So the deformation and stress are different under the circumstances.

Fig. 25. The stress of viewing plunger from the rotor lateral

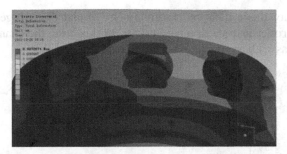

Fig. 26. The deformation of viewing plunger from the rotor lateral

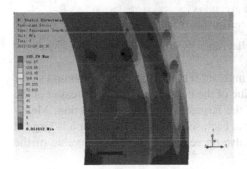
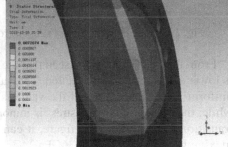

Fig. 27. The stress of the roller 8 contact stator

Fig. 28. The deformation of the roller 8 contact stator

5   Analysis of the Contact Stress

The maximum contact stress could be calculated by using Hertz theory. The equation is expressed by

$$\delta_{max} = \sqrt{\frac{NE\sum\rho}{2\pi l(1-\mu)^2}} \tag{1}$$

Where N is the force of roller contact with inner surface of the stator, $N=4463.8N$; E is the elastic modulus of roller and stator, $E=195GPa$; ρ is the distance of contact point of roller and stator and the center of roller; l is the length of roller, $l=24mm$; μ is Poisson's ratio of roller and stator, $\mu=0.3$;

The following equations can be obtained by the structure of water hydraulic motor. They are expressed by

$$\rho = \frac{\rho_0 + R \cdot \cos\alpha}{\cos\beta} \tag{2}$$

$$\beta = \arcsin\frac{R \cdot \sin\alpha}{\rho} \tag{3}$$

Substituting the available data into formula (1), (2) and (3) can get the maximum contact stress of roller and stator inner surface. In this case, the change of the maximum contact stress with argument as shown in Fig. 29.

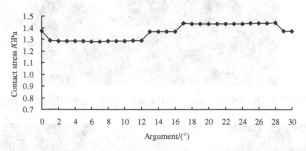

Fig. 29. The maximum contact stress of roller contact with inner surface of the stator

From the Fig. 29, it can be seen that the maximum contact stress of roller and stator is lower at the acceleration section. The value of the contact stress has increased at the uniform velocity section and the value is the largest at deceleration section. Meanwhile, the change of contact stress is minor at each section and the variation is large at transition section. From the figure, it can also be seen that the average value of the maximum contact stress is 1.36GPa.

Equivalent stress is also called Von Mises, it mainly inspects the stress difference of material in all directions. The relationship between equivalent stress and the principal stress can be represented as:

$$\sigma_e = \sqrt{\frac{1}{2}\left[(\sigma_1 - \sigma_2)^2 + (\sigma_2 - \sigma_3)^2 + (\sigma_3 - \sigma_1)^2\right]} \tag{4}$$

This contact stress is different from the calculated contact stress which is the local stress of the contact region when the two objects are squeezed. In FEA software, the CONTACT part of the tool have special simulation-CONTACT TOOLS.

Figure 30 to 33 shows the contact stress of 2, 4, 6 and 10 of roller. Figure 34 and 35 show the contrast figure which are the contact stress between rotor 8 and stator.

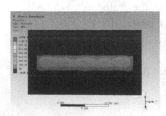  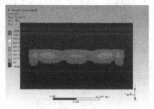

Fig. 30. The contact stress of roller 2   **Fig. 31.** The contact stress of roller 4   **Fig. 32.** The contact stress of roller 6

  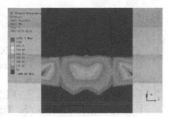

Fig. 33. The contact stress of roller 10   **Fig. 34.** The contact stress of roller 8   **Fig. 35.** The contact stress of surface of the stator

Through the above figure, it can be seen that the distribution of contact stress of roller is approximate rectangular plane distribution, and the maximum value is located at the center line of the contact area of the rectangle, which is consistent with the condition of the Hertz theory. And through numerical comparison can be found, the contact stress values of the roller are relatively small, it can only reach about 700MPa. And the stator surface of contact stress is significantly larger, the maximum value is 1231.7MPa (1.232GPa). Meanwhile, Hertz theory is used to calculate the contact stress, the maximum contact stress between the two surfaces is 1.36GPa. This shows that the simulation value of water hydraulic motor is in agreement with the theoretical value, and the percent of contact area is up to 91%.

6   Conclusion

- Using virtual prototype technology, the structure of water hydraulic motor was designed and the model of low speed high torque water hydraulic motor was established. And the working principle and structure characteristics of the water hydraulic motor were introduced.
- Through the study of the dynamic analysis of the motor assembly, the force variation law of the rolling body and the stator was obtained. The position of momentum mutation of the roller was got, and the stress distribution of the rolling body contact with the stator along the normal of the roller was gained. Through analyzing of kinematic pair of the water hydraulic motor, it can provide an effective reference for optimal design of the motor.
- The contact stress between the roller and stator was solved by the Hertz theory, and the location of the maximum stress and the maximum stress value were got. And the results show that it does not exceed the material allowable stress. This also verified the rationality of the structural design of the motor.
- Through the statics analysis of model in FEA software, the overall and local equivalent stress and deformation was got and the contact stress condition between the stator and the roller was obtained. The numerical simulation results are contrasted with the theoretical ones, and the validity of the numerical simulation method is testified.

References

1. Tang, Q., Zhao, G., Feng, X.: Research on the effects of port plate structure on the water hammering of pure water axial piston hydraulic pump. J. Machine tool & hydraulics **37**, 77–79 (2009)
2. Li, Y., Shi, G., Chen, Z.: FEM analysis of connecting-rod in low speed high torque hydraulic motors with ANSYS. Transactions of the Chinese Society for Agricultural Machinery **38**(11), 137–139 (2007)
3. Wang, B., Wang, C., Jianjun, F.: Strength analysis of hydraulic rotary vane actuators based on ANSYS Workbench. Machine Tool & Hydraulics **43**(7), 168–171 (2015)
4. Xing, L., Xie, H., Zhao, L., et al.: Modeling and simulation for virtual pilot machine of cam curve radial piston hydraulic motor. Ship Electronic Engineering **32**(10), 86–89 (2012)

Calculation of the Magnetic Field of the Permanent Magnet Using Multi-domain Differential Quadrature

Jun-wei Chen, Le-tong Ma, Bo Zhang$^{(\boxtimes)}$, and Han Ding

State Key Laboratory of Mechanical System and Vibration, School of Mechanical Engineering,
Shanghai Jiao Tong University, Shanghai 200240, China
cjwbuaa@163.com, m835527262@126.com, {b_zhang,hding}@sjtu.edu.cn

Abstract. On the basis of the multi-domain differential quadrature, this paper formulates the equation of the magnetic field of the permanent magnet (PM), which is widely used in precision engineering. The multi-domain, including the PM domain and the air domain, is generated by dividing the whole computational domain of the PM. By applying the differential quadrature rule in each domain, the unknown magnetic vector potential can solved, and then the magnetic field distribution is obtained. Numerical results show that the magnetic field obtained by the multi-domain differential quadrature method is accurate and efficient, which is validated by the comparison with the finite difference method and the finite element method.

Keywords: Permanent magnet · Multi-domain · Differential quadrature

1 Introduction

Devices based on permanent magnets (PMs) have been widely used in precision engineering, such as PM actuators [1,2] and PM brakes [3]. In these precision applications, accurate and efficient prediction of the magnetic field of the PMs is required to model the dynamic performance of the devices. To calculate the magnetic field, the traditional analytical method seems to be simple to implement. For example, the semi-analytical method, which derives from the Fourier series expansion, is widely utilized in linear PM synchronous motors. However, in this method, length of the PM arrays is assumed finite, and thus the end effect is neglected. In the case that a singular PM is utilized, the equivalent current model is available. But it greatly depends on the shape of the PM with loss of accuracy.

Due to the above disadvantages, an alternative way is to use the numerical methods. The traditional numerical methods, e.g., the finite difference method (FDM) and the finite element method (FEM), have been employed in problems of magnetic fields for decades. However, the FDM usually requires a great many points to achieve accurate solutions and is lack of geometrical flexibility in fitting irregular boundary shapes. Although the FEM is more powerful and versatile to handle problems involving complex geometries and inhomogeneous media, as a low-order scheme, it is difficult to achieve high accurate solutions.

© Springer International Publishing Switzerland 2015
H. Liu et al. (Eds.): ICIRA 2015, Part I, LNAI 9244, pp. 181–189, 2015.
DOI: 10.1007/978-3-319-22879-2_17

Compared to the FDM and the FEM, the differential quadrature method (DQM) [4,5] yields numerical results of high-order accuracy by using fewer discrete points. This advantage has been widely shown in a great many applications presented in [6], such as the complicated vibration problems of beams and plates [7,8], modeling and analysis of transmission lines [9,10], transient analysis of frequency-dependent inter-connecting circuits [11] and so on. In this paper, by applying the differential quadra-ture rule in multi-domains, including the PM domain and the air domain, the equation of the magnetic field of the PM is reformulated. Then, the advantages of the proposed method are showed by the comparison with the FDM and the FEM.

2   Modeling

2.1   Differential Quadrature Rule

In the DQM, the values of the derivatives at each sampling grid point are expressed as weighted linear sums of the function values at all sampling grid points within the considered domain. Consider a function $f(x)$ with respect to a variable x in $[-1,1]$, where n Chebyshev sampling grid points are employed, i.e.,

$$x_i = -cos\left(\pi \frac{i-1}{n-1}\right) \quad i = 1,2,...,n \tag{1}$$

The first order derivatives of $f(x)$ at the sampling grid points are given as

$$f'(x_i) = \sum_{j=1}^{n} G_{i,j}^{(1)} f(x_j) \quad i = 1,2,...,n \tag{2}$$

where $G_{i,j}^{(1)}$ represents the weighting coefficient of the first order derivative. It can be computed by the Lagrange's interpolation function $L_j(x)$ [12] explicitly as

$$G_{i,j}^{(1)} = L_j'(x_i) \tag{3}$$

So the first order weighting coefficient matrix is obtained as

$$\mathbf{G}^{(1)} = \begin{pmatrix} G_{1,1}^{(1)} & \cdots & G_{1,n}^{(1)} \\ \vdots & \ddots & \vdots \\ G_{n,1}^{(1)} & \cdots & G_{n,n}^{(1)} \end{pmatrix} \tag{4}$$

Similarly, the k-th order derivatives $f^{(k)}(x)$ at the sampling grid points are given as

$$f^{(k)}(x_i) = \sum_{j=1}^{n} G_{i,j}^{(k)} f(x_j) \quad i = 1,2,...,n \tag{5}$$

where $G_{i,j}^{(k)} = L_j^{(k)}(x_i)$. And the second order weighting coefficient matrix is [5]

$$\mathbf{G}^{(2)} = \begin{pmatrix} G_{1,1}^{(2)} & \cdots & G_{1,n}^{(2)} \\ \vdots & \ddots & \vdots \\ G_{n,1}^{(2)} & \cdots & G_{n,n}^{(2)} \end{pmatrix} \tag{6}$$

where

$$\begin{cases} G_{i,j}^{(2)} = 2\left(G_{i,j}^{(1)} G_{i,i}^{(1)} - \dfrac{G_{i,j}^{(1)}}{x_i - x_j} \right) & i,j = 1,2,...,n \quad i \neq j \\[4mm] G_{i,i}^{(2)} = -\displaystyle\sum_{j=1, j\neq i}^{n} G_{i,j}^{(2)} & i = 1,2,...,n \end{cases}$$

2.2   Magnetic Field of the PM

Consider a rectangular PM in the open domain. For simplicity, the computational domain is selected large enough compared with the dimension of the PM. Fig. 1 shows the computational domain of a quarter of the rectangular PM. As showed in Fig. 1, the area enclosed by ABCD represents a quarter of the PM, and the area enclosed by AEKF represents the whole computational domain.

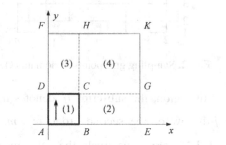

Fig. 1. Computational domain of a quarter of the rectangular PM

The PM is assumed to have a linear demagnetization characteristic, and is fully and homogeneously magnetized along the vertical direction. The equation for the magnetic field in both the PM and the air is [13]

$$\frac{\partial^2 A}{\partial x^2} + \frac{\partial^2 A}{\partial y^2} = 0 \tag{7}$$

where A represents the z component of the magnetic vector potential.
The associated interface and boundary conditions are

$$\begin{cases} B_n^{(PM)} = B_n^{(air)}, \ H_t^{(PM)} = H_t^{(air)}, & \text{along BC and CD} \\ B_n^{(PM)} = B_n^{(air)} = 0, & \text{along AF} \\ B_t^{(PM)} = B_t^{(air)} = 0, & \text{along AE} \\ B_n^{(air)} = B_t^{(air)} = 0, & \text{along EK and FK} \end{cases} \tag{8}$$

where B represents the magnetic flux density, H represents the magnetic field intensity, and the subscripts n and t denote the normal and the tangential components of the associated variable, respectively.

To apply the multi-domain DQM, the whole computational domain is divided into four domains, domain (1) of the PM and other three domains of the air around it, as illustrated in Fig. 1. In domain (k), $n_k \times m_k$ Chebyshev sampling grid points are applied, whose numbering order is showed in Fig. 2. Then Eq. (7) for region (k) can be rewritten as

$$\left(\mathbf{G}^{(2)}_{m_k,j} \mathbf{L}^{(k)r}_i + \mathbf{G}^{(2)}_{n_k,i} \mathbf{L}^{(k)c}_j \right) \mathbf{A}^{(k)} = 0, \quad i = 1,2,...,n_k \quad j = 1,2,...,m_k \quad\quad (9)$$

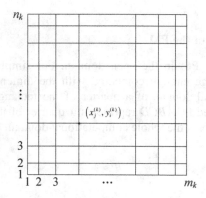

Fig. 2. Sampling grid points of domain (k)

where k denotes the k-th region, the subscript ij denotes the value at $(x^{(k)}_j, y^{(k)}_i)$, $\mathbf{G}^{(2)}_{m_k,j}$ represents the j-th row of the second order $m_k \times m_k$ weighting coefficient matrix $\mathbf{G}^{(2)}_{m_k}$, $\mathbf{L}^{(k)r}_i$ and $\mathbf{L}^{(k)c}_j$ are respectively the i-th row and j-th column index matrix and expressed as

$$\mathbf{L}^{(k)r}_i = \underbrace{\left[\mathbf{O}_{m_k} \quad \cdots \quad \mathbf{O}_{m_k} \quad \underset{i}{\mathbf{I}_{m_k}} \quad \mathbf{O}_{m_k} \quad \cdots \quad \mathbf{O}_{m_k} \right]}_{n_k}$$

$$\mathbf{L}^{(k)c}_j = \underbrace{\begin{bmatrix} \mathbf{E}_j & & \\ & \ddots & \\ & & \mathbf{E}_j \end{bmatrix}}_{n_k}$$

where $\mathbf{O}_{m_k}$ is the $m_k \times m_k$ null matrix, $\mathbf{I}_{m_k}$ is the $m_k \times m_k$ identity matrix, and

$$\mathbf{E}_j = \underbrace{\left[0 \quad \cdots \quad 0 \quad \underset{j}{1} \quad 0 \quad \cdots \quad 0 \right]}_{n_k}$$

Besides, the unknown vector $\mathbf{A}^{(k)}$ is

$$\mathbf{A}^{(k)} = \begin{bmatrix} \mathbf{A}_1^{(k)} & \mathbf{A}_2^{(k)} & \cdots & \mathbf{A}_{n_k}^{(k)} \end{bmatrix}^T$$

where

$$\mathbf{A}_i^{(k)} = \begin{bmatrix} A_{i1}^{(k)} & A_{i2}^{(k)} & \cdots & A_{im_k}^{(k)} \end{bmatrix}$$

By introducing the $n_k m_k \times n_k m_k$ matrix $\mathbf{D}^{(k)}$, Eq. (9) is simplified as

$$\mathbf{D}^{(k)}\mathbf{A}^{(k)} = \mathbf{0} \tag{10}$$

For the multi-domain, Eq. (10) is combined as

$$\mathbf{DA} = \mathbf{0} \tag{11}$$

where

$$\mathbf{D} = diag(\mathbf{D}^{(1)}, \mathbf{D}^{(2)}, \mathbf{D}^{(3)}, \mathbf{D}^{(4)})$$

$$\mathbf{A} = \begin{bmatrix} \left(\mathbf{A}^{(1)}\right)^T & \left(\mathbf{A}^{(2)}\right)^T & \left(\mathbf{A}^{(3)}\right)^T & \left(\mathbf{A}^{(4)}\right)^T \end{bmatrix}^T$$

It is noted that $\mathbf{B}^{(k)} = \nabla \times \mathbf{A}^{(k)}$, and the constitutive relation of the PM is

$$\mathbf{B}^{(1)} = \mu_0(\mu_r \mathbf{H}^{(1)} + \mathbf{M}_r) \tag{12}$$

where μ_0 is the permeability of air, μ_r is the relative permeability of the PM, $\mathbf{M}_r$ is the remanent magnetization. Using the differential quadrature rule, all the boundary conditions and the interface conditions are written as

$$\mathbf{D}_b \mathbf{A} = \mathbf{F} \tag{13}$$

where $\mathbf{D}_b$ and $\mathbf{F}$ are the coefficient matrix and the corresponding load vector, respectively. Consequently, Eqs. (11) and (13) can be solved simultaneously for the field distribution of the rectangular PM by the least square method.

3   Validation and Discussion

To validate the proposed method, the FDM and the FEM are also applied, as well as the analytical method [14,15], which is used as the benchmark. Fist, the results obtained by the FDM and the multi-domain DQM are compared, which are demonstrated in Fig. 3 and Fig. 4. Obviously, the solutions obtained by the multi-domain differential quadrature with 49×29 sampling grid points match very well with the analytical solutions. However, the solutions obtained by the FDM with 69×39 sampling grid points are still not accurate enough. This phenomenon is apparent near

$x = 6mm$ for B_x and near $x = 0$ for B_y. Although the accuracy of the FDM can be improved theoretically by applying more sampling grid points, it is not available because it will run out of memory under the computational condition listed in Table 1.

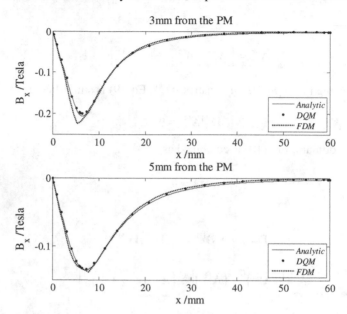

Fig. 3. Comparison of B_x between the multi-domain DQM and the FDM

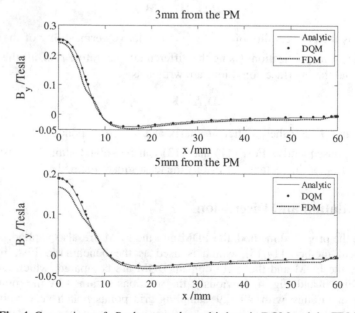

Fig. 4. Comparison of B_y between the multi-domain DQM and the FDM

Further, the FEM is also applied by Ansys through the element plane 53. The results are demonstrated in Fig. 5. For clarity, the results ranging from 40mm to 60mm are not shown. In Fig. 5, Ansys1 denotes the case that 1628 elements are applied, and Ansys2 denotes the case that 3640 elements are applied. Correspondingly, the number of nodes in the two cases of Ansys1 plane 53 and Ansys2 plane 53 is 5037 and 11145, respectively. In the multi-domain DQM, $49 \times 29 = 1421$ sampling grid points are applied. Obviously, the result obtained by the multi-domain DQM matches very well with the analytical solutions. However, the results obtained by Ansys1 with 1628 elements, using plane 53, are not accurate enough. There is evident disagreement with the analytical solution near the extreme values, both for B_x and B_y. When the number of elements is increased to 3640, namely, the cases of Ansys2, the disagreement gets smaller but is still larger than that of the multi-domain DQM. This is obvious in the B_y figure near the extreme values and in the range from 20mm to 40mm.

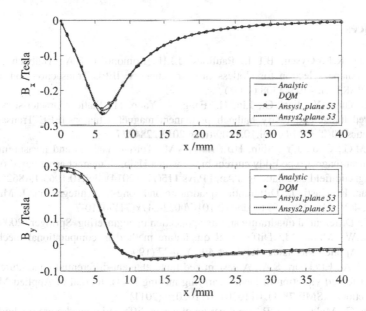

Fig. 5. Comparison of the magnetic flux density between the multi-domain DQM and Ansys

Table 1. Memory allocation for the present study

maximum possible array	801 MB (8.403e+008 bytes)
memory available for all arrays	1424 MB (1.493e+009 bytes)
memory used by MATLAB	336 MB (3.519e+008 bytes)
physical Memory (RAM)	2985 MB (3.130e+009 bytes)

4   Conclusion

In this paper, the differential quadrature is applied in multi-domain to the calculation of the magnetic field of the PM. Due to its high-order scheme, the proposed method is accurate and efficient even using fewer discrete points. This is validated by the numerical comparison with the FDM and the FEM. So the multi-domain differential quadrature method is an effective method for the calculation of the magnetic field due to the PM-based devices in engineering applications. In addition, the differential quadrature is applicable in the field analysis of irregular-shaped PMs and the dynamic analysis of electro-magnetic problems, which will be studied in the future.

Acknowledgements. This work was supported by the National Natural Science Foundation of China under grant 51375312, for which the authors are grateful.

References

1. Meessen, K.J., Gysen, B.L.J., Paulides, J.J.H., Lomonova, E.A.: Halbach permanent magnet shape selection for slotless tubular actuators. IEEE Transactions on Magnetics **44**(11 PART 2), 4305–4308 (2008)
2. Jin, P., Yuan, Y., Jian, G., Lin, H., Fang, S., Yang, H.: Static characteristics of novel air-cored linear and rotary halbach permanent magnet actuator. IEEE Transactions on Magnetics **50**(2) (2014). doi:10.1109/tmag.2013.2284277
3. Park, M.G., Choi, J.Y., Shin, H.J., Jang, S.M.: Torque analysis and measurements of a permanent magnet type Eddy current brake with a Halbach magnet array based on analytical magnetic field calculations. J Appl Phys **115**(17) (2014). doi:10.1063/1.4862523
4. Bellman, R., Casti, J.: Differential quadrature and long-term integration. J. Math. Anal. Appl. **34**(2), 235–238 (1971). doi:10.1016/0022-247x(71)90110-7
5. Shu, C.: Differential quadrature and its application in engineering. Springer (2000)
6. Bert, C.W., Malik, M.: Differential quadrature method in computational mechanics: A review. Applied Mechanics Reviews **49**(1), 1–27 (1996)
7. Jafari, A.A., Eftekhari, S.A.: A new mixed finite element-differential quadrature formulation for forced vibration of beams carrying moving loads. Journal of Applied Mechanics, Transactions ASME **78**(1), 0110201–01102016 (2011)
8. Karami, G., Malekzadeh, P.: Application of a new differential quadrature methodology for free vibration analysis of plates. International Journal for Numerical Methods in Engineering **56**(6), 847–868 (2003)
9. Xu, Q., Mazumder, P.: Accurate modeling of lossy nonuniform transmission lines by using differential quadrature methods. IEEE Transactions on Microwave Theory and Techniques **50**(10), 2233–2246 (2002)
10. Tang, M., Mao, J.: A differential quadrature method for the transient analysis of multiconductor transmission lines. In: 2008 International Conference on Microwave and Millimeter Wave Technology Proceedings, ICMMT 2008, pp. 1423–1426 (2008)
11. Tang, M., Mao, J.F., Li, X.C.: Analysis of interconnects with frequency-dependent parameters by differential quadrature method. IEEE Microwave and Wireless Components Letters **15**(12), 877–879 (2005)

12. Liu, J., Wang, X.W.: An assessment of the differential quadrature time integration scheme for nonlinear dynamic equations. Journal of Sound and Vibration **314**(1–2), 246–253 (2008). doi:10.1016/j.jsv.2008.01.004
13. Furlani, E.P.: Permanent Magnet and Electromechanical Devices, Materials, Analysis, and Applications Academic press series in electromagnetism. Academic Press (2001)
14. Halbach, K.: Design of permanent multipole magnets with oriented rare earth cobalt material. Nuclear Instruments and Methods **169**(1), 1–10 (1980)
15. Lee, M.G., Gweon, D.G.: Optimal design of a double-sided linear motor with a multi-segmented trapezoidal magnet array for a high precision positioning system. Journal of Magnetism and Magnetic Materials **281**(2–3), 336–346 (2004)

Optimization and Design of Exiting Coil in Giant Magnetostrictive Actuator

Xiaohui Gao, Yongguang Liu, and Zhongcai Pei[✉]

School of Automation Science and Electrical Engineering, Beihang University, Beijing, China
hgaoxiaohui@126.com, {lyg,peizc}@buaa.edu.cn

Abstract. Exiting coil as the center of the electromagnetism conversion is mainly to provide driving magnetic field for the giant magnetostrictive actuator (GMA) and control its output displacement through inputting different current. Therefore, the structural parameters of the exiting coil are the key factors to improve the electromagnetic conversion efficiency and make GMA effect sufficiently. This paper designs and optimizes parameters of the exiting coil according to the designing principles that are high uniformity and intensity of the magnetic field near giant magnetostriction materials (GMM) rod, minimum heat loss and compact-sized through analyzing the working conditions of the GMA. Then, the magnetic circuit with the optimized excitation coil is analyzed based on Ansoft. The result indicates that the distribution of magnetic field is more uniform and the uniformity rate is raised to be 99.35%.

Keywords: Giant magnetostrictive actuator (GMA) · Exiting coil · Ansoft · Uniformity · Optimization and design

1 Introduction

Since the 1970s, GMM as a kind of strategic functional materials got a fast development and is widely employed in the aerospace engineering, ocean, geology, measuring instrument, machinery, national defence and medical technology through taking advantage of its own features such as Joule effect, Villari effect, Wiedemann effect, Matteucci effect, ΔE effect and so on[1-4]. A large number of experimental results show that the coupling field of electricity-magnetism- machinery- hot and nonlinearities significantly exist in the GMA between its input current and output displacement which will destroy the property seriously [5-8]. The exciting coil as the only driving field is the source of the electromagnetic conversion and its parameters are the key factors to improve the efficiency and closely related to the frequency characteristic and stability control. The GMM can provide good anti-pressure and weaker tensile. If the GMM rod is placed in an inhomogeneous magnetic field, on the one hand the harmonic frequency output will appear as the higher harmonics in the process of measurement, on the other hand the mechanical property of the GMA and material characteristics of the GMM will be seriously damaged as the inconformity of the tension and compression between magnetic domain. Therefore, the uniformity of the magnetic field intensity plays an important role

© Springer International Publishing Switzerland 2015
H. Liu et al. (Eds.): ICIRA 2015, Part I, LNAI 9244, pp. 190–200, 2015.
DOI: 10.1007/978-3-319-22879-2_18

in the GMA [9]. Many researches have been carried out to design the exciting coil and made some achievement, but more focus is on the Magnetic intensity, heat loss and comprehensive research of high uniformity, strong intensity, small heat loss and compact-sized has never been researched [10-12]. Therefore, this paper design and optimize the structure parameters of the exciting coil based on all factors, which will significantly improve the property of the GMA.

2   Structure of GMA

Structure diagram of GMA is shown in Fig. 1. The working process can be described as follows: the exciting coil can produce driving magnetic field through adjusting the current and the GMM rod exports displacement. The magnetostrictive coefficient can be raised by applying pre-pressure relying on preloading spring. The exciting coil can also produce the bias magnetic field to eliminate the frequency-doubling phenomenon of the GMA on the basis of material characteristics of the GMM [14]. The conclusion summarized by linear constitutive equation of GMM (1) indicates that magnetic intensity is an important factor to influence the GMA. The source of magnetic intensity is exciting coil. Therefore, structural parameters of the exciting coil are particularly important.

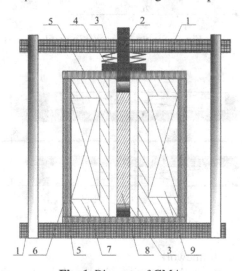

Fig. 1. Diagram of GMA

1- Non-magnetizer cover; 2- Output link; 3-Preload spring; 4-Magnetizer; 5- Permeability cover; 6- Permeable wall; 7- Coiler framework; 8-GMM rod; 9- Exciting coil.

$$\varepsilon = \frac{\sigma}{E_y^H} + d_{33}H$$

$$B = d_{33}\sigma + \mu^\sigma H$$

(1)

3   Optimization and Design of Exciting Coil

3.1   Magnetic Intensity Produced by Exciting Coil

The Magnetic intensity on the central axis of the single-layer exciting coil can be calculated by formula 2 based on Biot-Savart law[13].

$$H = \frac{1}{2} n_1 I(\cos \beta_1 - \cos \beta_2) \tag{2}$$

B is Magnetic intensity; n_1 is coil winding per unit length; I is input current; β_1 and β_2 are both angle which is related to the location in the center axis.

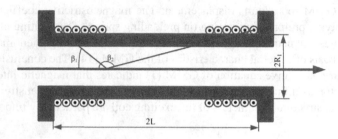

Fig. 2. Single-layer exciting coil

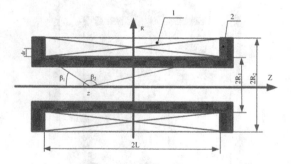

Fig. 3. Multilayer exciting coil

1- Exciting coil; 2-Coiler framework.
In order to calculate the magnetic intensity of the multilayer exciting coil, the coordinate system is established in Fig. 3. The magnetic intensity can be achieved by formula 3 in the radius r.

$$dH = \frac{\mu_0}{2} n_1 n_2 I \left(\frac{L+z}{\sqrt{(L+z)^2 + r^2}} + \frac{L-z}{\sqrt{(L-z)^2 + r^2}} \right) dr \tag{3}$$

n_2 is coil winding per unit thickness.

$$n_1 n_2 = \frac{N}{2L(R_2 - R_1)} \qquad (4)$$

The magnetic intensity in the location z can be calculated by formula 5

$$
\begin{aligned}
H_z &= \int_{R_1}^{R_2} dr \\
&= \int_{R_1}^{R_2} \frac{1}{2} n_1 n_2 I \left(\frac{L+z}{\sqrt{(L+z)^2 + r^2}} + \frac{L-z}{\sqrt{(L-z)^2 + r^2}} \right) dr \\
&= \frac{1}{2} n_1 n_2 I \left\{ (L+z) \ln \frac{R_2 + \sqrt{(L+z)^2 + R_2^2}}{R_1 + \sqrt{(L+z)^2 + R_1^2}} + (L-z) \ln \frac{R_2 + \sqrt{(L-z)^2 + R_2^2}}{R_1 + \sqrt{(L-z)^2 + R_1^2}} \right\}
\end{aligned}
\qquad (5)
$$

The distribution of magnetic intensity in the Z axis is shown in Fig. 4. It indicates that the magnetic intensity appears symmetric distribution along the midpoint of the exciting coil and its value decreases with distance growing from the center position. The maximum magnetic intensity is shown in formula 6 when z is 0.

$$H_{max} = n_1 n_2 I L \ln \frac{R_2 + \sqrt{L^2 + R_2^2}}{R_1 + \sqrt{L^2 + R_1^2}} \qquad (6)$$

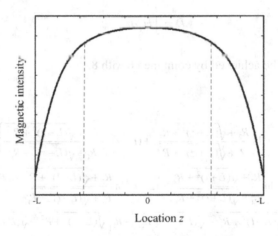

Fig. 4. Distribution of magnetic intensity

3.2   Uniformity of Magnetic Intensity

The Fig. 4 indicates that the magnetic intensity declines sharply near the end of exciting coil. In order to improve the uniformity of the coil, the GMM rode can be placed according to the Fig. 5, which can avoid end effect. This paper put forward a formula 7 to describe the uniformity of the magnetic field. The smaller value shows more uniformity of the magnetic field.

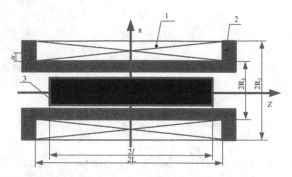

Fig. 5. Location of GMM rod

1- Exciting coil; 2-Coiler framework; 3-GMM rod.

$$C = \int_{l}^{-l} |H_z - H_{max}| dz \tag{7}$$

In order to calculate easily, the formula 7 can be changed to 8 since the value of H_{max} is the biggest. The bigger value shows more uniformity of the magnetic field.

$$D = \int_{l}^{-l} H_z dz \tag{8}$$

The formula 9 can be achieved by combined 6 with 8.

$$
\begin{aligned}
D &= \int_{l}^{-l} H_z dz \\
&= \int_{l}^{-l} \frac{1}{2} n_1 n_2 I \left\{ (L+z) \ln \frac{R_2 + \sqrt{(L+z)^2 + R_2^2}}{R_1 + \sqrt{(L+z)^2 + R_1^2}} + (L-z) \ln \frac{R_2 + \sqrt{(L-z)^2 + R_2^2}}{R_1 + \sqrt{(L-z)^2 + R_1^2}} \right\} dz \\
&= \frac{n_1 n_2 I}{4} \left\{
\begin{aligned}
&(L+l)^2 \ln \frac{R_2 + \sqrt{(L+l)^2 + R_2^2}}{R_1 + \sqrt{(L+l)^2 + R_1^2}} - (L-l)^2 \ln \frac{R_2 + \sqrt{(L-l)^2 + R_2^2}}{R_1 + \sqrt{(L-l)^2 + R_1^2}} \\
&+ R_2 \left[\sqrt{(L-l)^2 + R_2^2} - \sqrt{(L+l)^2 + R_2^2} \right] - R_1 \left[\sqrt{(L-l)^2 + R_1^2} - \sqrt{(L+l)^2 + R_1^2} \right]
\end{aligned}
\right\}
\end{aligned}
\tag{9}
$$

3.3   Heat loss of the Exciting Coil

Temperature is an important factor to influence the performance of GMA, while the source of heat is mainly resistance and eddy produced by exciting coil. In order to reduce the heat on the premise of meeting magnetic intensity, the power loss of the exciting coil can be achieved by formula 10 based on literature 15. The bigger value of G can produce smaller heat loss based on formula 10.

$$P_l = \frac{\rho_w R_1 H^2}{cG^2} \tag{10}$$

ρ_w is resistivity of exciting coil; c is a coefficient of exciting coil's cross-sectional area; G is a coefficient of exciting coil's shape.

$$G(\alpha,\beta) = \frac{1}{5}\sqrt{\frac{2\pi \cdot \beta}{\alpha^2 - 1}} \ln\left(\frac{\alpha + \sqrt{\alpha^2 + \beta^2}}{1 + \sqrt{1 + \beta^2}}\right) \tag{11}$$

3.4   Optimization and Design of Exciting Coil's Parameters

In order to design the parameters of the exciting coil, this paper puts forward a formula 12 through synthetically analyzing the high uniformity and intensity of the magnetic field near GMM rod, minimum heat loss and compact-sized. The P is to evaluate the advantages of the exciting coil and its bigger value shows better structural parameters.

$$P = \frac{D^2 G H_0}{V} \tag{12}$$

V is the volume of exciting coil.

$$V = \pi(R_2^2 - R_1^2)L \tag{13}$$

In order to easily design the parameters of the exciting coil, this paper introduces the shape factors α and β, while the value α and β can be achieved by $\alpha = R_2 / R_1$, $\beta = L / R_1$, $\gamma = l / R_1$. The formula 6, 9, 12, 13 can be changed to 14, 15, 16, 17.

$$H_0(\alpha,\beta) = n_1 n_2 IR_1 \beta \ln\frac{\alpha + \sqrt{\beta^2 + \alpha^2}}{1 + \sqrt{\beta^2 + 1}} \tag{14}$$

$$D(\alpha,\beta) = \frac{n_1 n_2 IR_1^2}{4}\left\{\begin{array}{l}(\beta+\gamma)^2 \ln\frac{\alpha+\sqrt{(\beta+\gamma)^2+\alpha^2}}{1+\sqrt{(\beta+\gamma)^2+1}} - (\beta-\gamma)^2 \ln\frac{\alpha+\sqrt{(\beta-\gamma)^2+\alpha^2}}{1+\sqrt{(\beta-\gamma)^2+1}} \\ +\alpha\left[\sqrt{(\beta-\gamma)^2+\alpha^2} - \sqrt{(\beta+\gamma)^2+\alpha^2}\right] - \left[\sqrt{(\beta-\gamma)^2+1} - \sqrt{(\beta+\gamma)^2+1}\right]\end{array}\right\} \tag{15}$$

$$V = \pi R_1^3 \beta(\alpha^2 - 1) \tag{16}$$

$$P(\alpha,\beta) = \frac{D(\alpha,\beta)^2 H_0(\alpha,\beta)G(\alpha,\beta)}{V(\alpha,\beta)} \tag{17}$$

The three-dimensional and contour line diagrams are shown in Fig. 6 and Fig. 7 which show that α is the monotonic functions of P. When α is increased, P becomes bigger. When $\beta=3.5$, D becomes the maximum. Therefore, $\beta=3.5$ is preliminarily confirmed.

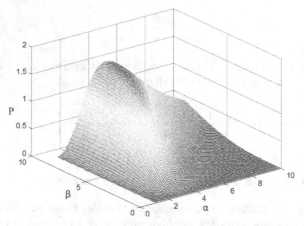

Fig. 6. Three-dimensional diagram of $P(\alpha, \beta)$

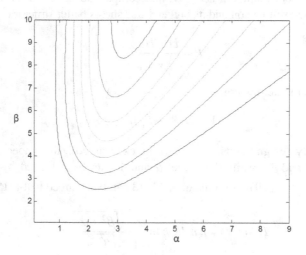

Fig. 7. Contour line diagram of $P(\alpha, \beta)$

The magnetic intensity is decided by turns and current of the exciting coil according to Ampere's circulation theorem. The current is limited by power supply and its square is proportional to power loss. Therefore, the current should not be too large. If the current is too small, turns becomes too much, which will pump up the volume and increase the inductance of exciting coil to reduce dynamic response of GMA. The current is preliminarily selected 10A.

$$NI = Hl_{coil} \tag{18}$$

N is turns of exciting coil; l_{coil} is the length of exciting coil.

The diameter of the wire can be selected by formula 19, when cross sectional area is circle [15].

$$d_w \geq 1.13 \sqrt{\frac{I_{max}}{J}} \qquad (19)$$

d_w is the diameter of bare wire; I_{max} is the maximum of current ; J is current density and selected2~4A/mm^2 in the long-term work[16]. So, J is 4A/mm^2 an $d_w \geq 1.79$mm. d_w is 1.81mm according to domestic specification of copper wire and external diameter of copper wire is 1.9mm.

The preliminary length of exciting coil: $2L=2\beta R_1 = 2 \times 3.5 \times 20 = 140$mm ;

$$n_1 = \frac{1}{k_\eta d} \qquad (20)$$

k_η is row coefficient of exciting coil; d is diameter of copper wire.
Turns of single row :

$$N_1 = 2n_1 L = = \frac{2L}{k_\eta d} = \frac{140}{1.05 \times 1.9} = 70.2 \qquad (21)$$

Considering the manufacturability of twisting: $N_1 = 71$.
The final length of exciting coil:

$$2L = \frac{N_1}{n_1} = N_1 k_\eta d = 71 \times 1.05 \times 1.9 = 141.6mm \qquad (22)$$

The excellent linearity scope of GMM rod is 200~1000Oe. The maximum magnetic intensity produced by exciting coil is 400Oe that is half of working scope considering frequency doubling. The formula 6 is changed to 23.

$$R_2 = \frac{\left[(R_1 + \sqrt{L^2 + R_1^2}) \, e^{\frac{H}{n_1 n_2 IL}} \right]^2 - L^2}{2(R_1 + \sqrt{L^2 + R_1^2}) \, e^{\frac{H}{n_1 n_2 IL}}} \qquad (23)$$

$$n_2 = \frac{1}{k_\beta d} \qquad (24)$$

k_β is overlay coefficient of exciting coil; d is diameter of copper wire.
$R_2 = 41.7$mm according to formula 22, 23, 24.
Turns of overlapping rows :

$$N_2 = \frac{R_2 - R_1}{k_\eta d} = \frac{41.7 - 20}{1.15 \times 1.9} = 9.9 \qquad (25)$$

Considering the manufacturability of twisting: $N_2 = 10$.

The external radius of exciting coil:

$$R_2 = k_\eta dN_2 + R_1 = 42\text{mm} \tag{26}$$

Turns of exciting coil:

$$N = N_1 N_2 = 71 \times 10 = 710 \tag{27}$$

The structural parameters of exciting coil are shown in Tab.1.

Table 1. Structural parameters of exciting coil

External radius R_2	Initial radius R_1	Lengthen 2L	Turns of overlapping rows N_2
42mm	20mm	141.6mm	10
Diameter of bare wire d_w	**Diameter of wire** d	**Turns** N	**Turns of single row** N_1
1.81mm	1.9mm	710	71

4   Magnetic Circuit Based on Ansoft

GMA can be simplified to 2D plane mode on the basis of its 3D axisymmetric structure. This paper analyzes the magnetic circuit of GMA by finite element analysis software Ansoft Maxwell 14.0. In order to evaluate the magnetic fields uniformity, this paper proposes the uniformity rate ε that is achieved by formula 28. The ε is closer to 1, the magnetic field is more uniform.

$$\varepsilon = (1 - \frac{H_{max} - H_{min}}{H_{max}}) \times 100\% \tag{28}$$

The finite mode is established and meshed by Ansoft Maxwell14.0.The exciting current is 5A that is half of maximum and boundary condition is balloon. The magneticfield distribution and magnetic intensity in the axis are respectively shown in Fig. 8 and Fig. 9 which indicate that the minimum and maximum magnetic intensity are respectively 30.63KA/m and 30.83KA/m. the uniformity rate is 99.35%.

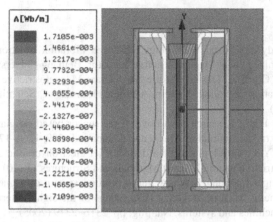

Fig. 8. Magnetic-field distribution

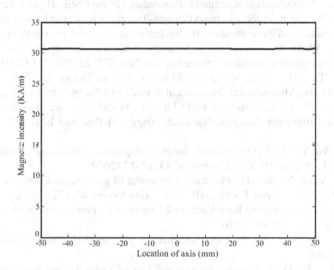

Fig. 9. Magnetic intensity in the axis

5   Conclusion

This paper optimizes and designs the structural parameters of exciting coil based on high uniformity, strong intensity, small heat loss and compact-sized. The Conclusions can be summarized as follows:

1. The end effect is produced by exciting coil and magnetic intensity declines sharply near both ends.
2. The design principles of exciting coil is provided and calculation steps are described in detail.
3. The length of GMM rod is shorter than exciting coil.
4. The properly magnetizer in the magnetic circuit can improve the property of GMA.

References

1. Zhang, X.: Study on Crucial Technology of Active Vibration Isolation System Based on Structure-Optimized Giant Magnetostrictive Actuator. Beihang University, Beijing (2008)
2. Mei, D., Pu, J., Chen, Z.: Design of Rare Earth Giant Magnetostrictive Actuator for Ultra-precision Vibration Isolation System. Chinese Journal of Scientific Instrument **25**, 766–769 (2004)
3. Jia, Z.-Y., Liu, H.-F.: Research on a novel force sensor based on giant magnetostrictive material and its model. Journal of Alloys and Compounds **509**, 1760–1767 (2011)
4. Zhu, Y.C., Li, Y.S.: Development of a deflector-jet electrohydraulic servo-valve using a giant magnetostrictive material. Smart Materials and Structures **11** (2014)
5. Cao S.Y., Zheng J.J., Wang B.W.: A micro-position system of a giant magnetostrictive actuator. In: Cheng, M., Lin, H. (eds.) Proceedings of the eighth international conference on electrical machines and systems, Nanjing, pp. 2316–2319 (2005)
6. Li, L., Yan, B., Zhang, C.: Influence of Frequency on Characteristic of Loss and Temperature in Giant Magnetostrictive Actuator. Proceedings of the CSEE **31**, 124–129 (2011)
7. Zhang, C.: Research on the Electric-Magnetic–Thermal Characteristics of Giant Magnetostrictive Actuator and its Application. Harbin Institute of Technology, HaeBin (2013)
8. Yan, R.-g., Wang, B.-w., Cao, S.-y.: Magneto-Mechanical Strong Coupled Model for a Giant Magnetostrictive Actuator. Proceedings of the CSEE **23**, 107–111 (2003)
9. Yang, B., Tao, H., Bonis, M., et al.: Magnetic Circuit Design for Terfenol-D Driven Magnetostrictive Mini-actuator. Mechanical Science and Technology **24**, 24 (2005)
10. Xue, G., He, Z., Li, D.: Magnetic Field Intensity Model for Giant Magnetostrictive Rod and Coil Optimization Analysis. Nanotechnology and Precision Engineering **12**, 85–90 (2014)
11. Zhao, Z., Wu, Y., Gu, J.: Optimization for giant magnetostrictive actuator based on genetic algorithm. Journal of Zhejiang University. **43**, 13–17 (2009)
12. Wang, X., Yang, X., Sun, H.: Finite element model of giant magnetostrictive actuator and its magnetic field analysis. China Civil Engineering Journal **45**, 172–176 (2012)
13. Clark, A.: Magnetostrictive Rare Earth-Fe2 Compounds: Ferromagnetic Materials. North-Holland Pub, Amsterdam (1980)
14. Liang, C., Qin, G., Liang, Z.: Electromagnetism, pp. 299–302. People's Education Press, Beijing (1980)
15. Engdahl, G.: Design Procedures for optimal Use of Giant Magnetostrictive Material in Magnetostrictive Actuator Applications. In: The 8th International Conference on New Actuators, pp. 554–557. Bremen, Germany (2002)
16. Wang, J.: Standard Handbook for Electrical Engineers. China Machine Press, Beijing (2002)

Mechanism and Impedance Control of Pneumatic Servo Driving System for Resistance Spot Welding

Bo Wang[✉], Tao Wang, Wei Fan, and Yu Wang

School of Automation, Beijing Institute of Technology, Beijing 100081, China
{wangbo231,wangtaobit,fanwei,wangyu}@bit.edu.cn

Abstract. In this paper, a mechanism of pneumatic servo driving system for resistance spot welding is designed. Impedance control is adopted to control the position and output force of pneumatic servo system simultaneously. It can achieve contact stability in the transition process from free motion to contact. In order to increase the performance of force tracking, a PID force controller is also used after steady contact. The validity of proposed method of impedance control is verified through computer simulation results.

Keywords: Mechanism · Pneumatic servo driving system · Impedance control · Contact · Force tracking

1 Introduction

Resistance spot welding is a main form in automobile body assembly process. It completes about more than 90% of assembly body work [1]. Pneumatic driving mechanism was widely used in traditional spot welding applications. Electrode force was controlled by cylinder pressure. Because pressure control was an open loop control, electrode force could not be adjusted with working conditions. Also large impact between the electrode and workpieces affected environment of workshop and service life of the electrode. Servo driving mechanism for spot welding is a new type drive mechanism developed in recent years [2-3]. It achieves high precision positioning of welding gun and flexible welding force control. As upgrading product of pneumatic driving mechanism, pneumatic servo driving mechanism has advantages of simple structure, low cost, fast and convenient maintenance comparing with electric servo gun [4-5]. At present main pneumatic elements manufactures in the world begin to develop related products.

Previous pneumatic servo control research mainly focus on position servo control. Research on pneumatic force servo control is relatively less [6]. As for pneumatic servo driving mechanism, piston position and output force are all needed to control at the same time. Impedance control is a one of the most important methods of controlling the interaction between a manipulator and an environment [7-8]. Some scholars have used it in the study of pneumatic force servo systems [9-10].

In this paper, a pneumatic servo driving mechanism designed for resistance spot welding is presented. Impedance control is adopted to control the force between the electrode and workpieces. Large impact can be greatly reduced by using this method. A PID force controller is also used after steady contact to improve force tracking

© Springer International Publishing Switzerland 2015
H. Liu et al. (Eds.): ICIRA 2015, Part I, LNAI 9244, pp. 201–209, 2015.
DOI: 10.1007/978-3-319-22879-2_19

accuracy. This paper is divided into five sections. In section 1, brief information about the subject is presented. In section 2, designed pneumatic servo driving mechanism is described in detail. Also mathematical model of the system is given. In section 3, proposed method based on impedance control is introduced. In section 4, computer simulation results are presented with comparison to pure position control. In section 5, conclusions are given.

2   Mechanism and Modeling of Pneumatic Servo Driving System

2.1   Mechanism of Pneumatic Servo Driving System

Three dimensional structure diagram and photograph of designed pneumatic servo driving mechanism for spot welding are as shown in Fig. 1 and Fig. 2. Pneumatic servo driving system consists of three main parts: mechanical structure, pneumatic circuit and electrical circuit. Pneumatic circuit includes a booster cylinder and four solenoid valves. Electrical circuit includes a force sensor, a position sensor and a controller.

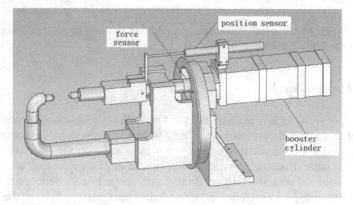

Fig. 1. Three dimensional structure diagram of pneumatic servo driving mechanism

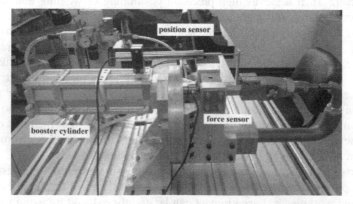

Fig. 2. Photograph of pneumatic servo driving mechanism

2.2   Modeling of Pneumatic Servo Driving System

Control structure of pneumatic servo driving system is as show in Fig. 3. The system consists of a booster cylinder, four solenoid valves, a position sensor, a force sensor, a computer and a data acquisition card. By controlling the states of solenoid valves, pressures in chambers of a booster cylinder can be adjusted. Therefore piston position and output force change.

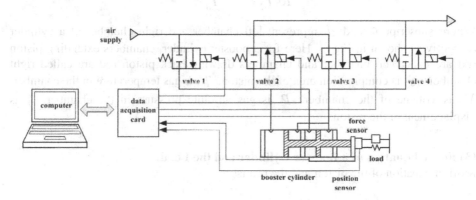

Fig. 3. Control structure of pneumatic servo driving system

(1) Mass Flow Equation of a Solenoid Valve.

Mass flow rate of a solenoid valve G is approximately equivalent to that through an orifice. Rough equation is

$$G = A_e p_1 \frac{\phi}{\sqrt{T_1}} \tag{1}$$

$$\phi = \begin{cases} \sqrt{\dfrac{k}{R}(\dfrac{2}{k+1})^{\frac{k+1}{k-1}}} & \dfrac{p_2}{p_1} \le \left(\dfrac{2}{k+1}\right)^{\frac{k}{k-1}} \\ \sqrt{\dfrac{2k}{k-1}\cdot\dfrac{1}{R}\left[\left(\dfrac{p_2}{p_1}\right)^{\frac{2}{k}} - \left(\dfrac{p_2}{p_1}\right)^{\frac{k+1}{k}}\right]} & \dfrac{p_2}{p_1} > \left(\dfrac{2}{k+1}\right)^{\frac{k}{k-1}} \end{cases}$$

Where A_e is effective area of the valve, p_1 is upstream pressure, p_2 is downstream pressure, T_1 is upstream temperature, k is adiabatic index, R is gas constant.

(2) Continuous Equation of a Booster Cylinder.
On the basis of state equation and energy equation

$$G_c = \frac{V_c}{RkT_c}\dot{p}_c + \frac{p_c S_c}{T_c R}\dot{x} \tag{2}$$

$$G_d = \frac{V_d}{RkT_d}\dot{p}_d - \frac{p_d S_d}{T_d R}\dot{x} \tag{3}$$

Where subscript c and d represent left chamber and right chamber of a cylinder respectively shown in Fig. 3. Here for a booster cylinder, chambers extending piston rod are called left chamber and chambers drawing back piston rod are called right chamber. S is compression area of the piston, T is gas temperature in the chamber, V is volume of the chamber, p is gas absolute pressure in the chamber, x is displacement of the piston.

(3) Force Equation of a Booster Cylinder and the Load.
Motion equation of the piston and the rod is

$$M\ddot{x} = S_c p_c - S_d p_d - p_a(S_c - S_d) - F_f - F_L \tag{4}$$

Where M is mass of the piston, rod and load, p_a is atmosphere pressure, F_f is the friction, F_L is the external load.

(4) Friction Model of a Booster Cylinder.
Friction of a booster cylinder greatly increases comparing with an ordinary cylinder. Considering common Stribeck friction model is discontinuous at zero velocity [11], following continuously differentiable model is used for a booster cylinder.

$$F_f = ((F_c + (F_s - F_c)*e^{-|v/v_s|})*\Lambda + f_v|v|)*\text{sgn}(v) \tag{5}$$

Where F_s is maximum static friction, F_c is Coulomb friction, f_v is viscous friction coefficient, v is velocity of the piston, v_s is Stribeck velocity. Here $\Lambda = 2/(1+e^{-\gamma|v|})-1$, $\gamma > 0$ [12]. γ is set to be 1000 according to simulation results.

Friction of a booster cylinder is tested by using a meter-out circuit. Based on tested data, friction model of a boosting cylinder can be gotten by using curve fitting method.

$$F_f = \left((48.72 + (207.346 - 48.72)*e^{\frac{-|v|}{0.027}})*(\frac{2}{1+e^{-1000*|v|}}-1) + 360*v\right)*\text{sgn}(v) \tag{6}$$

3   Impedance Control of Pneumatic Servo System

3.1   Impedance Control

Impedance control is a method to regulate mechanical impedance parameters of the manipulator in a desired value according to a given task. Impedance function is realized by the relationship between force and position/velocity error [13]. Desired dynamic impedance behavior relating the motion to external force due to contact with the environment can be expressed as equation (7).

$$m_d(\ddot{x}-\ddot{x}_r)+b_d(\dot{x}-\dot{x}_r)+k_d(x-x_r)=-(F-F_r) \tag{7}$$

Where m_d, b_d, k_d are target inertia, target viscosity and target stiffness. x is position of the manipulator. x_r is reference position. F is external force of the environment acting on the manipulator. F_r is reference force.

3.2   Impedance Control of Pneumatic Servo System

Impedance control for pneumatic servo system is designed as shown in Fig. 4. Here x_n is position adjustment, x_d is input position command. A PID force controller is also used when pneumatic servo system steady contacts with rigid workpieces to improve steady force tracking accuracy.

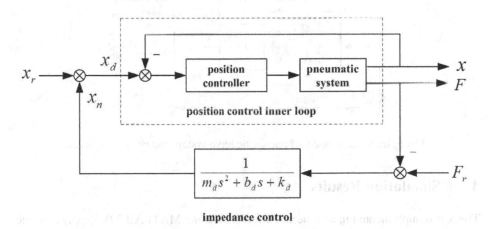

impedance control

Fig. 4. Block diagram of impedance control

In order to meet the needs of contact stability in transition process from free motion to contact, m_d, b_d, k_d are selected as follows according to [14].

$$\xi_d = \frac{b_d}{2\sqrt{k_d m_d}} \tag{8}$$

$$k = \frac{k_e}{k_d} \gg 1 \tag{9}$$

$$\xi_d \geq 0.5\left(\sqrt{1+2k} - 1\right) \tag{10}$$

Where k_e is equivalent stiffness of rigid workpieces.

3.3   Interaction Model of Pneumatic Servo System and Rigid Workpieces

Interaction model of pneumatic servo system and rigid workpieces is as shown in Fig. 5. Here x_e is position of workpieces. Rigid workpieces and force sensor are regarded as a linear spring and a thin plate with mass ignored. Interaction force depends on the size of deformation. Expression of it is as shown in equation (11).

$$F = \begin{cases} k_e(x - x_e) & x > x_e \\ 0 & x \leq x_e \end{cases} \tag{11}$$

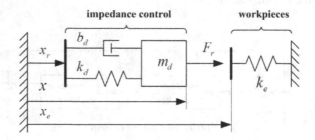

Fig. 5. Interaction model of pneumatic servo system and rigid workpieces

4   Simulation Results

The test is implemented in a numerical simulation using MATLAB 7.0. Supply absolute pressure p_s is 0.6 MPa. Operating frequency of solenoid valves is 50 Hz. The weight of piston, piston rod and load mass is 4 kg. The distance between pneumatic servo system and rigid workpieces is 0.035 m. The stiffness of workpieces is 300000 N/m. The parameters of a booster cylinder are set according to 1321.80.25.80.60.HBSP made by Pneumax corporation (Italy). The parameters of solenoid valves are set according to MHE4-MS1H-3/2G made by FESTO corporation (Germany).

Two sets of simulations are carried out to validate proposed control. In order to illustrate the performance of proposed control, it is compared with pure position control. In the first set of simulation, reference position is set to be 0.045 m and reference output force is set to be 2000 N. In the second set of simulation, reference position is set to be 0.045 m and reference output force is set to be a sinusoidal signal. The amplitude of sinusoidal signal is 500 N. The phase of it is 270 degrees. Frequency of it is 0.5 Hz. The offset of it is 1000 N.

Fig. 6 and Fig. 7 show the position and output force tracking results for step references of pneumatic servo system. According to the results shown in Fig. 6, pure position control has higher position tracking accuracy comparing with impedance control. Steady position tracking error of pure position control is 0.0005 m, which is about 1.1 % of reference position. Steady position tracking error of impedance control is 0.0035 m, which is about 7.7 % of reference position. According to the results shown in Fig. 7, pure position control can not control output force effectively. Force tracking error is 960 N, which is 48 % of reference force. Yet steady output force tracking error of impedance control is 100 N, which is about 5 % of reference output force. Impedance control realizes coordinated control of position and force at expense of position precision.

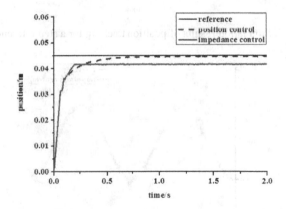

Fig. 6. Simulation results of position tracking for a step reference

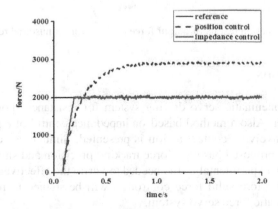

Fig. 7. Simulation results of output force tracking for a step reference

Fig. 8 shows position tracking results for a step reference of pneumatic servo system. Fig. 9 shows output force tracking results for a sinusoidal reference of pneumatic servo system. According to the results shown in Fig. 9, steady output force tracking error is 90 N. Impedance control has high output force tracking accuracy for a sinusoidal reference force. Yet it has large deviation for a step reference position.

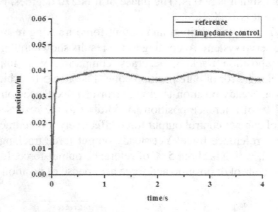

Fig. 8. Simulation results of position tracking for a step reference

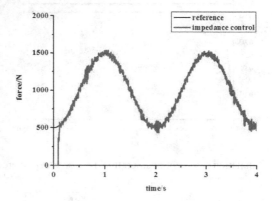

Fig. 9. Simulation results of output force tracking for a sinusoidal reference

5   Conclusions

A mechanism of pneumatic servo driving system for resistance spot welding is designed in this paper. Also a method based on impedance control of a pneumatic servo system for tasks involving contact action is presented. Simulation results show that impedance control proposed has high force tracking precision and small overshoot. In our future research, experimental test is needed to verify the effectiveness of proposed impedance control. More valid force controllers will be studied to improve the performance of pneumatic force servo system.

Acknowledgements. The authors extended their thanks to the National Natural Science Foundation of China and the State Key Laboratory of Fluid Power and Mechatronic Systems of China. The work is supported by the National Natural Science Foundation of China under Grant No. 51375045. This work is also supported by the State Key Laboratory Program of Fluid Power and Mechatronic Systems of China under Grant No. GZKF-201214.

References

1. Zhang, X., Chen, G., Zhang, Y., Lai, X.: Improvement of resistance spot weldability for dual-phase (DP600) steels using servo gun. Journal of Materials Processing Technology **209**, 2671–2675 (2009)
2. Zhang, X., Chen, G., Zhang, Y.: On-line evaluation of electrode wear by servo gun in resistance spot welding. International Journal of Advanced Manufacturing Technology **36**(7), 681–688 (2008)
3. Niu, B., Chi, Y., Zhang, H.: Electrode clamping force regulation of servo gun mounted on resistance spot welding robot. In: Proc. IEEE/ASME International Conference on Advanced Intelligent Mechatronics, pp. 576–582 (2008)
4. Nouri, B.M.Y., Al-Bender, F., Swevers, J., Vanherck, P., Van Brussel, H.: Modelling a pneumatic servo positioning system with friction. In: Proc. American Control Conference, pp. 1067–1071 (2000)
5. Scavarda, S., Thomasset, D.: Modeling and control of electropneumatic systems: an overview of recent French contributions. In: Proc. International Conference on Control, pp. 1462–1467 (1996)
6. Richer, E., Hurmuzlu, Y.: A high performance pneumatic force actuator system: Part I- nonlinear mathematical model, Journal of Dynamic Systems Measurement, and Control **122**(3), 416–425 (2000)
7. Hogan, N.: Impedance control: An approach to manipulation: Part I-III. ASME Journal of Dynamic System, Measurement, and Control **107**(1), 1–24 (1985)
8. Tsuji, T., Terauchi, M., Tanaka, Y.: Online learning of virtual impedance parameters in non-contact impedance control using neural networks. IEEE Transactions on Systems, Man, and Cybernetics-Part B: Cybernetics **34**(5), 2112–2118 (2004)
9. Tadano, K., Kawashima, K.: Development of a master slave system with force sensing using pneumatic servo system for laparoscopic surgery. In: Proc. IEEE International Conference on Robotics and Automation, pp. 947–952 (2007)
10. Richardson, R., Brown, M., Bhakta, B., Levesley, M.: Impedance control for a pneumatic robot-based around pole-placement, joint space controllers. Control Engineering Practice **13**, 291–303 (2005)
11. Meng, D., Tao, G., Ban, W., Qian, P.: Adaptive robust output force tracking control of pneumatic cylinder while maximizing/minimizing its stiffness. Journal of Central South University **20**(6), 1510–1518 (2013)
12. Zhao, W., Ren, X.: The fast terminal sliding mode control of dual-motor driving servo systems with friction. Journal of Harbin Institute of Technology **46**(3), 119–123 (2014). (in Chinese)
13. Jung, S., Hsia, T.C., Bonitz, R.G.: Force tracking impedance control of robot manipulators under unknown environment. IEEE Transactions on Control Systems Technology **12**(3), 474–483 (2004)
14. Surdilovic, D.: Contact stability issue in position based impedance control: theory and experiments. In: Proc. International Conference on Robotics and Automation, pp. 1675–1680 (1996)

Design of an Equivalent Simulation Platform for PMLSM-Driven Positioning System

Chao Liu, Hui Wang, Jianhua Wu, and Zhenhua Xiong[✉]

State Key Laboratory of Mechanical System and Vibration, School of Mechanical Engineering, Shanghai Jiao Tong University, Shanghai 200240, China
{aalon,351582221,wujh,mexiong}@sjtu.edu.cn

Abstract. This paper focuses on setting up an equivalent simulation platform for high-precision positioning system driven by permanent magnet linear synchronous motor (PMLSM). The transfer function from input voltage to output velocity is firstly developed by fully taking into account system characteristics, which contains not only a first-order low-frequency model but also a series of high-frequency resonance models. Subsequently, a reduced-order model is obtained according to the actual machine features and the relevant model parameters are identified by fitting the amplitude-frequency characteristics based on the low-frequency characteristics and former three natural frequencies. After introducing the external disturbances that cannot be definitely modeled, the accurate equivalent simulation platform of PMLSM-driven positioning system in Matlab/Simulink environment is established. Simulation and experiment studies validate the effectiveness of the proposed simulation platform due to the comparative analysis of tracking performance, overshoot and positioning time.

Keywords: Simulation platform · Dynamic modeling · Mechanical resonance modes · PMLSM-driven positioning system

1 Introduction

For a capable PMLSM-driven positioning table with satisfactory static and dynamic characteristics, a high-performance control algorithm is indispensable for achieving satisfactory performances, such as positioning time, positioning accuracy and disturbance suppression [1]. In the real application, experiments are directly carried out on the practical machine to verify the effectiveness of various control algorithms. However, since the control parameters are commonly tuned by trial and error method, it could be time-consuming for fulfilling the demand of satisfactory performances [2,3]. Therefore, the accurate equivalent simulation platform should be explored for shortening the development period of control algorithms design.

In order to implement the control algorithms conveniently, a simple dynamic model is usually established for PMLSM-driven positioning table based on rigid body assumption [4,5]. However, the equivalent simulation platform based on

© Springer International Publishing Switzerland 2015
H. Liu et al. (Eds.): ICIRA 2015, Part I, LNAI 9244, pp. 210–220, 2015.
DOI: 10.1007/978-3-319-22879-2_20

the simplified model cannot effectively reflect the actual system characteristics, since the PMLSM-driven positioning system is a typical flexible multibody system composed of rigid and flexible bodies. Therefore, some researches have paid attention to the influence of flexible modes. A modeling approach of multi-mode vibration coupled motion within wide frequency range was proposed for the high-performance motion system by analyzing the dynamic features with high velocity, high acceleration and high accuracy [6]. Chen et al. [7] proposed a novel physical model for the linear motor driven stage by taking into account nonlinearities and high-frequency structural flexible modes caused by the flexibility of ball bearings between the stage and linear guideways. Liu et al. [8] proposed a reduced-order modeling based on a number of frequency response test data, which considers time delay, first-order low-frequency model and three significant resonance models. However, equivalent simulation platforms based on the above-mentioned modeling methods are not suitable for the PMLSM-driven positioning system in this paper, which is verified by the comparative analysis of simulations and experiments.

Therefore, with the aim of obtaining an accurate equivalent simulation platform, the transfer function from input voltage to output velocity is firstly given by analyzing the system characteristics, which contains a first-order low-frequency model and a series of high-frequency resonance models. Secondly, a reduced-order model is obtained by taking into account system features and its parameters are identified by sine-sweep experiments. And then a simulation platform in Matlab/Simulink environment is established after introducing the external disturbances that cannot be definitely modeled, such as inertia perturbation, force ripple, nonlinear characteristics, uncertainties, high-frequency noises and quantization error. Simulation and experiment studies verify the effectiveness of the proposed simulation platform by virtue of the comparative analysis of tracking performance, overshoot and positioning time.

The remainder of this paper is organized as follows: in Section 2, the PMLSM-driven positioning system is firstly introduced and subsequently system model is discussed in detail. In Section 3, the model parameters are identified firstly and then a simulation platform is established. In Section 4, simulations and experiments are implemented to validate the effectiveness of the proposed simulation platform. Section 5 gives the conclusion.

2 Modeling of PMLSM-Driven Positioning System

2.1 PMLSM-Driven Positioning System

The layout view of servo mechanical system is shown in Fig. 1, which consists of a PC-based dSPACE controller, a PMLSM-driven table, two APS30 amplifiers, a set of cooling devices and so on. The PC-based dSPACE controller is used for computational intensive tasks associated with execution of relevant control algorithms. The PMLSM-driven table is a high-acceleration XY positioning platform for high-end wire bonder. Each axis is equipped with a non-contact linear optical encoder as the feedback device. The hardware resolution of each non-contact linear optical encoder is 0.4 μm (0.1 μm after quadruplication). In this paper, Y-axis is chosen as the study object, whose former three natural frequencies are tested as 251 Hz, 507 Hz and 719 Hz.

Cooling devices APS30 amplifiers PMLSMs-driven table

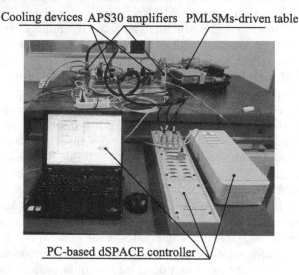

PC-based dSPACE controller

Fig. 1. PMLSM-driven positioning system

2.2 Modeling of PMLSM-Driven System

Fully taking into account of the above-mentioned PMLSM-driven system characteristics, the transfer function of Y-axis from input voltage u to output velocity v can be expressed as

$$G_v(s) = \frac{V(s)}{U(s)} = G_{pl}(s) + \sum_{i=1}^{n} h_i G_{pr_i}(s) \qquad (1)$$

where $V(s)$ is the Laplace transformation of v, $U(s)$ is the Laplace transformation of u, $G_{pl}(s)$ is the first-order low-frequency model, h_i is the weighting coefficient and $G_{pr_i}(s)$ is the ith resonance model.

Usually, the PMLSM-driven system can be assumed to be a rigid-body system during the low-frequency modeling process. So the dynamic model of Y-axis is written as

$$m\ddot{x} = k_t u - F_d, \qquad (2)$$

where m is the mass of the slider, $\ddot{x}$ is the acceleration, k_t denotes the thrust coefficient and F_d is the external disturbances.

Due to negligible Coulomb friction of the above-mentioned PMLSM-driven positioning system, viscous friction is only extracted from F_d. And then, the established model becomes

$$m\ddot{x} + B\dot{x} = k_t u - F_{dd}, \qquad (3)$$

where B is the viscous coefficient, $\dot{x}$ is the velocity and F_{dd} is the external disturbances besides Coulomb friction.

Denote $\alpha = m/k_t$, $\beta = B/k_t$ and $f_{dd} = F_{dd}/k_t$, the Eq.(3) can be expressed as

$$\alpha\ddot{x} + \beta\dot{x} = u - f_{dd}. \tag{4}$$

Substitute $\tau = \alpha/\beta$ and $k = 1/\beta$ into the Eq.(4), the first-order low-frequency model is written as

$$G_{pl}(s) = \frac{k}{\tau s + 1}. \tag{5}$$

However, the PMLSM-driven positioning system is actually not a rigid-body system but a typical flexible multibody system composed of rigid and flexible bodies. Therefore, to model the system accurately, the influence of high-frequency flexible modes caused by the system mechanism and mounting table itself should be taken into account. Based on the analysis of Y-axis characteristics, the influence of each high-frequency flexible mode can be equivalent to a mass-spring-damper system. Hence, the ith resonance model $G_{pr_i}(s)$ can be expressed as

$$G_{pr_i}(s) = \frac{\omega_i^2}{s^2 + 2\xi_i\omega_i s + \omega_i^2}, \tag{6}$$

where ξ_i and ω_i are the damping ratio and resonance frequency of $G_{pr_i}(s)$, respectively. In this paper, the mechanical resonance modes are discussed to establish the high-frequency resonance models and the unknown resonant modes are temporarily defined as high-frequency external disturbances.

3   Construction of Equivalent Simulation Platform

3.1   Model Parameters Identification

With the advantages of simplicity and effectiveness, sine-sweep method is often used for identifying the system model parameters in servo and drive areas. The excitation signal of sine-sweep method is a continuous sinusoid with frequency variation, which can be described as $u(t) = a\sin(\omega(t))$, a denotes the amplitude of sinusoid and $\omega(t)$ is the time-varying sweep frequency. In the real application, $\omega(t)$ can be obtained by setting the values of target time, initial frequency and target frequency. After the values of relevant four parameters are given, sin-sweep experiments are implemented on the Y-axis to obtain the output velocity, as is shown in Fig. 2. In the experiments, the amplitude and target time of sinusoidal excitation signal ranging from 20 Hz to 2000 Hz are set as 0.1 V and 10 s, respectively.

The frequency-domain characteristics of Y-axis are described by the dashed line as shown in Fig. 3. Since there are many crests and troughs in the relevant amplitude-frequency and phase-frequency characteristics, it can be concluded that the PMLSM-driven Y-axis is a multi-mode coupled system, which contains a low-frequency mode and a series of high-frequency resonance modes.

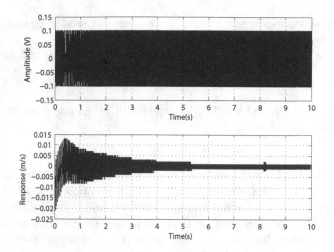

Fig. 2. Sinusoidal excitation signal and velocity response

In accordance with the analysis in Section 2 and Y-axis characteristics, the low-frequency mode can be modeled as Eq.(5) and the high-frequency resonance modes can be divided into mechanical resonance modes and unknown resonant modes. The mechanical resonance modes can be modeled as Eq.(6) and the unknown resonant modes are temporarily defined as high-frequency external disturbances. Next, the least squares method is adopted to identify the model parameters in Eq.(1) by fitting the amplitude-frequency curve on different frequency ranges.

By fitting the low-frequency amplitude-frequency curve, the first-order low-frequency model is obtained as

$$G_{pl}(s) = \frac{1378.5}{0.1032s + 1}. \tag{7}$$

Similarly, three mechanical resonance models and weighting coefficients are determined by fitting the high-frequency amplitude-frequency curve based on the former three natural frequencies. The weighting coefficient h_1 is 3.041 and the first mechanical resonance model $G_{pr_1}(s)$ is

$$G_{pr_1}(s) = \frac{2.517 \times 10^6}{s^2 + 521.7s + 2.517 \times 10^6}. \tag{8}$$

The weighting coefficient h_2 is 0.1956 and the second mechanical resonance model $G_{pr_2}(s)$ is

$$G_{pr_2}(s) = \frac{1.017 \times 10^7}{s^2 + 104.6s + 1.107 \times 10^7}. \tag{9}$$

The weighting coefficient h_3 is 0.0128 and the third mechanical resonance model $G_{pr_3}(s)$ is

$$G_{pr_3}(s) = \frac{2.045 \times 10^7}{s^2 + 6.553s + 2.045 \times 10^7}. \tag{10}$$

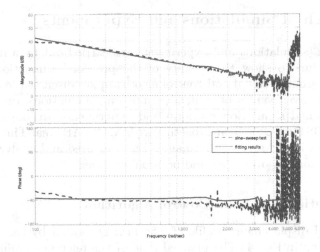

Fig. 3. Comparative analysis of sweep test and its fitting results

Accordingly, the model parameters in Eq.(1) can be determined and the frequency-domain characteristics are described by the solid line as shown in Fig. 3. From the figure, the comparative analysis demonstrates that the proposed modeling method is capable of effectively fitting the amplitude-frequency curve of sine-sweep test.

3.2   Equivalent Simulation Platform

Based on the identified model $G_v(s)$, an equivalent simulation platform is here designed for the PMLSM-driven positioning system in Matlab/Simulink environment. With a view to simulate the PMLSM-driven Y-axis accurately, the external disturbances which is difficult to be definitely modeled are also taken into account.

Firstly, inertia perturbation, force ripple, nonlinear characteristics and uncertainties are denoted as low-frequency disturbances and subsequently introduced into the established simulation platform. In this paper, the thrust force of low-frequency disturbances is described as a sinusoid with amplitude of 0.001 V and frequency of 50 Hz. Secondly, high-frequency noises are introduced into the established simulation platform. The thrust force of high-frequency noises is described as a sinusoid with amplitude of 0.0001 V and frequency of 1000 Hz. Furthermore, the quantization error is also taken into account, which is set as 0.1 μm. Till now, the equivalent simulation platform of PMLSM-driven positioning system are successfully established. And then, the effectiveness of equivalent simulation platform will be explored and discussed in the next section.

4   Research of Simulations and Experiments

In this section, simulations and experiments are carried out by a model-based control algorithm to show the validity of the proposed simulation platform. Firstly, the referential signal and control algorithm are discussed in detail. And then, with the given values of control parameters, simulations are carried out on the established simulation platform and experiments are implemented by the dSPACE DS1103 DSP board together with MATLAB Real-Time Workshop on the Y-axis. Finally, servo performances are compared and analyzed, such as tracking performance, overshoot and positioning time.

4.1   Referential Signal and Control Algorithm

The referential displacement profile of 2.54 mm generated by the asymmetric S-curve is shown in Fig. 4. The planned time of the reference profile is 12 ms. The planned peak velocity is 0.392 m/s while the maximum planned acceleration and deceleration are 9.1 g and 7.2 g, respectively.

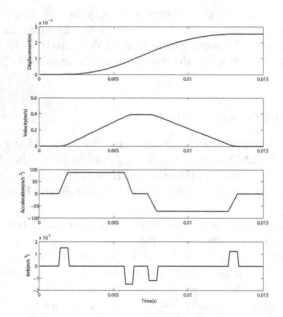

Fig. 4. The referential displacement profile

The model-based control algorithm is shown in Fig. 5, which is composed of a pole-placement proportion-derivative (PD) controller, a model-inverse feed-forward controller and a disturbance observer. Based on the above analysis, the model parameters of $G_v(s)$ can be accurately obtained. However, $G_v(s)$ cannot be directly implemented for the control algorithm by taking fully account of its calculation and implementation within the sampling time of 0.1 ms. Hence, the established model $G_v(s)$ should be reduced. To address the issue, $G_v(s)$ is replaced by the first-order low-frequency model, which is subsequently designed for the model-based control algorithm.

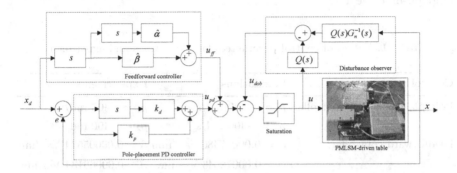

Fig. 5. The model-based control algorithm

The feedforward controller is designed for improving the tracking performance. With the model inverse method, the parameters $\hat{\alpha}$ and $\hat{\beta}$ can be determined by

$$\hat{\alpha} = \hat{\tau}/\hat{k}, \tag{11}$$

$$\beta = 1/\hat{k}. \tag{12}$$

where $\hat{k}$ and $\hat{\tau}$ are the identified values of k and τ in Eq.(5), respectively.

The proportion-derivative controller is implemented by the pole-placement design method for stabilizing the controlled system. The parameters k_p and k_d can be calculated by

$$k_p = p_1 p_2 \hat{\tau}/\hat{k}, \tag{13}$$

$$k_d = -(1 + (p_1 + p_2)\hat{\tau})/\hat{k}, \tag{14}$$

where p_1 and p_2 are the given pole values, respectively.

The disturbance observer is adopted for suppressing the external disturbances. Based on the platform characteristics, the low-pass filter $Q(s)$ is designed in the form that the denominator of $Q(s)$ is of third order and the numerator of $Q(s)$ is of first order, which can be expressed as

$$Q(s) = \frac{3\tau_1 s + 1}{\tau_1{}^3 s^3 + 3\tau_1{}^2 s^2 + 3\tau_1 s + 1}, \tag{15}$$

where τ_1 is the time constant of $Q(s)$.

218   C. Liu et al.

4.2   Servo Performances Comparison

Simulations are carried out on the established simulation platform with the values of control parameters, which is shown in Table 1. Experiments are implemented on the PMLSM-driven Y-axis and the values of control parameters are also given in Table 1.

Table 1. Values of control parameters for simulations and experiments

Control units	Simulations	Experiments
Pole-placement PD controller:	$p_1 = -400$ rad/s	$p_1 = -400$ rad/s
	$p_2 = -400$ rad/s	$p_2 = -400$ rad/s
Feedforward controller:	$\hat{\alpha} = 0.00007486$ Vs^2/mm	$\hat{\alpha} = 0.00005544$ Vs^2/mm
	$\hat{\beta} = 0.00072543$ Vs/mm	$\hat{\beta} = 0.00060134$ Vs/mm
Disturbance Observer:	$\tau_1 = 0.0005$ s	$\tau_1 = 0.0005$ s

The tracking performances are first analyzed under the high-performance control algorithm with different platforms. The tracking performance of equivalent simulation platform is shown in Fig. 6 and the maximum tracking error is 2.8 μm. Fig. 7 illustrates the tracking performance of the PMLSM-driven Y-axis and the maximum tracking error is 2.0 μm.

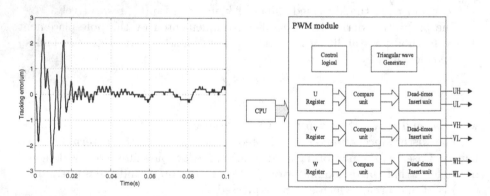

Fig. 6. Tracking errors of equivalent simulation platform

Fig. 7. Tracking errors of the PMLSM-driven Y-axis

The overshoots of equivalent simulation platform and PMLSM-driven Y-axis are secondly analyzed. Fig. 8 illustrates the overshoots of equivalent simulation platform and the maximum overshoot is 1.4 μm. The maximum overshoot of the PMLSM-driven Y-axis is 1.3 μm, as is shown in Fig. 9.

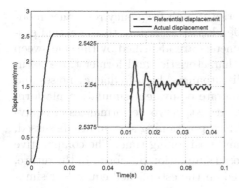

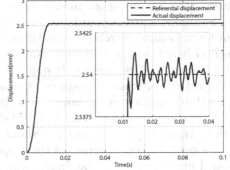

Fig. 8. Overshoot and positioning time of equivalent simulation platform

Fig. 9. Overshoot and positioning time of the PMLSM-driven Y-axis

We also compare the positioning times of equivalent simulation platform and PMLSM-driven Y-axis. Positioning time is defined as the time from the beginning of the displacement profile to the moment that the tracking error fully converges to the positioning error band while the actual displacement is close proximity to the referential displacement. Here, the position error band is set to 2.5 μm by reference to the requirement of the semiconductor manufacturing equipments. The positioning time of equivalent simulation platform is shown in Fig. 8 and the positioning time is 11.6 ms. Fig. 9 illustrates the positioning time of the PMLSM-driven Y-axis and the positioning time is given in Table 2.

Table 2. Servo performances with different platforms

Performances	Simulation platform	PMLSM-driven Y-axis
Maximum tracking errors:	2.8 μm	2.0 μm
Maximum overshoots:	1.4 μm	1.3 μm
Positioning times:	11.6 ms	11.4 ms

From Table 2, it can be seen that servo performances of equivalent simulation platform are very close to those of the PMLSM-driven Y-axis, especially maximum overshoot and positioning time. Therefore, it can be concluded that an effective equivalent simulation platform is successfully established in this paper for exploring the PMLSM-driven positioning system.

5 Conclusion

In this paper, an equivalent simulation platform is successfully built for PMLSM-driven positioning system. Based on the dynamic characteristics of rigid and flexible bodies, the system model is firstly discussed in detail, which is composed of a first-order low-frequency model and a series of high-frequency resonance models. Secondly, a reduced-order model is developed according to the actual system features and the relevant model parameters are identified by the sine-sweep method based on the frequency-domain characteristics and former three natural frequencies. After the low-frequency disturbances, high-frequency noises and quantization error are introduced, the accurate equivalent simulation platform in Matlab/Simulink environment is built. At last, servo performances of equivalent simulation platform and PMLSM-driven positioning system are discussed, such as tracking performance, overshoot and positioning time. The comparative results verify the effectiveness of the proposed simulation platform. In the future, control algorithms could be firstly explored on the established equivalent simulation platform and subsequently carried out on the actual system for shortening the development period.

Acknowledgments. This research was supported in part by National Key Basic Research Program of China under Grant 2013CB035804, Shanghai Economic and Information Technology Commission (No. CXY-2013-21), National Natural Science Foundation of China under Grant 51120155001 and U1201244.

References

1. Ding, H., Xiong, Z.: Motion stages for electronic packaging design and control. IEEE Robotics & Automation Magazine **13**(4), 51–61 (2006)
2. Sato, K., Shimokohbe, A., et al.: Characteristics of practical control for point-to-point (ptp) positioning systems: Effect of design parameters and actuator saturation on positioning performance. Precision Engineering **27**(2), 157–169 (2003)
3. Tursini, M., Parasiliti, F., Zhang, D.: Real-time gain tuning of pi controllers for high-performance pmsm drives. IEEE Transactions on Industry Applications **38**(4), 1018–1026 (2002)
4. Tesfaye, A., Lee, H.S., Tomizuka, M.: A sensitivity optimization approach to design of a disturbance observer in digital motion control systems. IEEE/ASME Transactions on Mechatronics **5**(1), 32–38 (2000)
5. Kim, B.K., Park, S., Chung, W.K., Youm, Y.: Robust controller design for ptp motion of vertical xy positioning systems with a flexible beam. IEEE/ASME Transactions on Mechatronics **8**(1), 99–110 (2003)
6. Liu, Q., Qi, C., Yuan, S., Ou, G.: Modeling and control of hi-performance motion system. In: 7th World Congress on Intelligent Control and Automation, WCICA 2008, pp. 455–460. IEEE (2008)
7. Chen, Z., Yao, B., Wang, Q.: Adaptive robust precision motion control of linear motors with integrated compensation of nonlinearities and bearing flexible modes. IEEE Transactions on Industrial Informatics **9**(2), 965–973 (2013)
8. Liu, Z.Z., Luo, F.L., Azizur Rahman, M.: Robust and precision motion control system of linear-motor direct drive for high-speed xy table positioning mechanism. IEEE Transactions on Industrial Electronics **52**(5), 1357–1363 (2005)

Development and Implementation of Modular FPGA for a Multi-Motor Drive and Control Integrated System

Xuan Zeng, Chao Liu, Xinjun Sheng$^{(\boxtimes)}$, Zhenhua Xiong, and Xiangyang Zhu

State Key Laboratory of Mechanical System and Vibration, School of Mechanical Engineering, Shanghai Jiao Tong University, Shanghai 200240, China
{zengxuan,aalon,xjsheng,mexiong,mexyzhu}@sjtu.edu.cn

Abstract. This paper focuses on the development and implementation of modular FPGA for a multi-motor drive and control integrated system, which is used for enhancing the operating and economical efficiency in industrial applications. In this paper, several IP components are firstly developed for reading the values of motor current and position signals simultaneously due to FPGA parallel processing capabilities. Secondly, field orientation control module is implemented for multi-motor closed-loop control. After that, multi-group PWM waves are generated by the modular FPGA at the same time. Finally, the effectiveness of the proposed FPGA structure is verified by the experiments, which are carried out on the self-designed multi-motor drive and control integrated system.

Keywords: Drive and control integrated system · FPGA · IP components · Motor drives · Nios II · Motion control

1 Introduction

With the rapid development of motor control and micro-electronic technologies, the demand of multi-motor drive and control integrated system has become a major trend. Several motors must be controlled simultaneously in industrial applications such as CNC machines, industrial robots and so on. The standard motion control system is usually set up by a motion controller and multiple drivers. This architecture not only occupies a large hardware space but also consumes a lot of computing resources. To deal with the issue, a feasible solution is explored and developed based on the fixed-point digital signal processor (DSP) and field programmable gate array (FPGA) [1,2]. Zhao et al.[3] presented a motion control board with architecture of DSP and FPGA. The earliest FPGA is mainly used to assist DSP chip, offloading them from the low level control layers to reduce the required execution time [4]. Compared with FPGA, DSP requires a higher sample-rate to control several motors. Therefore, more computing resources of DSP are devoted to repeating algorithms. As a result, the number of motors is limited for DSP.

© Springer International Publishing Switzerland 2015
H. Liu et al. (Eds.): ICIRA 2015, Part I, LNAI 9244, pp. 221–231, 2015.
DOI: 10.1007/978-3-319-22879-2_21

Recently, FPGA has received considerable improvements and become an alternative solution for the realization of digital control system, which is previously dominated by general-purpose microprocessor system [5,6]. With the advantages of programmable hard-wired feature, faster time-to-market, shorter design cycle, embedding processor, lower power consumption and higher density for the implementation of motion control system [10], the embedded system integration is improved greatly [7]. Chan et al. [8] designed a novel distributed-arithmetic-based PID controller algorithm with FPGA. FPGA-based controllers offer advantages such as high speed, complex functionality and low power consumption. Castro et al. [9] developed a new FPGA based platform for controlling a multi-motor electric vehicle.

Embedded processor IP and application IP are now be able to be developed and downloaded into FPGA to construct a system-on-a-programmable-chip (SoPC) environment, allowing users to design a SoPC module by mixing hardware and software integrated in one FPGA chip. With the highly parallel and reliable features of FPGA, several motors are able to be controled simultaneously by a single chip. In this paper, a multi-motor drive and control integrated system realized in FPGA is presented. There are four motors controlled by this system. The complete system is composed by several diving submodules. Four encoder modules are used to obtain motor position simultaneously. The current module, field-oriented-control (FOC) module, PWM module and Nios II processor are integrated in SoPC. The current module acquires current automatically. FOC module and PWM module compute four motors parameter simultaneously.

The organization of this paper is given as follows. In Section 2, the overview of multi-motor drive and control integrated system is presented. In Section 3, the approach for module design by using FPGA technology and software development is discussed in detail. In Section 4, experiments are conducted on the multi-motor drive and control integrated system to test the proposed modules. Conclusions are shown in Section 5.

2   System Description

The block diagram of multi-motor drive and control integrated system is shown in Fig. 1, where the FPGA chip (EP4CE10E22I7) is the main element of multi-motor drive unit. Feedback signals such as position, current and alarm signals are continuously collected and transmitted to FPGA during the control process, and finally sent to PC. The information interface module is integrated in FPGA to acquire position and current data. This information module contains one current submodule and four encoder submodules.

In order to complete the motor-control task, the interface for encoders and ADC chips is then designed in FPGA. The Nios II processor is responsible for motor vector control. The other modules contacting host PC, DSP and etc are used for system control. When the system is doing work, several motors are controlled by FPGA. In this way, multi-motor drive and control integrated system hardware becomes smaller and cheaper than traditional system.

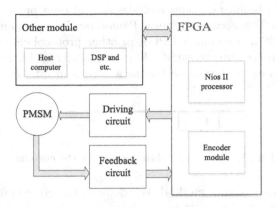

Fig. 1. The block diagram of multi-motor drive and control integrated system

3   Development of FPGA Modules

The structure of SoPC system in FPGA is shown in Fig. 2. During the FPGA working the encoder module automatically measures data of motor. After the current module receives multi-motor current information, the interrupt signal is generated and notify the Nios II processor. Nios II processor arranges modules to realize FOC and generates the corresponding PWM wave. This method enables the FPGA to control multi-motor system and improve the computational efficiency.

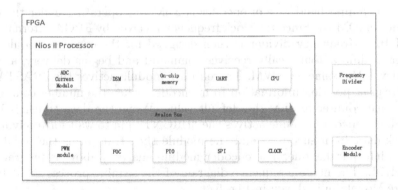

Fig. 2. The structure of SoPC system

3.1   Encoders Modules Development

Compared with incremental encoders, there are much more complicated structure for absolute encoders and the format of output data is still not unified.

Each standard of absolute encoder includes special ones made by Japan manufacturers such as Tamagawa, Yaskawa, Panasonic, Europe manufacturers such as EnDat and BiSS and so on. A FPGA or other protocol chip is then be used for communication decoding to satisfy different standards. Therefore, the FPGA encoder module must be developed for the absolute encoder.

Fig. 3. The request data format for the encoder

The question-answering method is adopted for communications in the encoder of Tamagawa. The encoder works in a receiving mode in default situation. When other components send the request command to it, the encoder sends back the corresponding data like position and alarm information after analyzing the command. The format of request command send to the encoder is shown as Fig. 3 and the format of returned data is shown as Fig. 4. Every field in the returned data contains 10 bits and different codes in the control field (CF) means different requests. The encoder returns corresponding information depending on the codes in CF.

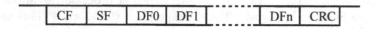

Fig. 4. The return data format from the encoder

The maximum baud rate of encoder is 2.5M bps operating under the clock frequency of 20M Hz. Since the clock frequency received by FPGA external clock is 50M Hz, a frequency divider is then designed for the encoder module. The encoder module automatically receives command and begins delivery, and the frequency is the same as PWM. The encoder module actives the "RASTART" pin to request position information from encoder, when the number counted by the internal counter reaches the default value. While the data transmission is finished, the encoder module actives the "DRCNT" pin to wait for receiving the feedback data from encoders, and the module then transforms the serial data received into parallel one. The encoder module analyzed above is programmed with VHDL. The module sequence diagram of the sending process simulated in software ModelSim is shown in Fig. 5.

3.2 Current Module

When the motor running, values of three phase currents satisfy the formula:

$$i_v + i_u + i_w = 0 \tag{1}$$

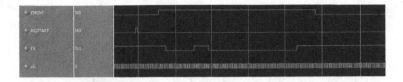

Fig. 5. Sequence diagram of the sending process in the ModelSim

Three-phase current values is calculated after measured the U-phase current and V-phase current. There are 8 channels of current data collected from four permanent magnet synchronous motors (PMSMs), and the U-phase and V-phase currents are collected respectively by two ADC78H89 chips. Traditional ADC interface is realized in processors such as DSP, which consumes much time to collect current values of 8 channels. Therefore, ADC interface is developed as a separate module in FPGA to make full use of its hardware-parallel features and not disturb the operation of Nios II processor. The current module is developed as shown in Fig. 6.

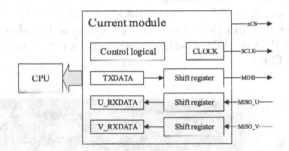

Fig. 6. Current module structure

3.3 PWM Waves Generation Module

After collecting the current and position information, FOC and PWM module complete the calculation and generate the corresponding PWM wave. According to the different switching order, the PWM wave is generated based on the different algorithm. At present, the PWM algorithm is generally classified into five-section modulation and seven-section modulation. The main difference between five-section modulation and seven-section modulation is the partitioned way of zero-vector. Each switch of seven-section modulation need to be started and stopped once during a PWM cycle. The five-section modulation releases one switch transistor during a PWM cycle. However, considering the existence of harmonic distortion, seven-section modulation usually obtains better performance. Therefore, the seven-section modulation is used in this system.

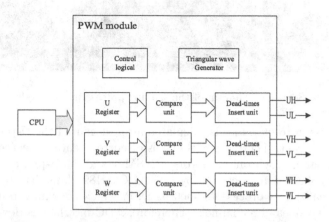

Fig. 7. Structure of PWM module

Dead time must be set to avoid the bridge arm short-circuit fault of the thyristor rectifiers during the running of three-phase inverter-bridge. This fault can cause over expenditure of switch transistor currents, and the whole electrical system may broke down in the worst case. According to the above requirement, the structure of PWM module is designed as shown in Fig. 7, and the simulation result is show in Fig. 8.

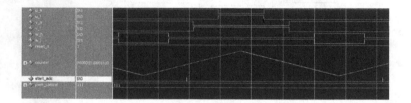

Fig. 8. Simulation of PWM generator

3.4 Nios II Processor Software Development

The Nios II processor is then utilized after the construction of current module, FOC module and PWM module is built. Nios II soft-core processor is a flexible and efficient embedded processor with customization peripherals and instruction. Due to the high real-time requirements of some hardware modules, the Nios II soft-core processor integrates these functions, which reduces the computing time and improves the performance. In order to realize multi-PMSM control, the code in Nios II processor must contains following functions:

1) Communicate with other processors;
2) Configure parameters in PMSM control;
3) Read the data of current and position in PMSM;
4) Complete the velocity-loop and torque-loop control.

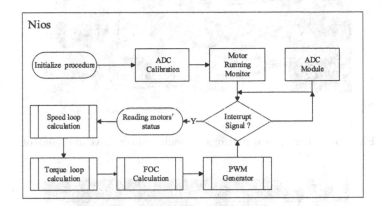

Fig. 9. The flowchart of multi-PMSM control in Nios II processor

Fig. 9 shows the flowchart of multi-PMSM control in Nios II processor. Nios core begins to read parameters for PMSM control from dual-port ram after FPGA was electrified. Then the hardware automatically detects the information such as the current data value. After that, the system initializes the motor state monitor, and waits for the interrupt signal which is generated by the current module. Once the system receives the interrupt signal, the PI controller parameters of speed and current controller are updated and the new current value is calculated out. Each PWM module generates the corresponding PWM wave to control the PMSM depend on the new value. Meanwhile, the Nios II processor waits for the next interrupt signal.

4   Experimental Studies

The overall experimental system is depicted in Fig. 10, and it includes four PMSM modules, one AC-DC Rectification module, one 24 V DC power and a multi-motor drive and control system. The power voltage of PMSM is 2000 W and the rating speed of PMSM is 3000 rpm.

To verify the performance of encoder module in Fig. 5, an experiment is conducted in corresponding circuit, the logic analyzer sampling the data of "DI" signal, "R/W" signal "DRCNT" and "RO" signal. The results are shown in Fig. 11, the behavior of "DRCNT" and "DI" signal is the same as that simulated in Modelsim, and the "RO" data is obtained from the encoder correctly. Motor running tests under unloaded and zero-speed conditions is shown in Fig. 12. The position value obtained form the encoder is stable at 130921.

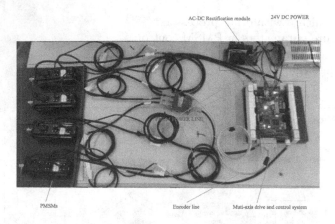

Fig. 10. Experimental platform of multi-motor drive and control

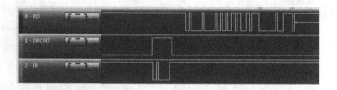

Fig. 11. Encoder reading signal wave

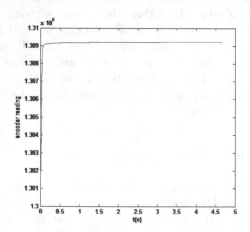

Fig. 12. Absolute encoder reading

Meanwhile, the logic analyzer samples the data of "MOSI", "SCLK", "nCS", "MISOU" and "MISOV" pins in current module. "MISOU" and "MISOV" are the U-phase value and V-phase value. The result is shown in Fig. 13, and the current value is thus able to be read by the module integrated in FPGA.

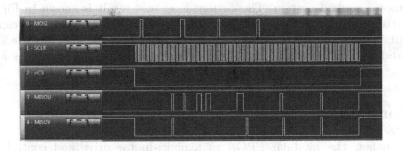

Fig. 13. Current reading signal wave

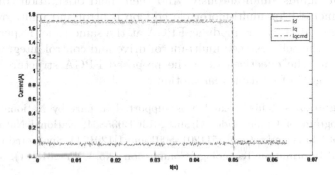

Fig. 14. Current loop step test

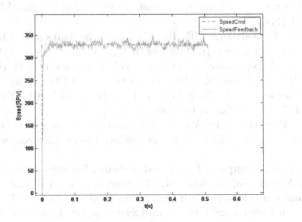

Fig. 15. Speed loop step test

A step response test is then conducted to verify the feasibility of overall multi-motor control system. The current-loop test result is shown in Fig. 14. The current command value is 1.7 A, and the rise-time value of step response is about 0.875 ms. The speed loop test result is shown in Fig. 15. The speed command value is 348 rpm, and the rise-time value of step response is about 20 ms.

5   Conclusions

In this paper, the modular FPGA of a multi-motor drive and control integrated system is successfully developed and implemented, which is responsible for acquiring relevant data and calculating motion control parameters. Firstly, several IP components are developed for reading the values of motor current and position signals simultaneously. And then, field orientation control module is implemented for multi-motor closed-loop control and multi-group PWM waves are generated by the modular FPGA at the same time. Experiments are conducted on the self-designed multi-motor drive and control integrated system to demonstrate the effectiveness of the proposed FPGA structure in light of achieving the goal of multiple-axis motion control.

Acknowledgements. This research was supported in part by National Key Basic Research Program of China under Grant 2013CB035804, National Natural Science Foundation of China under Grant 51120155001 and U1201244. Sponsored by Shanghai Economic and Information Technology Commission (No. CXY-2013-21).

References

1. Chen, H., Yeh, S., Hwang, J., Chen, M.: Development of dsp-based servo drive for permanent-magnet synchronous motors. In: Proc. 22nd Symp. Elect. Power Eng., pp. 417–421 (2001)
2. Zhou, Z., Li, T., Takahashi, T., Ho, E.: FPGA realization of a high-performance servo controller for pmsm. In: Nineteenth Annual IEEE Applied Power Electronics Conference and Exposition, APEC 2004, vol. 3, pp. 1604–1609. IEEE (2004)
3. Zhao, H., Zhu, L., Xiong, Z., Ding, H.: Design of a FPGA-based NURBS interpolator. In: Jeschke, S., Liu, H., Schilberg, D. (eds.) ICIRA 2011, Part II. LNCS, vol. 7102, pp. 477–486. Springer, Heidelberg (2011)
4. Tzou, Y.Y., Hsu, H.J.: FPGA realization of space-vector pwm control ic for three-phase pwm inverters. IEEE Transactions on Power Electronics **12**(6), 953–963 (1997)
5. Perdikaris, G.A.: Computer Controlled Systems. Springer (1991)
6. Samet, L., Masmoudi, N., Kharrat, M., Kamoun, A.: A digital pid controller for real time and multi loop control: a comparative study. In: 1998 IEEE International Conference on Electronics, Circuits and Systems, vol. 1, pp. 291–296. IEEE (1998)

7. Liu, X.W., Zhai, C., Yan, F.: The controlling and driving system of multiple motors based on fpga in lamost. Machinery & Electronics **6**, 008 (2008)
8. Chan, Y.F., Moallem, M., Wang, W.: Design and implementation of modular fpga-based pid controllers. IEEE Transactions on Industrial Electronics **54**(4), 1898–1906 (2007)
9. de Castro, R., Araújo, R.E., Oliveira, H.: Control in multi-motor electric vehicle with a fpga platform. In: IEEE International Symposium on Industrial Embedded Systems, SIES 2009, pp. 219–227. IEEE (2009)
10. Zwolinski, M.: Digital system design with VHDL. Pearson Education (2003)

Analytical Model for Flexure–Based Proportion Parallel Mechanisms with Deficient–Determinate Input

Jihao Liu, Weixin Yan, Yanzheng Zhao[✉], and Zhuang Fu

State Key Lab of Mechanical System and Vibration, Robotics Institute,
Shanghai Jiao Tong University, Shanghai 200240, China
liujihao.1@163.com, yzh-zhao@sjtu.edu.cn

Abstract. This paper presents the design and the specific analysis of a novel spatial symmetrical compliant mechanism with closed - loop subchains which is capable of driving the piezoelectric valve-less micropump in high frequence. This design of the topological configuration evolving from the bridge-type magnification mechanism amplifies the longitudinal deformation of the piezo-ceram actuators in parallel to the transverse displacement generation in dozens of times. In order to further estimate the performance of displacement amplifi-cation, the elastic beam theory, screw theory and the Pseudo Rigid Body Model (PRBM) are adopted to determine the ideal displacement amplification ratio. The FEM simulation is carried out to cross validate the design of the flexure-based mechanisms.

Keywords: Flexure hinges · Kinematic theory · Displacement amplification ratio · Symmetrical parallel compliant mechanism

1 Introduction

The valve-less micropump is one of the popular embedded components in the micro-fluidics system referring to devices and methods for controlling and manipulating pre-cise volume of fluid flows with scales of less than a millimeter [1-2]. The micropump is employed to support the driving force to mobilize the fluid (liquid or gas) through the micro-channels. By taking advantage of the micropump systems, many applications have been proposed in the medical, the chemical and the field of precision industry, such as micrototal analysis systems, implantable Insulin delivery systems and so on. There is increasing interest for the miniature climbing robot employing suction cups with the piezoelectric micropumps delivering the gas to generate the negative pressure.

Various approaches are used to optimize the performance of flow control from dif-ferent perspectives including check vlaves, diffuser/nozzles, topological configuration of pump chamber and peristaltic motion. Ivano Izzo [3] proposes a design methodolo-gy of no-moving-part (NMP) valve profile based on the definition of the efficiency

Y. Zhao—This work is supported by National Nature Science Foundations of China (under Grant 61273342, 6110510, 61473192 and U1401240).

H. Liu et al. (Eds.): ICIRA 2015, Part I, LNAI 9244, pp. 232–243, 2015.
DOI: 10.1007/978-3-319-22879-2_22

parameter of whole micropump in two limit load conditions (max pressure head and max flow rate). Comparing to the conventional diffuser NMP valve, there experimentally exists increment of the performance of the micro-valve under the same operative conditions driven by the same piezoelectric actuator. Anders [4] presents the design of a valve-less planar fluid pump with two pump chambers in the anti-phase which indicates the pump flow is more than twice as high as the classical configuration. Ki Sung Lee [5] demonstrates a novel micropump with a stepwise chamber, which enhances the rectifying efficiency by means of reducing the dead volume.

Due to the compressive property of the gas, the valve-less micropump in the suction cup requires a powerful actuator that could perform the outstanding feature of high frequency. D.H. Wang [6] models and tests the performance based on the characteristic of the circular piezoelectric unimorph actuator that has disadvantage of bearing strength under the high frequency. The voice coil motor or permanent magnets could actuate the film micropump with high frequency and large stroke, but they will occupy more space [7]. The piezostack is a typical actuator for precision motion and positioning with the remarkable characteristics including high force, high stiffness, high resolution and fast response. But one of the major drawbacks of conventional piezostack is small deformation stroke generation compared with its own dimension. The flexure-based compliant bridge-type mechanism is one of the classic magnification mechanisms for the piezostack [8].

In this work we propose a novel flexure-based coupled compliant mechanism with the combination features merging the bridge-type mechanism with the symmetrical parallel mechanism. The symmetrical compliant mechanism has the excellent features including no friction losses, no wear, no hysteresis and strong bearing strength. The configuration degree of freedom of this design is calculated using the reciprocal screw theory. According to the elastic beam theory, Pseudo Rigid Body Model (PRBM) and the kinematic theory, the proportion ratio is formulated. The finite element method is adopted to cross validate of the design in terms of the stress concentration and harmonic response under a certain frequency close to the first-order modal of the compliant mechanism.

2 Configuration Design

As demonstrated in Fig. 1, the suction cup for the miniature climbing robot employs a piezoelectric valve-less micropump in which the micro pump chamber and the suction cup are respectively etched in the opposite surfaces of the circular silicon substrate of 0.5mm thickness. Meanwhile, the design adopts a membrane of 50μm thickness in PMMA as the micropump film. The configuration of magnification mechanism evolves from the bridge-type mechanism merging the structural feature of the symmetrical parallel compliant mechanism.

As shown in Fig. 2, it detailedly demonstrates the magnification mechanism made up of same laminations of the flexure-based proportion parallel mechanisms in the configuration of rotational symmetry. The sub-chains rod with symmetric flexure hinges self-adapts subject to the station's motion along with the deformation displacement of the piezoelectric actuator and transmits radial displacement from the moving station to the lever mechanism. The axial output displacement part linking with tree rods in parallel, which adopts two flexure hinges in both ends, produces the axial displacement generation and driving force acting on the mciropump film. This configuration evolves from the bridge-type mechanism that is one of popular flexure-based compliant amplification mechanism in modern industry, and has function to convert and amplify the radial deformation displacement of the conventional piezostack in parallel into the axial displacement generation. In order to decrease the shaft drifting, this magnification is essential to coupling the inputs in the symmetric configuration.

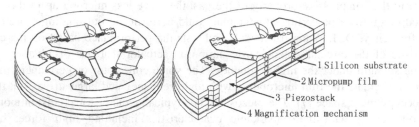

Fig. 1. Configuration view of the miniature suction cup

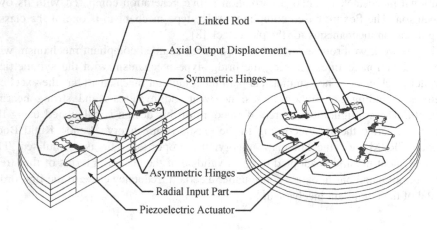

Fig. 2. The schematic view of the magnification

The silicon bulk micromachining technology is one of MEMS processing technic with high precision in the dust-free room. Compared with conventional Electrical Discharge Machining (EDM), it could be able to decrease even eliminate the machining error and guarantee quality. Meanwhile, the single crystal silicon (SCS) becomes more and more popular as structural material for the micro sensors and actuators in the automobiles [9-10]. One drawback in the applications in SCS is the mechanical property of silicon is orientation dependent. The convenient approach to overcome it is using <111> orientated silicon wafer whose mechanical property is isotropic. Compared with the metal in micro scale applications, SCS performs better in both material mechanics property and chemically inert, as shown in Table. 1. Besides SCS is a fragile material that performs no yield phenomenon.

Table 1. Various materials

Material	Material Mechanics Property			
	Yield Strength (GPa)	Young's Modulus (10^2GPa)	Density (g/cm^3)	Thermal Expansion ($10^{-6}/°C$)
Diamond	53	10.35	3.5	1.0
SiC	21	7.0	3.2	3.3
Si	7.0	1.9	2.3	2.33
Steel	4.2	2.1	7.9	12
Stainless steel	2.1	2.0	7.9	17.3
TiC	20	4.97	4.9	6.4

The flexure hinge made in SCS possesses enhanced performance in repeated accuracy and force bearing and performs excellent features including no yield, no rust and no backlash.

3   Modeling and Analysis

The flexure-based proportion parallel compliant mechanism evolves from the configuration of a symmetrical parallel mechanism with 6- PRRRR type subchains, in which all of the prismatic joints are active joints. The piezoelectric actuators embedded in the magnification mechanism are equivalent to prismatic joints of which one couples the adjacent. The kinematics of flexure-based magnification mechanism is formulated based on the model of the asymmetric parallel mechanism with closed loop subchains.

3.1   The Configuration Degree of Freedom

J Mrico [11] points out that mobility criteria based on the well-known Kutzbach-Grübler criterion would be possible to fail to derive the correct number of degrees of freedom of parallel mechanism. And according to group theoretical derivation, the Kutzbach-Grübler criterion is modified to provide the correct number of degrees of freedom for wider classes of parallel manipulators by resorting to the Lie algebra, $se(3)$, also known as screw algebra of the Euclidean group; $SE(3)$.

$$M = d(n - g - 1) + \sum_{i=1}^{g} f_i + v - \zeta \qquad (1)$$

where d represents the order of the mechanism, and $d = 6 - \lambda$, λ represents the number of the common constraints of a mechanism, n represents the total number of the members in the mechanism, g represents the total number of the kinematic pairs of the mechanism, f_i represents the DOF of the ith kinematic pair, v represents the number of redundancy constraints of the independent loops of a mechanism except the common constraints, ζ represents the number of passive degrees of freedoms of mechanism.

Based on PRBM, the flexure hinge could be represented by a revolute pair with the torsional spring. One subchain of the flexure-based parallel mechanism is shown in Fig. 3, which is also divided into two parts in serial, of which *Part 1* consists four revolute pairs and two prismatic pairs that couple with one in other subchain respectively.

Based on reciprocal screw theory, the kinematic motion of one subchain is formulated by screw system, while the constrain forces acting on this subchain could be presented by reciprocal screw system. The physical meaning of the screw system relatively has nothing with the location of the origin. Hence, we set up coordinate systems for *Part 1* and *Part 2* respectively. In computation of the configuration degree of freedom of spatial parallel mechanism with closed-loop subchains, a general kinematic pair takes place of one closed-loop subchain, and the whole degree of freedom of the parallel mechanism synthetizes all general kinematic pairs of all subchains.

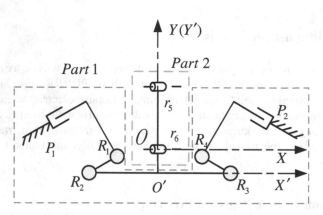

Fig. 3. The configuration model of the subchain

In *Part 1*, six paring elements are presented based on the screw theory in the coordinate system $O'X'Y'$, where Z axis is determined according to the right-hand rule. R_1, R_2, R_3, R_4 represents revolute pairs that rotate around Z axis, then the screw system is formulated in follow:

$$R_1 : \mathcal{S}_1 = (0 \quad 0 \quad 1; \quad b_1 \quad a_1 \quad 0) \tag{2}$$

$$R_2 : \mathcal{S}_2 = (0 \quad 0 \quad 1; \quad 0 \quad a_2 \quad 0) \tag{3}$$

$$R_3 : \mathcal{S}_3 = (0 \quad 0 \quad 1; \quad b_1 \quad -a_1 \quad 0) \tag{4}$$

$$R_4 : \mathcal{S}_4 = (0 \quad 0 \quad 1; \quad 0 \quad -a_2 \quad 0) \tag{5}$$

And the piezoelectric actuators are simplified as symmetric prismatic joints with respect to Y axis, and generate linear deformations motion pointing to the center of the fix platform along the radial directions. Hence, the screw system for the prismatic pairs is derived yields:

$$P_1 : S_5 = (0 \quad 0 \quad 0; \quad \frac{\sqrt{3}}{2} \quad \frac{1}{2} \quad 0) \tag{6}$$

$$P_2 : S_5 = (0 \quad 0 \quad 0; \quad -\frac{\sqrt{3}}{2} \quad \frac{1}{2} \quad 0) \tag{7}$$

According to the screw algebra, the reciprocal produce of the reciprocal screw system and the screw system is equal to zero, which means there exists no work by the constrain force:

$$S \circ S^r = 0 \tag{8}$$

where S^r denotes the reciprocal screw system of constrain force acting on the sub-chain of the mechanism; S denotes the screw system of the kinematics pairs.

Hence, we can derive the constrain force of **Part 1** in the reciprocal screw.

$$S_1^r = (0 \quad 0 \quad 1; \quad 0 \quad 0 \quad 0); \tag{9}$$

$$S_2^r = (0 \quad 0 \quad 0; \quad 1 \quad 0 \quad 0), \tag{10}$$

$$S_3^r = (0 \quad 0 \quad 0; \quad 0 \quad 1 \quad 0); \tag{11}$$

where $S_1^{\prime r}$ indicates a constrain force along Z axis, S_2^r indicates a moment around X axis; S_3^r indicates a moment around Y axis. According to the reciprocal screw system, we can equivalently derive the general kinematic pairs of **Part 1** in screw system:

$$S_1^g = (0 \quad 0 \quad 1; \quad 0 \quad 0 \quad 0); \tag{12}$$

$$S_2^g = (0 \quad 0 \quad 0; \quad 1 \quad 0 \quad 0); \tag{13}$$

$$S_3^g = (0 \quad 0 \quad 0; \quad 0 \quad 1 \quad 0) \tag{14}$$

where S_1^g indicates rotation around Z axis; S_2^g indicates translation along X axis; S_3^g indicates translation along Y axis. Combining the screw system of **Part 2**, we can obtain the reciprocal screw system S_i^r of one parallel mechanism's sub-chain, which indicates a constrain moment along the Y axis in the coordinate system for the model of the subchain:

$$r_5 : \mathcal{S}_7 = (1 \quad 0 \quad 0; \quad 0 \quad h_1 \quad -b_3) \tag{15}$$

$$r_6 : \mathcal{S}_7 = (0 \quad 0 \quad 1; \quad 0 \quad 0 \quad 0) \tag{16}$$

$$\mathcal{S}_i^r = (0 \quad 0 \quad 0; \quad 0 \quad 1 \quad 0) \tag{17}$$

Hence, the moving platform is constrained by moments produced by three symmetric subchains. These three moments parallel with their own fixed platforms respectively, and are no parallel to each other. So the maximum number of the linearly independence is 2, and there exists one redundancy constrain $v = 1$.

So we can obtain the configuration degree of freedom of the parallel mechanism, according to the modified Kutzbach Grübler's formula:

$$M = 6(14 - 15 - 1) + 15 + 1 = 4 \tag{18}$$

In this design, the number of the inputs U is less than the configuration degree of the freedom M, this spatial hybrid mechanism is regarded as one with deficient-determinate input.

3.2   The Ideal Displacement Amplification Ratio

Referring to Fig. 3, *Part 1* is regarded as input subchain that could transmit the planar motion and *Part 2* is output subchain in the analysis model for the amplification ratio based on the kinematics theory.

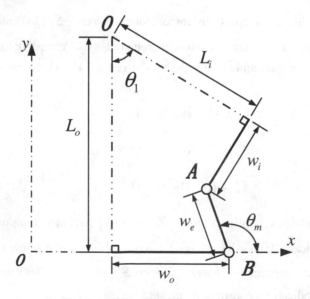

Fig. 4. Rigid model of input sub-chain

Input Subchain Proportion Ratio

The function of the input sub-chain is to realize the plane transmission of radial deformation displacements of piezoelectric actuators into output subchain. Given the ideal conditions that the deformation input keeps synchronous in both the amplification and driving frequency and the input subchain could be simplified further as shown in Fig. 4, we solve input subchain of the kinematic analysis in means of rigid model.

$$L_o = L_i \cos\theta_1 + w_i \sin\theta_1 + w_e \sin\theta_m \tag{19}$$

$$L_o = \frac{1}{2}L_i + \frac{\sqrt{3}}{2}w_i + w_e \sin\theta_m \tag{20}$$

where L_i denotes the radial distance from the input part to the center of the laminate along the input displacement; L_o denotes the radial distance from the position of outer flexure hinge to the center along motion of the linked rob in plane; A and B stands for flexure hinge respectively. Considering the range of θ_m, we can derive the amplify ratio:

$$\cos\theta_m = \frac{L_i \sin\theta_i - w_i \cos\theta_i - w_o}{w_e} \tag{21}$$

By differential on Equation (21), we can obtain amplified ratio of **Part 1**, and it yields:

$$\frac{dL_o}{dL_i} = \frac{1}{2} + w_e \cos\theta_m \frac{d\theta_m}{dL_i} \tag{22}$$

According to Equation (19) and Equation (20), Equation (22) becomes:

$$\frac{dL_o}{dL_i} = \frac{1}{2} - \frac{\sqrt{3}}{2}\cot\theta_m \tag{23}$$

$$\frac{dL_o}{dL_i} = \frac{1}{2} \pm \frac{\sqrt{3}}{2} \cdot \frac{1}{\sqrt{\left(\frac{\sqrt{3}}{2}\frac{L_i}{w_e} - \frac{1}{2}\frac{w_o}{w_e} - 1\right)^2 - 1}} \tag{24}$$

Output Subchain Proportion Ratio

Comparing to four-bar linkage and Scott-Russell mechanisms, the bridge-type mechanism has a compact structure and a large displacement amplification. According to the analysis by Ke-qi, the displacement amplification ratio of bridge-type mechanism is not related to the material and thickness of the structure, and it is not related to the length of dimension of the structure when the horizontal distance between the centers of hinges is fixed. [8]

The output subchain evolving from the bridge type mechanism is formulated in the same kinematic model. Hence the amplification ratio is formulated based on the elastic beam theory:

$$R_{amp} = \frac{3h(l+L)}{t^2 + 3h^2} \tag{25}$$

where, referring to Fig. 5, t represent the thickness of the flexure hinge, l represents the length of the flexure hinges, $L+l$ represents the horizontal distance between the centers of hinges, h represents the vertical distance between the centers of hinges.

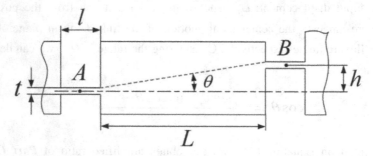

Fig. 5. The quarter of elastic model of bridge-type mechanism

The Proportion Ratio of Flexure-Based Mechanism
Combining Equation (21-22) and Equation (25), we can obtain the amplify ratio of the laminate:

$$R_m = \left| \frac{dh}{dL_i} \right| = \frac{dL_o}{dL_i} R_{amp} \tag{26}$$

where, $0 \leq \theta_m \leq \pi$. Based on Equation (24), it becomes:

$$R_m = \frac{3h(l+L)}{2(t^2 + 3h^2)} \left(1 \pm \frac{\sqrt{3}}{\sqrt{\left(\frac{\sqrt{3}}{2}\frac{L_i}{w_e} - \frac{1}{2}\frac{w_o}{w_e} - 1\right)^2 - 1}}\right) \tag{27}$$

According to Equation (27), we can obtain the optimal amplify ratio by optimizing parameters in the configuration.

4   Finite Element Analysis

We employ finite element analysis to verify the performance of the flexure-based laminate from the aspects of both harmonic respond and stress intensity in commercial software ANSYS.

We set up finite element model for one piece of laminate with 2.2 μm in the amplitude of radial input displacements. We could observe amplitude of axial output deflection according to the harmonic response analysis, as shown in Fig. 6. It indicates the first-order modal is a little more than *15KHz*, no matter whether the laminate bears the axial load. Hence, this displacement magnification can provide strong driving force for micro-pumps in high frequency.

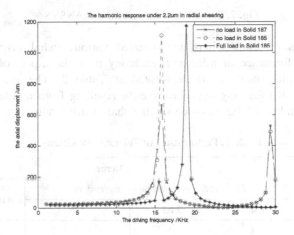

Fig. 6. The harmonic response analysis inANSYS

Considering the performance of the compliant mechanism dependent upon the stress concentration in the flexure hinges, we simulated the stress intensity in the laminate in the case that 2.2 μm in the amplitude of input displacements with 15kHz close to the first-order modal, and the center of laminate loaded 444.6*N/m* closing to the equivalent deflection stiffness of a piece of Parylene membrane with 50 μm in the thickness. As demonstrated in Fig. 7, the maximum stress of 0.533GPa, took place in the asymmetric flexure hinges in one piece of laminate. It indicates the safe factor (SF) of the spatial compliant-parrallel mechanism would be much more than 2.5 with respect to the yield strength of SCS.

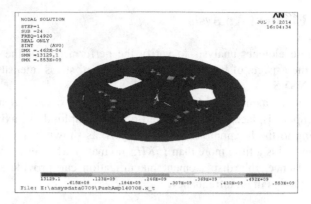

Fig. 7. Stress intension simulition inANSYS

After a series of trials where we inputted various radial displacements, we observed the performance of tolerant asynchrony from the aspect of error ratio of axial output displacement, as demonstrated at Table 2. The magnification has excellent feature of tolereting asynchrony error resulting from inpute amplitudes in parrallel, which indicated there existed smaller shaft drifting ratio.

Table 2. Performance of Tolerant asynchrony

| No. | Error | | | |
| | Error ratios of radial input amplitudes | | | Axial displacement |
	1 st	2 nd	3 th	
1	-5%	0	0	-1.66%
2	-5%	0	5%	0
3	0	0	0	0
4	0	0	5%	2.65%
5	0	0	10%	2.98%
6	0	5%	10%	4.3%

5   Conclusion

In order to improve the performance of piezoelectric actuators for film micro-pump, this paper presents a novel spatial hybrid compliant-paralleled mechanism also treated as a displacement magnification with excellent features including large amplify ratio, strong bearing, no friction and no backlash. We employ finite element analysis to verify the performance of the flexure-based laminate from the aspects of both harmonic respond and stress intensity by means of the finite element analysis in commercial software ANSYS. Meanwhile, this design possess low shaft drifting ratio according to the outstanding performance in toleration error resulting from amplitude of inputs in parallel.

Acknowledgment. This work is supported by National Nature Science Foundations of China (under Grant 61273342, 6110510, 61473192 and U1401240).

References

1. Lee, K.S., Kim, B., Shannon, M.A.: An electrostatically driven valve-less peristaltic micro-pump with a stepwise chamber. Sensors & Actuators A Physical. **187**(6), 183–189 (2012)
2. Stone, H.A., Stroock, A.D., Ajdari, A.: Engineering flows in small devices. Annual Review of Fluid Mechanics. **36**, 381–411 (2004)
3. Izzo, I., Accoto, D., Menciassi, A., et al.: Modeling and experimental validation of a piezoelec-tric micropump with novel no-moving-part valves. Sensors & Actuators A Physical. **133**(1), 128–140 (2007)
4. Olsson, A., Stemme, G., Stemme, E.: A valveless planar fluid pump with two pump cham-bers. Sensors & Actuators A Physical. **47**(1), 549–556 (1995)
5. Lee, K.S., Kim, B., Shannon, M.A.: An electrostatically driven valve-less peristaltic mi-cro-pump with a stepwise chamber. Sensors & Actuators A Physical. **187**(6), 183–189 (2012)
6. Wang, D.H., Huo, J.: Modeling and Testing of the Static Deflections of Circular Piezoelec-tric Unimorph Actuators. Journal of Intelligent Material Systems & Structures. **21**(16), 1603–1616 (2010)
7. Olaru, R., Petrescu, C., Hertanu, R.: A novel double-action actuator based on ferrofluid and permanent magnets. Journal of Intelligent Material Systems & Structures. **23**, 1623–1630 (2012)
8. Qi, K., Xiang, Y., Fang, C., et al.: Analysis of the displacement amplification ratio of bridge-type mechanism. Mechanism & Machine Theory. **87**, 45–56 (2015)
9. Jadaan, O.M., Nemeth, N.N., Bagdahn, J., Sharpe, W.N.: Probabilistic Weibull be-havior and mechanical properties of MEMS brittle materials. Journal of materials science. **37**, 4087–4113 (2003)
10. Petersen, K.E.: Silicon as a mechanical material. Proceedings of the IEEE. **70**(5), 420–457 (1982)
11. Rico, J.M., Aguilera, L.D., Gallardo, J., Rodriguez, R., Orozco, H., Barrera, J.M.: A group the-oretical derivation of a more general mobility criterion for parallel manipulators. Proceedings of the Institution of Mechanical Engineers - Part C. **220**(7), 969–987 (2006)

Study on the Electric-pneumatic Pressure Valve Driven by Voice Coil Motor

Tao Wang[✉], Jie Wang, Bo Wang, and Jinbing Chen

School of Automation, Beijing Institute of Technology, No 5 South Zhongguancun Street,
Haidian District, Beijing 100081, People's Republic of China
wangtaobit@bit.edu.cn

Abstract. Concerning the growing demand of high-speed response of electric-pneumatic proportional valve, in this paper, we developed an electric-pneumatic pressure valve system driven by voice coil motor (VCM). We build the mathematical model of the system and performed experiments for validation. Satisfied results of control of pressure were obtained owing to the excellent characteristics of VCM. And for further optimization we designed an active disturbance rejection controller (ADRC), whose superiority over PID controller has been validated through data analysis.

Keywords: Electric-pneumatic proportional valve · Voice coil motor · Modeling · Pressure characteristics

1   Introduction

Electric-pneumatic proportion/servo technology, which can realize the real-time precise control of pressure or flow satisfying the flexibility requirement of automation equipment [1], has become a development trend in the field of pneumatic drive [2].

In recent years, a special form of direct drive linear motor, VCM, has been widely used in different kinds of high speed, frequency and precision positioning motion system, especially suitable for high-speed periodic round motion [3][4]. A VCM-DDV produced by Yuken has reached 450Hz for bandwidth and 30L/min for rated flow [5], while 350Hz and 40L/min for another product from Parker [6]. In this essay we developed an electric-pneumatic pressure proportional valve system where the good performance of VCM has been fully showed and applied.

2   Algorithm

As shown in Fig. 1, the valve is mainly composed of three parts, including the pilot, the main valve and the controller.

H. Liu et al. (Eds.): ICIRA 2015, Part I, LNAI 9244, pp. 244–254, 2015.
DOI: 10.1007/978-3-319-22879-2_23

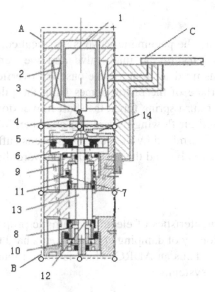

Fig. 1. Structure of the valve system

A-the pilot B- the main valve C- the controller
1-stator 2-rotor 3-flapper 4-nozzel 5-piston 6-backpressure chamber 7-spring of overflow valve 8-spring of positive valve 9-overflow valve pocket 10-positive valve pocket 11 overflow valve seat 12-positive valve seat 13- spool 14-orifice

2.1   The Pilot

As an important component of the system, this mechanism is like a converter that converts electrical signal into pneumatic signal along with the amplification of signal to control the movement of the spool [7]. When adjusting the distance between nozzle and flapper by giving different electrical signals to the VCM, the backpressure acting on the spool of valve can be greatly affected.

a. The VCM

Voice coil motor, with simple structure, high speed and fast response [8], is the ideal power source for low friction, high or low speed working condition. For this task, we chose a VCM produced by Winner as the drive device and a chip named L298N produced by SGS as the drive core chip.

b. The nozzle and flapper

In this sensitive mechanism even an extremely small increase of the distance between nozzle and flapper, that is x, can cause a great drop of backpressure-p_c, which assures that the valve is able to provide fast and precisely controlled air flows through the actuator chambers [9].

For our design, the diameter of fixed orifice (d_o) and of nozzle (d_n) are chosen as 0.0012m and 0.0016m, and the diameter of the flapper is 0.05m.

2.2   The Valve

As the command element, the pneumatic valve is a critical component of the actuator system. With the symmetry of the inlet valve and the vent valve, the pressure-feedback main valve has no dead zone on the onset of overflow [10]. Exterior force acting on the upper surface of the piston comes from the pilot chamber, when it is sufficient to restrict the return spring force, the shaft moves downward opening intake valve and the compressed air flowing inward one chamber. When the force from the lower surface which is determined by the outlet pressure is sufficient, the shaft moves upward opening overflow valve and the air flowing outward through the exhaust path.

2.3   The Controller

Due to the inherent characteristics of electric-pneumatic proportional system such as dead zone and non-stationary of damping coefficient, it's hard to achieve good control effect by PID algorithm. Thus an ADRC, which is more robust and adaptive, is proposed and applied to the system.

3   Modeling

3.1   Model of the VCM

According to voltage equation and force equation, the simplified mathematical model of VCM can be expressed directly by [11]

$$\frac{Y(s)}{U(s)} = \frac{B_g l}{mLs^3 + (K_l L + mR)s^2 + (K_l R + B_g^2 l^2)s} \tag{1}$$

Where L is the inductance in the circuit, R is the resistance, m is the mass of coil and rod, and K_l is the load resistance coefficient which satisfices $F_l = K_l v$.

In fact, L can be neglected as it's very small. By calculating, the model of the VCM we chose can be presented by

$$\frac{V(s)}{U(s)} = \frac{1}{0.00002954s^2 + 0.0513s + 5.29} \tag{2}$$

3.2   Pressure-flow Equation

According to ISO6358, the calculation formula of flow through small holes in ideal condition can be described by Sanville formula [12], followed as

$$G_m = \begin{cases} \dfrac{1}{5} S_{em} \rho_m p_1 \sqrt{\dfrac{T_0}{T_1}} & \dfrac{p_2}{p_1} \le b \\[4mm] \dfrac{1}{5} S_{em} \rho_m p_1 \sqrt{\dfrac{T_0}{T_1}} \cdot \sqrt{1 - \left(\dfrac{p_2/p_1 - b}{1-b}\right)^2} & \dfrac{p_2}{p_1} > b \end{cases} \tag{3}$$

Where G_m is the mass flow through a hole, S_{em} is the contraction area of hole, for orifice we have $S_{eo} = \pi c_o d_o^2/4$, and for nozzle we have $S_{en} = \pi c_n d_n x$, ρ_m is the density of fluid, p_1 is the upstream pressure, p_2 is the downstream pressure, T_0 is the temperature of fluid in standard reference atmosphere, T_1 is the temperature of upstream fluid, b is the critical pressure ratio.

If the downstream to upstream pressure ratio is smaller than a critical value b, the flow will attain sonic velocity (choked flow) and will only depend linearly on the upstream pressure. If the ratio is larger than b, the mass flow depends nonlinearly on both pressures.

Based on the flow equations of orifice and nozzle, the relationship between back-pressure and the distance between nozzle and flapper is show as

$$p^* = \begin{cases} \left(b_0 + \sqrt{b_0^2 - \left(1 + (1-b_0)^2 \dfrac{y^2}{k^2}\right)(2b_0 - 1)}\right) \Big/ \left(1 + (1-b_0)^2 \dfrac{y^2}{k^2}\right) \cdots p_c \in [b_o p_s, p_s] \\[5mm] \dfrac{k}{y} \cdots p_c \in \left[\dfrac{p_a}{b_n}, b_o p_s\right] \\[5mm] \left(\sqrt{\dfrac{p_a^2 b_n^2}{p_s^2} + (1-2b_n)\left(\left(\dfrac{p_a}{p_s}\right)^2 + \dfrac{(1-b_n)^2 k^2}{y^2}\right)} - \dfrac{p_a}{p_s} b_n\right) \Big/ (1-2b_n) \cdots p_c \in \left[p_a, \dfrac{p_a}{b_n}\right] \end{cases} \tag{4}$$

Where $p^* = \frac{p_c}{p_s}$, $y = \frac{4x}{d_n}$, $k = \frac{c_o d_o^2}{c_n d_n^2}$, p_s is the source pressure, p_c is the backpressure, c_o and d_o are the contraction area ratio and diameter of fixed orifice, c_n and d_n are the contraction area ratio and diameter of nozzle.

3.3　Force Balance Equation

With a fast and precise adjustment of the backpressure from the pilot, the valve provides accurate flow control as its force status changes with the change of back-pressure. Together, air pressure, spring pressure, damping force and Coulomb force decide the motion of the valve spool, followed as

$$M\ddot{x} = (p_c - p_a)A_0 - (p_o - p_a)A_1 - k_f(x + x_0) - c\dot{x} - F_c[sgn(\dot{x})] - Mg \tag{5}$$
$$M\ddot{x} = (p_o - p_a)A_1 - (p_c - p_a)A_0 + k_f(x_0 - x) - c\dot{x} - F_c[sgn(\dot{x})] + Mg \tag{6}$$

Where M is the mass of the movable parts, x is the displacement of the spool, x_0 is the preset spring compression at the equilibrium position, p_o is the pressure of exit, A_0 is the area of the upper surface, A_1 is the area of the lower surface, k_f is the stiffness of spool spring, F_c is the Coulomb friction force, c is viscous damping coefficient.

4   Design of the Controller

ADRC is a feedback linearization controller based on classic PID controller but without the inherent defects. By observation and compensation, the nonlinear and uncertain problems can be effetely overcome binging good dynamic performance. To improve the control effect of system, an ADRC, which is consist of TD, ESO and NLSEF three parts [13], is designed replacing the PID controller.

a. Calculation formula for TD

$$
\begin{cases}
v_1(k+1) = v_1(k) + hv_2(k) \\
v_2(k+1) = v_2(k) + hfh \\
fh = fhan(v_1(k) - v, v_2(k), r, h_0)
\end{cases} \tag{7}
$$

By parameter setting, we get the factor r=60 and the step h_0=0.005.

b. Calculation formula for ESO

$$
\begin{cases}
e(k) = z_1(k) - y(k) \\
z_1(k+1) = z_1(k) + h(z_2(k) - \beta_1 e(k)) \\
z_2(k+1) = z_2(k) + h(z_3(k) - \beta_2 fal(e(k), \alpha_1, \delta) + bu(k)) \\
z_3(k+1) = z_3(k) - h\beta_3 fal(e(k), \alpha_2, \delta)
\end{cases} \tag{8}
$$

Where

$$
fal(e, \alpha, \delta) =
\begin{cases}
e / \delta^{1-\alpha}, |e| \le \delta \\
|e|^\alpha \, sign(e), |e| > \delta
\end{cases} \tag{9}
$$

The results of parameter setting are as follows:

$$\beta_1 = 20, \beta_2 = 24, \beta_3 = 20, \alpha_1 = 0.5, \alpha_2 = 0.25, \delta = 5.$$

c. Calculation formula for NLSEF

$$
\begin{cases}
e_1(k+1) = v_1(k+1) - z_1(k+1) \\
e_2(k+1) = v_2(k+1) - z_2(k+1) \\
u_0(k+1) = k_p fal(e_1(k+1), \alpha_{01}, \delta_1) + k_d fal(e_2(k+1), \alpha_{02}, \delta_1) \\
u(k+1) = u_0(k+1) - \dfrac{z_3(k+1)}{b}
\end{cases} \tag{10}
$$

The results of parameter setting are as follows:

$$\alpha_{01} = 0.75, \alpha_{02} = 1.4, \delta_1 = \delta = 5, k_p = 15, k_d = 5, b = 2.$$

5   Simulation

a. Simulation of Nozzle and Flapper Model
The x-p_c curve when the inlet pressure is 0.6MPa is shown in Fig. 2.

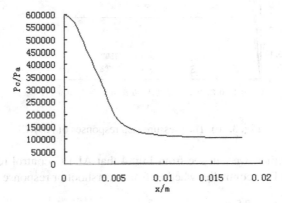

Fig. 2. The x-p_c curve of nozzle and flapper

The static characteristic curve of simulation, where the p_c is proportional to x somehow, is almost consistent with the one from theory, confirming the correctness of the model and parameters.

b. Simulation of Pressure Control System
Fig. 3 (a), (b) give the step responses of PID controller and ADRC based on different outlet pressures respectively.

It can be seen that the output pressures of both PID controller and ADRC trend to be steady, but the former's response time is about 0.3s along with significant overshoot, and the latter has been optimized significantly with overshoot almost zero, response time shortened to 0.1s.

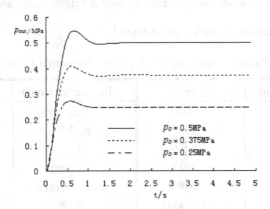

Fig. 3. (a). The pressure step responses of PID controller

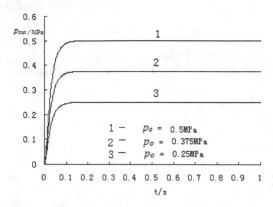

Fig. 3. (b). The pressure step responses of ADRC

When put together, we can see from Fig. 4 that ADRC control results are significantly better than PID controller whether from overshoot or response time.

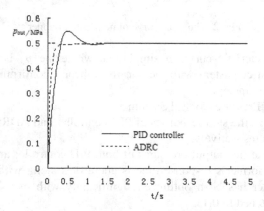

Fig. 4. Comparison curve of step responses of PID controller and ADRC

The relevant parameters are shown in table 1.

Table 1. parameters of E/P pressure proportional valve

Parameters(unit)	value	Parameters(unit)	value
c_o	0.7	c_n	0.7
p_s(MPa)	0.6	p_a(MPa)	0.1
b_o	0.528	b_n	0.528
T_0(K)	293	x_0(m)	0.002
d_o(mm)	0.999	d_n(mm)	0.269
k_f(N/M)	0.00002	F_c(N)	0.000025
ρ(kg/m^3)	1.18	M(kg)	0.01
A_0(mm^2)	100	A_1(mm^2)	1

6   Experiments and Results

The control circuit consisting of IPC, data acquisition card (PCLS - 818 and PCI - 1720), valve, air source handling components, and pressure sensor (PSE - 510) is shown in Fig. 5. The software platform uses Lab Windows/CVI to achieve the pressure real-time control.

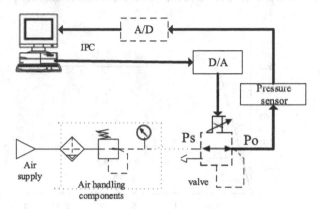

Fig. 5. Control system experiment setup

6.1   The Input and Output Static Characteristic of Pressure

The static characteristic of pressure, which is shown in Fig. 6, is the static relationship between the output pressure and the control voltage when the load channel is closed, namely the output flow.

Output pressure is proportional to control signal with good linearity. The experiment result matches well with the simulation result, but there are still some small deviations, which may be caused by the mechanical error as well as the error between the actual situations and ideal conditions.

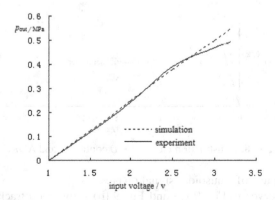

Fig. 6. Construction curve of input and output static characteristic

6.2   Dynamic Response Characteristics of Pressure

a. dynamic step response of pressure

The contrast curve of experiment and simulation under the PID control algorithm on condition that the inlet pressure is 0.6MPa, the outlet pressure is 0.5MPa, and the volume of the load container is 10ml is shown in Fig. 7. The simulation curve and test curve match perfectly, which verifies the correctness and feasibility of the mathematical model we build.

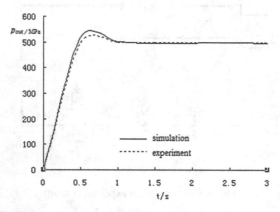

Fig. 7. Step response curves from experiment and simulation

From the construction curve of PID controller and ADRC in Fig. 8, we can see that the overshoot and response time of the former curve is about 4% and 0.25 s, while the latter curve shows zero overshoot and 0.15s response time, greatly improved the control effect.

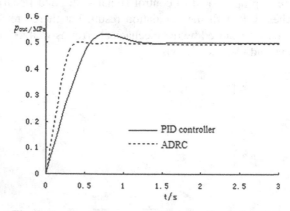

Fig. 8. Construction curve of PID controller and ADRC

b. The tracking test of sinusoidal signal on pressure

The tracking curves in Fig. 9 (a) and Fig. 9 (b) show good tracking ability of the control system. The output waveform and the input signal can basically coincide at low frequency, especially in the peak and valley. Differences of the curve at high frequency are a 1/10-cycle lag of phase and an amplitude declining.

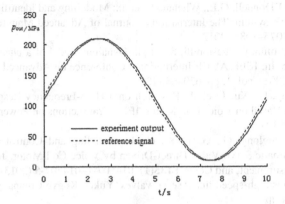

Fig. 9. (a) The sinusoidal tracking experimental curve with the frequency of 0.1 Hz

the reference signal,

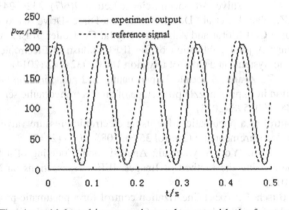

Fig. 9. (b) The sinusoidal tracking experimental curve with the frequency of 10Hz

7   Conclusions

In this article we developed a VCM driven pressure proportional valve system and its detailed mathematical model. The proposed model, which can be used to develop high performance pressure controller for applications in automated industry, is verified by the excellent agreement found when compared with the experimental data. We also developed an ADRC, whose superiority of performance is showed in results obtained both by simulation and experiment, greatly improved the control effect of the system.

References

1. Harris, P.G., O'Donnell, G.E., Whelan, T., et al.: Modelling and identification of industrial pneumatic drive system. The International Journal of Advanced Manufacturing Technology 58(9/12), 1075–1086 (2012)
2. Sorli, M., Figliolini, G., Pastorelli, S., et al.: Dynamic model of a pneumatic proportional pressure valve. In: IEEE/ASME International Conference on Advanced Intelligent Mechatronics (AIM2001), vol. 1, pp. 630–635 (2001)
3. Guo, H., Wang, D., Xu, J., et al.: Research on a High-Frequency Response Direct Drive Valve System Based on Voice Coil Motor. IEEE Transactions on Power Electronics 28(5), 2483–2492 (2013)
4. Baoren, L., Longlong, G., Gang, Y., et al.: Modeling and Control of a Novel High-Pressure Pneumatic Servo Valve Direct-Driven by Voice Coil Motor. Journal of Dynamic Systems, Measurement, and Control 135(1), 014507-1–014507-5 (2013)
5. LSV(H)G series high-speed linear servo valves. Yuken Kogyo Company, Ltd., Minato-ku. www.yuken.co.jp
6. Hydraulics proportional valve series DFplus. Parker Inc., Cleveland. http://www.parker.com/literature/
7. Wang, C., Li, Q., Ding, F., et al.: Design, Analysis and Experiment of GMM-Based Nozzle-Flapper Servo Valve. Advanced Science Letters 4(6/7), 2426–2430 (2011)
8. Wu, S., Jiao, Z., Yan, L., et al.: Development of a Direct-Drive Servo Valve With High-Frequency Voice Coil Motor and Advanced Digital Controller. IEEE/ ASME transactions on mechatronics: A joint publication of the IEEE Industrial Electronics Society and the ASME Dynamic Systems and Control Division 19(3), 932–942 (2014)
9. Li, S., Aung, N.Z., Zhang, S., et al.: Experimental and numerical investigation of cavitation phenomenon in flapper-nozzle pilot stage of an electrohydraulic servo-valve. Computers & Fluids 88, 590–598 (2013)
10. Situm, Z.: Control of a pneumatic drive using electronic pressure valves. Transactions of the Institute of Measurement and Control 35(8), 1085–1093 (2013)
11. Choi, J.-k., Lee, H.-i., Yoo, S.-y., et al.: Analysis and Modeling of a Voice-Coil Linear Vibration Motor Using the Method of Images. IEEE Transactions on Magnetics 48(11), 4164–4167 (2012)
12. Lambeck, S., Busch, C.: Exact linearization control for a pneumatic proportional pressure control valve. In: 8th IEEE International Conference on Control and Automation (ICCA) (2010)
13. Zhang, Y., Fan, C., Zhao, F., et al.: Parameter tuning of ADRC and its application based on CCCSA. Nonlinear dynamics 76(2), 1185–1194 (2014)

Biomechatronics in Bionic Dexterous Hand

EMG Onset Detection Based on Teager–Kaiser Energy Operator and Morphological Close Operation

Dapeng Yang[✉], Qi Huang, Wei Yang, and Hong Liu

State Key Laboratory of Robotics and System, Harbin Institute of Technology,
Harbin 150080, People's Republic of China
yangdapeng@hit.edu.cn

Abstract. As a typical biomedical signal, the electromyography (EMG) is now widely used as a human-machine interface in the control of robotic rehabilitation devices such as prosthetic hands and legs. Immediately detecting and eliciting of a valid EMG signal are greatly anticipated for ensuring a fast-response and high-precision EMG control scheme. This paper utilizes two schemes, Teager-Kaise Engergy (TKE) operator and Morphological Close Operation (MCO), to improve the accuracy of the onset/offset detection of EMG activities. The TKE operator is used to amplify the EMG signal's amplitude change on the initiation/cessation phases, while the MCO is adopted to filter out the false positives of the binary sequence obtained by the fore TKE operation. This method is simple and easily to be implemented. After selecting appropriate filtering parameters (T_1, T_2 and j), it can achieve precise onset detection (absolute error <10ms) over a variety of signal-to-noise ratios (SNR) of the biomedical signal.

Keywords: Myoelectric signal · Onset detection · Teager-Kaiser Energy · Morphological close operation

1 Introduction

Accurate onset/offset detection of a muscle activity is of great interest in the research of biomechanics [1], motor control [2], gait and posture initiation analysis [3], etc. This kind of kinetic information can be detected using camera, inertial sensors, accelerometers, or directly from biomedical signals collected over skeleton muscles. The myoelectric signal, also termed as electromyography (EMG), is such a biomedical signal that contains rich information about the user's motion intention. A study shows that the gait initiation, toe-off and heel-strike, can be detected using EMG on healthy subjects 130-260ms in advance, which provides sufficient time for controlling a lower-limb prosthesis [4]. Similarly, if a hand grasp initiation can be detected using EMG beforehand, then a fast-response prosthetic hand control can also be realized.

Acquiring a length of faithful EMG signal for feature extraction is a prerequisite for controlling advanced prosthetic hands. Since the EMG signal is weak, stochastic, and easily influenced by the environment, making precise onset detection is still challenge. Even experienced physicians in the hospital might introduce different levels of detection errors [5]. An accurate, automatic EMG onset detection has been long

© Springer International Publishing Switzerland 2015
H. Liu et al. (Eds.): ICIRA 2015, Part I, LNAI 9244, pp. 257–268, 2015.
DOI: 10.1007/978-3-319-22879-2_24

expected in the neuroscience research such as neural decoding, motor analysis, and control of various rehabilitation devices (exoskeleton, prosthetic leg and hand, etc.).

In the field of myoelectric control, a fast response of the control system (in other words, low control latency) is greatly expected for ensuring intuitive control feelings [6]. Accurate onset detection of an EMG activity is desired in either traditional amplitude-coding methods [7] or advanced pattern recognition (PR)-based control schemes [8, 9]. For the latter, a moving window is generally applied on the real-time EMG signal for extracting the specific pattern of the predefined motions. In this context, without accurate onset detection, this sliding window would elicit some unfaithful data for classification. It would result in a number of unexpected classification errors on the motion initiation stages [10] that largely affect the prosthetic hands' usability [11]. Of last resort, the majority voting [12, 13] could be a feasible solution to smooth these errors; however, it will introduce remarkable control latency. To improve the control accuracy and shorten the control delay especially for a dynamic motion control, the accurate onset detection then becomes necessary.

Simple implementation can utilize the EMG signal's standard deviation (STD) or absolute mean value (MAV) together with a threshold decision [14]; however, selecting a proper threshold [15] is always a difficult work. Moreover, these methods commonly require a processing window (finite moving average, FMA) that significantly reduces the control system's response rate. Other means, like double-threshold detector [15], wavelet transformation [16], statistical strategy [2], Kalman smoother [17] and maximum likelihood estimator [18], either need prior knowledge about the EMG model or require intensive calculation, which are not suitable for real-time practice.

The motivation of this paper is to propose a serial of procedures that can accurately detect the EMG onset and thus make a faithful EMG signal segmentation for pattern recognition. Studies on this issue have verified that the Teager-Kaiser Energy (TKE) operator [19, 20], which was firstly proposed in speech signal processing, can help improve the detection accuracy of the muscle activity onset in either simulation signals [21] or real EMG signals [22]. As an incremental study, this paper focuses on the intrinsic characteristics of a valid muscle contraction and applies a post-processing (Morphological Close Operation, MCO, a standard image processing algorithms) to leverage the TKE method's capability.

2   Materials and Methods

2.1   EMG Signal Simulation

To evaluate our method, a set of EMG signals was firstly simulated that contains the exact onset/offset time points of a muscle contraction. Notice that a muscular contraction comprises both isotonic and isometric contraction stages, the generation process of EMG should also be a dynamic one that contains both transient and steady-state phases. In other words, a proper EMG model should produce dynamic EMG signals including complete phases of onset, attack, and decay, within which various levels of the maximum voluntary contraction (MVC) are applied. However, most of the current sophisticated EMG models [23, 24] can only generate steady-state EMG signals with a constant level of muscle contraction, which are not suitable in our study. Studies have shown that the EMG signal's variance is a robust index for measuring the contraction level of a skeleton muscle [25]. To facilitate our study without loss of generality, this paper directly modeled the EMG signal as a Gaussian noise process with a ramp-shape variance in its initialization/depression stage [26]. The simulated EMG signal initiates from a standard background variance ($\sigma^2_{\text{noise}} = 1$), bursts at time t_0, and then dynamically changes its variance to σ^2_{signal} within a short time duration τ. Unlike the approach that superimposes steady-state EMG signals and background noise to implement EMG burst and various signal-to-noise ratios (SNR) [18], our simulated EMG signal contains accurate onset point t_0, transient EMG within τ, as well as a length of steady-state EMG rightly following the burst.

The signal to noise ratio (SNR) in the simulated EMG was configured as [26]

$$\text{SNR} = 10 \cdot \log_{10}\left(\frac{\sigma^2_{\text{signal}}}{\sigma^2_{\text{noise}}}\right)[\text{dB}],\tag{1}$$

where the σ^2_{noise} and σ^2_{signal} are the variances of the background noise and the steady-state EMG signal, respectively. Considering the bandwidth of the EMG signal, each simulated EMG was filtered using a high-pass filter of 20~500Hz (10-order Butterworth). The sampling frequency of the simulated signal is set to 1000Hz. At last, we simulated a total of 4000 EMG signal segments; each segment is 1000ms in length, with onset time t_0 locating on 500ms~600ms, transient duration τ lasting for 5ms~30ms and τ belonging to 10~20dB (all in uniform random distribution).

2.2 Teager-Kaiser Energy Operator

For a given digital signal $x(n)$, the TKE operator ψ can be written as [19]:

$$\psi[x(n)] = x^2(n) - x(n+1)x(n-1)\tag{2}$$

Since the output of the TKE operator is highly correlated to the signal's amplitude and normalized frequency, the change of EMG amplitude in the initial stage will be elevated that the onset point of the EMG can be easily labeled. This fact has been noticed by several studies [21, 22] and a normalized procedure can be rewritten as

$$\tilde{x}(n) = x(n) - \frac{1}{N}\sum_{i=1}^{N}x(i)$$

$$\psi(n) = \tilde{x}^2(n) - \tilde{x}(n+1)\tilde{x}(n-1), n = 1, 2, \cdots, M, \cdots, N,$$

$$\mu_0 = \frac{1}{M}\sum_{i=1}^{M}\psi(i), \delta_0 = \left\{\frac{1}{M-1}\sum_{i=1}^{M}(\psi(i) - \mu_0)^2\right\}^{1/2}, Th = \mu_0 + j \cdot \delta_0$$

$$s(n) = \text{sign}(\psi(n) - Th), \hat{t}_0 = \min(n \mid s(n) = 1), n = 1, 2, \cdots, M, \cdots, N$$

(3)

where $\tilde{x}(n)$ is the zero-mean signal, $\psi(n)$ is the TKE operator signal, N is the length of the signal, M is a proper length of the background noise, Th is the threshold, μ_0 and δ_0 are the mean and standard derivation of the background noise, j is the scale factor for δ_0, $s(n)$ is the state function with "1" representing the active state and "0" representing the null state, and sign(*) is the revised sign function (when input variable>0, output=1; otherwise, output=0). We applied this procedure on all of the simulated EMG signals with recommended j values (6-8) given in [21]. A sample of the TKE operation can be found in Fig. 1-a)~Fig. 1-b).

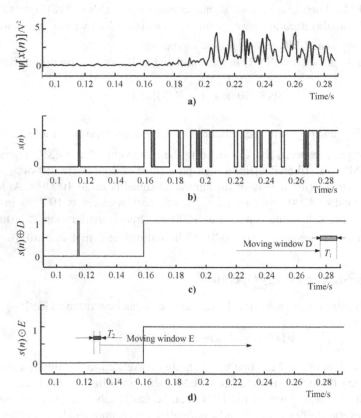

Fig. 1. The EMG onset detection using TKE and MCO: a) raw EMG data; b) state signal after TKE; c) morphological dilation operation with window D; d) morphological erosion operation with window E.

After the TKE operation, the raw EMG signal was transformed into the binary sequence $s(n)$. In most cases, this sequence was filled with scattering values of "1" and "0" that make a true active state of the EMG signal hard to be interpreted. Since the EMG signal may be seriously interfered with the background noise (typically when SNR<10), some low-level EMG activities, especially those on the contraction's initialization stage, might be regarded as null signals (false negatives) that the estimated onset point would lag behind the actual moment (delayed detection). For solving this, one way was to choose a low threshold Th by reducing the value of the scale factor j. On the other hand, a low threshold would take some spike noises in the signal as active EMG signals. When the spike noise randomly took place prior to the true onset time, an advanced detection error could also be produced (false positives). Because of those delayed or advanced detection artifacts (different from the motion artifacts that severely interference EMG signals), a faithful EMG onset/offset points cannot obtained through simply picking up the first "1" and last "1" in the $s(n)$.

2.3   Morphological Close Operation

According to the nature of a valid muscle contraction, we made a post-filtering scheme to the binary signal $s(n)$. We introduced two parameters, the *least contraction switching time* (T_1), defined as the minimum length of time duration needed for switching (or initiating) a contraction (typically, 25~30ms from spinal cord to muscle), and the *least contraction duration* (T_2), defined as the minimum length of a valid motion activity (typically, >100ms). Then, according to these two parameters, we applied a two-step filtering scheme on the state function $s(n)$ to rectify the intermittent sequence. Different from the previous study [27], we adopted a different filtering order and implemented our algorithm using a standard procedure from image processing, Morphological Close operation (MCO, first dilation and then erosion), to eliminate the onset artifacts.

The first filter (dilation filter) was used to connect any possible true active EMG state with an interval less than T_1. It is presumed that the skeleton muscle cannot contract too often with an idle interval less than T_1. The second filter (erosion filter) was used to eliminate any possible false active EMG state with duration less than T_2, which was constructed on a hypothesis that the muscle cannot contract too fast with duration less than T_2. To sum up, the parameter T_1 is the feasible interval between two active EMG state, while the parameter T_2 is the reasonable duration of a true active EMG state. Both can be determined according to specific application scenarios (the muscles, the motions, etc.).

Since these two filters adopt some morphological characters of the EMG signal, we term this post-processing procedure as the morphological operation hereafter. From the view of image processing, the onset detection is similar to the edge detection in a two-dimension image [28]. The first filter is like a morphological dilation operation (MDO), while the second filter is like a morphological erosion operation (MEO). These two operations together, termed as "morphological close" operation (MCO), first fills out the inner gaps (dilation) and then deletes outer noise points (erosion) in an image. Specifically, the MDO adopted a window D with length of T_1. The window D moved along the binary sequence $s(n)$ and all "0" intervals between the "1" values within D are changed into "1", as shown in Fig. 1-c). At the same time, the MEO adopted a window E with length of T_2, and all "1" intervals between "0" values within E were replaced by "0", as shown in Fig. 1-d). This two filtering scheme were denoted as $s(n) \oplus D$ and $s(n) \ominus E$ in the figure, respectively. After these two operations (together termed as MCO), $s(n)$ was changed to $s'(n)$ that the onset time $\hat{t}_0$, active state and the offset time of the EMG signal can be clearly interpreted.

2.4   Real EMG Verification

We also tend to validate the methods on real EMG signals. These signals were collected from four healthy subjects on the forearms using differential EMG electrodes (Danyang commercial modular electrodes, band filter of 20~500Hz, sampling frequency of 1000Hz, the same to the simulated signals). A sum of four bipolar electrodes were used and separately placed on the muscle of extensor pollicis brevis, extensor indicis proprius, extensor digitorum, and extensor digiti quinti proprius. The EMG signals were collected while the subjects were intending to grasp four different objects (a cup, an orange, a key and a screw nut), which requested four basic grasp patterns (cylinder grasp, spherical grasp, lateral grasp and tripod tip) [29]. When an object was presented, the subject was instructed to preshape their fingers from a neutral posture and then to compass it. The EMG signals were collected rightly on the grasp preparation phases. Our onset detection was used to processing the EMG signals rightly on the hand preshaping phase, for recognizing four grasp patterns for controlling a multi-fingered prosthetic hand. Each subject was requested to perform 10 replicas of a grasp (nearly 3 seconds in length). A total of 160 segments (4 subjects×4 patterns×10 replicas) of 4-channel EMG signals was collected. When necessary, the onset points within these raw EMG signals were manually labeled by an experienced expert.

We compared our methods with four traditional methods: 1) standard deviation (STD)-based method; 2) maximum mean absolute (MAV)-based method; 3) Hodges' method [14], and 4) Lidierth's method [27]. These methods extract EMG features (STD, MAV, etc.) using a shift window. We use M to represent the length of the calibration window, W the length of the shift window, O the stepping length, and h the scale factor. After adjustment, the parameters used in these methods are $j=7$ in TKE-MCO, $M=1000$, $W=20$, $O=10$, $h=3$ in the STD, MAV and Hodges' methods, respectively.

We defined the detection error t_e as

$$t_e = \hat{t}_0 - t_0 \tag{4}$$

where $\hat{t}_0$ is the predicted onset time, and t_0 is the actual onset time. An estimation of the probability density function (PDF) [30] of the detection error received by all methods was measured, to give a comprehensive prospect on the overall performance.

We took $t_e \in [-10\ 10]$ ms as successful detection, while t_e beyond this range as false detection. Then, the detection success rate p_e for all simulated data can be defined as

$$p_e = D_C / D_T \times 100\% \tag{5}$$

where D_C is number of the successfully detected onset points, D_t is the total number of onset points.

3   Results

We first examine the TKE procedure on all simulated signals. An estimation of the PDF is shown in Fig. 2-a. Details on the PDF's of the advanced error ($t_e < 10$) and delayed error ($t_e > 10$) obtained by different methods are also given in the figure.

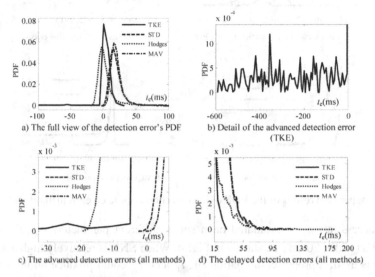

a) The full view of the detection error's PDF

b) Detail of the advanced detection error (TKE)

c) The advanced detection errors (all methods)

d) The delayed detection errors (all methods)

Fig. 2. Comparison of the PDF of the detection error t_e (without MCO)

From Fig. 2-a), the TKE method obtains a relatively tight distribution of t_e compared with the other three methods. The value of t_e according to the peak is more close to the origin ($t_e \approx 1.3$ ms). For Hodges method, since the starting point of the processing window is used as the onset point, the detection errors mostly lead the real onset point ($t_e \approx -3.1$ ms). From Fig. 2-b), the advanced errors appear broadly before the origin ($0 \sim 600$ms), largely due to the spike noise ahead the onset. The STD- and MAV-method get the shortest advanced error ($\min t_e > -5$ ms), as shown in Fig. 2-c). On the other hand, the Hodges method has the largest detection delay ($\max t_e < 175$ms), while the TKE has the smallest ($\max t_e < 35$ ms). It clearly shows that the processing window can effectively reduce the bad effect of spike noise; on the other hand, the window could also increase the detection delay.

We then apply the post-processing MCO method ($T_1 = 50$ms, $T_2 = 25$ms) to the same dataset, with PDF of t_e shown in Fig. 3. On the side of advanced errors (Fig. 3-a), the TKE-MCO method becomes more precisely ($\min t_e > -50$ ms), while the other three methods (STD-, MAV- and Lidierth) retain their performance. On the side of delayed errors, only the Lidierth method tends to deteriorate. In Fig. 3-d, we also present some detection errors on different T_1 values. It shows that when T_1 decreases from 25ms to 10ms, the advanced error tends to reduce while the delayed error tends to increase.

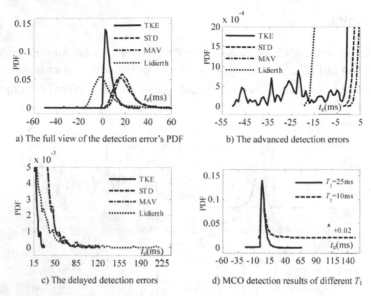

a) The full view of the detection error's PDF

b) The advanced detection errors

c) The delayed detection errors

d) MCO detection results of different T_1

Fig. 3. Comparative results of the PDF of the detection error te on each method (with MCO)

The success rate p_e is evaluated on a set of different parameters (j, T_1, T_2) with two SNR's (10dB, 20 dB), as shown in Fig. 4. When SNR is relatively high and a small j value is used, a proper combination of T_1 and T_2 could produce an unevenly distributed p_e, as show in Fig. 4-a). Generally, the p_e drops as T_1 increases (85% at $T_1 = 50$ ms, $T_2 = 5$ ms) , and ascends as T_2 increases (99.88% at $T_1 = 10$ ms, $T_2 = 50$ ms). In Fig. 4-b, the p_e becomes much plain as j increases from 5 to 6. The highest value 99.8% is at $T_1 = 10$ ms, $T_2 = 50$ ms. When SNR drops from 20dB to10dB, the p_e tends to decline, with the best value of 76.%. In this case, on selection of proper T_1 and T_2, selecting a small j value can improve p_e (highest, 83.5%, Fig. 4-d).

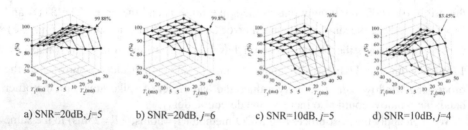

a) SNR=20dB, j=5   b) SNR=20dB, j=6   c) SNR=10dB, j=5   d) SNR=10dB, j=4

Fig. 4. The detection precision of TKE on different parameters

These results clearly indicate that the proposed TKE-MCO method can precisely detect the EMG onset point, only when the parameters of j, T_1 and T_2 are set properly. In general, a small j (say, 4~5) should be applied if the simulated EMG signal has a low SNR typically around 10dB. The T_1 should be as short as possible but longer than a normal EMG interval to fill the gaps within true EMG activities; while T_2 should be as long as possible but shorter than a valid contraction to resist any possible spike noise in the signal.

Result on a sample of real EMG signals is shown in Fig. 5. When MCO is applied, all methods can successfully detect the active states of the real EMG. Since the TKE makes the $s(n)$ signal more sensitive to amplitude variance, the TKE-MCO method gives the most accurate detection. Our experiments show that, within a normal muscle contraction, the T_1 is less than 30ms; in addition, the duration of an active state for each grasp is not less than 500ms. For the MAV- and STD-method, since they are naturally immune to spike noise, the T_1 can be large enough to compass all active state points ($s(n) = 1$). The Lidierth method produced more advanced detection error ($t_e < 0$), while the STD and MAV methods produced more detection latency ($t_e > 0$). These results are consistent to the results acquired on simulated EMG signals.

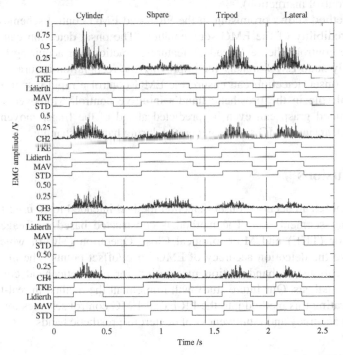

Fig. 5. Comparison of the onset detection methods on hand preshaping EMG ($T_1 = 50$ ms, $T_2 = 100$ ms in the Lidierth, STD, MAV method. $T_1 = 30$ ms, $T_2 = 100$ ms, j=7 in the proposed TKE+MCO method).

We labeled the most forward onset point (OP) in all channels as the overall onset point (OOP) of the multi-channel EMG signals. And, if the OOP is ahead of the other channels' OPs more than 100ms, we claimed it ill detection. Experiments indicated that the probability of ill onset detection is below 1% through all of the collected EMG signals.

4   Discussions

In the field of myoelectric control, a fast response of the control system (in other words, low control latency) is greatly anticipated for ensuring an intuitive control feeling [6]. For the PR-based EMG control schemes [9], the misclassifications on motion transition stages are now still a big issue that largely affects the advanced prosthetic hands' usability [11]. Those misclassifications, mostly appear in transient EMG signals generated by dynamic muscle contractions, may be largely attributed to the unreliable signal segmentation by the moving window (MW) [10]. Considering the response of the control system, it is better to restrict these misclassifications in a proper way for achieving a faster and safer control [31]. A feasible solution is to use the majority voting [12, 13], which is applied on a serial of classification outputs for increasing the control signal's credibility and consistency; however, the control latency of this method would be also significantly increased (several classifications output only a single control instruction).

The new method shows promising in the PR-based EMG control schemes for improving the credibility of the EMG segmentation. The onset detection can be integrated into the control scheme (ahead of feature extraction) for acquiring both transient and steady-state EMG signals for improving the recognition accuracy [32]. As well, accurate onset detection can ensure the EMG control a rapid response, which plays a critical role in the prosthetic hand's intuitive control. Moreover, a pattern about the intended grasp can even be predicted ahead of the finger movements by processing the transient EMG signals generated during hand preshaping.

5   Conclusions

This work proposes an accurate onset detection method intended to be used for faithful EMG signal segmentation. The method is constructed based on Teager-Kaiser Energy operator (TKE) and Morphological Close Operation (MCO), which in all largely promote the detection accuracy of EMG onset/offset points. The method has potential applications on hand motion recognition for prosthetic hand control and human motion analysis. Our future study will implement the method in real-time and consider integrating this method into the PR-based EMG control schemes for improving the control accuracy and intuitiveness of dexterous prosthetic hands.

Acknowledgments. This work is in part supported by the National Program on Key Basic Research Project (973 Program, No. 2011CB013306), the National Natural Science Foundation of China (No. 51205080), the China Postdoctoral Science Foundation Funded Project (No. 2013M540276, 2014T70316), and the Heilongjiang Postdoctoral Fund (No. LBH-Z13082).

References

1. Luca, C.J.D.: The use of surface eletromyography in biomechanics. Journal of Applied Biomechanics **13**, 135–163 (1997)
2. Staude, G.H.: Precise onset detection of human motor responses using a whitening filter and the log-likelihood-ratio test. IEEE Trans. Biomed. Eng. **48**, 1292–1305 (2001)
3. Maeda, Y., Tanaka, T., Nakajima, Y., Shimizu, K.: Analysis of Postural Adjustment Responses to Perturbation Stimulus by Surface Tilts in the Feet-together Position. Journal of Medical and Biological Engineering **31**, 301–305 (2011)
4. Wentink, E.C., Beijen, S.I., Hermens, H.J., Rietman, J.S., Veltink, P.H.: Intention detection of gait initiation using EMG and kinematic data. Gait & Posture **37**, 223–228 (2013)
5. Difabio, R.P.: Reliability of Computerised Surface Electromyography for Determining the Onset of Muscle Activity. Phys. Ther. **67**, 43–48 (1987)
6. Englehart, K., Hudgins, B.: A Robust, Real-Time Control Scheme for Multifunction Myoelectric Control. IEEE Transaction on Biomedical Engineering **50**, 848–854 (2003)
7. Zecca, M., Micera, S., Carrozza, M.C., Dario, P.: Control of multifunctional prosthetic hands by processing the electromyographic signal. Critical Reviews in Biomedical Engineering **30**, 459–485 (2002)
8. Scheme, E., Englehart, K.: Electromyogram pattern recognition for control of powered upper-limb prostheses: State of the art and challenges for clinical use. Journal of Rehabilitation Research and Development **48**, 643–660 (2011)
9. Oskoei, M.A., Hu, H.: Myoelectric control systems—A survey. Biomedical Signal Processing and Control **2**, 275–294 (2007)
10. Hargrove, L., Losier, Y., Lock, B., Englehart, K., Hudgins, B.: A real-time pattern recognition based myoelectric control usability study implemented in a virtual environment. In: 29th Annual International Conference of the IEEE on Engineering in Medicine and Biology Society. EMBS 2007, pp. 4842–4845. IEEE Press, New York (2007)
11. Peerdeman, B., Boere, D., Witteveen, H., in't Veld, R.H., Hermens, H., Stramigioli, S., Rietman, H., Veltink, P., Misra, S.: Myoelectric forearm prostheses: State of the art from a user-centered perspective. Journal of Rehabilitation Research and Development **48**, 719–737 (2011)
12. Oskoei, M.A., Hu, H.: Support Vector Machine-Based Classification Scheme for Myoelectric Control Applied to Upper Limb. IEEE Trans. Biomed. Eng. **55**, 1956–1965 (2008)
13. Bitzer, S., van der Smagt, P.: Learning EMG control of a robotic hand: towards active prostheses. In: Proceedings of the IEEE International Conference on Robotics and Automation, pp. 2819–2823. IEEE Press, New York (2006)
14. Hodges, P.W., Bui, B.H.: A comparison of computer-based methods for the determination of onset of muscle contraction using electromyography. Electromyography and Motor Control-Electroencephalography and Clinical Neurophysiology **101**, 511–519 (1996)
15. Bonato, P., D'Alessio, T., Knaflitz, M.: A statistical method for the measurement of muscle activation intervals from surface myoelectric signal during gait. IEEE Trans. Biomed. Eng. **45**, 287–299 (1998)
16. Merlo, A., Farina, D., Merletti, R.: A fast and reliable technique for muscle activity detection from surface EMG signals. IEEE Transactions on Biomedical Engineering **50**, 316–323 (2003)
17. Lee, J., Shim, H., Lee, H., Lee, Y., Yoon, Y.: Detection of onset and offset time of muscle activity in surface EMGs using the Kalman smoother. In: World Congress on Medical Physics and Biomedical Engineering 2006 IFMBE Proceedings, pp. 1103–1106

18. Xu, Q., Quan, Y.Z., Yang, L., He, J.P.: An Adaptive Algorithm for the Determination of the Onset and Offset of Muscle Contraction by EMG Signal Processing. IEEE Trans. Neural Syst. Rehabil. Eng. **21**, 65–73 (2013)
19. Teager, H.M.: Evidence for Nonlinear Sound Reduction Mechanisms in the Vocal Tract, pp. 241–261. Kluwer Acad Publ (1990)
20. Lemyre, C., Jelinek, M., Lefebvre, R.: New approach to voiced onset detection in speech signal and its application for frame error concealment. In: IEEE International Conference on Acoustics, Speech and Signal Processing. ICASSP 2008, pp. 4757–4760. IEEE, Las Vegas (2008)
21. Li, X.Y., Zhou, P., Aruin, A.S.: Teager-Kaiser energy operation of surface EMG improves muscle activity onset detection. Annals of Biomedical Engineering **35**, 1532–1538 (2007)
22. Solnik, S., DeVita, P., Rider, P., Long, B., Hortobágyi, T.: Teager-Kaiser Operator improves the accuracy of EMG onset detection independent of signal-to-noise ratio. Acta Bioeng Biomech **10**, 65 (2008)
23. Hamilton-Wright, A., Stashuk, D.W.: Physiologically based simulation of clinical EMG signals. IEEE Trans Biomed Eng **52**, 171–183 (2005)
24. Duchêne, J., Hogrel, J.-Y.: A Model of EMG Generation. IEEE Trans. Biomed. Eng. **47**, 192–201 (2000)
25. Phinyomark, A., Quaine, F., Charbonnier, S., Serviere, C., Tarpin-Bernard, F., Laurillau, Y.: EMG feature evaluation for improving myoelectric pattern recognition robustness. Expert Systems with Applications **40**, 4832–4840 (2013)
26. Staude, G., Flachenecker, C., Daumer, M., Wolf, W.: Onset Detection in Surface Electromyographic Signals: A Systematic Comparison of Methods. EURASIP Journal on Applied Signal Processing **2001**, 67–81 (2001)
27. Lidierth, M.: A Computer based Method for Automated Measurement of the Periods of Muscular Activity from an EMG and its Application to Locomotor Emgs. Electroenceph. Clin. Neurophysiol. **64**, 378–380 (1986)
28. Schluter, J., Bock, S.: Improved Musical Onset Detection with Convolutional Neural Networks, pp. 6979–6983. Florence, Italy (2014)
29. Taylor, C.L., Schwarz, R.J.: The anatomy and mechanics of the human hand. Artificial limbs **2**, 22–35 (1955)
30. Bowman, A.W., Azzalini, A.: Applied Smoothing Techniques for Data Analysis. Oxford University Press, New York (1997)
31. Englehart, K., Hudgins, B., Parker, P.A.: A Wavelet-Based Continuous Classification Scheme for Multifunction Myoelectric Control. IEEE Trans. Biomed. Eng. **48**, 302–311 (2001)
32. Yang, D., Zhao, J., Jiang, L., Liu, H.: Dynamic hand motion recognition based on transient and steady-state EMG signals. International Journal of Humanoid Robotics **9**, 11250007 (2012)

Analysis of Human Hand Posture Reconstruction Under Constraint and Non-constraint Wrist Position

Li Jiang, Yuan Liu, Dapeng Yang[✉], and Hong Liu

State Key Laboratory of Robotics and System,
Harbin Institute of Technology, Harbin 150080, China
yangdapeng@hit.edu.cn

Abstract. As a compactness unit, the human hand shows high versatility and sophisticated grasp functionality. How to design a robot hand replicating the human grasp posture is a challenging task. Mechanical implementation of postural synergies provides new hope for resolving this problem. Generally, these posture synergies are extracted from a large data set consisting of a variety of hand grasp postures that can be reconstructed through synergies with an acceptable error. In the daily life, people can successfully grasp an object within a grasp tolerance (acceptable scope of relative position between human hand and objects). However, the relative position between human hand and objects is almost ignored in previous studies on the reconstruction of human grasp postures. In this paper, we tend to analyze the difference of the reconstruction of hand postures in two different scenarios: constraint and non-constraint wrist positions. The principal component analysis (PCA) is applied to the posture data sets acquired under two different data-acquisition paradigms, with steady and varying wrist-object position, respectively, for reconstructing the hand postures. The reconstruction differences between these two data-collection paradigms are analyzed. The information transmission rates given by different number of PCs are qualified. The distributions of the first four PCs elements under the two paradigms are also presented, respectively. Our results show that the specific hand postures in changing relative position within grasp tolerance can be faithfully reconstructed only when the relative position between the human hand and the object are fully considered.

Keywords: Postural synergies · Posture reconstruction · Grasp · Prosthetic hand

1 Introduction

Human hand is highly versatile in its interactions with the environment. Replicating the sophisticated grasp function of human hand is a difficult task in robotic area. In terms of posture reconstruction, the first main problem we should solve is which postures we should reconstruct. Many human hand grasp type studies were presented. Schlesinger et al. made the first attempt to construct the human grasp posture taxonomy contained six categories: cylindrical, tip, hook, palmar, spherical and lateral [1]. In 1956, Napier suggests that the human grasp posture can be divided into power grasp and precision grasp [2]. In studying the grasps required for manufacturing task, Cutcosky

H. Liu et al. (Eds.): ICIRA 2015, Part I, LNAI 9244, pp. 269–281, 2015.
DOI: 10.1007/978-3-319-22879-2_25

established a comprehensive and detailed taxonomy. The taxonomy is begun with division of power grasp and precision grasp. Then the detailed taxonomy is carried on according to the object shapes [3]. Feix et al. reviewed a large number of grasp taxonomies and provide a comprehensive taxonomy contained 33 grasp postures [4]. Zheng et al. had an investigation of grasp type and frequency in daily household and machine shop tasks by the recording of a head-mounted camera. In grasp frequency results for daily household tasks, the top three grasp postures are medium wrap, power sphere and lateral pinch [5]. The corresponding grasp objects to the first three grasp postures are cylinder, sphere and prism. The 19 hours of video with over 9000 grasp instances from two housekeepers and two machinists was analyzed by Bullock et al. Their goal is to find small, versatile sets of human grasp to span common objects. The grasp span was defined and the first three versatile grasp postures with maximum score for the housekeeper were medium wrap, power sphere and lateral pinch [6].

For the design of prosthetic hand, high grasping versatility, compactness and affording are needed. Recent research in neuroscience illustrates that hand postures can be reconstructed by small number of principle components. Santello et al. reported that the first two principle components account for ~84% of the variance over 57 objects grasp postures. The principle components are called eigengrasps. The posture reconstruction method is called postural synergies [7]. Based on the research, some robot hands were designed via mechanical implementation of postural synergies [8, 9, 10, 11, 12, 13, 14, 15].

In mechanical implementation of postural synergies, building the human hand posture data set is the base. However, the object shape is the only effect factor on the previous research of the human hand gesture taxonomy and mechanical implementation of posture synergies. In fact, in daily grasp tasks the relative positions between human hand and objects are changing and not always in perfect status. Therefore, the reconstruction of grasp posture in changing relative position (within a grasp tolerance) is particularly important. There is a question to be raised: whether the posture reconstruction method not considering relative position can well represent human grasp posture in changing relative position condition? The aim of this paper is to compare the posture reconstruction results in changing relative positions and in perfect status. The difference of these two reconstruction methods is analyzed in the condition of changing relative position.

This paper is organized as follows: Section 2 details the experimental protocol, data collection, data analysis and evaluation of reconstruction difference; Section 3 presents the detail experiment results; Section 4 performs the observation and discussion of experiment; Section 5 concludes the study.

2   Material and Methods

2.1   Participants

Ten healthy subjects of right-handed (24~27 years old, 8 men and 2 women) volunteered to take part in the experiments. Each subject is of good health and has no history of neurological or motor disorders. All participants provided informed consent before the experiments, as required by the Declaration of Helsinki. The all experiments received IRB approval of Harbin Institute of Technology.

2.2   Experimental Protocol

The experiment was divided into two parts.

In first part, each subject was asked to grasp and lift 6 different objects, twice. The objects were chosen to span the hand postures of different grasp types. The shape of objects we chose was based on the Zheng [5] and Bullock[6] research results. The grasp postures corresponding to cylinder, sphere and prism have high grasp frequency and maximum versatile score. Six different objects were used in first part and second part of experiment (as shown in Fig. 1 and Fig. 2): two spheres (varying in diameter), two cylinders (varying in diameter) and two prisms (varying in length). The size of grasping object is shown in Table 1. The subjects had to:

1. Place two hands on the start areas which were demarcated by red tape.
2. Reach and grasp the objects in a natural and self-think perfect way with the right hand and hold this grasp gesture for 3 seconds.
3. Lift up the objects in an arbitrary height.
4. Put down the objects in the object position area which were demarcated by red tape.
5. Return the grasping hand to the start area.

Lifting up the objects was used to ensure that subject can move objects successfully by the gesture. In total, 12 trails (1 subject× 6objects× 2repeats) across all six objects were performed by each subject over a period of ~10 min.

Table 1. Size of the six grasping objects

Object		Size
Sphere	Large	Diameter 80mm
	Small	Diameter 60mm
Cylinder	Large	Diameter 60mm; height 200mm
	Small	Diameter 40mm; height 200mm
Prism	Large	Length:80mm;width:40mm;height:100mm
	Small	Length:40mm;width:40mm;height:100mm

In second part, the subject was asked to grasp 6 different objects in 27 different relative positions ($3x \times 3y \times 3z$), twice. The relative position between human hand and object is defined as the relative position between center of human wrist and object center of gravity in this paper. The subject sat in front of the table. The elbow and wrist rested on a flat surface, the forearm was horizontal, the arm was oriented in the parasagittal plane passing through the shoulder, and the hand was in a semipronated position. Right wrist through the anti-static wrist strap secured to the stationary bracket, which was utilized to constrain the wrist position. The grasped object target position on the plane of the table is shown in Fig. 2. The different relative heights (high, medium, low) between hand and objects were obtained by adjusting anti-static wrist strap height.

The object target position (shown in Fig. 2) was obtained after trying to successfully grasp objects in different position. The distance between the vertical lines and between the horizontal lines in the object target position area were 6cm and 4.5cm. The relative height between adjacent heights was 3cm.

Each subject was asked to grasp 6 different objects in 27 different relative positions ($3x\times3y\times3z$) between human hand and objects. Each object was grasped twice. In total, 324 trails (1 subjects$\times$ 6objects$\times$ 27 relative distances$\times$ 2repeats) across all six objects were performed by each subject over a period of ~2 h. Subjects were instructed to grasp the object with their right hand and place the object in target position with their left hand. Once subject completed grasp, he was asked to hold the gesture 3 seconds. No gesture constrains were given, grasp gesture and wrist posture were entirely decided by subjects under the premise of stable, nature and comfortable grasping. No explicit constraints on movement velocity were given.

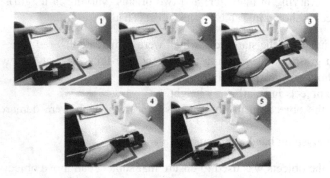

Fig. 1. Experimental protocol (the first part)

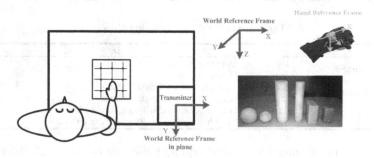

Fig. 2. Experimental protocol (the second part)

2.3   Record hand and Wrist Postures

The static wrist and hand postures were recorded in two hold grasp posture time points (3 seconds hold time) of first and second experiment part by CyberGlove (Virtual Technologies, Palo Alto, CA) and Fastrack™ Polhemus 3D motion-tracking system (acquisition rate: 100 Hz, positional accuracy: 0.8 mm RMS, rotational accuracy: 0.15° RMS; Roby-Brami et al. 2000). The Polhemus receiver was attached to CyberGlove. The wrist posture is given by Eular angles of azimuth, elevation and roll. The following joint angles were measured: flexion-extension of proximal interphalangeal (pip) and metacarpo-phalangeal (mcp) joints of digit II-V, as well as the inter-phalangeal (ip) and metacarpo-phalangeal (mcp) joints of the thumb (I), abduction/adduction (abd) of and opposition (rot) of the carpo-metacarpal joint of the thumb. The four abduction/adduction (abd) angles of the metacarpo-phalangeal (mcp) joints of digit II-V were deduced from three angular measures of finger spread. The glove was individually calibrated with several predefined postures by a C++ platform we developed. The wrist and hand posture data also was recorded by the C++ platform.

2.4   Data Analysis and Statistics

In this paper, two kinds of construction methods of human grasp posture data set were presented. The first kind of human grasp posture data set was constructed by the method like the first part of experiment. This is also the traditional method of recording human grasp posture data without considering the relative position between objects and human hand. The object shape was the only effect factor on hand posture and the different relative position between objects and hand was not considered. Subjects grasped the objects in a perfect status. For convenience of description, we used HPD-PS to represent human grasp posture data set in a perfect status. However, in the daily grasp the relative position between objects and hand is different especially for the prosthetic hand user. In this case, the grasp in our daily life can't be perfect like the first part of experiment. Therefore, we proposed the second kind of construction method of human grasp posture data set (the second part of experiment). In our construction method of human grasp posture data set, the relative positions between human hand and object were considered in three orthogonal directions (X, Y, and Z). For convenience of description, we used HPD-RP to represent human grasp posture data set in different relative position. The second kind of human grasp posture data set was constructed through the method like the second part of experiment. Principal component analyses (PCA) were applied to HPD-PS and HPD-RP for the reconstruction of human hand grasp posture. The first reconstruction method of human grasp posture was through PCA over HPD-PS, and we call it RCM-PS. The second reconstruction method of human grasp posture was through PCA over HPD-RP, and we call it RCM-RP. In RCM-RP, the object shape and the different relative positions between objects and hand were both considered as the effect factors on grasp posture.

The grasp posture data was averaged over the two repeats across two trails. The HPD-PS is a $[60 \times n]$ matrix and the HPD-RP is a $[1620 \times n]$ matrix. The n is the grasp posture kinematic DOF. Principal component analyses (PCA) were separately applied to HPD-PS and HPD-RP (the value of n can be set to 15 DOF (hand posture), 18 DOF (hand and wrist posture) and 11 DOF (hand posture without thumb) recorded from the first and second part of experiment).

The single one grasp posture of one subject can be defined as $Q_i = [q_{i1} \quad q_{i2} \quad \cdots \quad q_{in}]^T \in R^n$, n is the grasp posture kinematic DOF. N is defined as the number of hand posture and the wrist posture data set contained hand orientation and the digit configuration. Therefore posture data set can be represented as $Q = [Q_1 \quad Q_2 \quad \cdots \quad Q_N]^T \in R^{N \times n}$.

Therefore, hand posture and the wrist posture data set can be described by a small number of principal components:

$$Q \approx \tilde{Q} = \begin{bmatrix} s_{11} & s_{12} & \cdots & s_{1d} \\ \vdots & \vdots & \vdots & \vdots \\ s_{i1} & s_{i2} & \cdots & s_{id} \\ \vdots & \vdots & \vdots & \vdots \\ s_{N1} & s_{N2} & \cdots & s_{Nd} \end{bmatrix} \begin{bmatrix} e_1^T \\ e_2^T \\ \vdots \\ e_k^T \end{bmatrix} + \begin{bmatrix} \overline{q}_{11} & \cdots & \overline{q}_{1n} \\ \overline{q}_{21} & \cdots & \overline{q}_{2n} \\ \vdots & \cdots & \vdots \\ \vdots & \cdots & \vdots \\ \overline{q}_{N1} & \cdots & \overline{q}_{Nn} \end{bmatrix} \tag{1}$$

$$= S \times E + \overline{Q}$$

$$\overline{q}_{*j} = \frac{1}{N} \sum_{i=1}^{N} q_{ij} \tag{2}$$

$\overline{q}_{*j}$ is the average of joint angle of j-th kinematic DOF across N postures. $j = 1, 2, \cdots, n$. The vectors e_i are the eigenpostures, $e_k \in R^{n \times 1}$, $k = 1, 2, \cdots, d$. d is the driving DOF of grasp posture after dimension reduction by PCA. The values s_{ij} are scalar weights which is the controlled variable on eigenposture.

2.5   The Evaluation of Reconstruction Difference

In order to compare the difference between RCM-PS and RCM-RP, difference was evaluated in three aspects: information transmitted by PCs over different DOF (11 DOF, 15DOF or 18 DOF), element distribution of first four PCs and reconstruction angle difference. Each aspect details as follows:

Information Transmitted by PCs Over Different DOF (11 DOF, 15DOF or 18 DOF)

The posture data set Q also can be expressed as by singular value decomposition:

$$Q = [u_1 \, u_2 \cdots u_r] \text{diag}\{\lambda_1, \lambda_2, \cdots \lambda_r\} [v_1 \, v_2 \cdots v_r]^T \tag{3}$$

where $u_i \in R^{N \times 1}$, $i = 1, 2, \cdots, r$, $\lambda_1 \geq \lambda_2 \geq \cdots \geq \lambda_r$ are the singular values of matrix Q, $v_j \in R^{n \times 1}$, $j = 1, 2, \cdots, r$, $d \leq r \leq n < N$.

The reconstruction posture can be expressed as [13]:

$$\tilde{Q} = \sum_{i=1}^{d} \lambda_i u_i v_i^T \tag{4}$$

The information transmission rate can be written as:

$$T = \sum_{i=1}^{d} \lambda_i^2 / \sum_{i=1}^{n} \lambda_i^2 \tag{5}$$

This metric has been presented by Eckart and Young [16]. In robotic hand design area, T quantified anthropomorphic motion capability of robot finger motion [13].

Element Distribution of First Four PCs
Principal component analyses (PCA) were applied to HPD-PS and HPD-RP for the reconstruction of human hand grasp posture. All PCs were normalized. Then, the difference of PCs elements can be seen on radar chart.

Reconstruction Angle Difference
The reconstruction angle difference of each subject was separately calculated. For one subject, the reconstruction posture by RCM-PS contained six postures (each reconstruction posture corresponds to each object). However, the reconstruction posture by RCM-RP contained 162 postures (1 subject×6objects×27 relative distances). Each reconstruction posture corresponds to both relative position and object.

The reconstruction angle difference of RCM-PS and RCM-RP was obtained by four steps.

1. In order to confirm that posture reconstruction not considering the relative position can't well represent the grasp posture within grasp tolerance, we made the absolute difference between reconstruction posture by RCM-PS and HPD-RP. Then, reconstruction posture by RCM-RP was also made difference with HPD-RP. The reconstruction difference corresponding to both relative position and object by RCM-PS and RCM-RP was obtained.
2. In order to easily show results, reconstruction difference got from step1 in three relative heights was averaged. The reconstruction difference in nine relative positions (3X× 3Y) of horizontal plane was obtained.
3. The reconstruction difference got from step 2 is the difference of each subject. Then, we averaged ten subjects reconstruction difference got from step 2.
4. All 15 hand joint reconstruction angle errors of reconstruction difference got from step 3 in corresponding relative position were averaged. Reconstruction angle difference by first three PCs of RCM-PS and RCM-RP in nine relative positions of horizontal plane was obtained.

3   Result

3.1   Information Transmitted by PCs over Different DOF (11 DOF, 15DOF or 18 DOF) Across RCM-PS and RCM-RP

Information transmitted by PCs of different DOF (11 DOF, 15DOF or 18 DOF) across two posture reconstruction methods is shown on Fig. 3. The number of postures in HPD-PS (60 postures) is less than number of postures in HPD-RP (1620 postures). However, the information transmitted by PCs of hand posture (15 DOF) across RCM-PS and RCM-RP was highly consistent. This indicates that the information transmitted by PCs over hand postures (15DOF) is not sensitive to the increase in the number of hand postures. On the contrary, the information transmitted by different number of PCs of hand posture without thumb (11 DOF) and hand and wrist posture (18 DOF) across RCM-PS and RCM-RP was not consistent. To the reconstruction of hand posture without thumb (11DOF), the information transmitted ratio within first 5 PCs by RCM-RP was higher than by RCM-PS. To the reconstruction of hand and wrist posture (18DOF), information transmitted by PCs through RCM-RP was lower than through RCM-PS.

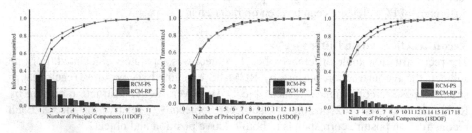

Fig. 3. Information transmitted by PCs got through RCM-PS and RCM-RP

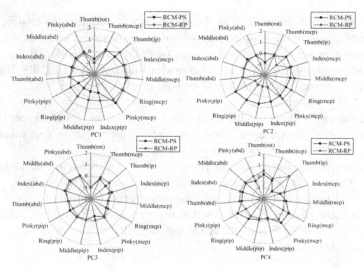

Fig. 4. Element distribution of first four PCs got by RCM-PS and RCM-RP

3.2   Element Distribution of First Four PCs got by RCM-PS and RCM-RP

Fig 4 shows each of joint angles changing for a unit change in the amplitude of the first to fourth PCs on hand postures (15DOF). Fig. 4 shows element distribution of the first four PCs got by RCM-PS and RCM-RP. It can be shown that only PC3 got by RCM-PS and RCM-RP overall had a high fitting degree. The PCs abd joint element got by RCM-PS and RCM-RP also had a high fitting degree. To the two different posture reconstruction methods, the PCs element of ip joint and opposition joint of thumb both had a large deviation. The pip joints element of first two PCs got by RCM-PS and RCM-RP had a large deviation and approximately had the opposite value. The elements of PCs are used to provide the coupling coefficient of hand joint in mechanical implementation of postural synergies. The results shown in Fig. 4 confirms that mechanical implementation on RCM-PS can't replicate the grasp posture in the condition of changing relative position.

3.3   Reconstruction Angle Difference of RCM-PS and RCM-RP

Based on the evaluation method of reconstruction angle difference, reconstruction angle difference by first three PCs of RCM-PS and RCM-RP in nine relative positions of horizontal plane are shown in Fig. 5. The X1, X2 and X3 in Fig. 5 represent X positional deviation from left to right. The Y1, Y2 and Y3 in Fig.5 represent Y positional deviation from distal to proximal.

As show in Fig. 5, a significant difference can be seen that posture reconstruction angle difference of the RCM-RP is lower. The mean angle differences of RCM-PS across all positions and objects were $6.0 \pm 1.1°$ (mean $\pm$ SD°). However, the mean angle differences of RCM-RP across all positions and objects were $19.2 \pm 3.3°$ (mean $\pm$ SD°).

Based on the reconstruction angle difference of Fig. 5, a problem can be raised: whether the objects or relative position has a higher effect on reconstruction angle difference by RCM-RP and RCM-PS? Therefore, based on results of Fig. 5, reconstruction angle differences by RCM-RP and RCM-PS on six objects and on nine relative position of horizontal plane were calculated. The results are shown in Fig. 6. Fig. 6 (A) shows the reconstruction angle difference on six different objects. For RCM-PS, the reconstruction angle difference in sphere and cylinder of large size was lower than in sphere and cylinder of small size (Fig. 6 (A)). The reconstruction angle difference in prism of large size was lower than in prism of small size. The reconstruction angle difference of grasp prism posture by RCM-RP was higher than other two kinds of shape objects. Using RCM-RP to reconstruct the human hand posture, the reconstruction angle difference of grasp sphere and cylinder posture was not obviously affected by size and shape.

The posture reconstructed angle difference by RCM-RP and RCM-PS on nine relative positions of horizontal plane is shown in Fig. 6 (B). If we use the RCM-PS as the hand posture reconstruction method, it can be seen that reconstruction angle differences in Y2 position were all lower than in other Y positions. Therefore, the changing relative position in Y position had an obvious effect on reconstruction angle difference. An obvious result we can see in Fig. 6 is that reconstruction angle difference of RCM-RP was lower than RCM-PS.

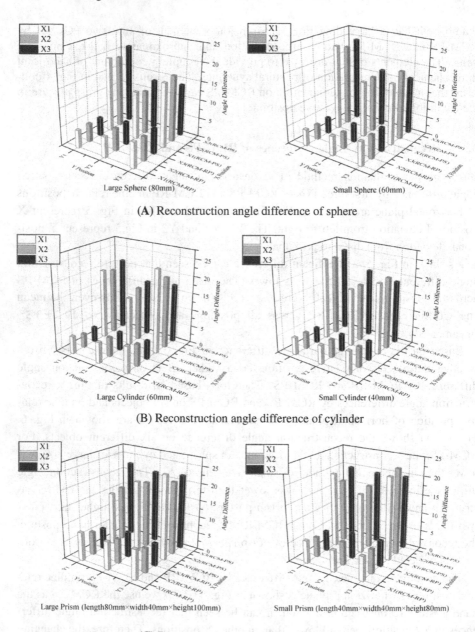

(A) Reconstruction angle difference of sphere

(B) Reconstruction angle difference of cylinder

(C) Reconstruction angle difference of prism

Fig. 5. Reconstruction angle difference by first three PCs of RCM-PS and RCM-RP in nine relative positions of horizontal plane

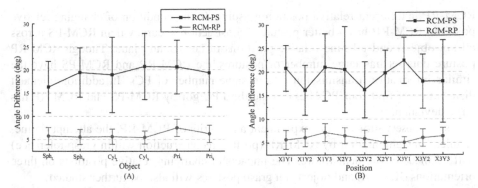

Fig. 6. Reconstruction angle difference by RCM-PS and RCM-RP on six objects and on nine relative positions of horizontal plane

4   Discussion

All the experiment results lead to the following three observations.

— Observation 1: To the separate reconstruction of hand posture (15DOF), the RCM-RP and RCM-PS have the similar information transmission rate in same number of PCs.
— Observation 2: Across first four PCs only PC3 got by RCM-PS and RCM-RP overall had a high fitting degree. The results shown in Fig.4 confirm that mechanical implementation on RCM-PS can't replicate the grasp posture in the condition of changing relative position.
— Observation 3: If the human grasps the objects in different relative positions, RCM-RP has a lower reconstruction angle error than the traditional hand posture reconstruction method (RCM-PS). The changing relative position in Y position has an obvious effect on reconstruction angle difference by RAM-PS. However, for RCM-RP, the object and relative position have a minimal impact on reconstruction angle difference.

From the above observations, three merits of RCM-RP can be obtained: 1) similar information rate with RCM-PS on the posture reconstruction of the hand posture (15DOF) (as shown in Fig. 3); 2) low reconstruction angle difference in changing relative positions (as shown in Fig. 5 and Fig. 6); 3) both object shape and relative position have minimal impact on reconstruction angle difference (as shown in Fig. 6).

5   Conclusion and Future Work

This research presents an experimental study on human grasp posture reconstruction method while considering the relative position between human hand and objects. Results give the posture reconstruction method comparison between RCM-PS and

RCM-RP in different relative position grasping. In the condition of changing relative position, RCM-RP has a better posture reconstruction accuracy than RCM-PS across all six objects and all nine relative positions of horizontal plane. Though RCM-RP posture data set has large number of postures, the RCM-RP and RCM-PS have the similar information transmission rate in same number of PCs. In addition, element distribution of first four PCs besides the third PC got by RCM-PS and RCM-RP has large deviation.

As the observations of the experiment and merits of RCM-RP, the attempt on mechanical implementation by RCM-RP (posture reconstruction within grasp tolerance) will be made in the near future. The impacts of changing relative positions on three orientations (X, Y, Z) and objects on grasp postures will also be further studied.

Acknowledgments. This work is supported in part by the National Basic Research Program (973 Program) of China (Grant No. 2011CB013306) and National Natural Science Foundation of China (51205080).

References

1. Schlesinger, G.: Der mechanische Aufbau der kunstlichen Glieder, Ersatzglieder und Arbeitshilfen für Kriegsbeschädigte und Unfallverletzte (1919)
2. Napier, J.R.: The prehensile movements of the human hand. Journal of Bone and Joint Surgery 38(4), 902–913 (1956)
3. Cutkosky, M.R.: On grasp choice, grasp models, and the design of hands for manufacturing tasks. IEEE Transactions on Robotics and Automation 5(3), 269–279 (1989)
4. Feix, T., Pawlik, R., Schmiedmayer, H.-B., Romero, J., Kragić, D.: A comprehensive grasp taxonomy. In: Robotics, Science and Systems Conference (RSS). Seattle, Washington, USA (2009)
5. Zheng, J.Z., De La Rosa, S., Dollar, A.M.: An investigation of grasp type and frequency in daily household and machine shop tasks. In: IEEE International Conference on Robotics and Automation (ICRA), Shanghai, China, pp. 4169–4175 (2011)
6. Bullock, I.M., Feix, T., Dollar, A.M.: Finding small, versatile sets of human grasps to span common objects. In: IEEE International Conference on Robotics and Automation (ICRA), Karlsruhe, Germany, pp. 1060–1067 (2013)
7. Santello, M., Flanders, M., Soechting, J.: Postural hand synergies for tool use. The Journal of Neuroscience 18(23), 10 105–10 115 (1998)
8. Brown, C.Y., Asada, H.H.: Inter-finger coordination and postural synergies in robot hands via mechanical implementation of principal components analysis. In: IEEE/RSJ International Conference on Intelligent Robots and Systems (IROS), San Diego, CA, USA, pp. 2877–2882 (2007)
9. Xu, K., Liu, H., Du, Y., Sheng, X., Zhu, X.: Mechanical implementation of postural synergies using a simple continuum mechanism. In: IEEE International Conference on Robotics and Automation (ICRA), Hong Kong, China, pp. 1348–1353 (2014)
10. Xu, K., Liu, H., Du, Y., Zhu, X.: Design of an underactuated anthropomorphic hand with mechanically implemented postural synergies. Advanced Robotics 28(21), 1459–1474 (2014)

11. Li, S., Sheng, X., Liu, H., Zhu, X.: Design of a myoelectric prosthetic hand implementing postural synergy mechanically. Industrial Robot: An International Journal **41**(5), 447–455 (2014)
12. Rosmarin, J.B., Asada, H.H.: Synergistic design of a humanoid hand with hybrid dc motor-sma array actuators embedded in the palm. In: IEEE International Conference on Robotics and Automation (ICRA), Pasadena, CA, USA, 773–778 (2008)
13. Sun, B., Xiong, C., Chen, W., Zhang, Q., Liu, M., Zhang, Q.: A novel design method of anthropomorphic prosthetic hands for reproducing human hand grasping. In: IEEE International Conference of Engineering in Medicine and Biology Society (EMBC), Chicago, USA, pp. 6215–6221 (2014)
14. Chen, W., Xiong, C., Liu, M., Liu, M.: Characteristics analysis and mechanical implementation of human finger movements. In: IEEE International Conference on Robotics and Automation (ICRA), Hong Kong, China, pp. 403–408 (2014)
15. Catalano, M.G., Grioli, G., Farnioli, E., et al.: Adaptive synergies for the design and control of the Pisa/IIT SoftHand. The International Journal of Robotics Research **33**(5), 768–782 (2014)
16. Eckart, C., Young, G.: The approximation of one matrix by another of lower rank. Psychometrika **1**, 211–218 (1936)

Three-Dimensional Simultaneous EMG Control Based on Multi-layer Support Vector Regression with Interactive Structure

Wei Yang, Dapeng Yang$^{(\boxtimes)}$, Yu Liu, and Hong Liu

State Key Laboratory of Robotics and System, Harbin Institute of Technology,
Harbin 150080, People's Republic of China
yangdapeng@hit.edu.cn

Abstract. In this paper, a novel three-dimensional (3D) simultaneous myoelectric (electromyography, EMG) control scheme established on multiple layers of support vector regression (SVR) with an interactive structure was proposed. For choosing a proper set of the three degrees of freedom (DOF's), a variety of DOF combinations (e.g., flexion/extension of the thumb and fingers, wrist pronation/supination, etc.) were compared in terms of their regression accuracy. The effort to drive a particular DOF for achieving a given motion with a specific strength was quantified through the root mean square (RMS) of the multi-channel signals, and then used to train a three-layer SVR model. An interactive structure was specially introduced in the model for improving the learning efficiency and control performance by taking advantage of the prior, supplementary regression knowledge. Both offline evaluation (regression criterions) and online experiments (3D target positioning) were conducted to verify our method's efficacy.

Keywords: Simultaneous control · EMG · SVR · Fitts' law · Pattern regression

1   Introduction

The surface electromyography (sEMG) is a kind of synthetically electrical reaction consisting of motor unit action potentials (MUAPs) that can be detected through the differential electrodes placed on the skin over the muscles. It is typically a biomedical, available signal that contains rich information about the motion intentions of the users, and has been widely used in neuropathy detection, motion analysis, and control of neural prostheses and robotic exoskeletons.

The EMG has been employed in the control of prosthetic hands since 1940s. Through modulating the amplitude of the EMG signals, amputees can voluntarily open or close their prosthetic hand [1]. Compared with the past, the structure of prosthetic hand has been largely transformed that a large variety of dexterous hand prostheses gifted with multiple functions, motorized fingers, and degrees of freedom (DOF's) start to be available in the market [2]. However, a suitable method for controlling these hands is still missing [3]. Since pattern recognition was introduced in the EMG control several

© Springer International Publishing Switzerland 2015
H. Liu et al. (Eds.): ICIRA 2015, Part I, LNAI 9244, pp. 282–293, 2015.
DOI: 10.1007/978-3-319-22879-2_26

decades ago [4], the schemes for controlling the prosthetic devices has been merely changed (normally consisting of data segmentation, feature extraction and classification) [5-7]. Although numerous motions related to fingers, wrist, and grasp patterns can be faithfully decoded in ideal laboratory conditions, it is pointed out that there is still a giant gap between research and real practice of the prosthetic hands [8, 9]. For those subjects with congenital disease or amputation, an unavoidable problem is that the available EMG signals on their stump are highly limited. Thus, the pattern recognition-based control is generally difficult to implement and a long term of training should be performed for the adaption between those subjects and the control system. On the other hand, the pattern recognition-based method only promises a single-DOF control once a time, which is far different from the physiological human motor control (multi-DOF simultaneous control). A new kind of methods that can achieve simultaneous control over multiple DOF's was urgently needed [10].

Recently, researchers start to pay close attentions on a type of 2-dimensional simultaneous EMG control methods that has been proved to be more flexible and intuitive than common pattern recognition-based methods. Jiang and his colleges [11] try to obtain a proportional output (for controlling the velocity or force of the finger) by making a comparison to the training data within the motion class predicted by pattern recognition. However, this method is predisposed to fail to give simultaneous outputs of multiple DOF's. Ameri, et al. [12, 13] take advantage of virtual reality (VR) to establish the relationship between EMG signals and continuous outputs of motions, by instructing the subjects to contract muscles according to the cursor's displacement displayed on the screen. The multi-DOF control performance is largely determined by the qualified relationship between the EMG signals and the cursor's continuous motions, in other words, the capability of the subject repeatedly conducting specific muscular contractions according to the cursor movement.

In this paper, we intend to extract the DOF-specific information (e.g. motor effort, as the training target) of a set of 3-DOF simultaneous motions directly from the EMG signals. Several methods are employed and compared for acquiring the desired outputs of training data. An interactive structure is introduced to the support vector regression (SVR) model for distinguishing similar motions. Online controlling experiments base on a task of tracking a 3D target are conducted, and the Fitts' law [14] from graphic design is introduced for evaluating our method's validness.

2   Materials and Methods

2.1   DOF Configuration and Motions

It was found that the human control their bodily joints in varied degrees of capabilities [15]. Suitable combination of DOF's needed to be selected first for obtaining preferred performance of the 3D simultaneous control. Different combinations of hand-wrist DOF's were compared. Assuming that DOF_i ($i=1, 2, 3$) represents the ith DOF, and each DOF has three states of the action (backward, neutral and forward) indicated as $DOF_i(j)$ ($j=-, 0, +$). The different combinations of the 3-DOF's are shown in Table1.

In total, for a given combination there were 27 different states (3 states on 3 DOF's, 3^3). Each state was termed as a motion expressed by ($\{DOF_1(*), DOF_2(*), DOF_3(*)\}$, *=-, 0, +). The force elicited by each DOF needed to be kept same for a control of accordance in the data acquisition.

Table 1. Five 3-DOF combinations of the hand and wrist

Index	DOF_1	DOF_2	DOF_3	Example
1	Flexion/extension of the thumb	Flexion/extension of the index finger	Flexion/extension of the middle finger, ring finger, and little finger	
2	Flexion/extension of the thumb	Flexion/extension of the index finger, middle finger	Flexion/extension of the ring finger, little finger	
3	Flexion/extension of the thumb	Flexion/extension of the index finger	Flexion/extension of the little finger*	
4	Wrist Flexion/extension	Wrist Abduction/Adduction	Wrist Supination/Pronation	
5	Wrist Flexion/extension	Wrist Abduction/Adduction	Hand close/open	

Note: * means the middle finger and ring finger were bound together and kept idle in the motions.

2.2   Data Acquisition Setup

Six commercial EMG electrodes (Otto Bock, 13E200=50) [16], which were placed on the antagonist muscles with regard to each DOF, were adopted to measure the EMG signals. This type of electrode directly outputted the on-board computed RMS of the EMG signal. The amplification of the electrodes were adjusted according to the maximum of signal (Grade 3~5, the magnification factor is 10,000~30,000), which gives a high resolution and an appropriate signal range. The EMG data acquired from finger motions generated by the first three DOF combinations (index 1~3 in Table 1) share the same positions of electrodes (extensor pollicis brevis, flexor pollicis longus, extensor indicis proprius, distal and proximal flexor digitorum superficialis, and extensor digiti quinti proprius). The electrodes in index 4 and 5 were differently configured (the wrist motions involved), with specific positions shown in Fig. 1.

The EMG data were sampled at 100Hz using a 12 bit A/D converter (NI USB-6008). For reducing the number of samples, every five consequent data were averaged and saved. Each acquisition session contained 27 trials; within each trial, a specific motion was collected. In data acquisition, the subject was requested to perform dynamic motion instead of a steady one for enriching the information of EMG signals [17]. Before data acquisition, a proper threshold (10%~20% of the maximum voluntary contraction,

MVC) was selected, according to the EMG amplitude in several collection rehearsals, to get rid of the unreliable samples that could be introduced by unintended or reverse movements. Each trial contained 200 samples, which needed related muscles to contract at least 10 seconds. Plenty of rest was given between trials for avoiding fatigue.

Every 45 minutes, we collected a session of EMG signals without removing the electrodes. The time used for collecting the data only occupied 15 minutes, while the rest (30 minutes) was used for the arm getting rest. To choose the proper DOF configuration, two sessions of EMG data (one for training, one for validation) were collected for each DOF combination in Table 1. In the following analysis, 7 sessions of the selected DOF combination were collected for examining the robustness of the control ability over a comparatively long term (about 5 hours).

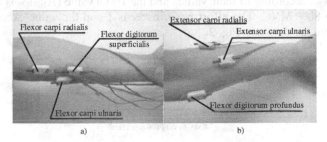

Fig. 1. Electrodes display for the combined hand and wrist DOF's (Left hand).

2.3   Data Processing

The acquired EMG signals (0~5V) were firstly normalized within each channel and the coefficients (the inverse of maximum voltage) were saved for future prediction. We intended to establish a relationship (regression) between the multi-channel EMG signals and the expected outputs of the three DOF's. The relationship consisted of two parts, quantifying the subject's intuitive information from the EMG signals and a map from this intuitive information to the output of DOF's. The intuitive information described the degree of subject's expectation of controlling DOF's and could be approximately quantified using the following two methods:

RMS-Based Method. The strong relativity between the root mean square (RMS) of the EMG signal and the exerted force of the muscle has been widely accepted [18]. Thus, we considered to use RMS to describe the degree of the muscular contraction as the intuitive information.

PCA-Based Method. Because the degree of voluntary contraction could describe the force of the muscles, the relationship among specific samples might be described by only one variable. Principle component analysis (PCA) was applied to obtain this relationship. Within each trial, the first principle component was calculated as the intuitive information for describing the degree of the muscular contractions. On the other hand, each DOF had only three states and could not distinguish the difference among the nine trials in the same state. And the intuitive information might be influenced by the other DOF's. Therefore, PCA was also calculated within the nine trials of a specific $DOF_i(j)$ (i=1,2,or 3; j= -,or +, the outputs when j equals to 0 is zero). These two results were denoted as PCA-based (within trial) and PCA-based (within DOF), respectively.

For the total 200 samples within each trial, we used y_k (k=1~200) to represent the intuitive information estimated through the methods above. We introduced a normalized output, $E_i \in (-1,1)$, to denote the degree of effort the subject elicited to control the ith DOF, and used $P_i \in (-1,1)$ to represent its corresponding value in the prediction procedure. A map from y_k to E_i was generally needed. Due to the high peak values produced by dynamical motions, a linear map would make the average absolute of E_i very small while the absolute value of prediction P_i often exceeds one. Thus, a nonlinear map from y_k to E_i was established in this paper. The y_k was sorted as $A_1 \leq A_2 \leq ... \leq A_{200}$ according to their values and the 160th value (A_{160}) was approximately mapped to 80% MVC. The effort never exceeds 1 and the value exceeds 0.8 were largely avoided, as shown in Fig. 2.

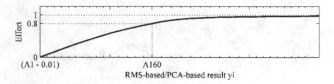

Fig. 2. The nonlinear map between the RMS-based /PCA-based values and the Effort.

As proposed in [12, 13], each DOF has an independent SVR model that the relationships among different DOF's were ignored. In this paper, the dimension of the feature for training specific SVR models was enlarged thus the information from the other DOF's could be utilized while training the model. The structure of the proposed method is shown in Fig. 3. Firstly the three SVR's was arranged from top to bottom according to their regression accuracy. Then, the first and the second SVR's outputs were introduced to the second and the third SVR's, respectively, to take fully advantage of the available regression experiences. Since this interactive structure of SVR's notices the relationships among the models, we expected it could somehow distinguish the similar motions, thus to improve the regression accuracy.

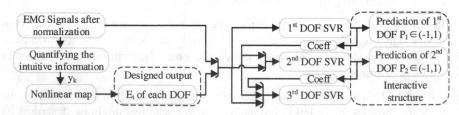

Fig. 3. Structure of the simultaneous regression model with interactive structure.

The SVM algorithm (ε-support vector regression, ε-SVR) was realized using the libsvm package [19]. The radial basis function (RBF) was adopted as the kernel type, the cost parameter C was set to 2, and the kernel parameter γ was set to 0.5, according to the grid search method.

2.4   Offline/Online Evaluations

The regression accuracy (RA), relative mean square error (RMSE), and average absolute error (AAE) were selected for evaluating the offline estimation accuracy. Besides, the rate of support vector (RSV) which can indicate the velocity of learning and the degree of overfitting was employed to evaluate the complexity of the model. Equations are listed in Table. 2. All calculations were performed in MATLAB environment (Version 8.0, Dual Core 3.4GHz, 4GB RAM).

In addition to offline evaluation, an online 3D simultaneous EMG control experiment over a moving cursor (along with the X-axis, Y-axis and its size) on the screen in front of the subject was also conducted. The graphic-user interface (GUI), as shown in Fig. 4, was compiled in LabVIEW (version 2012). The Fitts' law [14] was employed as the foundation of the experiment, and the metrics [13] applied in the experiment were shown on Table.3. The velocity of the cursor was controlled by the outputted prediction P_i, while the movement on the X/Y direction is controlled by wrist at the velocity of 200 pixels per second and the size of the cursor was controlled by hand open/close. The directions controlled by wrist were adjusted according to the subject's preference. In Fig. 4, the color of the cursor would change if the distance between the cursor and target was within W/2 for instructing the subject. If the cursor was kept on target for 1s, then this control trail was claimed successful and a new target would be generated according to the given task parameters and the position of the cursor. If the target was not reached within 15s, then the control was claimed unsuccessful and a new target would also be generated in the same way. Thus, the subject's attention could always keep following the cursor during the experiment. An online control experiment lasted 100s.

For practicing the control task, a 2D control over the cursor's position was conducted firstly before the 3D control. This 2D control was totally accomplished by the wrist motions without displaying the information of the third DOF. In other configurations, the 2D was the same to the 3D control.

Table 2. Equations of standards for evaluating the model.

Metric	Equation	Metric	Equation
RA	$RA_i = 1 - \dfrac{\sum\limits_{t=0}^{M}\left(P_i(t) - S_i(t)\right)^2}{\sum\limits_{t=0}^{M}\left(S_i(t) - \overline{S_i(t)}\right)^2}$	RMSE	$RMSE_i = \dfrac{\sum\limits_{t=0}^{M}\left(P_i(t) - S_i(t)\right)^2}{\sqrt{\sum\limits_{t=0}^{M}S_i(t)^2}\sqrt{\sum\limits_{t=0}^{M}P_i(t)^2}} \times 100\%$
AAE	$AAE_i = \dfrac{\sum\limits_{t=0}^{M}\left|P_i(t) - S_i(t)\right|}{M}$	RSV	$RSV_i = \dfrac{\text{number of support vector}}{\text{number of samples}} \times 100\%$

3   Results

3.1   Different Combinations of DOF's

For each combination in Table.1, the cross validation (one session for training, one session for predicting) on regression accuracy was made. The desired effort E_i was calculated through the RMS-based method. The obtained RA and RSV are shown in Table. 4. In terms of RA and RSV, the DOF combinations with Index of 3 and 5 perform better than the others. Due to its suitable placement of the electrodes (all near the elbow) and high control intuitiveness, we chose the 3-DOF configuration as wrist flexion/extension, wrist abduction/adduction and hand close/open in our following experiments.

Table 3. Metrics in the 3-D fitts' law experiment [13]

Metric	Description
Target distance (D)	The shortest distance between the starting position and the target.
Target width (W)	If the distance between cursor and target is smaller than W/2, it means the cursor reach the target.
Index of difficulty (ID)	Describing the difficulty of the task; equal to $\log_2(D/W+1)$.
Movement Time	The time the cursor needs to reach and keep in the target.
Completion Rate	Describing the task success rate; the percentage of the reached targets within 15 seconds.
Path Efficiency	Describing the control quality; the mean of the ratio between D and the distance cursor travelled.
Overshoot	Describing the control precision; the average times the target was reached but lost before finishing.

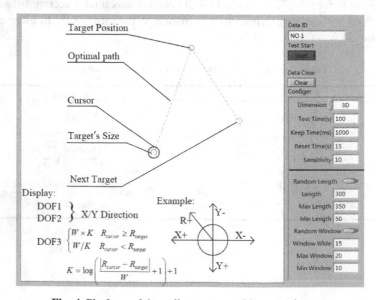

Fig. 4. Platform of the online target tracking experiments

3.2   Different Pre-processing Methods

Through cross validation, the regression performance acquired by different pre-processing methods is compared in Table 5. The RMS-based and PCA-based (within trial) pre-processing method share a superior performance. Because of the fewer support vectors and less calculations, the RMS-based method is employed for the following experiments.

Table 4. Performance comparison on different 3-DOF configurations

	Index	DOF_1	DOF_2	DOF_3
RA (mean±standard)	1	0.524±0.273	0.501±0.261	0.599±0.162
	2	0.629±0.159	0.691±0.183	0.700±0.172
	3	0.828±0.063	0.852±0.073	0.902±0.064
	4	0.631±0.207	0.635±0.208	0.638±0.198
	5	0.853±0.077	0.800±0.079	0.876±0.056
RSV (mean±standard)	1	49.7±3.6	46.6±1.0	39.0±1.5
	2	47.6±2.5	40.6±0.1	41.5±3.6
	3	39.2±0.2	29.4±3.6	24.4±3.4
	4	41.3±9.3	42.6±0.8	48.6±5.4
	5	33.1±3.2	39.4±0.3	29.0±2.4

Table 5. Comparison of different pre-processing methods (RMS- and PCA-based)

		DOF_1	DOF_2	DOF_3
RA (mean±standard)	RMS-based	0.852±0.043	0.647±0.096	0.890±0.037
	PCA-based(within trial)	0.850±0.042	0.653±0.095	0.883±0.040
	PCA-based(within DOF)	0.855±0.041	0.608±0.102	0.866±0.030
RSV (mean±standard)	RMS-based	32.3±3.5	48.8±2.7	26.6±3.7
	PCA-based(within trial)	34.5±3.2	49.1±3.1	28.1±4.2
	PCA-based(within DOF)	33.8±3.3	50.9±4.4	32.8±3.8

3.3   Non-interactive and Interactive Structure

The interactive structure in the order of DOF_3, DOF_1 and DOF_2 was employed according to their regression performance. The results obtained by the SVR with and without the interactive structure are compared in Table 6. The number of support vector decreases 12% and 21%, respectively, after introducing the interactive structure. The regression accuracy (RA) of DOF_2 increases 4%, while this value of DOF_1 declines a little.

For detailing the performance of the interactive structure, the predictions of 27 motions within a session were divided into three groups according to the number of active DOF's: 1) one active DOF's and the rest (No-Multi), 2) two active DOF's (2-Multi), and 3) three active DOF's (3-Multi). Then, the RA, RMSE and AAE were calculated with regard to each DOF (DOF_1, DOF_2, DOF_3, respectively), as shown on Fig. 5. It is found that, by introducing the interactive structure, the DOF_2 can obtain slightly better regression accuracy especially when few active DOF's are getting involved at the same time.

3.4   Online Experiment

Nine combinations of D and W were configured in the online control experiment. The SVR model with interactive structure were finally utilized on the platform, and the parameters were configured according to Table 7. The result of the metrics is shown in Table. 8. The relationships between ID and movement time are shown in Fig. 6. Segmental trajectory of the cursor controlled by the simultaneous EMG control scheme (ID= 4) are shown in Fig. 7 and Fig. 8.

Table 6. Comparison of the different structures (non- interactive and interactive)

	Without interactive structure			With interactive structure		
	DOF_1	DOF_2	DOF_3	DOF_1	DOF_2	DOF_3
RA	0.852	0.647	0.890	0.849	0.674	0.890
(mean±standard)	±0.043	±0.096	±0.037	±0.050	±0.109	±0.037
RSV (mean±standard)	32.3±3.5	48.8±2.7	26.6±3.7	28.5±3.3	38.7±2.4	26.6±3.7

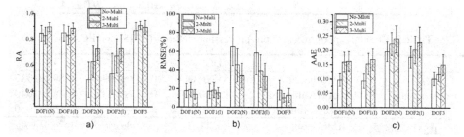

Fig. 5. Comparison of the regression performance according to a specific DOF (DOF_1, DOF_2 or DOF_3) with or without interactive structures (denoted as I and N, respectively)

4   Discussion

After comparing the offline performance among the first three combinations of DOF's, the one consisting of thumb, index and little finger while the other fingers were bound together performances better. Performance promotion was also found while replacing the wrist supination/pronation by hand open/close, which means that the appropriate separation of the DOF's would contribute to the performance. The RSV could reflect this degree of separation as well as regression overfitting while other conditions were kept same. An ideal combination of DOF's should have a high RA and a low RSV at the same time.

Table 7. Parameter configuration on each task

No.	1	2	3	4	5	6	7	8	9
D(pixel)	200	200	200	100	100	100	300	300	300
W(pixel)	20	30	40	20	30	40	20	30	40
ID	3.46	2.94	2.58	2.58	2.12	1.81	4.00	3.46	3.09

Table 8. Results of the matrics of the Fitts' law

	Complete Rate	overshoot	efficiency
2D Fitts' law	0.981	0.009	0.790
3D Fitts' law	0.943	0.175	0.692

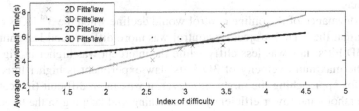

Fig. 6. The relationship between the movement time and index of difficulty

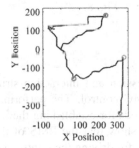
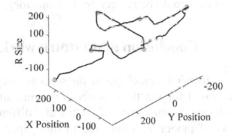

Fig. 7. Controlling trajectory of a 2-D targeting task (ID = 4)

Fig. 8. Controlling trajectory of a 3-D targeting task (ID = 4)

For estimating the driving effort of the specific $DOF_i(j)$, the PCA-based method was attempted for analyzing the influence of other DOF's within the nine trials with relation to $DOF_i(j)$. But the result showed that the influence could not be eliminated through PCA. However, the regulation of effort within trials could be calculated by the RMS-based or PCA-based method, since the proportional information within trials was more obvious.

The interactive structure was introduced in the model for distinguishing similar motions, which could be reflected by the decrease of RSV. The supernumerary feature predicted by other DOF's could differ the similar training data within the $DOF_i(j)$ and promote the performance of prediction. The interactive structure increased the feature by 14%, which was calculated in the inner product of the kernel function and seldom influenced the time cost. Meanwhile, the 12% reducing support vector directly decreased the complexity of model. This reduction could be obviously observed while predicting. However, while building the model, two times of predictions for all the samples should be introduced in and the reduction of time was not so evident.

For DOF_1, the improvement of regression was only reflected on the reduced number of support vectors because of the incomplete information only given by DOF_3. When complete information was given, as for DOF_2, the performance of regression would get a promotion in terms of both regression accuracy and complexity of the model.

According to the performance given by different groups divided by the number of active DOF's, the regression accuracy on No-Multi and 2-Multi (e.g. motions with few active DOF's) can be better improved by introducing the interactive structure. After comparing the RMSE and AAE, results also indicated that the prediction of the neutral state often retained a bias which could be reduced by employing the interactive structure.

The performance of the online control would decline when the 3rd DOF was added, indicating that the 3-DOF trajectory control was more difficult and less intuitive. The slope of 3D Fitts' law was less cliffy while the intercept was higher in Fig. 6. It was because the maximum velocity of 3D Fitts' law experiment is higher thus it needed less time for approaching the target; however, it was more difficult (indicated by the higher overshoot and lower efficiency) for reaching and keeping in the target. Moreover, the subject could hardly perceive the velocity relationship between R direction and X/Y direction on the display in real time. Thus, a severe change of velocity in R-direction could be noticed in the trajectory, as shown in Fig. 8.

5   Conclusion and Future work

A regression model based on support vector regression and interactive structure is proposed for the three-dimension simultaneous EMG control. The performances of different combinations of DOF's and different methods for establishing the ideal outputs of model are compared in this paper. Results show that the RMS of the EMG signals could approximately reflects the motor effort for driving the joints. The model has a better performance on distinguishing the difference between similar motions with an interactive structure. Online control experiment on 3D cursor tracking (positioning and zooming) finally validates our design through the metrics of Fitts' law.

Future work will focus on a strategy that can reduces the number of the training motions while keeping the online control performance. Experiments on a population of subjects including amputees will be conducted to obtain more faithful results. Finally, our method will be examined in the real-time control of multi-DOF prosthetic hands.

Acknowledgments. This work is in part supported by the National Program on Key Basic Research Project (973 Program, No. 2011CB013306), the National Natural Science Foundation of China (No. 51205080), the China Postdoctoral Science Foundation Funded Project (No. 2013M540276, 2014T70316), and the Heilongjiang Postdoctoral Fund (No. LBH-Z13082).

References

1. Zecca, M., Micera, S., Carrozza, M.C., Dario, P.: Control of multifunctional prosthetic hands by processing the electromyographic signal. Critical Reviews in Biomedical Engineering **30**, 459–485 (2002)
2. Belter, J.T., Segil, J., Dollar, A.M., Weir, R.F.: Mechanical design and performance specifications of anthropomorphic prosthetic hands: a review. J Rehabil Res Dev **50**, 599–618 (2013)
3. Castellini, C., Artemiadis, P., Wininger, M., Ajoudani, A., Alimusaj, M., Bicchi, A., Caputo, B., Craelius, W., Dosen, S., Englehart, K., Farina, D., Gijsberts, A., Godfrey, S.B., Hargrove, L., Ison, M., Kuiken, T., Markovic, M., Pilarski, P., Rupp, R., Scheme, E.: Proceedings of the first workshop on Peripheral Machine Interfaces: going beyond traditional surface electromyography. Frontiers in Neurorobotics **8**, Article 22: 21–17 (2014)
4. Hudgins, B., Parker, P., Scott, R.N.: A new strategy for multifunction myoelectric control. IEEE Trans. Biomed. Eng. **40**, 82–94 (1993)
5. Fougner, A., Stavdahl, O., Kyberd, P.J., Losier, Y.G., Parker, P.A.: Control of Upper Limb Prostheses: Terminology and Proportional Myoelectric Control: A Review. IEEE Transactions on Neural Systems and Rehabilitation Engineering **20**, 663–677 (2012)
6. Scheme, E., Englehart, K.: Electromyogram pattern recognition for control of powered upper-limb prostheses: State of the art and challenges for clinical use. Journal of Rehabilitation Research and Development **48**, 643–660 (2011)
7. Oskoei, M.A., Hu, H.: Myoelectric control systems—A survey. Biomedical Signal Processing and Control **2**, 275–294 (2007)
8. Ning, J., Dosen, S., Muller, K.R., Farina, D.: Myoelectric control of artificial limbs: is there a need to change focus? IEEE Signal Process. Mag. **29**, 148–152 (2012)
9. Peerdeman, B., Boere, D., Witteveen, H., in't Veld, R.H., Hermens, H., Stramigioli, S., Rietman, H., Veltink, P., Misra, S.: Myoelectric forearm prostheses: State of the art from a user-centered perspective. Journal of Rehabilitation Research and Development **48**, 719–737 (2011)
10. Ameri, A., Scheme, E.J., Kamavuako, E.N., Englehart, K.B., Parker, P.A.: Real-time, simultaneous myoelectric control using force and position-based training paradigms. IEEE Trans Biomed Eng **61**, 279–287 (2014)
11. Jiang, N., Lorrain, T., Farina, D.: A state-based, proportional myoelectric control method: online validation and comparison with the clinical state-of-the-art. Journal of Neuroengineering and Rehabilitation **11**, 11 (2014)
12. Ameri, A., Kamavuako, E.N., Scheme, E.J., Englehart, K.B., Parker, P.A.: Real-time, simultaneous myoelectric control using visual target-based training paradigm. Biomedical Signal Processing and Control **13**, 8–14 (2014)
13. Ameri, A., Kamavuako, E., Scheme, E., Englehart, K., Parker, P.: Support Vector Regression for Improved Real-Time, Simultaneous Myoelectric Control. IEEE Trans Neural Syst Rehabil Eng **22**, 1198–1209 (2014)
14. MacKenzie, I.S.: Fitts' law as a research and design tool in Human-Computer Interaction. Human-Computer Interaction **7**, 91–139 (1992)
15. Jones, L.A., Lederman, S.J.: Human Hand Function. Oxford University Press, Oxford (2006)
16. Otto Bock MYOBOCK 13E200=50 electrodes, http://www.ottobock.com/cps/rde/xchg/ob_us_en/hs.xsl/16573.html
17. Yang, D., Zhao, J., Jiang, L., Liu, H.: Dynamic hand motion recognition based on transient and steady-state EMG signals. International Journal of Humanoid Robotics **9**, 11250007 (2012)
18. De Luca, C.J., Adam, A., Wotiz, R., Gilmore, L.D., Nawab, S.H.: Decomposition of Surface Emg Signals. J Neurophysiol **96**, 1646–1657 (2006)
19. http://www.csie.ntu.edu.tw/~cjlin/libsvm

Idle Mode Detection for Somatosensory-Based Brain-Computer Interface

Xiaokang Shu, Lin Yao, Xinjun Sheng, Dingguo Zhang, and Xiangyang Zhu$^{(\boxtimes)}$

State Key Laboratory of Mechanical System and Vibration,
Shanghai Jiao Tong University, Shanghai, China
shuxk89@gmail.com
http://bbl.sjtu.edu.cn

Abstract. *Objective.* Idle mode detection is a vital problem to be solved in self-paced (asynchronized) BCI, because patients need to control the BCI system whenever he or she wants, rather than output commands according to the system cues which is essentially different between self-paced and synchronized BCIs. With the detection of idle mode, we can finally increase the performance of real-time BCI. *Approach.* In this work, we introduce a new experiment paradigm to research the difference between idle mode and task mode from those domains of time, frequency and spatial. The experiment is carried out with 14 volunteers from 19 to 26 years old. *Main results.* Off-line analysis shows significant differences of the distribution of EEG power spectrum between task mode and idle mode. When the subjects execute a left or right hand selective sensation, there appears an obvious ERD/ERS in the subject's contralateral cortex. However, there is no ERD/ERS during idle mode periods even with simultaneously applied vibration. With combining data set of left and right hand sensation for classifier calibration, we recognize idle mode from mixed task mode with TPR value of 86%, and this result is significantly higher than that TPR achieved from single task type (left or right hand sensation) with $p < 0.05$. *Significance.* The new proposed calibration method is demonstrated to be feasible for idle mode detection in SS-BCI. With the combination of two mental tasks, we can improve the BCI performance by increasing true positive rate, and limit false positive rate at the same level compared to the idle mode recognition from single task type.

Keywords: Self-paced BCI · Selective sensation · Idle mode detection · True positive rate · False positive rate

1 Introduction

Brain computer interface (BCI) has been proposed to help the motor-disabled patients establish a new pathway to communicate with external world [1], especially for those people suffering from amyotrophic lateral sclerosis (ALS) [2] and spinal cord injuries (SCI) [3]. Existing BCIs mostly operate only during specific

© Springer International Publishing Switzerland 2015
H. Liu et al. (Eds.): ICIRA 2015, Part I, LNAI 9244, pp. 294–306, 2015.
DOI: 10.1007/978-3-319-22879-2_27

periods predefined by the system. One disadvantage of these BCIs is subject must perform corresponding mental tasks according to the system cues, it is usually called synchronized BCI. However, a more practical BCI is required to be controlled by users freely, namely users can activate and control the system whenever they wish, this kind of BCI is called asynchronized (self-paced) BCI. In the later type, the system firstly needs to predict whether the subject is outputting a control command [4]. Two states should be classified in this step, task mode (TM) and idle mode (IM). In second step, if the system is in task mode, another classifier is used to predict the actual control commands. Then the predicted commands will be used to control the outside device, or the system should maintain inactive when idle mode is predicted.

Motor imagery is frequently used in asynchronized BCI as user's mental tasks. In this kind of study, spectral power variation [5][6] or event related potential [7] is extracted from raw EEG signal, and these features are strictly related to the simultaneously performed mental task. While the system continuously predicting the user's intention, a manual threshold value is used to determine the mental state. The detail information about this method can be found in [6]. Actually, the threshold is always optimized to produce a highest TPR or a lowest FPR. However, the limitation is the BCI with a lowest FPR will also have a very low information transfer rate (ITR), and it is supposed to make the subject feel frustrated. Another solution to detect mental state is pattern classification [8]. In this kind of BCI, IM is considered to be one individual class and should be classified from TMs. Actually, the IM can be extended to be an additional pattern to provide the user with more control commands. In [8], three mental tasks including left hand imagery, words association and rest are introduced to be the users' tasks and each of them is expected to evoke a specific pattern of EEG spectral power. However, results of this study show the pattern differences, especially between TM and IM, are not significant among most subjects. In this way, there are many 'BCI-illiteracy' of motor imagery based BCIs (MI-BCI) [9].

To improve the performance of 'BCI-illiteracy', we need to select some efficient mental tasks for those users. Currently, mental tasks like motor imagery, words association, auditory imagery, and spatial navigation have been proposed to modulate subject's cortical rhythm for BCI utilization. Among these tasks, motor imagery is most used, as it produces changes in EEG that occur naturally in movement planning and is relatively straightforward to detect [10]. However, 'BCI-illiteracy' could not change their cortical oscillation with motor imagery [9]. In the study of [11], authors explored three non-motor imagery mental tasks including sentence visualization, multiplication and object rotation as well as one motor imagery task of right hand extension. As a result, there seems no significant difference among these four mental tasks. Other than these independent mental tasks, one dependent mental task named selective sensation based on mechanical vibrotactile stimulation was proposed in [12], selective sensation based BCI (SS-BCI) has been proven to be better than motor imagery based BCI (MI-BCI), especially for 'BCI-illiteracy'.

In this work, we aim at improving the performance of idle mode detection by introducing a new calibration method for training of classification between idle mode and task mode to make the data set more separable. Common spacial pattern (CSP) algorithm is used for feature extraction and linear discriminate analysis (LDA) serves as the classifier. The experiment paradigm is corrected from that of [12], we have combined idle mode with left and right hand sensation to study which calibration method is more suitable for idle mode detection in SS-BCI. False positive rate (FPR) and true positive rate (TPR) [6] are calculated to evaluate the system's performance. For each subject, we analyse the EEG signal difference from those dimensions of temporal, spatial and frequency.

2   Materials and Methods

2.1   Subjects

14 subjects (8 males and 6 females) at all have successfully participated in the experiment, the subjects' age ranged from 19 to 26, and all right handed. They were all informed with the whole experiment process. This study has been approved by the Ethics Committee of Shanghai Jiao Tong University. All participants signed the informed consent forms before participating in the experiments.

2.2   EEG Recording and Stimulation Device

EEG signals were recorded using a SynAmps2 system(Neuroscan, U.S.A.). 64 channel quick-cap was used to collect 62 channel EEG signals, and the electrodes were placed according to the extended international 10/20 system. The reference electrode was located on the vertex [13], and the ground electrode was located on forehead. An analog bandwidth filter with 0.5 Hz to 70 Hz and a notch filter with 50 Hz to diminish power line interference were applied to the original signals, which were sampled at 250 Hz.

In this experiment, stimulation was applied to the wrist skins, with 175 Hz (resonant frequency of the stimulating device) sinusoidal carrier frequency, modulated with 27 Hz to induce flutter sense. In this stimulating configuration, two types of mechanical receptor (Pacinian corpuscles and Meissner corpuscles) were stimulated, which were especially sensitive to frequency above 100 Hz and 20 to 50 Hz respectively [21]. And these stimulation configurations would attract much cortical processing of afferent inflow compared to single frequency stimulation either with high or low frequency, with which only one type of mechanical receptors was stimulated.

Both left and right wrist skins were simultaneously stimulated, with equal amplitude and the same modulation frequency. The linear resonant actuators(10 mm, C10-100, Precision Microdrives Ltd. Typical Normalized Amplitude 1.4G) were used for vibrotactile stimulation. Electrical signal of 175 Hz sinusoidal carrier frequency modulated with 27 Hz sinusoidal frequency was produced via computer sound card, and amplified with audio amplifier to drive the actuators.

The amplitude of vibration was individually adjusted within the range of 0.5 times the device normalized amplitude to maximum amplitude of 11.3 um at resonant frequency, so that the subjects could feel the intense vibration with flutter sense, and it was modulated neither too small nor too strong that the subject could concentrate himself or herself on performing the predefined experimental task.

2.3 Experimental Paradigm

The experiment is designed as figure.1. Subjects were seated in front of a LCD displayer with hands and forearms rested on the armrest, and one vibrator was tied to each of the subject's wrists. Distance between the subject and screen was set as about 70 cm to ensure the subjects could see the displayed information clearly. During the experiment preparation, the subjects were informed of the experiment procedures and corresponding mental tasks. The subjects were required not to blink eyes frequently and no real movement during task periods.

At the beginning of each trial, there would be a white cross in the center of screen and a short time (200 ms) vibration was produced on the subject's wrists to remind the start of task. In the 3rd second, the system randomly gave a cue to introduce the subject to complete corresponding mental task. There were three kinds of system cue, red rectangle appearing in left, right and center, and each of them represented left hand sensation, right hand sensation and taking rest, respectively. Meanwhile, the vibrators started working and subjects should selectively sense the cued side vibration and ignore the opposite one. The cue disappeared after 1.5 seconds, while the cross and vibration continued until the 8th second. Then a rest of 3 ± 2 s would be taken before end of the trail. Totally 60 trials (20 for left hand sensation, 20 trial right hand sensation and 20 trials for idle mode) made up a session, and each subject needed to carry out 4 sessions. In this case, the experiment lasted less than 1 hour.

This experiment paradigm is similar to our former work [12], the most difference is we aimed at extending the conventional binary BCI to a three class BCI system by adding idle mode detection. Compared with the work of Huang [14], our experiment is based on selective sensation which has been proved to be more robust than motor imagery.

2.4 Decoding Algorithms

In this research, Common Spatial Pattern (CSP) and Linear Discriminant Analysis (LDA) have been introduced for spatial filtering and pattern classification, respectively. CSP is a most effective decoding algorithm in motor imagery based BCI, its mathematical basis is realized by simultaneous diagonalization of the covariance matrices for the two classes [15][16]. With this method, we can find some spatial directions which will be used for the raw data projection to maximize the power difference of two classes, and this information can be used for classification. LDA is a most used classifier in BCI literatures, and it is suitable

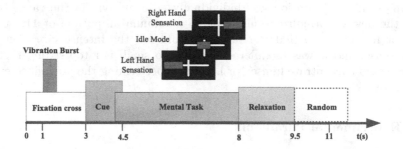

Fig. 1. Experiment paradigm design. At the beginning of a single trial, a black cross is displayed on the screen, and a short vibration is applied to notice the subject to focus on the screen. In 3rd second, a red block is displayed in left, right or center area of the screen to inform the mental tasks of left and right hand sensation, and idle mode, separately. The task continues until the 8th second, and is followed with a random period (1.5±2) time of rest.

for Gaussian distribution data, especially for a two-class classification problem. The detail implementation of LDA can refer to [17].

We analysed the event related desynchronization (ERD) and event related synchronization (ERS) variation within the dimensions of temporal, frequency and spatial. ERD and ERS are defined as the percentage of power decrease or power increase in a defined frequency band referred to the base line [18]. In our research, we aimed at having a better understand of brain activation during different mental tasks, then with this essential knowledge we may get a better performance of our BCI system. In order to emphasize the endogenous event related information, we focus on the time-frequency variation of two channels of C3 and C4 according to the international extended 10/20 system, and spatial analysis was done within the interval from 0.6s to 4s after the system cue. All of the feature analysis were done with a Matlab toolbox named FieldTrip [19], which was developed by Radboud University for advanced analysis of MEG, EEG and invasive electrophysiological data, especially for time-frequency analysis.

In order to study the effects of time and frequency parameters for pattern classification, different frequency bands and time segments are used in following analysis. In time domain, given the reaction time (from the appearance of the indicating cue to the actual mental performing) of subjects, we only analysed the data from time 1s to 5s after the appearance of each start cue. The time segment of [1 5]s is regarded as the whole trial and other three time segments of [1 4]s, [1 3]s, [1 2]s, [2 5]s, [2 4]s, [2 3]s, [3 5]s, [3 4]s and [4 5]s are also used for classification to learn whether the classification results are robust about time. In frequency domain, the normally used frequency band of [8 26] Hz is separated into lower α [8 10] Hz, upper α [10 13] Hz, α [8 13] Hz, and lower β [13 20] Hz, upper β [20 26] Hz, and β [13 26] Hz. Then each frequency band was combined with the 10 time segments, and $7\times10=70$ pairs of parameters are formed.

We carried out a trial-based 10-fold cross validation to evaluate the system's performance. For single-task idle mode classification, there were 80 trials of left/right hand sensation and 80 trials of idle mode for calculations, the data set were separated into 10 parts randomly and we picked up one of them (8 trials of each class) for testing, then the remaining trials (72 trials for one class) were used for training the classifiers. Each part of data should be used for testing and this step would be repeated for 10 times. For mixed-task idle mode classification, 50% (40 trials) of left hand sensation and 50% (40 trials) of right hand sensation were picked up to form the mixed-task type data set. Then the selected mixed-task trials would be classified from those 80 trials of idle mode tasks with 10-fold cross validation.

3 Results

3.1 Signal Feature Analysis of Different Mental Tasks

To have a better understanding of the signal feature difference between three types of mental task, a time-frequency-spatial analysis is conducted on the filtered data of each subject. The results of subject s1 are shown in figure.2. The left column represents band power variation along time and frequency corresponding to different mental tasks. The right column shows the spatial distribution of band power. The top and middle row are related to left and right hand sensation, while the bottom row is corresponding to idle mode. When the subject is performing left or right hand sensation, there is an obvious ERD phenomenal shortly after the start cue, and this power decreasing phenomenal appears both in α band and β band. From spatial distribution, we find the contralateral dominance exists. On the other hand, when the subject is performing left hand sensation, the ERD value of his right somatosensory cortex is larger than the left side. Otherwise, when the mental task is right hand sensation, the result is contrary. This result has been proven in former studies [12]. However, when we compare the bottom row with those two upper rows, there was no significant ERD phenomenal in neither domain of time, frequency or spatial. This difference could be further used for idle mode detection.

For pattern classification with features of band power, selection of frequency band is of great concern. As shown in figure.2, ERD phenomenal exists in both α and β bands, then which band has better performance for recognition needs to be decided. Actually, we have analysed power variation of all subjects and we find the frequency band of cortical activation is specific for every subject. Time-frequency analysis results of the first three subjects (S1, S2, and S3) are shown in figure.3. From left column of the figure, we can see that subject S2 shows significant ERD in [10 13]hz which is upper α, while subject S3 shows significant ERS in [10 12]hz, and these two features are both different from that of subject S1. In this way, frequency band optimization is going to be introduced in the following calculation.

Figure.4 has shown the topographic maps of both task mode and idle mode. Similar results with figure.2 are obtained considering the band power spatial distributions. When the subject is performing a selective sensation task (no matter

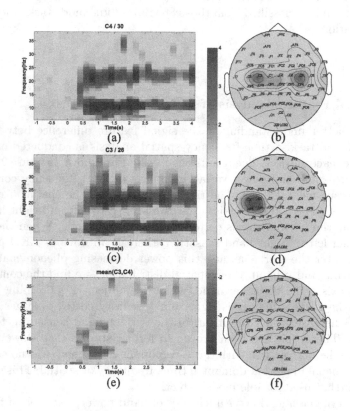

Fig. 2. Time-frequency-spatial analysis of EEG from subject S1. The left column represents band power variation along time and frequency corresponding to different mental tasks. The right column shows the spatial distribution of band power. The top and middle row are related to left and right hand sensation, while the bottom row is corresponding to idle mode. Time-frequency maps are draw with EEG data of channel C3 and C4. Topographic maps are drawn with time=[0.6 4]s and frequency=[10 13]hz

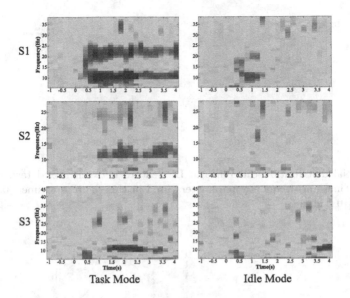

Task Mode   Idle Mode

Fig. 3. Time-frequency analysis of EEG from subject S1, S2 and S3. The left column represents spectral power variation corresponding to task mode and right column represents that of idle mode. It's obvious the activation frequency bands depends on individual characters.

right hand or left hand), the somatosensory cortex (both or contralateral hemisphere) is activated due to the subject's internal modulation during mechanical vibration. However, if the subject do not sense the vibration on purpose during idle mode task periods, corresponding cortex is not going to be activated. This phenomenal is suitable for recognizing idle mode from selective sensation tasks.

3.2   Evaluation of the BCI System

Evaluation method of 10-fold cross validation is introduced to predict the BCI system's performance. There are three groups of evaluation results which are corresponding to detecting idle mode from left hand sensation, from right hand sensation, and from mixed task, separately. For the idle detection of mixed tasks, 40 trials of left hand sensation and 40 trials of right hand sensation are picked up to form a new data set which is regarded as task mode. With these processes, we acquire the TPR values of 79%, 81% and 86% for each group with frequency optimization, as shown in figure.5. The filter and classifier used here were CSP and LDA.

With a T-test between these three groups of recognition results, we find there is no significant ($p > 0.5$) difference between group one and group two, but the third group is significantly higher ($p < 0.01$) than the first two groups. It means either of left hand sensation or right hand sensation has the same ability to be separated from idle mode. However, if these two tasks are combined and regarded

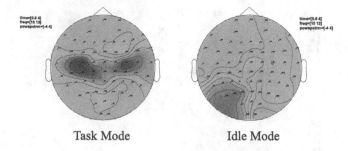

Task Mode   Idle Mode

Fig. 4. Spatial distribution difference of BP feature between mixed task mode and idle mode. Time and frequency parameters used in this figure are time=[0.6 4]s and frequency=[10 13]hz

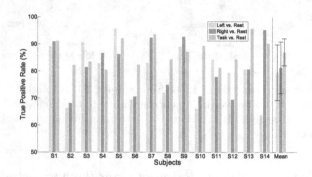

Fig. 5. True Positive Rate (TPR) of different task modes and idle mode. The blue bar represent TPR of idle mode and left hand sensation, the green bar represent TPR of idle mode and right hand sensation, while the brown bar represent TPR of idle mode and mixed task mode

as a mixed task type, the signal features are changed and ERD distribution is expanded to both sides of somatosensory cortex. In this way, those subjects with low TPR values will get improved. From figure.5, we can see the TPR values of subject s2, s6, s10 have been improved from below 70% to above 80%. In this way, the mixed task type is better for idle mode detection compared with the task type of single hand sensation.

3.3 Effects of Different Parameters

Latency of the BCI system is a limitation for daily use. In this study, we analyze the effects of data segment length on decoding results. This parameter is an important factor of the decoding latency. The segment length used in the above analysis is 4 s, and in this section we consider three segment lengths of 1 s, 2 s and 3 s. Given the reaction delay of subjects, we extract data from the 2nd

Table 1. True Positive Rate(TPR) and False Positive Rate(FPR) of different time segments between task mode and idle mode

Subject	1s		2s		3s	
	TPR	FPR	TPR	FPR	TPR	FPR
S1	0.85	0.29	0.86	0.26	0.90	0.23
S2	0.84	0.30	0.81	0.31	0.82	0.39
S3	0.80	0.28	0.83	0.28	0.81	0.30
S4	0.79	0.46	0.82	0.33	0.83	0.34
S5	0.89	0.19	0.94	0.16	0.92	0.20
S6	0.70	0.31	0.85	0.33	0.92	0.41
S7	0.85	0.26	0.92	0.10	0.93	0.09
S8	0.79	0.26	0.84	0.31	0.84	0.36
S9	0.89	0.14	0.85	0.23	0.90	0.20
S10	0.75	0.38	0.85	0.39	0.86	0.39
S11	0.79	0.35	0.83	0.25	0.87	0.16
S12	0.81	0.29	0.83	0.31	0.81	0.36
S13	0.91	0.20	0.93	0.16	0.95	0.25
S14	0.79	0.36	0.91	0.28	0.89	0.30
MEAN	0.82	0.29	0.86	0.26	0.88	0.28
STD	0.06	0.08	0.04	0.08	0.05	0.10

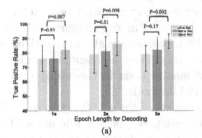 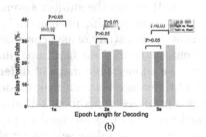

(a)   (b)

Fig. 6. Comparison of TPRs and FPRs for different epoch lengths and task types. The three groups are corresponding to segment length of 1s, 2s, 3s. Blue, green and brown bars represent idle mode detection results of left hand sensation, right hand sensation and mixed task type.

second after the system cue, it means the time window used in this analysis will be [1 2]s, [1 3]s, and [1 4]s, separately.

With a same decoding method used in section 3.2, we obtain three groups of TPR and FPR values which has been shown in table.1. The mean TPR values for each time segment are 0.82±0.06, 0.86±0.04 and 0.88±0.05, and mean FPR values are 0.29±0.08, 0.26±0.08 and 0.28±0.1. T-test show the results of TPR and FPR between three groups have no significant difference.

Another analysis is to compare TPR and FPR values among different segment lengths and task types. In this step, the idle mode detection advantage of mixed task type will be demonstrated. Nine TPR values have been acquired for 3 task types and 3 segment lengths, the result is shown in figure.6(a). There are three groups of bars, and each group represent one segment length parameter. In each group, blue and green bars mean the TPR calculated with decoding idle mode from left and right hand sensation, and brown bar is for idle mode detection from mixed task type. T-test shows that TPR value for mixed task type is significantly higher than those of the other two task types in three groups. However, the results of FPR which has been shown in figure.6(b) show no significant difference between different task types.

4   Discussion and Conclusion

In this paper, we have proposed a new experiment paradigm for idle mode detection. We aim at improving the decoding accuracy by combining different task types into a mixed type and predict it from idle mode. With this goal, we have analysed EEG signal characteristics among different mental tasks, especially compared to that of idle mode. The features we use in this work is band power, and the feature variations of subject S1 along time, frequency and spatial are displayed in figure.2. We find this experiment paradigm show significant difference between task mode and idle mode. In detail, the phenomenal of ERD/ERS appearing during the subject's task performing hasn't been found in the idle mode trials. It means the subject's somatosensory cortex hasn't been activated during rest. Actually, this phenomenal also exists in motor imagery based BCI experiment. [8] is a study of asynchronized BCI, researchers aim at detecting idle mode from the task of left hand imagination movement and words association. However, the topographic maps of different mental tasks in this study show no significant difference between task mode and idle mode.

As this proposed experiment paradigm consists of three tasks (two sensation tasks and one rest task), we need to predict rest task (namely idle mode) from the other two types of data set, rather than a binary classification. In the study of Farhad [11], research need to predict idle mode from four types of mental tasks including visualizing word, multiplication, object rotating and motor imagery. They simplify this problem by choosing the type of task with best performance of idle mode detection as the first step to determine the subject's real state, and further calculation will be carried out if idle mode not detected in the first step. In this way, the problem has been simplified to be a binary classification. However, in our proposed method, we combine the two task types of left and right hand sensation into a mixed task type, and classified from the idle mode tasks. From the topographic map of mixed task shown in figure.4, we find the ERD phenomenal exists in both sides of somatosensory cortex, and no activation with idle mode tasks.

With this proposed method, the TPR value of idle mode detection from mixed task type is 86% which is significantly higher than that of the idle mode detection from single task type (79% for left hand sensation and 81% for right hand sensation). This result is also much better than that of [4]. In [4], researchers sct up two classifiers and one was used for classifying task types while the left one was used to detect idle mode trials from mixed task mode which consisted of 3 kinds mental tasks (left or right hand motor imagery, foot motor imagery and tongue imagery). However, mean value of TPR is only 28.4 while FPR is 16.9.

With another analysis of the influence of different decoding parameters, we find the advantage of idle mode detection also exists when the decoding time window is changed from [1 5]s to [1 4]s, [1 3]s, and [1 2]s. The comparison result of TPR and FPR of different decoding time window is shown in figure.6. In three different time segments, TPR of mixed task type is always higher than that of single task type with $p<0.05$. Meanwhile, there is no significant difference between three time segments for FPR values. It means that our proposed method has the potential to be used in real-time BCI system in which the decoding time segment is always limited to no longer than 1 s.

One drawback of this study is there is no well-designed transition between task mode and idle mode. In former studies [6][4][20][11], one threshold value was employed to determine the subject's mental state. With this method, subjects can smoothly transfer from task mode to idle mode rather than a sudden transition. The value of threshold can be changed according to the subject's performance, therefore the FPR value can be limited to a very low level (even zero). In contrast, our FPR values are higher than 20%. However, TPR value will also decrease when we get a lower FPR, and it will discourage the subjects when the output commands have a too long latency.

Acknowledgment. We thank all volunteers for their participation in the study. We thank Prof. Pete B. Shull for the helpful suggestions. This work is supported by the National Basic Research Program (973 Program) of China (Grant No.2011CB013305), the Science and Technology Commission of Shanghai Municipality (Grant No.13430721600), and the National Natural Science Foundation of China (Grant No.51375296).

References

1. Wolpaw, J., Birbaumer, N., McFarland, D., Pfurtscheller, G., Vaughan, T., et al.: Brain-computer interfaces for communication and control. Clinical Neurophysiology 113(6), 767–791 (2002)
2. Murguialday, A.R., Hill, J., Bensch, M., Martens, S., Halder, S., Nijboer, F., Schoelkopf, B., Birbaumer, N., Gharabaghi, A.: Transition from the locked in to the completely locked-in state: a physiological analysis. Clinical Neurophysiology 122(5), 925–933 (2011)
3. Birch, G.E., Bozorgzadeh, Z., Mason, S.G.: Initial on-line evaluations of the lf-asd brain-computer interface with able-bodied and spinal-cord subjects using imagined voluntary motor potentials. IEEE Transactions on Neural Systems and Rehabilitation Engineering 10(4), 219–224 (2002)
4. Scherer, R., Lee, F., Schlogl, A., Leeb, R., Bischof, H., Pfurtscheller, G.: Toward self-paced brain-computer communication: navigation through virtual worlds. IEEE Transactions on Biomedical Engineering 55(2), 675–682 (2008)

5. Pfurtscheller, G., Solis-Escalante, T.: Could the beta rebound in the eeg be suitable to realize a brain switch? Clinical Neurophysiology 120(1), 24–29 (2009)
6. Townsend, G., Graimann, B., Pfurtscheller, G.: Continuous eeg classification during motor imagery-simulation of an asynchronous bci. IEEE Transactions on Neural Systems and Rehabilitation Engineering 12(2), 258–265 (2004)
7. Xu, R., Jiang, N., Mrachacz-Kersting, N., Lin, C., Asin, G., Moreno, J., Pons, J., Dremstrup, K., Farina, D.: A closed-loop brain-computer interface triggering an active ankle-foot orthosis for inducing cortical neural plasticity (2014)
8. Galán, F., Nuttin, M., Lew, E., Ferrez, P.W., Vanacker, G., Philips, J., Millán, J.D.R.: A brain-actuated wheelchair: asynchronous and non-invasive brain-computer interfaces for continuous control of robots. Clinical Neurophysiology 119(9), 2159–2169 (2008)
9. Blankertz, B., Sannelli, C., Halder, S., Hammer, E.M., Kübler, A., Müller, K.-R., Curio, G., Dickhaus, T.: Neurophysiological predictor of SMR-based BCI performance. NeuroImage 51(4), 1303–1309 (2010)
10. Curran, E., Sykacek, P., Stokes, M., Roberts, S.J., Penny, W., Johnsrude, I., Owen, A.M.: Cognitive tasks for driving a brain-computer interfacing system: a pilot study. IEEE Transactions on Neural Systems and Rehabilitation Engineering 12(1), 48–54 (2004)
11. Faradji, F., Ward, R.K., Birch, G.E.: Toward development of a two-state brain-computer interface based on mental tasks. Journal of Neural Engineering 8(4), 046014 (2011)
12. Yao, L., Meng, J., Zhang, D., Sheng, X., Zhu, X.: Selective Sensation Based Brain-Computer Interface via Mechanical Vibrotactile Stimulation. PloS One 8(6), e64784 (2013)
13. Teplan, M.: Fundamentals of EEG measurement. Measurement Science Review 2(2), 1–11 (2002)
14. Huang, D., Qian, K., Fei, D.-Y., Jia, W., Chen, X., Bai, O.: Electroencephalography (eeg)-based brain-computer interface (bci): A 2-d virtual wheelchair control based on event-related desynchronization/synchronization and state control. IEEE Transactions on Neural Systems and Rehabilitation Engineering 20(3), 379–388 (2012)
15. Fukunaga, K.: Introduction to statistical pattern recognition, 2nd edn., vol. 1, p. 2. Academic Press, New York (1990)
16. Ramoser, H., Muller-Gerking, J., Pfurtscheller, G.: Optimal spatial filtering of single trial EEG during imagined hand movement. IEEE Transactions on Rehabilitation Engineering 8(4), 441–446 (2000)
17. Fisher, R.A.: The use of multiple measurements in taxonomic problems. Annals of Eugenics 7(2), 179–188 (1936)
18. Graimann, B., Huggins, J., Levine, S., Pfurtscheller, G.: Visualization of significant ERD/ERS patterns in multichannel EEG and ECoG data. Clinical Neurophysiology 113(1), 43–47 (2002)
19. Oostenveld, R., Fries, P., Maris, E., Schoffelen, J.-M.: Fieldtrip: open source software for advanced analysis of meg, eeg, and invasive electrophysiological data. Computational Intelligence and Neuroscience 2011 (2010)
20. Solis-Escalante, T., Müller-Putz, G., Brunner, C., Kaiser, V., Pfurtscheller, G.: Analysis of sensorimotor rhythms for the implementation of a brain switch for healthy subjects. Biomedical Signal Processing and Control 5(1), 15–20 (2010)

A Control Strategy for Prosthetic Hand Based on Attention Concentration and EMG

Changcheng Wu, Aiguo Song[✉], and Peng Ji

School of Instrument Science and Engineering, Southeast University, Nanjing 210096, China
tgreatw@sina.com, a.g.song@seu.edu.cn

Abstract. In order to control the prosthetic hand following the user's intention, a control strategy based on the EMG signals and the user's attention concentration is proposed in this paper. A portable EEG device, MindWave, is employed to capture the user's attention concentration. In the procedure of motion recognition, the beginning and end of the user's action intent is discriminated by the user's attention concentration and the Willison amplitude of the EMG signals. The integrated EMG is used to estimate the grasp force and the opening-and-closing speed of the prosthetic hand. An EMG signal model is proposed to eliminate the interference between the two channels of the EMG signals. The Experiments are implemented to verify the proposed control strategy. The results indicate that the proposed strategy is of effectiveness.

Keywords: Prosthetic hand · Attention concentration · EMG signal model

1 Introduction

The human machine interface (HMI) is a system that establishes a connection between human and machines. The communication channels provided by HMI can help people manipulate a machine. Prosthetic hand is a kind of typical HMI device. For the people with upper limb amputation, prosthetic hand can help them manipulate a functional prosthetic hand to recover some hand functions. In the past decades, several HMIs have been investigated to meet the demands of the amputees. In these HMIs, Electromyography (EMG) based control of the prosthetic hands has received great attention due to its simple operation and in accordance with the operation habits of the natural hand.

A great many of experts and scholars have done a lot of studies in this area. The signal processing method such as waveform length, integrated EMG, variance, spectrum analysis, wavelet transform, AR model, and etc were widely used [1-4]. A supervised feature mining method and some other work have been done by Huang et al [6]. Zhang et al presented a preliminary study for the forearm functional movement recognition using low-density EMG [7]. Erik D. Engeberg proposed a biomimetic sliding mode controller for EMG prosthetic hand [8]. A mobile controller for EMG prosthetic hand with tactile feedback was proposed to improve the perceptibility of the prosthetic hand [9]. To reduce the effects of the individual difference on the EMG

© Springer International Publishing Switzerland 2015
H. Liu et al. (Eds.): ICIRA 2015, Part I, LNAI 9244, pp. 307–318, 2015.
DOI: 10.1007/978-3-319-22879-2_28

motion recognition, Wu et al proposed an EMG self-learning motion recognition method [10]. In addition, much research work have been conducted in[11-16].

Although much research work have been done, the existing prosthetic hand still does not very satisfy the amputees. In a real world situation, about 30 to 40 percent of amputees do not regularly use their prostheses [5]. In EMG based control prosthetic hand, just relying on the EMG signals to discriminate the users' motion intention may cause misjudgment for the reason that the EMG signals may also generate when arm is doing some other irrelevant action.

This paper proposes a control strategy based on the user's attention concentration and the EMG signals for the purpose of providing a more reliable and more precise prosthetic hand control. The Attention concentration and EMG signals are adopted to understand the user's control intent. In order to obtain a more accurate estimation of the grasp force and the opening-and-closing speed (OCS) of the prosthetic hand, a model of the EMG signals is established. And the purpose of establishing this model is to eliminate the influence between the two channels of the EMG signals. The experiments are implemented to verify the feasible of the proposed strategy.

The rest of the paper is organized as follows. A detailed description of the system is in section 2. Section 3 presents the proposed strategy. The experiments are performed in section 4, and the conclusions of the paper are in section 5.

2   System Configuration

This section describes the system configuration. The system consists of a prosthetic hand, an EEG signals acquisition device, and an EMG signals acquisition device.

Prosthetic Hand
The MPH-II is a one degree of freedom prosthetic hand, which is equipped with three force sensors (to measure the grasp force) and a position sensor (encoder). Wearing silicone glove, the MPH-II looks like a natural hand. The core of the control system is a single chip microcomputer (C8051F320, Silicon Laboratories) [10]. The control board has a USB interface by which data can be exchanged between the prosthetic hand's control board and the computer.

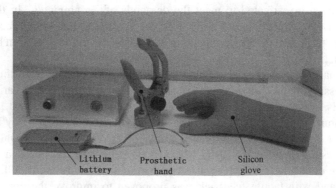

Fig. 1. The MPH-II EMG prosthetic hand

EEG Signals Acquisition Device

In this paper, the MindWave (MW) is employed to obtain the user's EEG signals. The MW is a NeuroSky product. It has the function of wireless Bluetooth communication. Using non-invasive EEG with a dry electrode at the forehead, the raw EEG signals, the current magnitude of 8 commonly-recognized types of the EEG signals and the user's brain states can be measured. And these data can be transmitted to the PC or other signal processing devices through the built-in Bluetooth module of the MW. Two user's brain state meters, attention and mediation, output by MW are comprised of a complex combination of artifact rejection and data classification methods. A study of testing these two meters has been conducted in [17]. The study results in [17] demonstrate that the meters are able to differentiate between certain mental states.

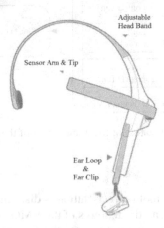

Fig. 2. The MindWave

EMG signals Acquisition Device

Two EMG sensors are used to acquire the EMG signals. A 10- bit A/D converter is used to digitize signals (the sampling ratio is 1 kHz). A Bluetooth module is utilized to transmit the EMG signals to the controller of the prosthetic hand or other signal processing devices. All the components of the EMG signals acquisition device are fixed on a ribbon (see Fig. 3) to make it convenient for users to wear, and the positions of the EMG sensors on the ribbon are adjustable because the best detecting position for different users are different.

Fig. 3. The EMG acquisition device

3   Control Strategy

The EMG prosthetic hand control strategy mainly includes the motion recognition and the controller as shown in Fig. 4.   In this paper, the user's attention concentration and the EMG signals are combined to recognize the user's motion intent. A Back-stepping controller with stiffness fuzzy observation which is proposed in authors' paper [15] is employed to realize the proportional control of the grasp force and the OCS of the prosthetic hand.

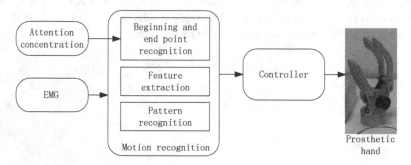

Fig. 4. The Visualization of the components of the control strategy

3.1   Motion Recognition

The common practice of the motion recognition is discriminating the starting point in the EMG signals and extracting the features of the EMG signals firstly, then conducting the motion recognition by using the extracted features. And the methods for the recognition of the beginning and end (BE) point are commonly the threshold method. However, just relying on the EMG signals to discriminate the BE point may cause misjudgment for the reason that the EMG signals may also generate when arm is doing some other irrelevant action such as swing arm. In this paper EEG signals and EMG signals are combined to recognize the user's motion intent. The user's attention concentration and the Willison amplitude of the EMG signals are combined to discriminate the BE point of the user's action intent.

3.1.1   Attention Concentration

In this paper the user's attention concentration data are obtained from MW. And the data updating rate is one time per-second.

The human's brain needs to deal with a lot of information at the same time, it may cause fluctuation of the human's attention concentration. In order to overcome the instability, the weight-Smith Gate is employed to modify the attention concentration data. The steps are as follows.

Step1: calculate the weight average value (*Ave*) of the two recent attention concentration data.

$$Ave(t) = \alpha a(t) + (1-\alpha) a(t-1), \alpha \in (0,1)$$

(1)

Where, $a(t)$ is the current data of the attention concentration, $a(t\text{-}1)$ is the previous attention concentration data, α is the weight. The updating rate of $a(t)$ is 1 Hz. The value of $a(t)$ is between 0 and 100.

Step2: select a mean threshold (MT).

Step 3: select two factors, k_1 and k_2, for the upping-edge threshold (UT) and the falling-edge threshold (FT). These two thresholds are calculated as follows:

$$\begin{cases} UT = k_1 \times MT \\ FT = k_2 \times MT \end{cases} \quad k_1 \in [1,2], k_2 \in [0,1] \tag{2}$$

The characteristic curve of this weight-Smith Gate is shown in Fig. 5. The red arrow and the blue arrow represent the direction of the switch on and switch off, respectively.

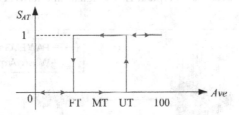

Fig. 5. The characteristic curve of this weight-Smith Gate

From the characteristic curve, we can know that: the sensibility and the stability are determined by MT, k_1 and k_2. Fig. 6 shows an example of the user's attention concentration, here $\alpha = 0.7$, $k_1=0.8$, $k_2=1.4$ and $MT=50$.

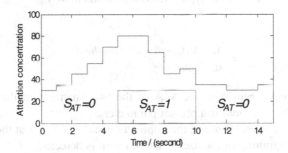

Fig. 6. The example of the user's attention concentration

3.1.2   Willison Amplitude of the EMG Signals

A pair of antagonistic muscles on the user's forearm is selected to place the two EMG sensors. And these two muscles are marked as muscle_1 and muscle_2, respectively. Under the rest state, the EMG signals captured from the arm show no changes in potential. And, the contracting of the muscle would lead to the changes in potential.

Willison Amplitude (WA) is a common characteristic parameter which can indicate the muscle contraction state. The definition is as follows:

$$WA = \sum_{j=0}^{N-1} \frac{1}{2}[1 + f(E_{i-j} - E_{i-j-1})] \tag{3}$$

$$f(x) = \begin{cases} 1, & |x| > threshold \\ 0, & other \end{cases} \tag{4}$$

where, E_i is the current sample data of the EMG signals, E_{i-j} is the jth previous sample data of the EMG signals, N is the window size.

The window size and the threshold will influence the performance of the calculation result. In this paper, the amplitude of the EMG signals which are measured under the rest state is adopted as the threshold. A typical EMG signals and its WA are shown in Fig. 7.

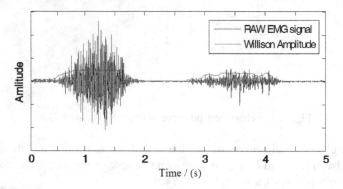

Fig. 7. The typical EMG signals and its WA

Define S_{WA} as follows,

$$S_{WA} = \begin{cases} 1, & WA_{E1} + WA_{E2} > Threshold \\ 0, & other \end{cases} \tag{5}$$

where WA_{E1} and WA_{E2} represent the WA of the two channels of the EMG signals respectively, and the threshold is a pre-set parameter.

$S_{WA}=1$ indicate the muscles are in the contraction state. If $S_{AT}=1$ at the same time, it indicates that the beginning of the user's action intent is detected.

3.2   Estimation of the Grasp Force and the OCS

In this paper, the IEMG is employed to estimate the grasp force and the OCS of the prosthetic hand. The IEMG is defined as follows:

$$IMEG(t) = \sum_{j=0}^{M-1} |E_{t-j}| \tag{6}$$

where, E_t is the current sample data of the EMG signals, E_{t-j} is the jth previous sample data of the EMG signals, M is the window size.

During the experimental process we observed that there was a correlation between the two channels of the EMG signals. In order to obtain more accurate estimation results, eliminating the interference between the two EMG signals is necessary. As for one degree of freedom prosthetic hand, the most commonly grasp force estimate method is to compute the difference of the two channels of the IEMG signals.

In this paper, a model of the two channels of the EMG signals is established as follows:

$$\begin{cases} X = S_{xx} + S_{xy} \\ Y = S_{yy} + S_{yx} \end{cases} \tag{7}$$

where, X and Y refer to the two channels IEMG of the EMG signals respectively, S_{xx} denotes the signal generated by the contraction of muscle_1 itself, S_{xy} denotes the disturbing signal generated by the contraction of muscle_2. Similarly, S_{yy} denotes the signal generated by the contraction of muscle_2 itself, and S_{yx} denotes the disturbing signal generated by the contraction of muscle_1.

Assuming that

$$\begin{cases} S_{xy} = \varphi_{xy}(Y) \\ S_{yx} = \varphi_{yx}(X) \end{cases} \tag{8}$$

Then, the model can be rewritten as follows:

$$\begin{cases} S_{xx} = X - \varphi_{xy}(Y) \\ S_{yy} = Y - \varphi_{yx}(X) \end{cases} \tag{9}$$

Through the calibration experiments, the functions φ_{xy} and φ_{yx} can be easily obtained. Owing to the spatial confined, the details of the calibration experiments and the performance evaluations of the proposed model will be described in a upcoming article.

Fig. 8(a) shows two channels of the amplified raw EMG signals. These signals are captured under a grasp action and a hand open action. Fig. 8(a) shows that there exists the interference between the two channels of the EMG signals. Fig. 8(b) and Fig. 8(c) show the IEMG and the decoupled IEMG, respectively. After the decoupling processing, the interference between the two channels of the IEMG is reduced significantly. And in this paper, these decoupled IEMG are used to estimate the grasp force and the OCS of the prosthetic hand.

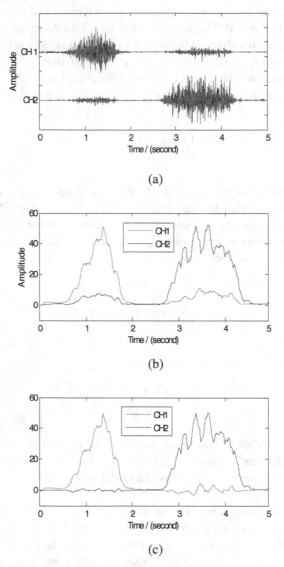

Fig. 8. (a) The raw EMG signals. (b) The integrate EMG. (c) The decoupled Integrate EMG.

The flow chart of the estimation is shown in Fig. 9. The output, F_d, is the estimation result. In free space, F_d reflects the OCS of the prosthetic hand, and in restricted space, F_d reflects the grasp force. k_s is scaling factor.

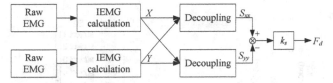

Fig. 9. The flow chart of estimating the grasp force and the OCS of the prosthetic hand

4   Experiments and Results

To verify the validity of the control strategy presented in this paper, the Lab-based experiments were carried out.

Seven non-amputee volunteers (four males, three females, aged from 22 to 28) were chosen to accomplish the experiments. The experimental scene is shown in Fig. 10. Before the experiments, the participants had informed of all the details of the experiments.

Fig. 10. The experimental scene

4.1   Attention Concentration Control Experiment

In this paper, the user's attention concentration is employed to discriminate the BE point of the user's action intent. The experiments verify the effectiveness of the proposed BE point discrimination method firstly. In this experiment, the participants are required to try to control their attention concentration according to the prompts. When a prompt occurs, the participant should try to improve their attention concentration as high as possible. The experiments last for 10 days. And the participant participate the test one time a day, each test consists 50 action prompts. During the experiments, the user's attention concentration data are recorded.

The time from the instant the prompt occurs to the instant the participant's attention concentration reaches to UT of the Smith Gate is recorded as shown in Fig. 11.

The results show that the participants can control their attention concentration to reach to UT of the Smith Gate in a short time, especially after a period of time training. The experiments indicate that the application of user's attention concentration in the BE point discrimination is feasible.

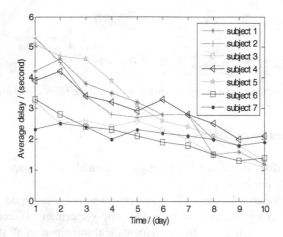

Fig. 11. The results of the attention concentration control experiments

4.2   Prosthetic Hand Control Experiment

After the 10 days attention concentration tests, the participants were required to control the prosthetic hand by using the system which described in section 2. This experiment consists two parts, one is to control the opening and the closing of the prosthetic hand according to the prompts, and the other is to control the prosthetic hand to grasp a paper cup with water according to the prompts. In paper cup grasp experiment, the participants were asked to grasp and lift the paper cup with water (see Fig. 12). In this process, the participant should keep water without overflowing or dropping.

Table 1 shows the results of the prosthetic hand's opening and closing control experiment. And the results of the paper cup grasping experiment are shown in table 2.

Table 1. The results of the prosthetic hand's opening and closing control experiment

Participant's number	Average success rate (%)
1	88
2	80
3	92
4	100
5	90
6	92
7	96

Fig. 12. The paper cup grasping experiment

Table 2. The results of the paper cup grasping experiment

Participant's number	Average success rate (%)
1	84
2	72
3	90
4	90
5	88
6	86
7	90

5   Conclusion

This paper proposes a control strategy for prosthetic hand, which captures the user's action intent from the combination of user's attention signals and EMG signals. The user's attention concentration and the WA of the EMG signals are employed to recognize the BE of the actions. The integrated EMG is calculated to estimate the grasp force and the OCS of the prosthetic hand. An EMG model has been established for the purpose of eliminating the interference between the two channels of the EMG signals. The experiments are implemented to verify the feasibility of the proposed strategy. The experimental results show that the proposed strategy is effective.

For the future work, we will try to discover a better feature from the EEG signals to indicate the user's motions, which will be easily accepted by the amputees.

References

1. Mahaphonchaikul, K., Sueaseenak, D., Pintavirooj, C., et al.: EMG signal feature extraction based on wavelet transform, Chiang Mai (2010)
2. Xie, H., Huang, H., Wu, J., et al.: A comparative study of surface EMG classification by fuzzy relevance vector machine and fuzzy support vector machine. Physiology Measurement **36**(2), 191–206 (2015)
3. Davidson, J.: A survey of the satisfaction of upper limb amputees with their prostheses, their lifestyles, and their abilities. Journal of hand therapy: official journal of the American Society of Hand Therapists **15**(1), 62–70 (2002)
4. Zaini. M.H.M., Ahmad, S.A.: Surgical and non-surgical prosthetic hands control: A review, Langkawi (2011)

5. Al Omari, F., Hui, J., Mei, C., et al.: Pattern Recognition of Eight Hand Motions Using Feature Extraction of Forearm EMG Signal. Proceedings of the National Academy of Sciences India Section A: Physical Sciences **84**(3), 473–480 (2014)
6. Han-Pang, H., Yi-Hung, L., Chun-Shin, W.: Automatic EMG feature evaluation for controlling a prosthetic hand using supervised feature mining method: an intelligent approach (2003)
7. Zhang, Z., Wong, C., Yang, G.: Forearm functional movement recognition using spare channel surface electromyography, Cambridge, MA, USA (2013)
8. Engeberg, E.D.: A physiological basis for control of a prosthetic hand. Biomedical Signal Processing and Control **8**(1), 6–15 (2013)
9. Hirata, T., Nakamura, T., Kato, R., et al.: Development of mobile controller for EMG prosthetic hand with tactile feedback, Budapest (2011)
10. Wu, C., Song, A., Zhang, H.: Adaptive fuzzy control method for EMG prosthetic hand. Chinese Journal of Scientific Instrument **34**(6), 1339–1345 (2013)
11. Antfolk, C., Bjorkman, A., Frank, S., et al.: Sensory Feedback from a Prosthetic Hand Based on Air-Mediated Pressure from the Hand to the Forearm Skin. Journal of Rehabilitation Medicine **44**(8), 702–707 (2012)
12. Engeberg, E.D., Meek, S.G.: Backstepping and Sliding Mode Control Hybridized for a Prosthetic Hand. IEEE Transactions on Neural Systems and Rehabilitation Engineering **17**(1), 70–79 (2009)
13. Yanagisawa, T., Hirata, M., Saitoh, Y., et al.: Real-time control of a prosthetic hand using human electrocorticography signals. Journal of Neurosurgery **114**(6), 1715–1722 (2011)
14. Cheng-Hung, C., Naidu, D.S., Perez-Gracia, A., et al.: A hybrid adaptive control strategy for a smart prosthetic hand, Minneapolis, MN (2009)
15. Wu, C., Song, A., Zhang, H., et al.: A Backstepping Control Strategy for Prosthetic Hand Based on Fuzzy Observation of Stiffness. Robot **35**(6), 686–691 (2013)
16. Wu, C., Song, A., Ling, Y., et al.: A control strategy with tactile perception feedback for EMG prosthetic hand. Journal of Sensors (In press, 2015)
17. NeuroSky, Inc. NeuroSky's eSense™ Meters and Detection of Mental State (2009)

Providing Slip Feedback for Closed-Loop Control of Myoelectric Prosthesis via Electrotactile Stimulation

Linjun Bao, Dingguo Zhang$^{(\boxtimes)}$, Heng Xu, and Xiangyang Zhu

State Key Laboratory of Mechanical System and Vibration,
Shanghai Jiao Tong University, Shanghai 200240, China
dgzhang@sjtu.edu.cn

Abstract. Introducing tactile feedback into closed-loop control of prosthetic hand has always been an issue to provide amputees with a kind of more use-friendly and efficient prosthesis. This paper describes a method to improve prosthetic closed-loop control performance via slipping feedback. Cognitive load task is introduced into the system to verify whether the slip feedback adding much attention. The results show that with slip and visual feedback, subjects could finish grasp task more stably and faster than that only with visual feedback. And there are little difference on cognitive load task between slip & visual feedback and visual feedback.

Keywords: Electrotactile · Slip feedback · Closed-loop control · Prosthesis

1 Introduction

Human hand is sophisticated and flexible, and there are a lot of feed forward and feedback mechanisms during the process of grasp and manipulation [1]. Utilizing this mechanism, able-bodied people can achieve a fast, stable and economic grasp. But for amputees, parts of the feedback system were damaged and this situation should be taken into consideration when researchers designed the ideal prosthesis. Nowadays the commercial prostheses such as ilimbultra (Touch Bionics, USA) and Michelangelo (Ottobock, German) have a good motion performance, but none of them could provide somatosensory feedback and users also cannot feel the stiffness, roughness or texture of objects. Lack of tactile feedback, the control of prosthesis tends to be passive, which may be one of the key reasons for the abandonment of the prosthesis [2].

In recent researches of prosthetic control, there are two main methods producing somatosensory feedback, the direct nerve stimulation [3] and the noninvasive sensory feedback such as force feedback [4], vibrotactile feedback [5], cutaneous electrotactile feedback [6][7] and hybrid feedback [8]. In terms of noninvasive somatosensory feedback, electrotactile is more similar to natural sensory feelings and includes richer sensory information. In recent studies of real-time closed-loop

© Springer International Publishing Switzerland 2015
H. Liu et al. (Eds.): ICIRA 2015, Part I, LNAI 9244, pp. 319–328, 2015.
DOI: 10.1007/978-3-319-22879-2_29

prosthetic control, most of them operated grasp task in a virtual system [6] or virtual reality environment [9], and the real/virtual prosthesis was controlled by a wheel [6], the force of finger joints [10], or the mainstream electromyography (EMG). The main feedback method was pressure feedback generated by modifying the stimulation strength or position. However, what is often neglected in prosthetic control is that slip feedback also contributes a lot in grasping, which has been indicated in the physiology [11]. Besides, most amputees have visual feedback assistance to manipulate the prosthesis, integrating too much sensory feedback in the system may induce additional cognitive load and have negative effect on prosthetic control [6]. Therefore, based on electrotactile, providing subjects slip feedback could offer slip and pressure information in the meantime. And it remains to be confirmed whether it is an efficient way to improve grasp performance with little cognitive load.

In this paper, we test the real-time closed-loop prosthetic control system with slip feedback via electrotactile stimulation. The slip feedback was reproduced by changing different channels and time orders of stimulation pulses. By modulating the strength of the stimulation pulse, the slip feedback carried the contact pressure information as well , achieving the sensory information integration. Besides, the stability, rapidity and efficiency as evaluation indexes were adopted to evaluate the prosthetic control performance. Our preliminary results indicated that it slipped faster when the contact pressure was smaller than the balanced force.

2   Materials and Methods

2.1   Subjects

Six able-bodied subjects (mean age 23.83 ± 0.98, six males) participated in this experiment. And all of them are right-handed. Before the experiment each participant was told about the procedure of the experiment and sighed an informed consent form. The study protocol was approved by the local Ethics Committee of Shanghai Jiao Tong University.

2.2   The SJTU-5a Hand

The SJTU-5a Hand (custom-made) is a kind of 6-DOF anthropomorphic hand with humanoid finger joint design and palm shape. The thumb has two DOFs and each of the other four fingers has one DOF, driven by DC motors via a lead screw nut and linkage mechanism. Besides controlling the motor action of the prosthetic hand, the microcontroller Arduino Mega2560 (Massimo Banzis Team, Italy) simultaneously receives the contact force data collected by the pressure sensor FSR. The FSR402 is a kind of piezoresistive thin-film pressure sensor (Interlink Electronics, USA) and has a little nonlinearity in the target force area. After amplification and noise reduction by the hardware, the force data is transferred to the PC at a frequency of 100Hz for postprocessing. Polynomial fitting is applied to calibrate the force sensor a better linearity in a range of 0~30N as shown in Fig .1.

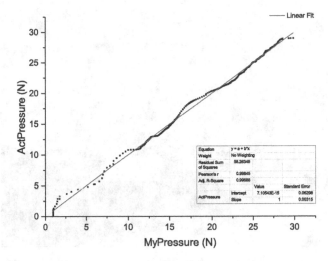

Fig. 1. Linear fit and force calibration for the pressure sensor.

2.3 Haptic Feedback

The SJTU-Stimulator 2.0 (custom-made) provides haptic feedback by stimulating the cutaneous tactile receptors. As shown in Fig.2, working in slip feedback mode, the stimulator activates the multichannel electrodes in a routine of 1-2-3-4-1. The stimulation frequency is set to 100Hz, in order to ensure the subjects can not discriminate the frequency of the pulse and they can only feel pressure instead of vibration. The amplitude of the stimulation pulse is set to 3mA. The strength of the electrotactile stimulation is modulated by changing the duration of the stimulation pulse (50us~1ms), and the speed of slipping is modulated by changing the frequency of changing the channels (1~200Hz).

2.4 Experimental Setup

The subjects controlled the prosthetic hand to interact with the real environment in real time through surface EMG (sEMG) and electrotactile feedback. The control system was composed of the following components: (1) SJTU-5a Hand, (2) DataLOG (Biometrics Ltd., UK), (3) driving and sensing controller, (4) SJTU-Stimulator 2.0, (5) multi-channel electrodes, (6) a laptop (host PC). The surface EMG was acquired at 1000 Hz though two electrodes which were placed on the flexor carpi radialis and extensor carpi ulnaris. The sampled EMG was processed by the following steps in turn: (1) 20-400 Hz fifth-order Butterworth band-pass filter, (2) 100 Hz fifth-order notch of comb filter which used to remove the electrical artifact from electrotactile, (3) full-waved rectification, (4) 10 Hz fifth-order Butterworth low-pass filter, (5) mean absolute value (MAV): the mean of 200ms analysis window and 50ms slide window [12]. The processed EMG was proportionally transformed into the equivalent grasp force order, and then it

Fig. 2. The multichannel electrodes. Each number represents a channel and each channel is independent. The upper copper pad is anode and the corresponding lower pad is cathode. All of the copper pads are covered with hydrogel to avoid the pricking caused by bad contact between electrodes and skin.

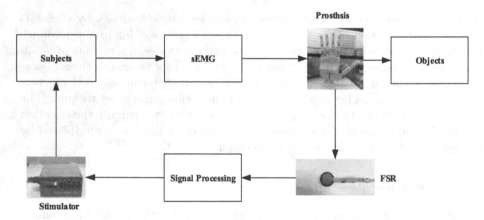

Fig. 3. The real-time control loop chart of prosthesis.

was sent to the prosthesis actuator. When the prosthesis held the object, the FSR attached to the finger tips sensed the contact pressure, and the host PC provided electrotactile feedback based on the pressure. As shown Fig.3, the real-time control loop operated at 20Hz.

2.5 Experimental Protocol

In order to evaluate the reliability and efficiency of prosthesis control with slip feedback, the prothesis control evaluation experiments were carried out.

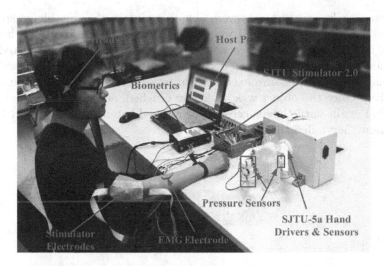

Fig. 4. The experiment protocol of prosthesis control method with slip feedback.

As shown in Fig.4, the subjects were instructed to sit beside a desk with theirs arms softly lying on the desk.

The experimenter started the grasp task and subjects were required to reach a suitable grasp force and maintain it for a while just like the process that able-bodied people grasp and manipulate objects. During grasp task, subjects always received visual feedback, in which they could watch the prosthesis action. But in different experiments, slipping feedback was selectable. In order to investigate the influence of the cognitive load, in some experiments subjects were required to randomly listen to different numbers and repeat it loudly and correctly while operating the grasp task [6]. If not, the grasp task must be tested again.

In this experiment, two kinds of stiffness and weight of four objects were measured and entered into the object database before the experiment. (See details in Table.1). In this paper, balanced force value was equal to the minimum value of the contact pressure when objects were able to be held in the air by prothetic hand, and the contact pressure was regarded as the crush pressure once objects were distinctly deformed or the obviously uneconomical pressure when grasping the hard object. The control flow diagram of the grasp task was shown in Fig.5.

Table 1. The attributes of the objects.

Number	Object	balanced force/mN	Crush Pressure/mN
1	empty polycarbonate bottle	300	4000
2	empty mineral water bottle	300	2800
3	polycarbonate bottle with water	1000	4000
4	mineral water bottle with water	1000	2800

During grasp task, once subjects were able to maintain the suitable grasp force more than five seconds, this trail was finished successfully. However, when the situation, in which the grasp force was less than the balanced force, lasted over five seconds, this object was broken because of slipping out of the prosthesis. And if the value of contact pressure exceeded the preset crushed pressure, the object was crushed. Once the object was broken or crushed, subjects were required to do this trail again until they succeeded or reached the maximum restart times, which meant they failed this trail.

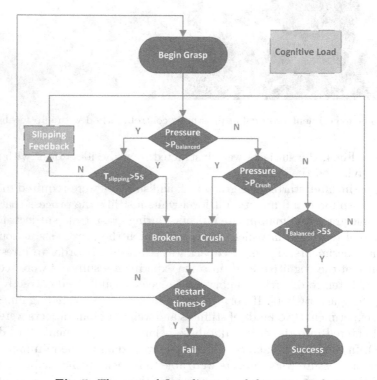

Fig. 5. The control flow diagram of the grasp task.

The experiment was conducted in following two steps.

Step 1. In the preliminary experiment, subjects should be familiar with the control system of the prosthesis, with/without slip feedback and with/without cognitive load.

Step 2. The formal experiments were divided into four experiments. **Experiment 1:** Grasp task with visual feedback, **Experiment 2:** Grasp task with visual and slip feedback, **Experiment 3:** Grasp task with visual feedback, combined with cognitive load, **Experiment 4:** Grasp task with visual and slip feedback, combined with cognitive load. Each experiment included 20 trails, 5 trails for each object and in a random order.

Fig.6 showed the contact pressure in a typical successful grasp task. The T_f was the total time spent to finish the grasp task, and the Fr indicated the fluctuation of the grasp force when subjects maintained it. Obviously, the success rate, T_f and F_r were three indexes to demonstrate whether the control method of the prosthesis was efficient.

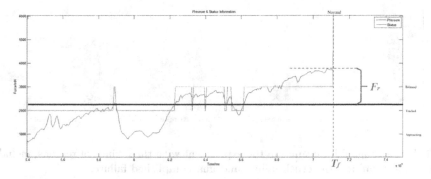

Fig. 6. The contact pressure and evaluation index in a successful grasp task.

3   Results and Discussion

As shown in Fig.7, compared vision feedback with vision & slip feedback, the times of object broken and crush decreased as the slip feedback induced. When adding cognitive load, the performance of both the visual feedback and vision & slip feedback became worse, and the times of object broken and crush increased. Even under the situation of cognitive load, providing vision & slip feedback failed less times than visual feedback. As shown in Fig.8, for the object 1,2 and 3, the T_f of slip feedback was slight shorter than visual feedback and had no obvious difference. But for object with small stiffness and large weight, visual feedback combined with slip feedback significantly spent less time than visual feedback alone. However, as shown in Fig.9, the F_r of vision & slip feedback has no significant difference with that of visual feedback.

The results of the prothesis control evaluation experiment indicated that, along with visual feedback, slip feedback truly improved the grasp performance of prosthesis. From Fig.7, we can see clearly that slip feedback reduced the failure times of grasp, which meant it could improve the stability of the control system. From Fig.8, faced with some difficult grasp task such as grasping heavy and soft object 4, the slip feedback helped to accomplish the task faster, which meant it could improve the rapidity of the control system. From Fig.9, we found no significant difference about the F_r, the reason may be that subjects did not fully understand the meaning of economic grasp and tend to use redundant grasp force to hold the objects. In future, we would refine the parameters of electrotacile feedback and try to find better coordinated prosthetic control system.

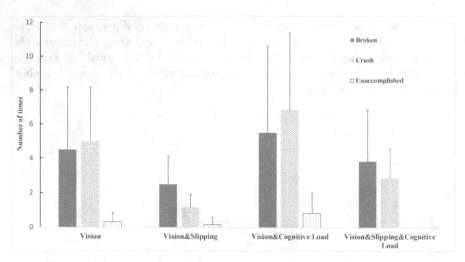

Fig. 7. The average failure times in the experiment with three kinds of reasons, including the broken failure, the crush failure and unaccomplished failure.

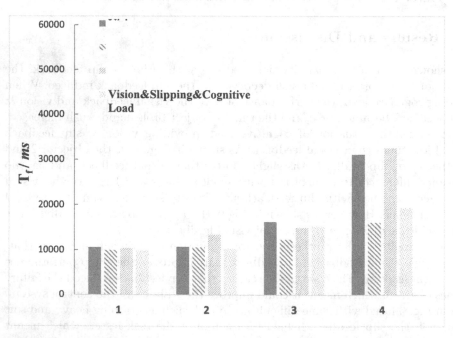

Fig. 8. The average accomplished time T_f in the experiment on four kinds of feedback, including visual feedback, visual and slip feedback, visual feedback & cognitive load, visual and slip feedback & cognitive load,1~4 indicate four different objects.

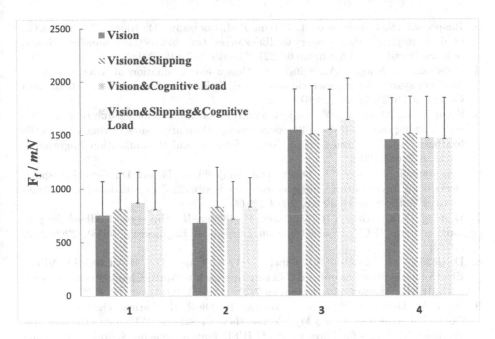

Fig. 9. The fluctuation of the grasp force F_r of in the experiment with four kinds of feedback, including visual feedback, visual and slip feedback, visual feedback & cognitive load, visual and slip feedback & cognitive load,1~4 indicate four different objects.

4   Conclusion

In this study, the real-time closed-loop control system of prosthetic hand improved by slip feedback achieved better performance than visual feedback only. Even though the difference was not significant, the experiment results indicate that slip feedback could work as an improved interface of prosthesis control. And the protocol of grasp and force maintainance may help to wake up the damaged motor and sensory system of amputees.

Acknowledgments. This work is supported by the National Basic Research Program (973 Program) of China (Grant No.2011CB013305), the National Natural Science Foundation of China (Grant No.51475292), and the Science and Technology Commission of Shanghai Municipality (Grant No. 14ZR1421300).

References

1. Johansson, R.S., Cole, K.J.: Sensory-motor coordination during grasping and manipulative actions. Current Opinion in Neurobiology **2**(6), 815–823 (1992)
2. Biddiss, E., Chau, T.: Upper-limb rosthetics: critical factors in device abandonment. American Journal of Physical Medicine & Rehabilitation **86**(12), 977–987 (2007)

3. Raspopovic, S., Capogrosso, M., Petrini, F.M., Bonizzato, M., Rigosa, J., Di Pino, G., et al.: Restoring natural sensory feedback in real-time bidirectional hand prostheses. Science Translational Medicine **6**(222), 222ra19–222ra19 (2014)

4. Pylatiuk, C., Kargov, A., Schulz, S.: Design and evaluation of a low-cost force feedback system for myoelectric prosthetic hands. JPO: Journal of Prosthetics and Orthotics **18**(2), 57–61 (2006)

5. Stepp, C.E., Matsuoka, Y.: Object manipulation improvements due to single session training outweigh the differences among stimulation sites during vibrotactile feedback. IEEE Transactions on Nerual Systems and Rehabilitation Engineering **19**(6), 677–685 (2011)

6. Jorgovanovic, N., Dosen, S., Djozic, D.J., Krajoski, G., Farina, D.: Virtual grasping: closed-loop force control using electrotactile feedback. Computational and Mathematical Methods in Medicine, **2014** (2014)

7. Damian, D.D., Arita, A.H., Martinez, H., Pfeifer, R.: Slip speed feedback for grip force control. IEEE Transactions on Biomedical Engineering **59**(8), 2200–2210 (2012)

8. D'Alonzo, M., Dosen, S., Cipriani, C., Farina, D.: HyVEHybrid Vibro-Electrotactile StimulationIs an Efficient Approach to Multi-Channel Sensory Feedback. IEEE Transactions on Haptics **7**(2), 181–190 (2014)

9. Ninu, A., Dosen, S., Muceli, S., Rattay, F., Dietl, H., Farina, D.: Closed Loop Control of Grasping With a Myoelectric Hand Prosthesis: Which Are the Relevant Feedback Variables for Force Control? IEEE Transactions on Neural Systems and Rehabilitation Engineering **22**(5), 1041–1052 (2014)

10. Saunders, I., Vijayakumar, S.: The role of feed-forward and feedback processes for closed-loop prosthesis control. J. Neuroeng. Rehabil. **8**(60), 1–12 (2011)

11. Johansson, R.S., Cole, K.J.: Grasp stability during manipulative actions. Canadian Journal of Physiology and Pharmacology **72**(5), 511–524 (1994)

12. Humbert, S.D., Snyder, S.A., Grill Jr., W.M.: Evaluation of command algorithms for control of upper-extremity neural prostheses. IEEE Transactions on Neural Systems and Rehabilitation Engineering **10**(2), 94–101 (2002)

Development of a Hybrid Surface EMG and MMG Acquisition System for Human Hand Motion Analysis

Weichao Guo, Xinjun Sheng, Dingguo Zhang, and Xiangyang Zhu[✉]

State Key Laboratory of Mechanical System and Vibration,
School of Mechanical Engineering, Shanghai Jiao Tong University,
Shanghai 200240, People's Republic of China
mexyzhu@sjtu.edu.cn

Abstract. Surface electromyography (EMG) is widely investigated in human-machine interface (HMI) by decoding movement intention to intuitively control intelligent devices. Mechanomyogram (MMG) is the mechanical vibration signal produced by contracting muscles, and is also useful tool for intuitive HMI. Combining EMG and MMG together would provide more comprehensive information of muscle activities. This paper presents a hybrid EMG and MMG acquisition system to capture these two signals simultaneously. Experiments were carried out to study the performance of the proposed hybrid sensor for human hand motion analysis, and the results indicated that EMG and MMG reflected muscle contractions from different aspect for their different time-frequency responses. Furthermore, the pattern recognition experiment results showed that the classification accuracy using combined EMG-MMG outperformed EMG-only by 20.9% and 14.6%, with mean absolute value (MAV) and power spectral density (PSD) features respectively. The findings of this study support and guide the fusion of EMG and MMG for improved and robust HMI.

Keywords: Surface EMG · MMG · Human-machine interface · Hand motion analysis

1   Introduction

Surface electromyography (EMG) representing the muscular activity driven by the motoneuron [1], is usually monitored to manifest muscle activation. Because of the capability of estimating movement intention, EMG is widely used as intuitive human-machine interface (HMI) to control intelligent devices, such as multifunction upper-limb prostheses [2,3]. It is based on the assumption that the EMG signals of muscle contractions can be mapped to the corresponding commands of external devices. In recent years, promising results have been reported for the EMG based HMI [4–6], however, there are still some difficulties hindering the practical usage of myoelectric control [7]. For instance, the surface EMG

© Springer International Publishing Switzerland 2015
H. Liu et al. (Eds.): ICIRA 2015, Part I, LNAI 9244, pp. 329–337, 2015.
DOI: 10.1007/978-3-319-22879-2_30

signal is susceptible to interference from sweat and skin impedance change with time [3,7–9], as a result, the control accuracy is deteriorated.

In contrast with EMG signals, mechanomyogram (MMG) signals are not affected by sweat and changes in skin impedance. Reflecting muscle activity in form of low-frequency ($<$ 50 Hz) vibration, MMG is the mechanical signal produced by contracting muscles [10]. The low-frequency vibrations generate displacements of about 500 nm on the skin surface when a muscle is contracted [11], and the accompanied MMG signals can be measured with accelerometers or microphones. Recently, MMG signals have been investigated for HMI applications, such as upper-limb prostheses control [11–13] and hand movement analysis [14–17]. However, the control accuracy of MMG looks inferior to EMG due to the considerably lower signal-to-noise ratios (SNR), and the external noise caused by movement artifact is known to introduce considerable interference to MMG measurement [12,18]. Thus, attenuating the movement artifact noise of MMG, and then combining the advantages of both EMG and MMG would be of great significance to facilitate the understanding of muscle activities. Moreover, fusion of EMG and MMG would be valuable for enhanced and robust HMI.

This paper presents a hybrid surface EMG and MMG acquisition system for hand gesture analysis. The hybrid system allows capture EMG and MMG signals simultaneously, and the movement artifact of MMG is eliminated by means of differential filter. Experimental tests are carried out to investigate the time-frequency responses of EMG and MMG to muscle contractions. Furthermore, pattern recognition based on EMG and MMG signals for five classes of hand motion is performed.

2   Materials and Methods

2.1   System Configuration of Hybrid EMG and MMG Sensor

The aim of the hybrid sensor was to detect high quality EMG and MMG signals simultaneously during muscle activation. To attenuate common-mode noise, especially the movement artifact caused by excess limb or body movement, differential amplification was adopted to capture EMG and MMG signals, as shown in Fig. 1. Instrumentation amplifier (INA326, Texas Instruments) was used for the differential amplification for its high common mode rejection ratio (CMRR) and input impedance. The differential EMG and MMG signals were bandpass filtered by 20–450 Hz [19] and 5–100 Hz [16,17] respectively to remove the unwanted noise. Combined with the bandpass filter, signals were further amplified with a gain of 500 and 50 for EMG and MMG respectively to improve the SNR. The further amplification was implemented by using the operation amplifier (AD8603, Analog Devices) chosen for its rail to rail property. Moreover, the analog EMG and MMG signals were digitalized by means of analog to digital conversion (ADC) with a sampling rate of 1 kHz, and then the digitized signals were wirelessly transmitted to the PC host for storage, displaying and further processing.

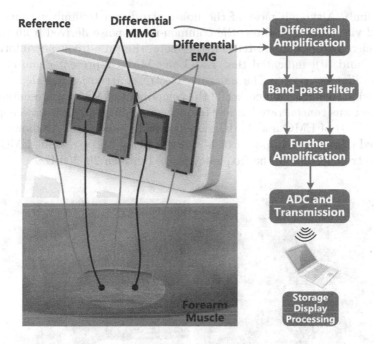

Fig. 1. The framework of hybrid EMG and MMG acquisition system that allows detect electrophysiological and mechanical activities of the underlaying muscle.

Micro control unit C8051F310 (Silicon Labs) was employed to manage the signal digitization and transmission, and the ADC was realized by AD7607 (Analog Devices).

2.2   Implementation of System Hardware and Tests

Fig. 2 shows the hardware implementation of the proposed hybrid EMG and MMG acquisition system. Dry gold-plated copper electrodes, with an inter-electrode spacing of 13 mm, were deployed to capture EMG signals for their stable and reliable contact with skin. In addition, MMG signals generated by the vibrations of muscle fibers were detected by a couple of MEMS accelerometers (ADXL203, Analog Devices Inc, USA) with an interval of 13 mm. ADXL203 was chosen for its small geometric dimensions (5 mm×5 mm×2 mm) and high sensitivity (1000 mV/g).

To test and compare the time-frequency domain responses, EMG and MMG signals from Flexor Digitorum Superficialis [FDS, Fig. 2 (b)] during three hand gestures (fist, wrist flexion and hand open) were collected. The signal waveform and time-frequency information of EMG and MMG signals were investigated, as shown in Fig. 3. Both EMG and MMG made responses to muscle contractions [Fig. 3 (a) and (c)], however, MMG signals showed lower SNR than EMG. The relative low SNR of MMG was likely attributed to the fact that accelerometers measured gross movement of the sensor itself, coupled with the movement of subject's limb. Although most of the noise introduced by limb movement was eliminated via differential filter, the common-mode noise derived from the couple of accelerometers still existed. Furthermore, the time-frequency information [Fig. 3 (b) and (d)] indicated that EMG and MMG represented muscle activities from different aspect. The muscular electrical activities revealed by EMG showed a wide frequency range, while the component of MMG representing muscle vibration was concentrated at low frequency. Additionally, the power spectral densities (PSD) of EMG and MMG signals during hand open, as shown in Fig. 4, indicated that the main power of MMG was below 100 Hz, while EMG had a higher spectrum power at the frequency range between 20–300 Hz.

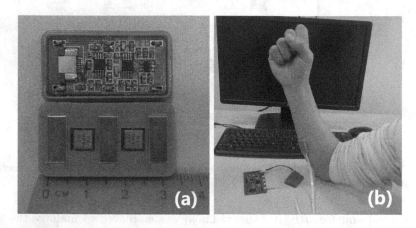

Fig. 2. Hardware implementation of the hybrid EMG/MMG sensor: (a) the hybrid sensor integrates the function of amplification and bandpass filter; (b) the hybrid sensor is attached over the Flexor Digitorum Superficialis (FDS) muscle via double-adhesive tapes.

3   Experimental Results and Discussion

Experiments were performed to verify the hybrid EMG/MMG system for human hand motion analysis and recognition. In the experiment, a hybrid sensor was placed over subject's FDS muscle [Fig. 2 (b)], and the subject was instructed to naturally hang his limb down and perform five classes of motions: wrist flexion (WF), wrist extension (WE), fist, hand open (HO) and rest, during which EMG and MMG signals were recorded simultaneously. In each trial, one motion was sustained for 5s and all motions were repeated once in a trial. Ten trials data were collected, and the subject was allowed for 2–5 minutes' rest between trials to avoid muscle fatigue.

To analyze EMG and MMG signals during the five hand motions, mean absolute value (MAV) of time-domain signals and average amplitude of PSD were extracted, expressed as equation (1) and equation (2) respectively. The radar plot of EMG and MMG representing the relative behaviour of five hand motions,

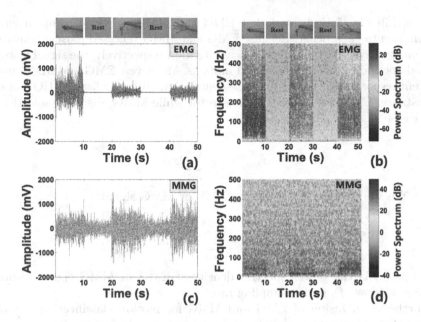

Fig. 3. Time-domain waveform and time-frequency information of EMG and MMG signals during muscle contractions: (a) the surface EMG responses to fist, wrist flexion and hand open; each motion is lasting 10 s and there is 10 s rest between contractions; (b) the time-frequency information of EMG signals computed using short-time Fourier transform (STFT); (c) the MMG responses synchronized with EMG; (d) the time-frequency information of MMG signals.

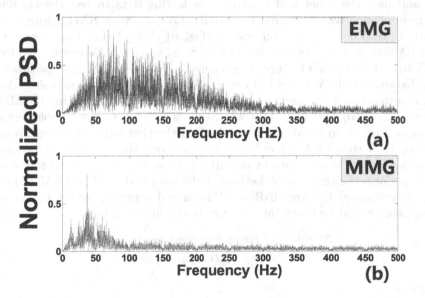

Fig. 4. The power spectral densities (PSD) of (a) EMG and (b) MMG signals during hand open.

as shown in Fig. 5, indicated that EMG and MMG provided complementary information to each other. Because EMG and MMG reflected muscle activation from electrophysiology and mechanical vibration respectively, the signal features showed specificity for different hand motions. Moreover, EMG and MMG made different responses to the same muscle contraction. Take the fist and HO motions for instance, EMG was more sensitive to fist while MMG was more sensitive to HO (Fig. 5).

$$x_{mav} = \frac{1}{N} \sum_{i=1}^{N} |x_i| \tag{1}$$

where N is the window size, x_i is the EMG or MMG signal.

$$x_{psd} = \sum_{i=1}^{fs/2} psd(i) \tag{2}$$

where $psd(i)$ is the power spectral density of EMG or MMG signal within a analysis window, fs is the sampling rate.

Furthermore, fusion of EMG and MMG for motion classification was also investigated. The motion classification had two stages: feature extraction and pattern recognition. Firstly, the EMG and MMG signals were segmented into a series of 300 ms windows with an overlap of 200 ms, and then MAV or PSD features were extracted from these windows. After feature extraction, pattern recognition was performed by means of training and testing. Half of the 10 trials data were used to train the linear discriminative analysis (LDA) classifier [20], and then the other half were used as testing data to test the classifier. The performance was evaluated by classification accuracy (CA), expressed as equation (3). The classification results (Fig. 6) showed that the average CA using EMG-only was improved from 66.7% to 87.6% for MAV feature and from 74.1% to 88.7% for PSD feature when combining EMG and MMG. As shown in Fig. 6 (a) and (c), the WF and HO were confused with other motions using EMG-only, however, these two motions were distinguishable via combining the MMG feature [Fig. 6 (b) and (d)]. This was likely attributed to the complementary information provided by MMG. It was noteworthy that only one channel sensor was used to distinguish five hand motions, therefore, the classification accuracy in this study was below state of the art [4]. However, the point of this study was to perform the comparison between EMG-only and combined EMG-MMG. In consideration of the low SNR of MMG signal captured by accelerometers, microphones would be taken into account in our further study.

$$CA = \frac{Number\ of\ correct\ testing\ samples}{Total\ number\ of\ testing\ samples} \times 100\% \tag{3}$$

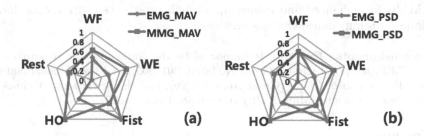

Fig. 5. Radar plots of EMG and MMG exhibiting the relative behaviour of five hand motions: (a) MAV features; (b) PSD features. The values of the features are normalized to the maximum of all five motions, and are averaged among ten trials.

66.7	WF	WE	Fist	HO	Rest
WF	34.6	0.8	0.4	72.9	0.0
WE	25.8	88.8	0.0	16.7	0.0
Fist	0.0	0.0	99.6	0.0	0.0
HO	39.6	8.3	0.0	10.4	0.0
Rest	0.0	2.1	0.0	0.0	100.0

(a)

87.6	WF	WE	Fist	HO	Rest
WF	65.4	4.6	0.4	7.9	0.0
WE	29.6	88.8	0.0	7.9	0.0
Fist	0.0	0.0	99.6	0.0	0.0
HO	5.0	5.4	0.0	84.2	0.0
Rest	0.0	1.3	0.0	0.0	100.0

(b)

74.1	WF	WE	Fist	HO	Rest
WF	51.7	0.0	9.2	55.8	0.0
WE	0.8	95.0	0.0	11.3	0.0
Fist	2.1	0.0	90.8	0.0	0.0
HO	45.4	3.3	0.0	32.9	0.0
Rest	0.0	1.7	0.0	0.0	100.0

(c)

88.7	WF	WE	Fist	HO	Rest
WF	77.5	0.4	9.2	10.4	0.0
WE	6.3	92.1	0.0	6.7	0.0
Fist	2.1	0.0	90.8	0.0	0.0
HO	14.2	6.7	0.0	83.0	0.0
Rest	0.0	0.8	0.0	0.0	100.0

(d)

WF   WE   Fist   HO   Rest

Fig. 6. Confusion matrix for five hand motions using different signals and features: (a) MAV feature of EMG, the average classification accuracy is 66.7%; (b) MAV feature of combined EMG and MMG, the average accuracy is 87.6%; (c) PSD feature of EMG, the average accuracy is 74.1%; (d) PSD feature of combined EMG and MMG, the average accuracy is 88.7%.

4 Conclusion

This paper presented and validated a hybrid EMG and MMG capturing system for human hand motion analysis. The proposed hybrid sensor system allowed detect EMG and MMG signals simultaneously, and could eliminate the limb movement artifact of MMG via differential filter. The experimental results indicated that EMG reflected muscle contraction in electrophysiology having a wide frequency distribution, while MMG was the mechanical vibration signal concentrating on low-frequency. Therefore, providing complementary information to each other, understanding of muscle activities was facilitated by combining EMG and MMG. The pattern recognition results of five hand gestures showed that the classification accuracy using combined EMG-MMG features was remarkable improved comparing to EMG alone. Our future work will focus on fusion of EMG and MMG for enhanced and robust upper-limb prostheses control, considering the long-term impedance changes and sweat.

Acknowledgments. This work is supported by the National Basic Research Program (973 Program) of China (Grant No.2011CB013305), and the National Natural Science Foundation of China (Grant No.51375296), and the Science and Technology Commission of Shanghai Municipality (Grant No. 13430721600).

References

1. De Luca, C.J., Adam, A., Wotiz, R., Gilmore, L.D., Nawab, S.H.: Decomposition of surface EMG signals. Journal of Neurophysiology **96**(3), 1646–1657 (2006)
2. Parker, P., Englehart, K., Hudgins, B.: Myoelectric signal processing for control of powered limb prostheses. Journal of Electromyography and Kinesiology **16**(6), 541–548 (2006)
3. Jiang, N., Dosen, S., Müller, K.R., Farina, D.: Myoelectric control of artificial limbsis there a need to change focus. IEEE Signal Process. Mag. **29**(5), 150–152 (2012)
4. Scheme, E., Englehart, K.: Electromyogram pattern recognition for control of powered upper-limb prostheses: State of the art and challenges for clinical use. Journal of Rehabilitation Research & Development **48**(6), 643 (2011)
5. Liu, J., Zhang, D., Sheng, X., Zhu, X.: Quantification and solutions of arm movements effect on sEMG pattern recognition. Biomedical Signal Processing and Control **13**, 189–197 (2014)
6. Pan, L., Zhang, D., Sheng, X., Zhu, X.: Improving myoelectric control for amputees through transcranial direct current stimulation (2015)
7. Farina, D., Jiang, N., Rehbaum, H., Holobar, A., Graimann, B., Dietl, H., Aszmann, O.: The extraction of neural information from the surface EMG for the control of upper-limb prostheses: Emerging avenues and challenges. IEEE Trans. Neural Syst. Rehabil. Eng. **22**(4), 797 (2014)
8. He, J., Zhang, D., Zhu, X.: Adaptive pattern recognition of myoelectric signal towards practical multifunctional prosthesis control. In: Su, C.-Y., Rakheja, S., Liu, H. (eds.) ICIRA 2012, Part I. LNCS, vol. 7506, pp. 518–525. Springer, Heidelberg (2012)
9. Young, A., Kuiken, T., Hargrove, L.: Analysis of using EMG and mechanical sensors to enhance intent recognition in powered lower limb prostheses. Journal of Neural Engineering **11**(5), 056021 (2014)
10. Courteville, A., Gharbi, T., Cornu, J.Y.: Mmg measurement: A high-sensitivity microphone-based sensor for clinical use. IEEE Transactions on Biomedical Engineering **45**(2), 145–150 (1998)
11. Silva, J., Heim, W., Chau, T.: A self-contained, mechanomyography-driven externally powered prosthesis. Archives of Physical Medicine and Rehabilitation **86**(10), 2066–2070 (2005)
12. Xie, H.B., Zheng, Y.P., Guo, J.Y.: Classification of the mechanomyogram signal using a wavelet packet transform and singular value decomposition for multifunction prosthesis control. Physiological Measurement **30**(5), 441 (2009)
13. Geng, Y., Chen, L., Tian, L., Li, G.: Comparison of electromyography and mechanomyogram in control of prosthetic system in multiple limb positions. In: 2012 IEEE-EMBS International Conference on Biomedical and Health Informatics (BHI), pp. 788–791. IEEE (2012)

14. Alves, N., Chau, T.: Uncovering patterns of forearm muscle activity using multi-channel mechanomyography. Journal of Electromyography and Kinesiology **20**(5), 777–786 (2010)
15. Alves, N., Chau, T.: Recognition of forearm muscle activity by continuous classification of multi-site mechanomyogram signals. In: 2010 Annual International Conference of the IEEE Engineering in Medicine and Biology Society (EMBC), pp. 3531–3534. IEEE (2010)
16. Sasidhar, S., Panda, S.K., Xu, J.: A wavelet feature based mechanomyography classification system for a wearable rehabilitation system for the elderly. In: Biswas, J., Kobayashi, H., Wong, L., Abdulrazak, B., Mokhtari, M. (eds.) ICOST 2013. LNCS, vol. 7910, pp. 45–52. Springer, Heidelberg (2013)
17. Alves, N., Chau, T.: Classification of the mechanomyogram: its potential as a multifunction access pathway. In: Annual International Conference of the IEEE Engineering in Medicine and Biology Society, EMBC 2009, pp. 2951–2954. IEEE (2009)
18. Silva, J., Chau, T.: Coupled microphone-accelerometer sensor pair for dynamic noise reduction in MMG signal recording. Electronics Letters **39**(21), 1496–1498 (2003)
19. Guo, W., Yao, P., Sheng, X., Liu, H., Zhu, X.: A wireless wearable sEMG and NIRS acquisition system for an enhanced human-computer interface. In: 2014 IEEE International Conference on Systems, Man and Cybernetics (SMC), pp. 2192–2197. IEEE (2014)
20. Hargrove, L.J., Scheme, E.J., Englehart, K.B., Hudgins, B.S.: Multiple binary classifications via linear discriminant analysis for improved controllability of a powered prosthesis. IEEE Transactions on Neural Systems and Rehabilitation Engineering **18**(1), 49–57 (2010)

Disturbance Observer–Based Fuzzy Control for Prosthetic Hands

Xiao-Bao Deng[1,2], Xiao-Gang Duan[1,2(✉)], and Hua Deng[1,2]

[1] State Key Laboratory of High-Performance Complex Manufacturing,
Central South University, Changsha 410083, China
xgduan@csu.edu.cn
[2] School of Mechanical & Electrical Engineering, Central South University,
Changsha 410083, China

Abstract. There exists uncertainties and environmental disturbances, such as mechanism friction and filling water into cups, etc., when a prosthetic hand grasps an object. Those disturbance can not be obtained easily because of the lack of sensors. Thus, a disturbance observe-based fuzzy control is presented for prosthetic hands. The environmental disturbance can be observed by using the proposed disturbance observer. The performance of the observer is analyzed. A fuzzy logical controller is then adopted to track the desired force. Finally, the proposed method is applied to controlling a single-freedom prosthetic hand. The experimental results show that the proposed method is effective.

Keywords: Prosthetic hand · Disturbance observer · Fuzzy control

1 Introduction

People who have amputated their extremities by labor accidents, traffic accidents or other afflictions have been desiring a prosthetic hand to help them. For a prosthetic hand, grasp force control is a key element. For example, to hold an egg without breaking it or to grasp a hammer without letting it fall [1]. Although there are some control methods that can improve grasp stability, the control of grasp force is still imperfect considering the system should be able to overcome mechanism friction and monitor external disturbances.

In fact, to generate enough force, high ratio gear and linkage mechanisms are used to compensate for small motors. These hardy mechanical components result in viscous friction and backlash in prosthetic hands [2]. These nonlinearities produce a very large dead-band of the motor. There are a lot of uncertainties in the system of prosthetic. Robust nonlinear control methods are wise choices to deal with the nonlinear friction and backlash inherent to prostheses. Some literatures show a sensory feedback control for initiating stable grasp [3,4]. However, the disturbance observer(DOB) is seldom applied in prosthetic hand , although the disturbance observer is used to many applications [5-8] since it has simple structure and is easy to understand. The reason may be the worry about robust stability of the control system with disturbance observer. But K.Ohnishi consider this problem and examine an analysis of parameters [9].

© Springer International Publishing Switzerland 2015
H. Liu et al. (Eds.): ICIRA 2015, Part I, LNAI 9244, pp. 338–347, 2015.
DOI: 10.1007/978-3-319-22879-2_31

On the other hand, facing unknown environment and being required to grasp a variety of objects, it is a good choice for prosthetic hands to use the disturbance observer to compensate external disturbance and self-uncertainty. A disturbance observer can provide   force controller with more information when lacking sensors. Since fuzzy logical control does not require a certain model and has a good effect on nonlinear, a fuzzy logical control is adopted to track the desired force considering uncertainty of prosthetic hand. Additionally, fuzzy controllers have been widely used for industrial processes due to their heuristic natures associated with simplicity and effectiveness for both linear and nonlinear system.[10-13].

Inspired by aforementioned researches, since the prosthetic hand is a highly complex and nonlinear system, we design disturbance observer-based fuzzy logical controller to compensate disturbances and keep grasping force stable. The model of output force and drive voltage is established, and then the performance of disturbance observer was analyzed. Also the work utilizes a TI-TM320F28335 DSP as the platform to control the force of a single DOF prosthetic hand with force sensory feedback. Experiments of the grasping force control in the DSP controller are realized and presented in this paper.

2   Problem Formulation

The basic structure of the prosthetic hand is as shown in Fig. 1. The basic mechanical structure of the prosthetic hand includes a precious metal-commutation DC micro motor, a gear train and a four-link system. A force sensor-force sensing resistor (FSR) is pasted on the surface of thumb finger. The dynamics of a prosthetic hand with single freedom can be described by [14]

$$M\ddot{\theta} + C\dot{\theta} + D\theta = Bu - lf \tag{1}$$

where, θ, $\dot{\theta}$ and $\ddot{\theta}$ denote the joint angle, angular velocity and angular acceleration, respectively; M is the equivalent inertia; C includes the effective damping of the motor and linkage system in addition to the back electromotive force terms; D is the system's stiffness; B is the quotient of the torque constant and armature resistance of the motor; u is the voltage input to the motor of the prosthetic hand; f is the force exerted by the environment, and l is the distance between the contact point and rotational joint.

A prosthetic hand is effected by designed constraints on weight, size, and power necessitate very small actuators. High ratio gear and linkage mechanisms are used to compensate for small motor to generate enough force for prosthetic hands. These hardy mechanical components result in viscous friction and backlash in prosthetic hand. These nonlinearities produce a very large dead-band of the DC motor. The uncertain is hard to obtain since lacking appropriate sensors.

Besides, the prosthetic hand is vulnerable to environmental disturbance in the grip of objects. for example a glass of water can be grasped stably in the normal environment. But when the cup was subjected to a sudden disturbance, if there is no enough force to cope with the current disturbance, it is likely to tilt or even drop. Although there are many control methods like sliding mode controller [4], PID controller [15], EMG proportional control [16], and so on, undesired grasping force still exists because prosthetic hands faces on unknown environment.

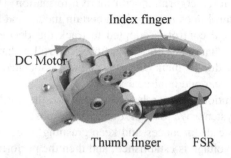

Fig. 1. Schematic structure of prosthetic hands

3   Control Methods

The overall control block diagram is as shown in Fig. 2, which includes a fuzzy controller and a disturbance observer. The disturbance observer is used to obtain disturbance information. It can give a voltage increment to motor directly to compensate a sudden disturbance. The fuzzy controller is used to track the desired force.

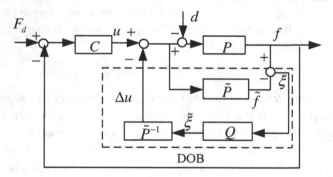

Fig. 2. Overall control block diagram

where F_d is the given input signal in the force closed-loop. d and Δu denote the external disturbances and estimation of voltage increment, respectively. P and $\tilde{P}$ represent mathematical model of the actual object and the nominal object, respectively. Q is the filter. C denotes the controller. f and $\tilde{f}$ are the actual output force and estimation of the nominal model, respectively. u is the output of C.

The whole control system mainly concludes the two parts. Their design is independent of each other according to the Fig. 2. By adjusting the drive voltage, jointly maintain the grip strength stability of a prosthetic hand.

3.1   Disturbance Observer

The DOB is placed in the control loop for a prosthetic hand to removed the distur-bance entirely and reduce the influence of disturbance on stability of grasping objects.

In the system design, the DOB can regard the system uncertain as disturbance, and have a good effect on the estimation and compensation of disturbance. In a certain error range, the actual model can be equivalent to its nominal model. Design mainly embarks from two aspects. First, when dead-band of the motor causes initial grip force too large, disturbance observer obtains the signal, a voltage is given to reduce motor driving voltage. Second, when detecting a external disturbance caused by de-crease of grip force, DOB will increase the drive voltage by giving a negative incre-ment of voltage Δu. Then the model will be established.

Suppose impedance between the prosthetic hand and environment is an elastic element. Thus, the contact force $\tilde{f}(t)$ in (1) is described by

$$\tilde{f}(t) = K_{ev}x(t) \tag{2}$$

where $x(t) = \tilde{x}(t) - x_0(t)$ and $\tilde{x}(t)$ denotes real value, $x_0(t)$ denote contract position when the prosthetic hand is initially contacted with the environment, K_{ev} is environ-ment stiffness.

The displacement $x(t)$ can be approximated by

$$x(t) = l\theta(t) \tag{3}$$

where l is the distance between the contact point and rotational joint. For simplicity, we consider that force is applied to only one direction. Based on impedance control [17], integrating (2) and (3), equation (1) becomes

$$\ddot{\tilde{f}}(t) + a_1\dot{\tilde{f}} + a_0\tilde{f} = cu \tag{4}$$

where $a_1 = C/M$, $a_0 = (D + l^2K_{ev})/M$ and $c = BlK_{ev}/M$. Equation (4) is a typi-cal second order system.

In a certain error range, the actual model can be equivalent to its nominal model. According to equation (4), one can get:

$$\Delta u = \frac{1}{c}\left[\ddot{\xi}(t) + a_1\dot{\xi} + a_0\xi\right] \tag{5}$$

Where Δu is the increment of voltage. $\tilde{\xi}$ is the deviation of the actual force f and the estimated force $\hat{f}$ after filtering. If $\tilde{\xi} = 0$, there is no disturbance. If $\tilde{\xi} > 0$, then $\Delta u > 0$. It is necessary to reduce drive force of motor. If $\tilde{\xi} < 0$, then $\Delta u < 0$. It is necessary to increase drive force of motor.

3.2   Fuzzy Controller

The disturbance observer can weaken the influence of disturbance and system uncer-tainty. However, in the control process of the whole system, according to the change of the force, a reliable input voltage u of motor must be guaranteed.

To ensure the stability of grasping force for a nonlinear prosthetic hand, we adopted a fuzzy PID controller [18,19] is as shown in Fig. 3.

The membership function (MFs) of FLC selected a standard triangular membership functions.

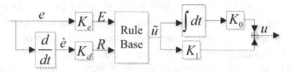

Fig. 3. The structure of fuzzy controller

In fact, rules are always finite in real-world application. According to the triangle membership function, we select N=3. Setting the{NL,NM,NS,ZR,PS,PM,PL} as input E and R as the fuzzy linguistic variables, namely {negative large, negative medium, negative small, zero, positive small ,positive middle, positive large}. Rule Base can be determined.

Thus, finite rules clearly show the saturation effect. The output model of the rule base becomes [20]:

$$\tilde{u} = sat(\sigma) = \begin{cases} sgn(\sigma) & |\sigma| > 1 \\ g(\sigma) & |\sigma| \le 1 \end{cases} \qquad (6)$$

with

$$g(\sigma) = \sigma + (1 - \gamma)(kh - \sigma) \qquad (7)$$

where $\sigma = E + R$.

So the mathematical models of main controller and reflex controller can be easily derived as:

$$u_m = K_m \tilde{u}_m = K_m sat(\sigma_m) \qquad (8)$$

$$u_r = K_0 \int sat(\sigma_r) dt + K_1 sat(\sigma_r) \qquad (9)$$

where $sat(\sigma)$ is given in (7).

4   The Characteristic of Disturbance Observer

According to Fig. 2, the mathematical model of disturbance observe by using Laplace transform is given:

$$\Delta u(s) = \frac{Q(s)}{\tilde{P}(s)[1 - Q(s)]} f(s) - \frac{Q(s)}{1 - Q(s)} u(s) \qquad (10)$$

where $u(s)$ and $f(s)$ are Laplace transforms of $u(t)$ and $f(t)$, respectively. To illustrate the role of DOB, one can obtains:

$$\Delta u(s) = G_u u(s) + G_d d(s) \tag{11}$$

where $d(s)$ is Laplace transforms of $d(t)$,

$$G_u(s) = \frac{\left[P(s) - \tilde{P}(s)\right]Q(s)}{\tilde{P}(s)\left[1 - Q(s)\right] + P(s)Q(s)}, \quad G_d(s) = \frac{Q(s)P(s)}{\tilde{P}(s)\left[1 - Q(s)\right] + P(s)Q(s)}.$$

When $P(s) = P_n(s)$, for this situation, one gets $\Delta u(s) = -Q(s)d(s)$. When $P(s) \neq P_n(s)$, Δu contains not only d, but also the equivalent interference of the system uncertainties.

Through DOB compensation, the output force can be expressed as:

$$f(s) = P(s)\left[u(s) - \Delta u - d(s)\right] \tag{12}$$

According to equations (7) and (8),

$$f(s) = G_{fu}(s)u(s) + G_{fd}d(s) \tag{13}$$

Where, $G_{fu}(s) = \dfrac{P(s)\tilde{P}(s)}{\tilde{P}(s)\left[1 - Q(s)\right] + P(s)Q(s)}$ $\quad G_{fd}(s) = \dfrac{P(s)\tilde{P}(s)[Q(s) - 1]}{\tilde{P}(s)\left[1 - Q(s)\right] + P(s)Q(s)}$

When $P(s) = P_n(s)$, one obtains:

$$f(s) = P(s)u(s) + P(s)\left[Q(s) - 1\right]d(s) \tag{14}$$

When $Q(s) \approx 1$, one can get $G_{fu}(s) \approx \tilde{P}(s)$, $G_{fd}(s) \approx 0$. Therefore, equation (10) is approximately correct. This shows that DOB can make the actual system performance close to the nominal system, which provides a strong robustness for the control system. When $Q(s) \approx 0$, one can get $G_{fu}(s) = P(s)$, $G_{fd}(s) = -P(s)$. This can reduce the increase of force when force is rise.

The performance of DOB depends on the design of the filter. The low frequency characteristics of the filter are close to 1, and the high frequency characteristic is close to 0 in order to reduce the uncertainty of the model. According to the literature [7,9], the following low-pass filter is adopted.

$$Q(s) = \frac{3\tau s + 1}{(\tau s + 1)^3} \tag{15}$$

By choosing different τ values, the different cut-off frequency can be obtained.

5   Experiments

The method of disturbance observer-based fuzzy control proposed in this paper is applied in a single-DOF prosthetic hand as shown in Fig. 4. The thumb and index fingers wear silicone gloves to avoid impact between fingers and objects. A force sensor-force sensing resistor (FSR) is pasted on the surface of the silicone glove at thumb finger's tip. The resistor of FSR on the thump tip changes sensitively when the prosthetic hand contacts the object. A circuit of the FSR, a constant resistor and a battery in series was used to obtain the voltage of FSR.

Control algorithms are realized on the Digital signal processor (DSP) F28335 with C language. In order to achieve the requirement of real time control, FSR signal was acquired at sample rate of 1000Hz. We adopted $\tau = 0.005$. The Parameters of fuzzy controller is given as follow by Cut and Try method. $K_e = 0.43$, $K_d = 0.25$, $K_0 = 0.002$, $K_1 = 0.15$. We conducted three group of experiments. As shown in Fig. 5, from left to right are metal cup, plastic cup and paper cup, respectively.

DC   Index finger

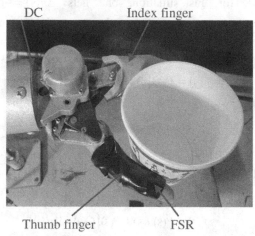

Thumb finger   FSR

Fig. 4. Test-bed of the single DOF prosthetic hand

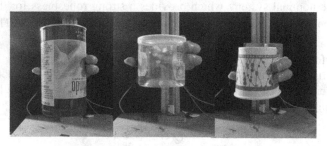

Fig. 5. Grasped objects.

When the object is caught stably, we put a external disturbance into the object. Then observe the state of the object and acquire the signals of FSR and voltage. Fig. 6, Fig. 7 and Fig. 8 are force and voltage signals of metal cup, plastic cup, and paper cup, respectively.

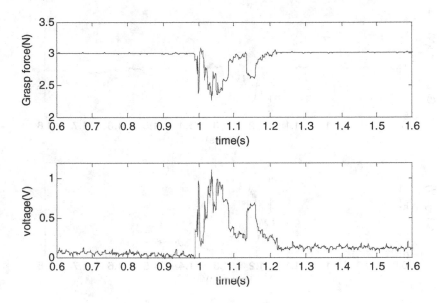

Fig. 6. Force signal of metal cup with DOB under disturbance

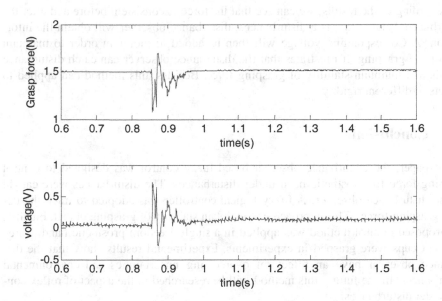

Fig. 7. Force signal of plastic cup with DOB under disturbance

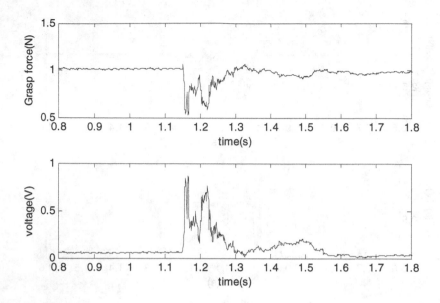

Fig. 8. Force signal of paper cup with DOB under disturbance

According to the results, we can see that the force is consistent before and after the disturbance. When there is a disturbances, disturbance observer will obtain the information. A Corresponding voltage will then be added to motor in order to maintain stability of grasping. It illustrates that the disturbance observer can catch disturbance signals and maintain stability of grasping force. Besides, this method can applied to objects of different rigidity.

6   Conclusion

In the paper, the disturbance observer-based fuzzy control was designed to control grasping force for prosthetic hand under disturbances. The disturbances were caught by the disturbance observer. A fuzzy logical controller was adopted to track the desired grasping force . This approach can maintain stability of grasping object. Finally , the proposed control method was applied in a single-freedom prosthetic hand. Three different cups were grasped in experiments. Experimental results show that the disturbance observer have an effect on overcoming uncertainties and environmental disturbances. In the future, this method will be researched in the aspect of reflex control under disturbances.

Acknowledgements. This study was supported by a grant from National Basic Research Program 973 of China (Grant Nos. 2011CB013302).

References

1. Castellini, C., Gruppioni, E., Davalli, A., Sandini, G.: Fine detection of grasp force and posture by amputees via surface electromyography. Journal of Physiology-Paris **103**, 255–262 (2009)
2. Engeberg, E.D., Meek, S.G.: Backstepping and Sliding Mode Control Hybridized for a Prosthetic Hand. IEEE Transactions on Neural Systems and Rehabilitation Engineering **17**, 70–79 (2009)
3. Wettels, N., Parnandi, A.R., Moon, J.-H., Loeb, G.E., Sukhatme, G.: Grip control using biomimetic tactile sensing systems. IEEE/ASME Transactions on Mechatronics **14**, 718–723 (2009)
4. Engeberg, E.D., Meek, S.G.: Adaptive Sliding Mode Control for Prosthetic Hands to Simultaneously Prevent Slip and Minimize Deformation of Grasped Objects. IEEE/ASME Transactions on Mechatronics **18**, 376–385 (2013). [1]
5. Chen, W.-H., Ballance, D.J., Gawthrop, P.J., O'Reilly, J.: A nonlinear disturbance observer for robotic manipulators. IEEE Transactions on Industrial Electronics **47**, 932–938 (2000)
6. Kent, A., Engeberg, E.D.: Robotic hand biomimicry: Lateral finger joint force and position feedback during contour interaction. In: 2011 IEEE International Conference on Robotics and Biomimetics (ROBIO), pp. 2245–2246 (2011)
7. Kempf, C.J., Kobayashi, S.: Disturbance observer and feedforward design for a high-speed direct-drive positioning table. IEEE Transactions on Control Systems Technology **7**, 513–526 (1999)
8. Tomizuka, M.: On the design of digital tracking controllers. Journal of dynamic systems, measurement, and control **115**, 412–418 (1993)
9. Kobayashi, H., Katsura, S., Ohnishi, K.: An analysis of parameter variations of disturbance observer for motion control. IEEE Transactions on Industrial Electronics **54**, 3413–3421 (2007)
10. Sugeno, M.: Industrial applications of fuzzy control. Elsevier Science Inc. (1985)
11. Zhu, D., Wu, C.: Fuzzy control of group elevators. In: 1993 IEEE Region 10 Conference on Proceedings of the Computer, Communication, Control and Power Engineering, TENCON 1993, vol. 4, pp. 304–307 (1993)
12. Po-Rong, C., Bor-Chin, W.: Adaptive fuzzy power control for CDMA mobile radio systems. IEEE Transactions on Vehicular Technology **45**, 225–236 (1996)
13. Tang, K., Man, K.F., Chen, G., Kwong, S.: An optimal fuzzy PID controller. IEEE Transactions on Industrial Electronics **48**, 757–765 (2001)
14. Engeberg, E.D., Meek, S.G., Minor, M.A.: Hybrid Force-Velocity Sliding Mode Control of a Prosthetic Hand. IEEE Transactions on Biomedical Engineering **55**(5), 1572–1581 (2008)
15. Cipriani, C., Zaccone, F., Micera, S.: On the Shared Control of an EMG-Controlled Prosthetic Hand: Analysis of User–Prosthesis Interaction. IEEE Transactions on Robotics, 170–184 (2008)
16. Fougner, A., Stavdahl, Ø., Kyberd, P.J., Losier, Y.G., Parker, P.A.: Control of upper limb prostheses: terminology and proportional myoelectric control – a review. IEEE Transactions on neural systems and rehabilitation engineering **20**(5), 663–677 (2012)
17. Kang, S.H., Jin, M., Chang, P.H.: A solution to the accuracy/robustness dilemma in impedance control. IEEE/ASME Transactions on Mechatronics **14**(3), 282–294 (2009)
18. Li, H.X., Gatland, H.B., Green, A.W.: Fuzzy variable structure control. IEEE Transactions on Systems, Man, and Cybernetics, Part B: Cybernetics **27**, 306–312 (1997)
19. Duan, X.-G., Li, H.-X., Deng, H.: Effective Tuning Method for Fuzzy PID with Internal Model Control. Industrial & Engineering Chemistry Research **47**, 8317–8323 (2008). 2008/11/05
20. Duan, X.-G., Li, H.-X., Deng, H.: Robustness of fuzzy PID controller due to its inherent saturation. Journal of Process Control **22**, 470–476 (2012)

A Method to Modify Initial Desired Force of Prosthetic Hand Based on Stiffness Estimation

Yi Zhang[1,2], Xiao-Gang Duan[1,2(✉)], and Hua Deng[1,2]

[1] State Key Laboratory of High-Performance Complex Manufacturing,
Central South University, Changsha 410083, China
xgduan@csu.edu.cn
[2] School of Mechanical and Electrical Engineering,
Central South University, Changsha 410083, China

Abstract. If desired force decoded by biological signals of prosthetic hand is too large, the soft object may produce undesired deformation or damaged. In order to reduce this phenomenon, the excessive grasping force needs to be limited. Based on contact-impact force model, a fuzzy logic system for stiffness estimation determined by contact force and its gradient is presented. In order to avoid undesired deformation, the maximum grasping force of the soft object is set up by experimental data. Then, a fuzzy control model is proposed to modify initial desired force. The experiment results are presented to demonstrate effectiveness of the proposed method.

Keywords: Prosthetic hand · Stiffness estimation · Fuzzy control

1 Introduction

Amputees expect to own prosthesis with a rich variety of functionality in order to deal with their daily lives. Although artificial hands can be built stronger and faster than the human hand, functions of such hands are still not to be compared with that of a natural hand. Because most kinds of prosthetic hands are controlled by electromyography signals or potentials from voluntarily contracted muscles within a person's residual limb on the surface of the skin. Such prosthetic hands are only available in both voluntary opening and closing movements due to the limited level of utilization of biological signals, the dexterity of these are still far away from human hands [1].

Humans are able to adjust grasping force very quickly, especially during tasks that need interactions with different environment. Such adaptability is credited to the characteristic of variable stiffness of muscles. Humans can adjust their activation levels of the agonistic and antagonistic muscles depending on the properties of the environment. When the two kinds of muscles are simultaneously activated, the stiffness of the fingers increases; hence, able to utilize tools in high precision. Besides, high stiffness also can provide larger grasping force. On the other hand, it can be observed that the fingers become more flexible and compliant when the hand grasping or manipulating the soft or fragile objects. However, on the contrary, the prosthetic

© Springer International Publishing Switzerland 2015
H. Liu et al. (Eds.): ICIRA 2015, Part I, LNAI 9244, pp. 348–358, 2015.
DOI: 10.1007/978-3-319-22879-2_32

hands are manufactured with a fixed stiffness. To deal with the problem, many high-cost commercially prosthetic hands, such as Otto Bock's Sensor Hand and Touch Bionics' i-Limb Ultra equipped with sensors to provide the low-layer control. The low-layer control is often based on position/velocity or force feedback from the pros-thetic hand to fulfil effective actions in auto grasp feature. It not only provide auto-power adjusting for adaptation to pre-defined grasp tasks, but can also solve many problems such as force overshoot, slippage and grasp instability through force or position feedback [2,3]. Rigid prosthetic hands with more users provide only basic structural support with limited function, which are especially not quite suitable for dexterous contact manipulation.

In order to improve motion performance, many researchers have put lots of effort in the recognition of movement patterns [4] and estimation of grasping force, position or velocity [5] by biological signals decoding. But that, biological signal decoding invariably introduces a latency between the time control the prosthetic hand and the time react to the amputee [6]. Unfortunately, the latency problem can negatively im-pact the amputee's experience and it's hard to resolve under current technology.

As a result, the principal focus of research work has been on the active force con-trol of prosthetic hand. In fact, the most important one for a prosthetic hand is the grasping force. When a prosthetic hand grasps an object, an excessive grasping force cause possible damage to some delicate objects. On the other hand, too little force may cause the instability of grasp. Therefore, an effective force control strategy is needed. The adaptive force control and impedance control technique applied in pros-theses shows very good application prospects [7]. Engeberg proposed an adaptive PID sliding mode controller based on measurement of the object stiffness [8]. The controller is effective to prevent unwanted force overshoot but could not correct the desired force determined by biological signal in gripping state so the softer object is prone to undesired deformation or damage.

In this paper, a fuzzy controller with estimate initial desired force is proposed. Based on contact-impact force model, a fuzzy logic system for stiffness estimation determined by contact force and its gradient is modeled. In order to avoid undesired deformation, the maximum grasping force of the soft object is set up by experimental data. Then, a fuzzy control model is proposed to modify initial desired force. The experiment results are presented to demonstrate effectiveness of the proposed method.

2   Problem Formulation

For a prosthetic hand, the control system is usually composed of two parts: an upper-layer control and a low-layer control. The input of the upper-layer control is biological signals of humans. Most of prosthetic hands are reproducing the force under upper-layer controller when grasping an object. However the motion and force information cannot decode completely, in addition, the delay times of up to 300ms. That is, accurate and real-time grasping force cannot be obtained from biological signal. Thus, bottom-layer control should be used in conjunction with upper-layer control to fulfil effective actions expected by amputees.

Prosthetic hands need to grab various objects with random uncertainty environment. Taking human hands as the reference, when a man try to grasp different objects with various shapes and degrees of stiffness, he generally tries to touch with objects and perceives the surface information about the object. After he /she perceives the stiffness of objects, his/her grasping force is affected. Therefore, the key problems are how to offer a quick way to estimate the object's stiffness and adjust the parameter of feedback controller by detected stiffness.

3 Estimation of Object Stiffness

The dynamics of a prosthetic hand can be described by [9]

$$M\ddot{\theta} + B\dot{\theta} + K\theta = \tau - J^{T}F \tag{1}$$

where, θ, $\dot{\theta}$ and $\ddot{\theta}$ are the joint angle, angular velocity and angular acceleration respectively; M is the equivalent inertia matrix; B denotes the effective damping matrix of the motor and mechanical structure; K is the system's stiffness matrix; τ is the is the actuator joint torques; J^{T} is Jacobian transpose matrix and F is the contact force exerted by the environment. The force and motion state of the prosthetic hand can be calculated from Eq. (1).

When the prosthetic hand contact with the object with a slow speed, Hertzan contact theory is especially suitable to the contact force calculation [10]. The contact-impact force model as follows:

$$F(\delta, \dot{\delta}) = \begin{cases} K_{i}\delta^{n} + C\delta^{n}\dot{\delta} & \delta \geq 0 \\ 0 & \delta < 0 \end{cases} \tag{2}$$

where K_{i} is the contact stiffness; C is the contact damping; n is an index coefficient; δ and $\dot{\delta}$ denote deformation displacement and deformation velocity, respectively.

For the stiffness estimation, a simple way is to calculate by the force and position measurements during the grasp. Hooke's law defined the stiffness as:

$$K_{i} = dF / dx \tag{3}$$

where x is the deformation during grasping.

It's important to note that the stiffness of object is not constant. In general, most objects take on complicated nonlinear stiffness characteristics. If the object is deformable, the stiffness depends upon the amount of deformation of the object as a function of the pressure. However, from Eq. (2), only an instantaneous stiffness can be gotten. More precise stiffness characteristics should be obtained by applying a set of forces to the objects and measuring their deflections. But this method too slowly to calculate, which is bad for rapid and steady grasp.

Suppose the prosthetic hand is close to the object at very low speed. From Eq. (2), δ and $\dot{\delta}$ with a slow increase or approaching zero when the finger touch the object in a very short time.   As the finger hit the object instantaneously, if the object is deformed, the instantaneous contact force is directly affected by the deformation. It is very easy to calculate the stiffness of the object with angle sensor and force sensor by Eq. (3). However, the majority of prosthetic hand is not equipped with angle sensor, and it is inconvenient to increase it.

Therefore, for the rapid estimation of contact stiffness, a simple method is proposed herein considering accuracy as well as efficiency. The gradient of grasping force is calculated by using only one force sensor as

$$dF_m = \frac{F_{m+1} - F_m}{T} \tag{4}$$

where $F_m = \frac{1}{3}\sum_{k=1}^{3} F_k$, F_k is the instantaneous value of the contact force collected by force sensor.

And then, the stiffness of the object is approximate estimated by the grasp force and its gradient with fuzzy reasoning. Thus, the initial desired force will be modified according to the stiffness of object to avoid deformation or damage.

4   Desired Force Estimation and Main Controller

In this section, we propose a fuzzy control model to modify initial desired force as shown in Fig. 1. In this controller, F_d^* is the decoding force by biological signals, F_{d0}, F_d and F denote estimated, initial desired force and real grasping force, respectively, $e = F_d - F$ is error, F_0 denotes initial contact force at the initial voltage u_0.

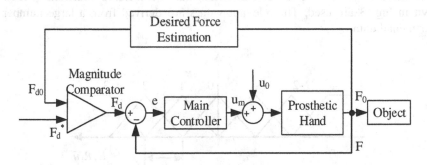

Fig. 1. The fuzzy logic-based control scheme of the prosthetic hand

The controller mainly includes two units, desired force estimation and main controller. The initial desired force is determined by logical judgment (Magnitude Comparator) for the first time to grasp as

$$F_d = \begin{cases} F_{d0}, & \text{if } F_{d0} \leq F_d^* \ \& \ F_{d0} \neq 0 \\ F_d^*, & \text{if } F_{d0} > F_d^* \ \text{or} \ F_{d0} = 0 \end{cases} \tag{5}$$

4.1   A. Fuzzy Logic System for Force Estimation

Considering that detection of the stiffness is vague, a fuzzy logic system is proposed to estimate desired grasping force when the prosthetic hand first contact with the object. The detail design of the fuzzy logic system was discussed in [11] and [12]. Here, only the structure of the fuzzy logic system is presented as shown in Fig. 2.

Generally, for a fuzzy logic system, the fuzzy rule adopts "If E is A_i and R is B_j then $\tilde{u}$ is G_{i+j}," where E, R and A_i, B_j $(i, j = -N, \cdots, -1, 0, 1, \cdots, N)$ denote input variables and input fuzzy sets, respectively; $\tilde{u}$ and G_{i+j} denotes the control action and output fuzzy set, respectively; N denotes the number of input fuzzy sets .

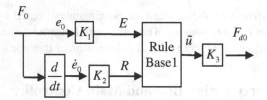

Fig. 2. Fuzzy logic system for desired force estimation

Practically, rules are always finite in real-world application, i.e., N is finite. Here, we select N=3, where the linguistic labels are negative large (NL), negative medium (NM), negative small (NS), zero (ZO), positive small (PS), positive medium (PM) and positive large (PL). The standard triangular membership functions (MFs), as shown in Fig. 3 are used. The relevant values are derived from a large number of experimental data.

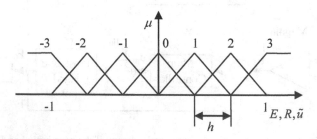

Fig. 3. Membership functions of input and output variable.

Thus, finite rules clearly show the saturation effect, as shown in Fig. 4 The output model of the rule base becomes [13] :

$$\tilde{u} = sat(\sigma) = \begin{cases} sgn(\sigma) & |\sigma| > 1 \\ g(\sigma) & |\sigma| \le 1 \end{cases} \tag{6}$$

with $g(\sigma) = \sigma + (1 - \gamma)(kh - \sigma)$ (7)

where $\sigma = E + R$.

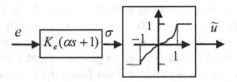

Fig. 4. Saturation of the rule base.

Thus, the estimated force can be easily derived as:

$$F_{d0} = K_3 \tilde{u} \tag{8}$$

4.2   Main Controller

Prosthetic hand has nonlinearities caused by the existence of motor dead bands, friction and large gear ratio [9]. This uncertainty can be described by fuzzy systems. FLC has two advantages, 1) it has robustness due to its inherent saturation; 2) it does not depend on accurate model of controlled object. Thus, a FLC is chosen as reflex controller as shown in Fig. 5.

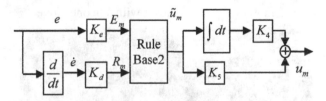

Fig. 5. Structure of the fuzzy main controller.

The model of the FLC is given as [14, 15]

$$u_m = u_e + K_5 sat(\sigma) \tag{9}$$

with $u_e = K_0 \int \tilde{u}_m dt = \alpha K_4 K_e \left(e + \dfrac{1}{\alpha} \int edt \right) + \Delta u$ (10)

where $\sigma_m = E_m + R_m = K_e e + K_d \dot{e}$; $\tilde{u}_m = sat(\cdot)$ is a saturation function, Δu is an error; K_e, K_d, K_0, K_4, α and β are positive constant.

5   Experiments and Results

In this section, a simple method of estimation of object stiffness by contact with its experimental verification is present.

5.1   Experimental Setup

The experiment setup used for detecting contact force and force gradients based control of prosthetic hand is presented in Fig. 6. The setup composed by prosthetic hand, force sensing resistor, data acquisition and grasped object. The prosthesis consists of a precious metal-commutation DC micro motor, a gear train and a four-link system. In the four-link system, the thumb finger and the index finger are coupled by a connecting rod. So the thumb and index fingers open and close simultaneously. The maximum force of the prosthetic hand grasping an object is about 15N. This prosthetic hand is similar to one described in [16]. Particularly, the thumb and index fingers wear silicone gloves to avoid impact between fingers and objects. A force sensor-force sensing resistor (FSR) is pasted on the surface of the silicone glove at thumb finger's tip. And a thin rubber covers the FSR to protect it. The resistor of FSR on the thump tip changes sensitively when the prosthetic hand contacts the object. We use a circuit of the FSR, a constant resistor and a battery in series and the voltage of FSR is detected. The voltage was calibrated by describing the force as a quantic polynomial function of the detected voltage. The data of FSR is collected by a real-time experimental platform based on LabVIEW and PXI. Control algorithms and the real-time (RT) data acquisition software module are realized on the upper computer by LabVIEW graphical programming language (G-language). The National Instruments (NI) PCI eXtensions for Instrumentation (PXI) system is controlled by the LabVIEW software.

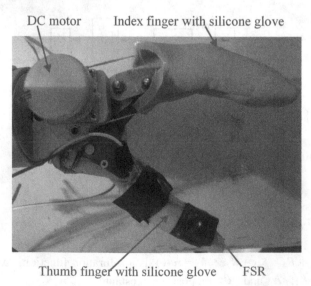

Fig. 6. The single DOF prosthetic hand with FSR.

5.2   Stiffness Estimation

In the experiment, ceramic cup, plastic bottle and paper bottle were used as grasped objects. These three objects have the same radius. The purpose of the experiment is to obtain the grasp forces in the same conditions. First of all, adjust the prosthetic fingers very close to the objects, but without contact with it. Next, set the same input voltage of prosthetic hand to grasp obiects. At the same time, recording data from FSR.

A group result of grasping force in an open-loop control at 1.2 voltages is shown in Fig. 7. The loading rates are related to the stiffness of the object, and the hard object required shorter time and larger force to reach the same grasping force than the soft one. According to the Hertzan contact theory as shown in Eq. (2), the soft object deformed less under the same stress, which is validated by the experiments.

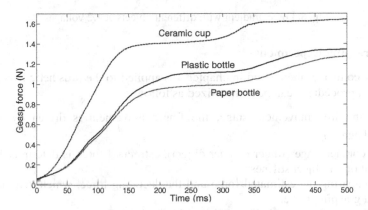

Fig. 7. Grasping force with different objects at 1.2 voltages.

The Fig. 8 shows force gradient relationship of three selected objects with different stiffness by Eq. (4). By comparing the Fig. 7 and Fig. 8, we can find that the force gradient is more able to reflect the stiffness of the object. Therefore, wc can use the information of the force and its gradient to calculate the stiffness of the object by using the method of fuzzy estimation, as shown in Fig. 1. By using this method, the information of the force and its gradient of different stiffness objects is obtained by a lot of experiments. At the same time, the maximal bearable grasping force of soft objects without excessive deformation has also been assessed. The parameters of fuzzy controller as shown in Fig. (1) can be determined by these experiment data.

However, it is difficult to distinguish the gradient information of plastic bottle and paper bottle. The method can only be used to identify objects that have a large stiffness difference. Moreover, it's interesting to find that the grasping force increased in very short time when the prosthetic fingers contact with the object. If the desired force is improper, the objects may have damaged in an instant.

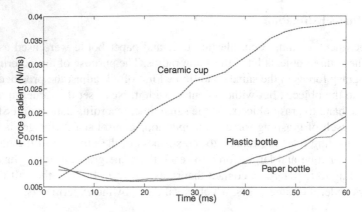

Fig. 8. Force gradient with different objects at 1.2 voltages.

5.3   Grasping Experiment

The fuzzy controller mentioned in chapter 4 is applied to the prosthetic hand, and the experiment procedures can be summarized as follows:

A) In the free movement stage, the finger movement is driven by an initial voltage u_0;

B) In contact stage (finger contact object), estimated force is determined by estimation of object stiffness.

C) Comparing the estimated force and the decoding force, which determines initial grasping force;

D) The main fuzzy controller realize stable grasping.

In addition, according to the experimental data, the parameters of fuzzy controller are set as $K_1=1$, $K_2=0.8$, $K_3=1.5$ and $K_e=0.5$, $K_d=0.2$, $K_4=0.01$, $K_5=0.2$, respectively. Suppose the decoding force $F_d^* = 2.5N$, the results of the experiments are as follow.

When the object is ceramic cup, the control performance is shown in the Fig.9, where the desired force is not changed.

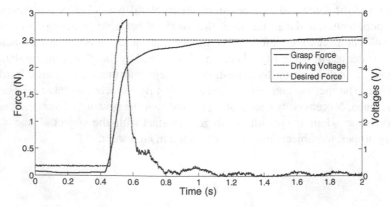

Fig. 9. Control performance for hard object

When grasping object is paper cup, the control performance is shown in the Fig. 10. In order to protect the cup from being damaged, the decoding force is modified to 1.4N after a brief contact. The experimental results demonstrate the feasibility of this approach.

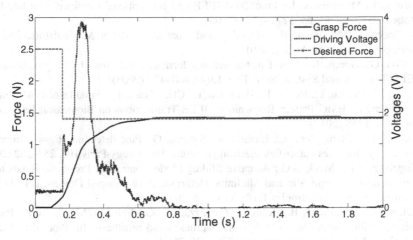

Fig. 10. Control performance for soft object

6   Conclusion

This paper presents a fuzzy control method to avoid undesired deformation of soft object for prosthetic hand.   According to the results of grasping force in an open-loop control experiments in this paper, it's believed that the grasping force and force gradient are able to reflect the stiffness of the object. In this way, based on contact-impact force model, a fuzzy logic system for stiffness estimation determined by contact force and its gradient is modeled. In order to avoid undesired deformation, the maximum grasping force of the soft object is set up by experimental data. Then, a fuzzy control model is proposed to modify initial desired force. The experiment results are presented to demonstrate effectiveness of the proposed method.

However the results have shown that objects with similar stiffness are difficult to be distinguished. Take the stiffness of prosthetic hand as a reference, only a far greater than it or far less than it can be identified. Therefore to improve the performance of the proposed method, other sensors such as position or angle will be considered in future work.

Acknowledgements. This study was supported by a grant from National Basic Research Program 973 of China (Grant Nos. 2011CB013302).

References

1. Carozza, M., Cappiello, G., Stellin, G., Zaccone, F., Vecchi, F., Micera, S., et al.: On the development of a novel adaptive prosthetic hand with compliant joints: experimental platform and EMG control. In: Proc. 2005 IEEE/RSJ International Conference on Intelligent Robots and Systems, pp. 1271–1276 (2005)
2. Sciavicco, L., Siciliano, B.: Modeling and Control of Robot Manipulators, 2nd edn. Springer-Verlag, New York (2000)
3. Dillon, G., Horch, K.: Direct neural sensory feedback and control of a prosthetic arm. IEEE Trans. Neural Syst. Rehabil. Eng. 13(4), 468–472 (2005)
4. Young, A., Smith, L., Rouse, E., Hargrove, L.: Classification of Simultaneous Movements using Surface EMG Pattern Recognition. IEEE Transactions on Biomedical Engineering 60, 1250–1258 (2013)
5. Castellini, C., Gruppioni, E., Davalli, A., Sandini, G.: Fine detection of grasp force and posture by amputees via surface electromyography. Physiology-Paris 103, 255–262 (2009)
6. Engeberg, E.D., Meek, S.G.: Adaptive Sliding Mode Control for Prosthetic Hands to Simultaneously Prevent Slip and Minimize Deformation of Grasped Objects. IEEE/ASME Transactions on Mechatronics 1(18), 376–385 (2013)
7. Scherillo, P., Siciliano, B., Zollo, L., Carrozza, M., Guglielmelli, M., Dario, P.: Parallel force/position control of a novel biomechatronic hand prosthesis. In: Proc. IEEE/ASME Int. Conf. Adv. Intell. Mechatron. pp. 920–925 (2003)
8. Andrecioli, R., Engeberg, E.D.: Adaptive sliding manifold slope via grasped object stiffness detection with a prosthetic hand. Mechatronics 8(23), 1171–1179 (2013)
9. Engeberg, E.D., Meek, S.G., Minor, M.A.: Hybrid Force-Velocity Sliding Mode Control of a Prosthetic Hand. IEEE Transactions on Biomedical Engineering 55, 1572–1581 (2008)
10. Herbert, R.G., McWhannell, D.C.: Shape and frequency composition of pulses from an impact pair. ASME Journal of Engineering for Industry 99, 513–518 (1977)
11. Li, H.X., Gatland, H.B., Green, A.W.: Fuzzy variable structure control. IEEE Transactions on Systems, Man, and Cybernetics, Part B: Cybernetics 27, 306–312 (1997)
12. Duan, X.-G., Li, H.-X., Deng, H.: Effective Tuning Method for Fuzzy PID with Internal Model Control. Industrial & Engineering Chemistry Research 47, 8317–8323 (2008)
13. Duan, X.-G., Li, H.-X., Deng, H.: Robustness of fuzzy PID controller due to its inherent saturation. Process Control 22, 470–476 (2012)
14. Duan, X.G., Deng, H., Li, H.X.: A saturation based tuning method for fuzzy PID controller. IEEE Transactions on Industrial Electronics 11(60), 5177–5185 (2013)
15. Mendel, J.M.: Uncertain Rule-Based Fuzzy Logic Systems: Introduction and New Directions. Prentice Hall PTR, London (2000)
16. Engeberg, E. D.: Human Model Reference Adaptive Control of a Prosthetic Hand. Intelligent & Robotic Systems, 1–16 (2013)

Preliminary Testing of a Hand Gesture Recognition Wristband Based on EMG and Inertial Sensor Fusion

Yangjian Huang[1], Weichao Guo[1], Jianwei Liu[1], Jiayuan He[1], Haisheng Xia[1], Xinjun Sheng[1], Haitao Wang[2], Xuetao Feng[2], and Peter B. Shull[1]([✉])

[1] State Key Laboratory of Mechanical System and Vibration, School of Mechanical Engineering, Shanghai Jiao Tong University, Shanghai 200240, People's Republic of China
pshull@sjtu.edu.cn
[2] Samsung R&D Institute of China, Beijing 100028, People's Republic of China
xuetao.feng@samsung.com

Abstract. Electromyography (EMG) is well suited for capturing static hand features involving relatively long and stable muscle activations. At the same time, inertial sensing can inherently capture dynamic features related to hand rotation and translation. This paper introduces a hand gesture recognition wristband based on combined EMG and IMU signals. Preliminary testing was performed on four healthy subjects to evaluate a classification algorithm for identifying four surface pressing gestures at two force levels and eight air gestures. Average classification accuracy across all subjects was 88% for surface gestures and 96% for air gestures. Classification accuracy was significantly improved when both EMG and inertial sensing was used in combination as compared to results based on either single sensing modality.

Keywords: Surface EMG · Inertial motion sensing · Human-machine interface · Hand gesture recognition

1 Introduction

Mobile devices such as smart phones, tablets and laptops are increasingly integral in our daily lives. Traditional interaction paradigms like on-screen keyboards can be an inconvenient and cumbersome way to interact with these devices. This has led to the advancement of bio-signal interfaces such as speech based and hand gesture based interfaces. Compared to speech input interfaces, hand gesture devices can be more intuitive in resembling physical manipulations related to spatial tasks such as navigation on a map or manipulating a picture [1].

Different technologies have been proposed for hand gesture recognition, including devices based on optical depth sensing [2,3], magnetic sensing [4], force sensitive resistors (FSR) [5], inertial motion sensing (IMU) [6–11] and electromyography (EMG) [12–14]. Depth sensing and magnetic sensing are susceptible to ambient light and magnetic disturbances and FSR may require unnatural

© Springer International Publishing Switzerland 2015
H. Liu et al. (Eds.): ICIRA 2015, Part I, LNAI 9244, pp. 359–367, 2015.
DOI: 10.1007/978-3-319-22879-2_33

hand gestures. IMUs have advantages in discriminating dynamic gestures involving direction change but are not as strong in discriminating gestures involving minimal movement like making a fist. The advantage of EMG sensing is detecting static gestures involving distinct muscle activations but can be sensitive to electrode placement and susceptible to sweat and skin impedance changes over time [15–18]. Thus combining inertial and EMG sensing could increase hand gesture classification accuracy by combining the strengths of each.

Previous research has combined inertial and EMG sensing by primarily placing sensors on the forearm [1,19–22]. Georgi et al. [1] used a system consisting of an IMU worn on the wrist and 16 EMG sensors worn on the forearm to classify 12 gestures, which achieved a recognition rate of 97.8% in session-independent and 74.3% in person-independent testing. Zhang et al. [21] used a three-axis accelerometer and 5 EMG sensors to classify 72 Chinese Sign Language (CSL) words at a 95.8% recognition rate. Chen et al. [22] improved recognition rates by 5-10% for 24 hand gestures with 2D-accelerometers and EMG sensors together as compared to using EMG sensors alone. Additionally, Wolf et al. [19] used a sleeve with an IMU and 8 EMG sensors and get 99% accuracy for 9 dynamic gestures. Though these studies provide promising results, for practical applications it may be inconvenient to place EMG sensors on the forearm.

This paper presents a hybrid recognition wristband based on surface EMG and inertial sensor fusion for hand gesture recognition. We aimed to combine EMG and inertial sensing in a practical wristband form with relatively high accuracy.

2   Materials and Methods

2.1   Hand Gestures

We choosed two kinds of gestures: four surface pressing gestures with the hand pressing against a flat, hard surface and eight air gestures which are performed in the air without any contact forces. For the surface gestures, we distinguished between two levels of pressing force. Gestures were selected to be familiar and intuitive to users, accustomed to interacting with mobile devices. The four surface gestures were: index finger press (T1), index finger, middle finger and ring finger pressing together (T3), all five fingers pressing together (T5), and making a fist (F), as shown in (Fig. 1). The eight air gestures were: okay sign (OK), peace sign (PS), hang loose (HL), finger snap (FS), thumbs up (TU), thumbs down (TD), turn palm over (TP), and walking fingers (WF), as shown in Fig. 2. The two force levels for the surface gestures were set to 10% of maximum voluntary contraction (MVC) and 50% of MVC.

2.2   Hybrid EMG and Inertial Sensing Wristband System Configuration

The aim of the hybrid system was to detect EMG and motion signals simultaneously. Dry gold-plated copper electrodes were deployed to capture EMG signals

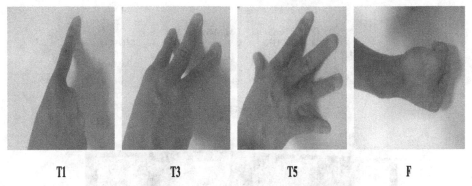

T1	T3	T5	F

Fig. 1. Four surface gestures: index finger press (T1), three finger press (T3), five finger press (T5), and making a fist (F).

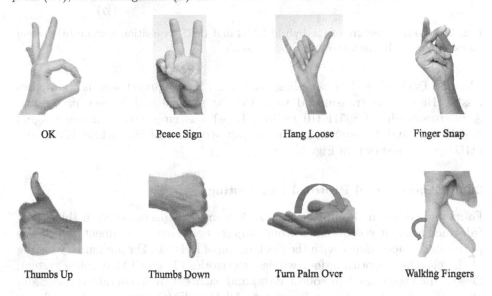

OK	Peace Sign	Hang Loose	Finger Snap
Thumbs Up	Thumbs Down	Turn Palm Over	Walking Fingers

Fig. 2. Eight air gestures: okay sign (OK), peace sign (PS), hang loose (HL), finger snap (FS), thumbs up (TU), thumbs down (TD), turn palm over (TP), walking fingers (WF).

for their stable and reliable contact with skin. To attenuate common-mode noise and crosstalk, differential amplification was adopted to capture EMG signals. An instrumentation amplifier (INA326, Texas Instruments) was used for differential amplification because of its high common mode rejection ratio (CMRR) and input impedance. The differential EMG signals were bandpass filtered at 20–450 Hz to remove unwanted noise. EMG signals were further amplified with a gain of 500 to improve the signal to noise ratio. Inertial motion sensors consisted of a 3-axis analog accelerometer (ADXL335, Analog Devices) and a dual-axis pitch and roll analog gyroscope (LPR4150AL, STMicroelectronics). EMG and motion signals were digitalized via analog to digital conversion AD7607

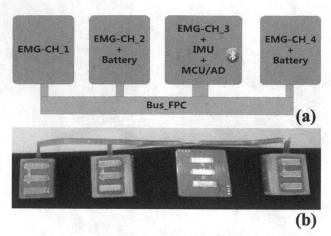

(a)

(b)

Fig. 3. Overall structure of the hybrid EMG and IMU acquisition system: (a) system block diagram; (b) hardware implementation.

(Analog Devices) with a sampling rate of 1 kHz. Digitized signals were wirelessly (Bluetooth) transmitted to a PC for storage and further processing. A microcontroller C8051F310 (Silicon Labs) was employed to manage signal digitization and transmission. The overall structure of the hybrid EMG and IMU system is shown in Fig. 3.

2.3   Experimental Protocol and Testing

Four able-bodied male subjects (20 to 30 years old) participated in this study. Informed consent was obtained from subjects before the experiment, which was performed in accordance with the Declaration of Helsinki. For preliminary testing of the classifier algorithm, four wireless electrodes (Trigno TM Wireless system, Delsys Inc.) were used to collect data, and each electrode contained one EMG channel and one 3-axis accelerometer. All four EMG sensors and the 3-axis

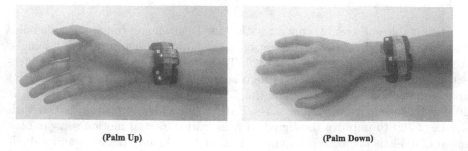

(Palm Up)   (Palm Down)

Fig. 4. Experimental setup: two electrodes were placed on the anterior muscles, and two were plced on the posterior muscles, 5 cm away from the edge of the palm

accelerometer on the top lateral wrist were used for classification. Electrodes were placed 5 cm away from the edge of the palm, of which two were on the anterior muscles, and two were on the posterior muscles (Fig. 4). For surface gestures recognition, each subject performed twenty trials, half of which were conducted at 10% MVC and another half were conducted at 50% MVC. For air gestures recognition, each subject performed ten trials with the moderate force. Each trial consisted of one repetition of the defined gestures, where each gesture sustained 5 seconds with 5 seconds rest between them. Half of the trials were used for training and the left half were used for testing.

2.4   Feature Extraction and Pattern Recognition

Motion classification was performed in two stages: feature extraction and pattern recognition. First, EMG and motion signals were segmented into a series of 200 ms windows with an overlap of 50 ms, which were used to extract the feature vectors. Each feature vector consisted of the EMG features (mean absolute value, waveform length, zero crossing and sign slope change) and acceleration features (mean value and standard deviation). The linear discriminant analysis (LDA) classifier was used for the pattern recognition.

3   Experimental Results and Discussion

Classification accuracy (CA) was computed as:

$$CA = \frac{Number\ of\ correct\ testing\ samples}{Total\ number\ of\ testing\ samples} \times 100\% \qquad (1)$$

Only EMG features were used in surface gesture recognition because of the minimal hand movements. Mean classification accuracies for T1, T3, T5 and F surface gestures were 86.17%, 85.68%, 91.20% and 87.60% respectively, as seen in (Fig. 5). For surface gestures at two force levels, the average CA for T1-10%, T3-10%, T5-10%, F-10%, T1-50%, T3-50%, T5-50% and F-50% were 82.99%, 82.79%, 93.90%, 85.71%, 89.74%, 86.88%, 94.81% and 92.21% respectively, as seen in (Fig. 6). Average CA across all subjects was 88.63% for all surface gestures in this study which is better than another studies [12] which reported a recognition rate of 79% and similar to the recognition accuracy of 94% reported by Kim et al [13]. Provided that the system was worn on the wrist where muscular activation signals are relatively lower, the classification accuracy in this study is promising.

Furthermore, classification performance based on EMG and inertial sensing fusion for eight air gestures is shown in Fig. 7. Average CA using EMG alone improved from 86.81% to 95.97% when combining EMG with inertial sensing. We expected EMG and inertial sensing to be complementary because they capture different aspects of movement. For static gestures such as OK, PS and HL which involve minimal movement, classification using EMG alone can be quite accurate, because movement features are negligible. Hand gesture recognition

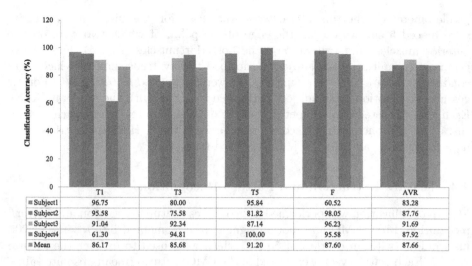

Fig. 5. Classification accuracy for four surface gestures (T1, T3, T5 and F) without force discrimination.

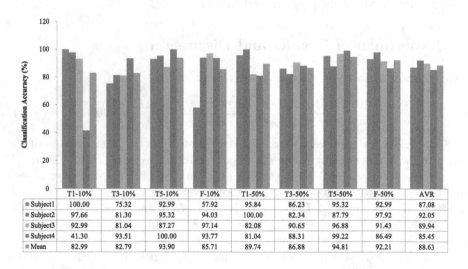

Fig. 6. Classification accuracy for four surface gestures at two force levels (T1-10%, T3-10%, T5-10%, F-10%, T1-50%, T3-50%, T5-50% and F-50%) for each subject.

based on motion signals for minimal movement gestures may result in a classification overfitting problem. For dynamic gestures such as TU, TD, TP and WF which involve large direction changes during rotation or translation, classifiers based on motion signals alone can discriminate gestures accurately. In this case, hand gesture recognition based on EMG alone might have difficulty identifying dynamic gestures that differ primarily in direction. Overall accuracy is higher

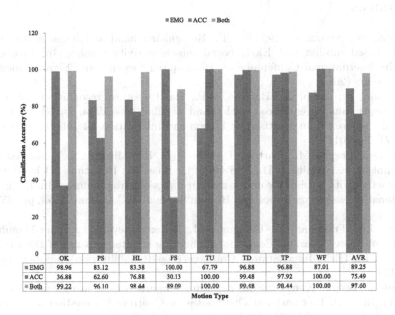

Fig. 7. Classification accuracy of eight air gestures (OK, PS, HL, FS, TU, TD, TP and WF) for one representative subject. For each gesture, classification accuracy is shown when using EMG or inertial sensing (ACC) individually, and in combination (Both).

when using EMG and inertial sensing together compared to performance using either modality alone.

4   Conclusion

This paper presents a wrist-based hand gesture recognition system that combines EMG and inertial motion sensing. A classification algorithm was validated via preliminary testing of four surface gestures at two force levels and eight air gestures. EMG sensing is particularly accurate for static hand gestures because of its ability to track muscle contractions. Inertial sensing is well suited for dynamic hand gestures due to its inherent ability to sense motion direction and amplitude changes. Used in combination these sensor signals can provide complementary information for higher classification accuracy. Classification accuracy using combined EMG-motion features was significantly improved as compared to EMG or motion sensing classification used in isolation.

Acknowledgments. This work was supported by the Samsung R&D institute of China (Beijing).

References

1. Georgi, M., Amma, C., Schultz, T.: Recognizing hand and finger gestures with IMU based motion and EMG based muscle activity sensing. In: Proceedings of the International Conference on Bio-inspired Systems and Signal Processing, pp. 99–108 (2015)
2. Trindade, P., Lobo, J., Barreto, J.P.: Hand gesture recognition using color and depth images enhanced with hand angular pose data. In: IEEE International Conference on Multisensor Fusion and Integration for Intelligent Systems, pp. 71–76 (2012)
3. Van Den Bergh, M., Carton, D., De Nijs, R., Mitsou, N., Landsiedel, C., Kuehnlenz, K., Wollherr, D., Van Gool, L., Buss, M.: Real-time 3D hand gesture interaction with a robot for understanding directions from humans. In: IEEE International Workshop on Robot and Human Interactive Communication, pp. 357–362 (2011)
4. Rowe, J.B., Friedman, N., Bachman, M., Reinkensmeyer, D.J.: The Manumeter: a non-obtrusive wearable device for monitoring spontaneous use of the wrist and fingers. In: IEEE International Conference on Rehabilitation Robotics, pp. 1–6 (2013)
5. Jeong, E., Lee, J., Kim, D.: Finger-gesture recognition glove using velostat (ICCAS 2011). In: 11th International Conference on Control, Automation and Systems. Number Iccas, pp. 206–210 (2011)
6. Ko, S., Bang, W.: A Measurement System for 3D Hand-Drawn Gesture with a PHANToM TM Device. Journal of Information Processing Systems 6(3), 347–358 (2010)
7. Muthulakshmi, M.: Mems Accelerometer Based Hand Gesture Recognition. International Journal of Advanced Research in Computer Engineering & Technology (IJARCET) 2(5), 1886–1892 (2013)
8. Wang, J.S., Chuang, F.C.: An accelerometer-based digital pen with a trajectory recognition algorithm for handwritten digit and gesture recognition. IEEE Transactions on Industrial Electronics 59(7), 2998–3007 (2012)
9. Benbasat, A.Y., Paradiso, J.A.: An inertial measurement framework for gesture recognition and applications. In: Wachsmuth, I., Sowa, T. (eds.) GW 2001. LNCS (LNAI), vol. 2298, pp. 9–20. Springer, Heidelberg (2002)
10. Hartmann, B., Link, N.: Gesture recognition with inertial sensors and optimized DTW prototypes. In: IEEE International Conference on Systems, Man and Cybernetics, pp. 2102–2109 (2010)
11. Wu, J., Pan, G., Zhang, D., Qi, G., Li, S.: Gesture recognition with a 3-d accelerometer. In: Zhang, D., Portmann, M., Tan, A.-H., Indulska, J. (eds.) UIC 2009. LNCS, vol. 5585, pp. 25–38. Springer, Heidelberg (2009)
12. Samadani, A.A., Kuli, D.: Hand gesture recognition based on surface electromyography. In: 36th Annual International Conference of the IEEE Engineering in Medicince and Biology Society, pp. 4196–4199 (2014)
13. Kim, J., Mastnik, S., André, E.: EMG-based hand gesture recognition for real-time biosignal interfacing. In: Proceedings of the 13th International Conference on Intelligent user Interfaces, IUI 2008, vol. 39, p. 30 (2008)
14. Saponas, T.S., Tan, D.S., Morris, D., Balakrishnan, R.: Demonstrating the feasibility of using forearm electromyography for muscle-computer interfaces. In: Proceeding of the Twentysixth Annual CHI Conference on Human Factors in Computing Systems, CHI 2008, p. 515 (2008)

15. He, J., Zhang, D., Zhu, X.: Adaptive pattern recognition of myoelectric signal towards practical multifunctional prosthesis control. In: Su, C.-Y., Rakheja, S., Liu, H. (eds.) ICIRA 2012, Part I. LNCS, vol. 7506, pp. 518–525. Springer, Heidelberg (2012)
16. Jiang, N., Dosen, S., Müller, K.R., Farina, D.: Myoelectric control of artificial limbs-is there a need to change focus. IEEE Signal Process. Mag. **29**(5), 148–152 (2012)
17. Farina, D., Jiang, N., Rehbaum, H., Holobar, A., Graimann, B., Dietl, H., Aszmann, O.: The extraction of neural information from the surface EMG for the control of upper-limb prostheses: Emerging avenues and challenges. IEEE Trans. Neural Syst. Rehabil. Eng. **22**(4), 797 (2014)
18. Young, A., Kuiken, T., Hargrove, L.: Analysis of using EMG and mechanical sensors to enhance intent recognition in powered lower limb prostheses. Journal of Neural Engineering **11**(5), 056021 (2014)
19. Wolf, M.T., Assad, C., Stoica, A., You, K., Jethani, H., Vernacchia, M.T., Fromm, J., Iwashita, Y.: Decoding static and dynamic arm and hand gestures from the JPL biosleeve. In: IEEE Aerospace Conference Proceedings (2013)
20. Li, Y., Chen, X., Tian, J., Zhang, X., Wang, K., Yang, J.: Automatic recognition of sign language subwords based on portable accelerometer and EMG sensors. In: International Conference on Multimodal Interfaces and the Workshop on Machine Learning for Multimodal Interaction on - ICMI-MLMI 2010, p. 1 (2010)
21. Zhang, X., Chen, X., Li, Y., Lantz, V., Wang, K., Yang, J.: A framework for hand gesture recognition based on accelerometer and EMG sensors. IEEE Transactions on Systems, Man, and Cybernetics Part A: Systems and Humans **41**(6), 1064–1076 (2011)
22. Chen, X., Zhang, X., Zhao, Z.Y., Yang, J.H., Lantz, V., Wang, K.Q.: Hand gesture recognition research based on surface EMG sensors and 2D-accelerometers. In: 2007 11th IEEE International Symposium on Wearable Computers, pp. 1–4 (2007)

Robot Actuators and Sensors

Robot Joint Module Equipped with Joint Torque Sensor with Disk-Type Coupling for Torque Error Reduction

Jae-Kyung Min, Hong-Seon Yu, Kuk-Hyun Ahn, and Jae-Bok Song[✉]

School of Mechanical Engineering, Korea University, Seoul, Korea
jbsong@korea.ac.kr

Abstract. Force control and collision detection of a robot were usually conducted using a 6 axis force/torque sensor mounted at the end-effector. This scheme, however, suffers from its high cost and inability to detect collisions at the robot body. As an alternative, joint torque sensors embedded in each joint were used, but they also suffered from various errors in torque measurement. To solve this problem, a robot joint module with an improved joint torque sensor is proposed in this study. In the proposed torque sensor, a cross-roller bearing and disk-type coupling are added to prevent the moment load from adversely affecting the measurement of the joint torque and to reduce the stress induced in the assembly process of the sensor. The performance of the proposed joint torque sensor was verified through various experiments.

Keywords: Joint torque sensor · Robot joint module · Disk-type coupling · Torque measurement · Moment load

1 Introduction

Recently, robots collaborating with humans in the same space have received considerable attention [1]. For this purpose, much research has been done in performing force control of a robot arm and collision detection based on the sensing of external forces. The robot arm is usually equipped with a 6 axis force/torque sensor at its end-effector for force control capability. This sensor enables contact control and collision detection at the end-effector, but it cannot respond to the collisions that occur at the robot body. In order to solve this problem, some researchers mounted joint torque sensors (JTS) in each joint of the robot arm, and estimated the forces and moments acting on the end-effector by the dynamic calculations [3-6]. The proposed methods are able to perform force control and collision detection including the robot body collision. However, there are some problems in measuring the torques from joint torque sensors. It is easy to be affected by unwanted loads such as the moment load at the joint [7], and the joint structure needs to be carefully designed because of the complex assembling structure.

Therefore, in this study, we propose a robot joint module with a built-in disk-type coupling that can be easily assembled and is also insensitive to an unnecessary load. The JTS performance of the joint module is verified by various experiments.

© Springer International Publishing Switzerland 2015
H. Liu et al. (Eds.): ICIRA 2015, Part I, LNAI 9244, pp. 371–378, 2015.
DOI: 10.1007/978-3-319-22879-2_34

The proposed module includes the components such as a speed reducer, JTS, bearing, and the output frame. This makes it easy to configure different types of robots by realizing different combinations of the proposed robot joint module. In this study, accurate torque measurement and thus force control and collision detection are achieved by designing the robot joint module with a disk type coupling which minimizes the performance degradation due to the crosstalk and assembly stress.

This paper is organized as follows, Chapter 2 describes the structure of the proposed robot joint module. Chapter 3 verifies that the robot joint module reduces the moment load and stress through various experiments. Finally, we draw conclusions in Chapter 4.

2   Robot Joint Modules

2.1   Joint Module Structure

A robot arm usually performs force control and collision detection based on the external forces measured by the force/torque sensor mounted at its end-effector. However, this method requires the use of an expensive sensor, and cannot detect collisions that occur at the robot body except at the end-effector. The calculation of the external forces based on the torques measured from the joint torque sensors (JTS) mounted at each joint enables force control and collision detection at the robot body. However, it is difficult to measure accurate torques because the moment load is applied to the joint, and the stresses from assembly degrade the JTS performance. Therefore, in this research, we design a joint structure containing a JTS which measures the accurate torque effectively, and the other joint components such as the motor and gear reducer.

The proposed robot joint module is composed of the timing pulley, harmonic drive, joint torque sensor and bearings. Also, the robot joint module is designed with a hollow center for smooth wiring of the robot. It is easy to construct various types of robots and verify the performance of force control for each module using the proposed robot joint module.

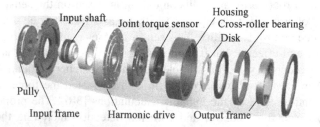

Fig. 1. Structure of robot joint module

A gear reducer, a JTS, and the output frame are connected in series as shown Fig. 2(a). When a load is applied to the output frame in this structure, the rotational torque between the frame and the gear reducer can be measured by the joint torque sensor. However, the load other than the rotational torque under consideration, such as the moment load, is applied to each joint depending on the robot arm's motion and posture. Because the moment load causes crosstalk, it degrades the performance of force control of the robot arm. Therefore, in this study, by inserting the cross-roller bearing which supports the loads in all the 5 axes except the axis of rotation into the robot joint module as shown as Fig. 2(b), only pure rotational torque can be transmitted to the joint torque sensor.

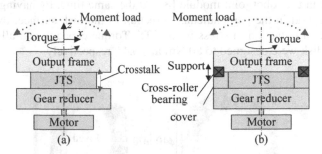

Fig. 2. Structure design: (a) previous design, and (b) proposed design to minimize crosstalk

2.2 Stress Reduction with Disk

As the output frame is in contact with both the JTS and the cross-roller bearing as shown in Fig. 3(a), the tolerance may cause the inclination between the JTS and the output frame. If one is to assemble these two sides using bolts by force, deformation occurs in the JTS. Since the principle of a JTS is to measure the torque by examining the structural deformation, the strain and stress caused by the assembly process lead to the torque measurement error. This error, which is different in size for every fabrication, deteriorates the reproducibility of torque measurement. In order to cope with this problem, we designed the connection structure which allowed the declination between the output frame and the JTS as shown in Fig. 3(b), which improved the torque measurement performance and reproducibility.

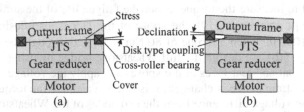

Fig. 3. (a) Stress generated by tolerance, and (b) coupling-type JTS design to avoid stress

The connection structure which allows the inclination should only transmit the rotational torque, excluding the other forces and moments to the JTS. In this study, we propose a built-in disk type coupling joint structure as shown in Fig. 4. The thin disk shape has low stiffness in a direction perpendicular to the surface, and high stiffness in the horizontal direction (i.e., on the disk surface). Due to these characteristics, the elastic deformation with low stiffness of the disk reduces the stress on the JTS caused by the inclination. Also, due to the high stiffness in the direction of rotation, the rotational torque is transmitted without loss. The disk is designed considering the size and specifications of the components of the robot joint module, and the disk is a plate spring-like material that possesses elasticity. In this study, a disk with the outer diameter of ϕ70 and the inner diameter of ϕ40 was designed. The disk has a 5×10^4 Nm/rad rotational stiffness to fit the rotational stiffness of the harmonic drive (5×10^4 Nm/rad) used in the robot joint module [8]. At the same time, by having the stiffness of 4.5×10^2 Nm/rad in the perpendicular direction to the disk surface, the disk effectively achieves a reduction in stress in the JTS. Torque capacity and allowable inclination of the disk were designed to 80 Nm and 0.4°, respectively.

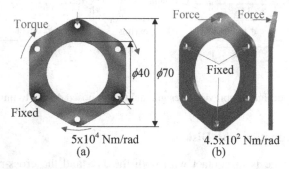

Fig. 4. Design of elastic disk: (a) torsional stiffness, and (b) axial stiffness

2.3   Joint Torque Sensor (JTS)

The joint torque sensor (JTS) inserted into the robot joint module must be sensitive to torque under consideration and insensitive to other forces or moments (i.e., moment loads). Therefore, the JTS is designed as a hub-spoke structure as shown in Fig. 5(a) and has a hollow inner hub for wiring. When the inner hub is fixed and torque is applied to the outer ring, the strain occurs on the spoke which is relatively sensitive than the hub or the outer ring. In this instance, strain gauges attached on the faces of the spokes are used to measure the torque. Since the fatigue life of the strain gauge is 10^7 times at ±1000 με strain rate [10], the spoke of the JTS is also designed to have a strain rate of ±1000 με when a torque of 80 Nm is applied through the FEM analysis. As shown in Fig. 5(b), we design the full Wheatstone bridge by using 8 strain gauges attached to the front and rear faces of the spokes and apply the voltage of E. When the resistance of the strain gauge is changed, it is able to calculate the torque by using the amplifier and the voltage difference from the two nodes of the Wheatstone bridge [9].

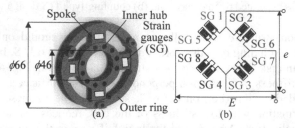

Fig. 5. Design of JTS: (a) hub-spoke structure, and (b) Wheatstone bridge

Detailed specifications of the prototype of the robot joint module developed in this study are shown in Table 1.

Table 1. Specifications of robot joint module

Feature	Specification	
	Weight	2.0 kg
	Dimension	ϕ110×59 mm
	Rated torque (speed ratio)	75 Nm(1:160)
	Torque sensing range	±80 Nm

3   Experiments and Results

3.1   Experiments of the Torque Measurement

To verify the performance of the proposed robot joint module, we performed the torque measuring experiments using the experimental setup shown in Fig. 6. A dual arm where the weight can be hung at each end was installed. We mate the center of mass of the dual arm and the rotation axis to compensate for the torque from the weight of the dual arm. By adding the weight of 12 kg to one end of the dual arm (0.55 m), we applied the torque of 66 Nm to the robot joint module and rotated the module to measure the joint torque according to the rotational angle. As shown in Fig. 6(b), after changing the direction of the robot joint module, we performed the same experiment by applying a moment load and rotating the robot joint module as above. The results of two experiments are shown in Fig. 7.

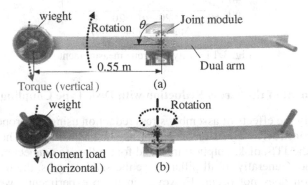

Fig. 6. Experimental setup to apply external load to robot joint module: (a) torque, and (b) moment load

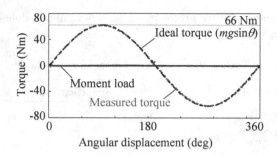

Fig. 7. Experimental results

In the experiment of Fig. 6(a), the ideal torque is given by

$$\tau_e = mgl \sin \theta \qquad (1)$$

Where θ is the angle of rotation, m and l are the mass and moment arm length, and g is the acceleration of gravity. As shown in Fig. 7, the torque measured from the JTS is effectively tracking the ideal torque. Moreover, by comparing the measured torque to the ideal torque, the linearity error of the proposed robot joint module is computed to be 0.48% as shown in Fig. 8, and it is comparable to that of commercial force/torque sensors (0.5 %). As a result of the experiment shown in Fig. 6(b), the torque measurement from the applied moment load is no more than 0.5 %, which means that the moment load is effectively supported by the proposed structure and poses insignificant effects to the measured rotational torque.

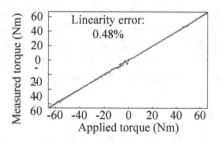

Fig. 8. Linearity of torque measurement

3.2   Experiment of the Stress Reduction with Disk Type Coupling

In order to verify the effects of assembly stress reduction using the proposed disk-type coupling, we constructed a test environment as shown in Fig. 9. The experimental setup includes the JTS, disk, output frame, and four pillars to connect base frame with the output frame. Generally, if all pillars have the same lengths, the torque error due to the inclination does not occur. However, in these experiments, we intentionally made the length differences in each case by 0.05 mm, 0.08 mm, and 0.2 mm, so that it generates the inclination of 0.03°, 0.05° and 0.13°, respectively. With the length difference in each case, the setup was assembled with different inclinations, and the corresponding torque measurements are listed in Table 2.

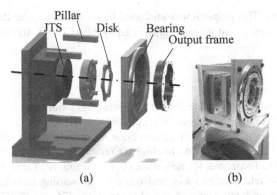

(a)   (b)

Fig. 9. Experimental setup to apply inclination error to JTS: (a) disassembled and (b) assembled.

Table 2. Experimental results of inclination error

Torque error	Inclination			
	0.0°	0.03°	0.05°	0.13°
without disk	0 Nm	1.2 Nm	3.6 Nm	5.5 Nm
with disk	0 Nm	0.3 Nm	1.1 Nm	1.8 Nm

As shown in Table 2, if the inclination increased from 0.03° to 0.13°, the torque error in a regular JTS increased from 1.2 Nm to 5.5 Nm. However, by using the proposed disk-type coupling structure, the torque error was reduced by 70 % compared to the sensor without a disk. Therefore, by using the proposed robot joint module structure containing the disk-type coupling, it is possible to profoundly reduce the torque error caused by assembly stress and inclination.

4   Conclusion

In this study, we proposed the robot joint module including the joint torque sensor and disk-type coupling. The following conclusions were drawn.

(1) It is easy to construct various types of robots using the proposed robot joint module including the joint torque sensor.
(2) The proposed robot joint module effectively supports the moment load occurring from the robot's motion and posture, and reduces the torque error due to the crosstalk down to 0.5%.
(3) By containing the disk-type coupling structure, the proposed robot joint module shows good reproducibility and accurately measures the torque which is not affected by the assembly stress.

Acknowledgements. This research was supported by the MOTIE under the Industrial Foundation Technology Development Program supervised by the KEIT No. 10048980 and 10038660.

References

1. Kaneko, K., Kanehiro, F., Kajita, S., et al.: Humanoid robot HRP 2. In: Proc. of the IEEE Int. Conf. on Robotics and Automation, pp. 1083–1090 (2004)
2. Gravel, P.D., Newman, S.W.: Flexible robotic assembly efforts at Ford motor company. In: Proc. of the IEEE Int. Sym. on Intelligent Control, pp. 173–182 (2011)
3. Tsetserukou, D., Tadakuma, R., Kajimoto, H., Kawakami, N., Tachi, S.: Development of a whole-sensitive teleoperated robot arm using torque sensing technique. In: Proc. of the IEEE Int. Conf. on Intelligent Robots and Systems, pp. 476–481 (2007)
4. Parmiggiani, A., Randazzo, M., Natale, L., Metta, G., Sandini, G.: Joint torque sensing for the upper-body of the iCub humanoid robot. In: IEEE/RAS Int. Conf. on Humanoid Robots, pp. 15–20 (2009)
5. Cho, C.N., Kim, J.H., Lee, S.D., Song, J.B.: Collision detection and reaction on 7 DOF service robot arm using residual observer. Journal of Mechanical Science and Technology **26**(4), 1197–1203 (2012)
6. Hirzinger, G., Sporer, N., Albu-Schaffer, A., Hahnle, M., Krenn, R., Pascucci, A., Schedl, M.: DLR's torque-controlled light weight robot III-are we reaching the technological limits now?. In: Proc. of the IEEE Int. Conf. on Robotics and Automation, pp. 1710–1716 (2002)
7. Siciliano, B., Khatib, O.: Handbook of Robotics. Springer-Verlag (2007)
8. SamickHDS Co. LTD. http://www.samickhds.co.kr
9. Aghili, F., Buehler, M., Hollerbach, J.: Design of a Hollow Hexaform Torque Sensor for Robot Joint. Int. Journal of Robotics Research **20**(12), 967–976 (2001)
10. CAS Co. http://www.globalcas.com

Development of a Hydraulic Artificial Muscle for a Deep-Seafloor Excavation Robot with a Peristaltic Crawling Mechanism

Mamoru Nagai[1(✉)], Asuka Mizushina[1], Taro Nakamura[1], Fumitaka Sugimoto[2], Kensuke Watari[2], Hidehiko Nakajo[2], and Hiroshi Yoshida[2]

[1] Department of Precision Mechanics, Faculty of Science and Engineering, Chuo University, 1-13-27 Kasuga, Bunkyo-ku, Tokyo 112-8551, Japan
{m_nagai,a_mizushina}@bio.mech.chuo-u.ac.jp,
nakamura@mech.chuo-u.ac.jp
[2] Japan Agency for Marine-Earth Science and Technology, 2-15 Natsushima-cho, Yokosuka, Kanagawa 237-0061, Japan
{sugimotof,watarik,nakajohh,yoshidah}@jamstec.go.jp

Abstract. In recent years, observations and explorations of the deep seafloor have been actively pursued. One goal of such explorations is to obtain the samples of seafloor mud and its inclusions. Mud that contains minerals and submarine microorganisms has great potential for studies in biology, geology, and marine science. To contribute to these efforts, we propose a robot using peristaltic crawling to excavate deep seafloor. The robot consists of three parts: excavation, propulsion, and extraction units. The propulsion actuator of the proposed robot must be able to function under water at high pressures. As the first stage in the development, we developed a subunit using an oil hydraulic artificial muscle intended for use in deep sea, and conducted a performance experiment under water pressure. Our results confirmed that the artificial muscle can be used in water pressures of up to 5 MPa.

Keywords: Earthworm robot · Seabed excavation · Artificial muscle

1 Introduction

Currently, various technical efforts to observe and study the ocean, including deep-sea water and deep seabeds, are being actively pursued [1][2]. In particular, sediments in the deep seafloor contain mineral resources and submarine microorganisms, and extracting them is expected to be a major goal of deep-sea explorations [3]. For example, within the exclusive economic zone of Japan, rare earth deposits with ultrahigh densities have been discovered in the seabed below the water depths of several thousand meters, and these deposits have growing potential as a future resource [4]. Furthermore, revealing the ecology of microorganisms living in the seabed offers potential to elucidate various mysteries in the vast biosphere.

© Springer International Publishing Switzerland 2015
H. Liu et al. (Eds.): ICIRA 2015, Part I, LNAI 9244, pp. 379–389, 2015.
DOI: 10.1007/978-3-319-22879-2_35

A common way to explore under the seabed is by using surface research vessels. Such vessels are routinely used for coring, drilling, etc. [5][6]. By drilling, heavy mud samples can be raised vertically from the deep areas of over a few thousands meters through a drill pipe extending from a research vessel. This procedure requires a long device with length equal to the depth being explored; consequently, costs increase quickly as the target depth increases. Coring is a simple method that can acquire a mud sample of up to a few dozen meters using a piston corer to pierce the seafloor. These two methods collect mud samples in columnar shapes. However, they impose a drawback in which the research area available at one time is limited to a point. This limitation means that these methods effectively sample the vertical direction but are not suitable for extension in the horizontal direction. To investigate the extent of a selected layer, for example the layer containing seabed mineral resources, we must increase the number of sampling spots and perform detailed investigations. Thus, the costs for sampling increase.

An alternative that can solve these problems is a small autonomous buried-type robot. However, the development of a sub-seafloor robot has received little attention. Omori et al. developed an excavation robot that uses peristaltic crawling for lunar and planetary exploration [7][8]. As the propulsion mechanism, this robot uses the peristalsis behavior of earthworms, and experiments have confirmed that peristaltic crawling is a useful propulsion mechanism for an excavation robot. However, their robot used a DC motor as the propulsion actuator. The disadvantage is that the motor needs pressure compensation when used in deep-sea water. This requirement increases the size of the system because it should be placed in an oil-filled container.

In this study, we propose the use of a hydraulic artificial muscle as the propulsion actuator to imitate the segments of an earthworm, thereby developing a deep-seafloor excavation robot. The main objective of the proposed robot is to explore under the seabed in deep seas by acquiring the samples of mud and its contents. Furthermore, this robot can serve as a platform for various observations and measurements such as acoustic and electromagnetic sounding in deep seas. Because no platform currently exists for penetrating the seabed, a robot that can excavate into the seafloor has the potential to advance future research in geology, marine biology, and other fields.

In this paper, we report a study on an artificial muscle as the propulsion actuator for peristaltic crawling in a deep-sea excavation robot. The propulsion actuator of the proposed robot must be able to function under water at high pressures. We developed a subunit using a hydraulic artificial muscle and conducted a performance experiment under water pressure. In this experiment, we confirmed the performance of subunits under water pressure of up to 5 MPa.

2   Peristaltic Crawling of Earthworms

In this section, we describe the peristaltic crawling of earthworms, which is used for propulsion of the proposed robot. Figure 1 shows the mechanism of earthworm locomotion by peristaltic crawling [9]. The earthworm's body is separated into numerous segments, each having the same structure. The earthworm moves by sequential contraction and extension of its segments beginning from its head. The segments become thick when contracting and grip the wall. Conversely, the segments become thin when extending, which enables the worm to reduce the contact area with the wall. Propagating the waves of contraction and extension from the head to the rear body section, the worm moves forward using friction with the wall.

This locomotion mechanism has some advantages. First, the earthworm eats soil at its head, conveys the soil through its body, and discharges waste from its rear. Adopting this mechanism to the robot, the robot can move forward without pressing excavated soils against the wall of the borehole. This advantage enables the robot to move and excavate without its motion being inhibited by excavated soils after the robot is completely buried. Second, this locomotion has a large contact area with the wall, which enables stable movement in soils. It also enables the robot to support the reaction force of the earth auger when excavating. Third, this robot needs only a small area for movement. As previously mentioned, body segments have less contact area when extending, so the influence of soil pressure received from the wall surface is reduced. We think that these features would also be useful in a deep-sea excavation robot, so we adopted peristaltic crawling as the propulsion mechanism in our robot.

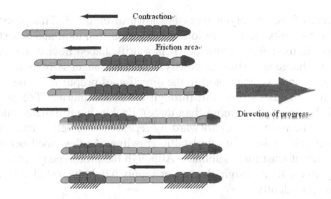

Fig. 1. Pattern of earthworm locomotion using peristaltic crawling

3   Concept of the Deep-Seafloor Excavation Robot

In this section, we describe the concept of the deep-seafloor excavation robot using peristaltic crawling. The proposed robot, housed in a launcher, would be dropped onto the seafloor from a research vessel. The launcher would support the robot body until the robot excavates and enters the soil. The entry of the robot into the seabed is illustrated in Figure 2. The activities of the robot during a typical mission would be as follows.

1) Excavate soils vertically to a depth of several meters from the launcher.
2) Bend the body to change from vertical to horizontal orientation.
3) Excavate soils horizontally, extracting and collecting mud samples and inclusions inside the body.
4) Discharge excavated soil from its rear.
5) Continue horizontal excavation, then again bend the body to become vertical for the return to the seafloor.

6)Rise to the sea surface to be collected by the research vessel.

By simultaneously performing these operations with multiple robots, efficient explorations would be realized at low cost.

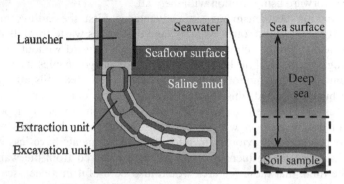

Fig. 2. Method of robot operation in seafloor

The structure of the prototype robot is shown in Figure 3. This proposed robot is composed of three units: an excavation unit, a propulsion unit, and an extraction unit. An earth auger is used for excavation. It is a drill-shaped device that can excavate, convey, and discharge soil simultaneously. The earth auger is mounted in the center of the robot body and excavates soil in the direction of propulsion. The excavated soil is conveyed into the extraction unit through the robot interior. The propulsion unit consists of multiple subunits (more than three). Each subunit imitates one segment of an earthworm. The robot moves forward by repeated contraction and extension motions of the subunits, mimicking peristaltic crawling. As described below, peristaltic crawling requires at least three subunits. Although the prototype robot has an external power supply, in the final design, the power supply would be installed inside the robot and operate independently.

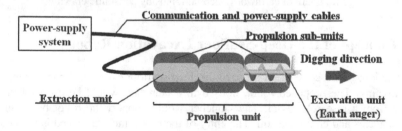

Fig. 3. Prototype of excavation robot

To realize these motions in a deep-sea environment under high water pressure, the actuator must be small, water resistant, and have high output. At the locations of the intended operations, the water pressure is very high. Thus, when small size and high output are required, motor actuators are typically used. However, a motor used in seawater needs an oil-filled compensator, which increases the system size. Therefore, we decided to use an artificial muscle as the propulsion actuator and hydraulic oil as the working fluid.

The artificial muscle used in the propulsion subunit is a straight-fiber-type [10]. A schematic of a straight-fiber-type artificial muscle is shown in Figure 4. The muscle has a tubular shape and is made of natural latex rubber that contains lengthwise carbon fiber layers in the axial direction. When pressure is applied, the muscle expands in the radial direction and contracts in the axial direction. Based on the non-linear dynamic characteristics model of artificial muscle [11] shown in Figure 5, theoretical value of muscle's inner volume V can be expressed as follow:

$$V = \frac{l_0 \pi}{4\phi_0^3}\left[d_0\phi_0^2 l_0 + \left(d_0^2\phi_0^2 + \frac{3}{4}l_0^2 \right)\sin\phi_0 - \phi_0 l_0 \cos\phi_0 (l_0 + d_0 \sin\phi_0) + \frac{l_0^2}{12}\sin 3\phi_0 \right]. \qquad (1)$$

where l_0 is the initial length, d_0 is the initial diameter. Shape change is regarded as an arc shape in this model. The radial expansion is used to grip the wall of the borehole and creates friction; the amount of contraction determines the propulsion displacement in one motion. These expansion and contraction cycles imitate one segment of an earthworm (cf. Figure 1).

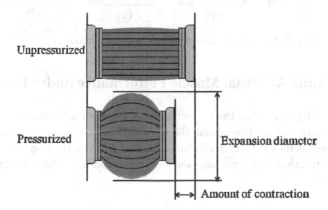

Fig. 4. Schematic of straight-fiber-type artificial muscle

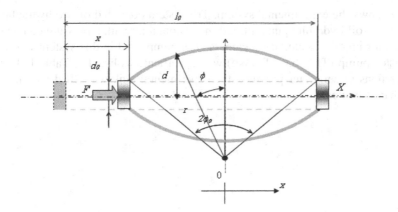

Fig. 5. Shape model of the artificial muscle viewed along the Z-axis

One example of the propulsion motion pattern of the robot is shown in Figure 6. In Panel (a), all subunits are in the expansion state, and then in Panel (b), the head subunit extends first. In Panel (c), the second subunit extends, while the head subunit simultaneously contracts, gripping the wall. In Panel (d), motion proceeds by repeated extensions and contractions of the subunits in sequence from the head to the rear of the robot. This peristaltic crawling enables the robot to move stably. The motion requires a minimum of three subunits, but changing the number of subunits and motion patterns enables us to change the propulsive force and speed.

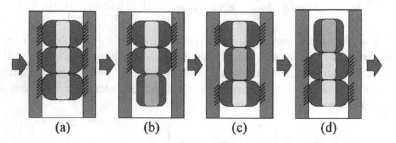

(a)   (b)   (c)   (d)

Fig. 6. Example of motion pattern in three subunits of robot.

4   Hydraulic Artificial Muscle Performance under Pressure

We experimentally evaluated the performance of the hydraulic artificial muscles under water pressure and confirmed that the units perform the same expansion and contraction motions as they do under atmospheric pressure. The relationships between the amount of contraction and flow rate and between expansion diameter and flow rate were evaluated.

4.1   Experimental System

Figure 7 shows the experimental system. The system consisted of two hydraulic muscles units, an oil hydraulic pump placed in the container with a pressure compensator, large strain gauges, a scale, and a camera. The pump used in this system was a small axial piston pump (TFH-630, Takako) with specifications listed in Table 1. To verify the operations under water pressure, the experimental tank was filled with tap water rather than seawater.

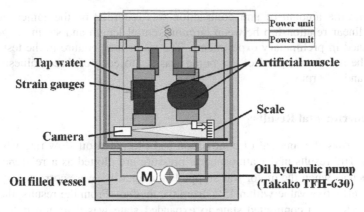

Fig. 7. Experimental water tank system

Table 1. Pump Specifications

Parameter	Value
Displacement (cm^3)	6.29
Maximum operating pressure (MPa)	21
Maximum speed (rpm)	3,000
Rotational direction	Bi-direction
Weight (kg)	2.26

The hydraulic artificial muscle units are shown in Figure 8. The rubber and working fluid used in these units were NBR and AeroShell Fluid 41, respectively. The artificial muscle was fixed to the metal flanges with metal fixtures. The natural dimensions of the units were a length of 350 mm, diameter of 110 mm, and length between crimping of 186 mm. To study the performance of only the hydraulic artificial muscle, the bellows were not mounted inside.

Fig. 8. Photo of hydraulic artificial muscle units in test tank

The pressure in the test water tank was increased from atmospheric pressure to 5 MPa (equivalent to 500 m depth) in increments of 1 MPa while both artificial muscle units alternately expanded and contracted. The flow rate of oil was estimated by the rotational speed of the pump, the expansion diameter was estimated by the strain gauges, and the amount of unit contraction was recorded by the camera and scale. A nearly linear relationship between circumferential length and strain gauge reading was obtained in preliminary experiments. The water temperature in the test tank was 285 K. The rotational speed of the pump was set to each of three values: 100 rpm, 150 rpm, and 200 rpm.

4.2   Experimental Results

Figure 9 shows the amount of contraction and estimated oil flow rate when Unit 1 expanded. The results under atmospheric pressure are plotted as a reference. Invalid values, such as when the motor was stopped momentarily, were corrected by the average as the effective value within 15% from the median. From the results, the cumulative flow rate from contracted state to expanded state was determined to be about 1450–1500 cm^3, independent of the external pressure. The data confirm that performance under 5 MPa of water pressure was the same as that under atmospheric pressure.

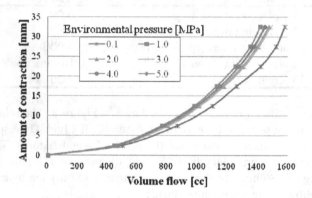

Fig. 9. Relation between contraction and volumetric flow rate at pressures of up to 5 MPa

Figures 10 and 11 show the expansion diameters and contraction percentages, respectively, as functions of the oil flow rate. Under the water pressures tested in this experiment, the units operated independently of the external water pressure. During contractions and expansions, the relationships between flow rate and expansion diameter differ due to hysteresis of the artificial muscles. The experiment confirmed that the actuator using a hydraulic artificial muscle can be operated at water pressures of up to 5 MPa. However, the increasing current drawn by the pump became a concern, and we stopped the performance experiment at 5 MPa. Therefore, further studies of the relationship between the current and water pressure and improvements in the pump system are necessary.

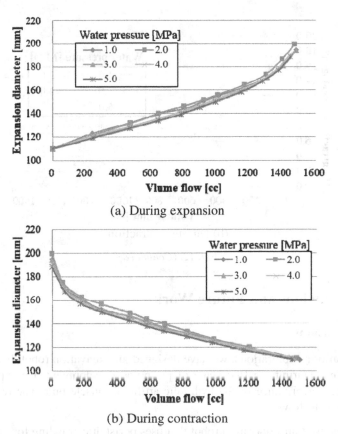

(a) During expansion

(b) During contraction

Fig. 10. Relationships between flow rate and expansion diameter during (a) expansion and (b) contraction at pressures of up to 5 MPa

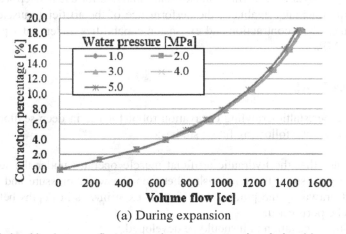

(a) During expansion

Fig. 11. Relationships between flow rate and percent contraction during (a) expansion and (b) contraction for pressures of up to 5 MPa

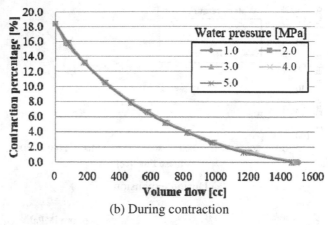

(b) During contraction

Fig. 12. (*Continued*)

5   Conclusions and Future Work

5.1   Conclusions

For deep-seafloor explorations, we have designed an excavation robot that uses peristaltic crawling. During the first stage of prototype development, we experimentally evaluated the performance of the hydraulic artificial muscle unit. The results can be summarized as follows:

1) We proposed an excavation robot that uses peristaltic crawling for deep-seafloor explorations.
2) We made a hydraulic artificial muscle unit intended for use in deep sea.
3) We experimentally evaluated the performance of the artificial muscle units and confirmed their contraction and expansion capabilities under water pressures of up to 5 MPa.

5.2   Future Work

To develop a peristaltic-crawling excavation robot for use in deep-seafloor explorations, we propose the following future studies:

1) To ensure that the hydraulic artificial muscle operates under the seabed, it is necessary to study the relationship between the water pressure and amount of current drawn by the pump. Then, further experiments at depths below 500 m should be performed.
2) A flexible excavation unit should be developed.

References

1. Nakajoh, H., Osawa, H., Miyazaki, T., Hirata, K., Sawa, T., Utsugi, H.: Development of work class ROV applied for submarine resource exploration in JAMSTEC. Oceans-Yeosu, pp. 1–5 (2012)
2. Yoshida, H., Hyakudome, T., Ishibashi, S., Sawa, T., Nakano, Y., Ochi, H., Watanabe, Y., Nakatani, T., Ota, Y., Sugesawa, M., Matsuura, M.: An autonomous underwater vehicle with a canard rudder for underwater minerals exploration. In: IEEE International Conference on Mechatronics and Automation (ICMA), pp. 1571–1576 (2013)
3. Yoshida, H., Aoki, T., Osawa, H., Ishibashi, S., Watanabe, Y., Tahara, J., Miyazaki, T., Itoh, K.: A deepest depth ROV for sediment sampling and its sea trial result. In: Underwater Technology and Workshop on Scientific Use of Submarine Cables and Related Technologies, pp. 28–33 (2007)
4. Kato, Y., Fujinaga, K., Nakamura, K., Takaya, Y., Kitamura, K., Ohta, J., Toda, R., Nakashima, T., Iwamori, H.: Deep-sea mud in the Pacific Ocean as a potential resource for rare earth elements. Nat. Geosci. 4, 535–539 (2011)
5. Wada, K.: Coring Technology to be applied in IODP NAnTroSEIZE. In: Oceans, MTS/IEEE Kobe Techno-Ocean, pp. 1–4 (2008)
6. Kyo, M.: Challenges to drill through seismogenic zone. In: Underwater Technology Symposium (UT), IEEE International, pp. 1–7 (2013)
7. Omori, H., Murakami, T., Nagai, H., Nakamura, T., Kubota, T.: Planetary subsurface explorer robot with propulsion units for peristaltic crawling. In: Proc. IEEE Int. Conf. Robot. Autom., pp. 694–654 (2011)
8. Omori, H., Nakamura, T., Yada, T.: Development of an underground explorer robot based on peristaltic crawling of earthworms In: Proceedings of 11th Int. Conf. on Climbing and Walking Robots and the Support Technologies for Mobile Machines, pp. 1053–1060 (2008)
9. Sugi, H.: Evolution of Muscle Motion. The University of Tokyo Press (1977, in Japanese)
10. Nakamura, T.: Experimental comparison between McKibben type artificial muscles and straight fibers type artificial muscles. In: SPIE International Conference on Smart Structures, Devices and Systems III (2006)
11. Nagai, S., Majima, T., Nakamura, T.: Motion control of instantaneous force for an artificial muscle manipulator with variable rheological joint. In: IEEE International Conference on Robotics and Biomimetics (ROBIO), pp. 402–407(2012)

Dynamic Characteristic Model for Pneumatic Artificial Muscles Considering Length of Air Tube

Shota Yamazaki$^{(\boxtimes)}$, Tatsuya Kishi, and Taro Nakamura

Department of Precision Mechanics, Chuo University, 1-13-27 Kasuga,
Bunkyo-ku, Tokyo 112-8551, Japan
s_yamazaki@bio.mech.chuo-u.ac.jp

Abstract. We developed an earthworm-type 15A pipe inspection robot with pneumatic artificial muscles (PAMs). The movement of the robot is performed by a drive unit that repeatedly expands and contracts the muscle. Pressurized air is supplied to the muscle via an air tube. The robot is required to change the specifications of the tube in accordance with the distance of inspection. When the specifications change, the responses of air transmission and the drive unit also change. To date, the optimum operating condition of the drive unit has been obtained experimentally, but this is an extremely time-consuming problem. Thus, we derive a dynamic characteristic model of the PAMs that considers the length of the air tube. Then, the usefulness of the model is confirmed by comparing its results to experimental values of the artificial muscle.

Keywords: Artificial muscle · Pipe inspection · Earthworm robot · Dynamic characteristic

1 Introduction

Currently, pipes are regularly inspected to prevent accidents caused by aging. However, it is difficult to inspect long pipes with narrow diameters; thus, we have developed a pipe inspection robot for 15A pipes (whose inner diameter is approximately 16 [mm]) that are difficult to inspect traditionally.

Our robot uses an earthworm-type pneumatic drive to move through a 15A pipe [1]. Figure 1 shows the appearance of the developed 15A pipe inspection robot. The movement of the robot is performed by a drive unit that repeatedly expands and contracts pneumatic artificial muscles (PAMs). Pressurized air is supplied to the muscle via an air tube. The robot is required to change the specifications of the tube in accordance with the distance of inspection. Because the robot inspects over long distances, a thin and long air tube is used. When the specifications of the tube change, the responses of air transmission and the drive unit also change. Thus, we have experimentally determined the most suitable operational conditions (input pressure and operating time) of the robot whenever the specifications changed. However, the experiments are extremely time consuming.

© Springer International Publishing Switzerland 2015
H. Liu et al. (Eds.): ICIRA 2015, Part I, LNAI 9244, pp. 390–401, 2015.
DOI: 10.1007/978-3-319-22879-2_36

To solve this problem, we need a dynamic characteristic model for the PAMs that considers the length of the air tube. Currently, only a dynamic characteristic model of the artificial muscles [2,3] exists. Thus, we derive the dynamic model of pressure response in the air tube in this study.

Herein, we first describe the artificial muscle; Next, we derive the dynamic model of pressure response in the air tube and develop a dynamic characteristic model for PAMs that considers the length of the air tube. Finally, to confirm the usefulness of the model, we compare its results to experimental values of the artificial muscle.

Fig. 1. 15A pipe inspection robot developed in this study

2   Straight-Fiber-Type Artificial Muscle

The drive unit of the robot used a straight fiber type artificial muscle [4] that has been studied by Shinohara et al. Figure 2 shows the appearance of the muscle. It is made of natural rubber latex and a micro-carbon fiber. When pressurized air is supplied to the muscle, the rubber expands, as shown in Fig. 2, but the fibers do not extend because carbon fiber is placed in the axial direction. Therefore, the muscle contracts axially and expands radially.

Figure 3 shows the operation of the robot with the straight-fiber-type artificial muscle. At first, the anterior of the muscle contracts; then, the contraction is sequentially transmitted to the rear portion of the muscle. In this case, because friction is generated between the contracted portion and the pipe, the expanded portion obtains a reaction force for further expansion.

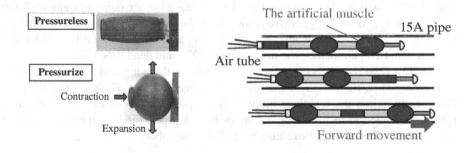

Fig. 2. Straight-fiber-type artificial muscle   **Fig. 3.** Operation of the robot

3 Reduced Model Representing Dynamic Characteristics of a Pneumatic Transmission Line

3.1 Reduced Model Representing Dynamic Characteristics of a Pneumatic Transmission Line

We focus on the reduced model used by Kagawa et al. [5] to represent the dynamic characteristics of a pneumatic transmission line. Equations (1) and (2) present the reduced model:

$$R_N > 3.64, \qquad G_1(s) = \frac{P_2(s)}{P_1(s)} = \frac{\omega_n^2 e^{-T_d s}}{s^2 + 2\zeta\omega_n s + \omega_n^2}, \tag{1}$$

$$R_N \leq 3.64, \qquad G_2(s) = \frac{P_2(s)}{P_1(s)} = \frac{e^{-T_d s}}{T_L s + 1}. \tag{2}$$

$R_N = cr^2/v\,L$: Parameters representing the characteristics of the pneumatic transmission line

P_1: Input pressure [MPa], P_2: Output pressure [MPa]
L: Length of the pipe [m], c: Speed of sound [m/s]
v: Kinetic viscosity [m^2/s], s: Laplace symbol [1/s]
r: Radius of the pipe [m], T_L: Time constant [s]
T_d: Dead time [s], ω_n: Natural angular frequency [rad/s]
ζ: Damping ratio

The model represented by Equations (1) and (2) has a characteristic parameter of the boundary of the pneumatic transmission line given by $R_N = 3.64$. If R_N is greater than 3.64, Equation (1) is approximated by a second-order delay system with three parameters: the natural angular frequency (ω_n), damping ratio (ζ), and dead time (T_d). Conversely, if R_N is 3.64 or less, Equation (2) is approximated by a first-order delay system with two parameters: the time constant (T_L), and dead time (T_d).

The 15A pipe inspection robot uses an air tube with an internal radius of 0.6 [mm]. Therefore, when the length of the tube is less than 2.2 [m], R_N is greater than 3.64. Conversely, when the length is more than 2.2 [m], R_N is 3.64 or less. Now, as the robot is intended to inspect long distances in 15A pipes, the length of the tube is 2.2 [m] or more. Therefore, we use Equation (2) for the robot.

3.2 Need of an Empirical Formula for Time Constant

The time constant (T_L) expressed in Equation (2) varies with the input pressure, length of the tube, and radius. Therefore, the time constant is typically determined experimentally each time the input pressure, length, or radius of the tube change; however, this is extremely time consuming. Thus, for each of the tube's parameters, we measure the change in the time constant and determine an empirical formula to predict its value.

4   Determining an Empirical Formula for Time Constant

To find an empirical formula for the time constant, we perform measurements of the pressure response in the air tube to determine its behavior.

4.1   Evaluation of Pressure Response in Air Tube

Figure 4 shows the experimental environment for evaluating the pressure response. At first, constant pressure is supplied by an air compressor. Next, the pressure value is input from a PC to proportional solenoid valve, and the commanded pressure is supplied from the proportional solenoid valve to the air tube. Then, the pressure response is evaluated by a sensor attached to the tip of the tube. The specifications of the air tube satisfy $R_N \leqq 3.64$. In addition, first, the tube is gently bent, and then, it is moved toward as linear of a configuration as possible.

The time constant included in the reduced model may be changed by the input pressure (P [MPa]), the length of the tube (L [m]), or the radius of the tube (R [mm]). Therefore, we focus on these three parameters and measure the pressure response when two parameters are fixed and the other is changed. Input pressures are 0.08, 0.12, and 0.16 [MPa], lengths of the tube are 10, 20, and 30 [m], and radius of the tube are 0.4, 0.6, and 1.25 [mm]; the experiment is performed in 27 patterns.

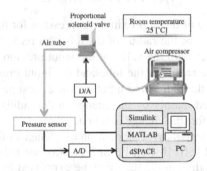

Fig. 4. Experimental environment for evaluating pressure response

4.2   Experimental Results

Figure 5 shows the change in the time constant with varying lengths of the tube. Because the time constant is directly proportional to the length of the tube, we believe that it can be approximated by a straight line. The plots in Fig. 5 indicate the experimental values, and solid lines indicate the approximate expressions.

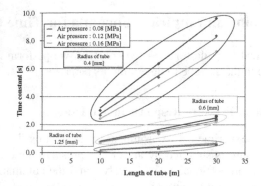

Fig. 5. Change in the time constant with varying lengths of the tube

4.3   Determining the Empirical Formula for Time Constant

From the experimental results shown in Fig. 5, an empirical formula for the time constant is obtained. The time constant is expressed as an empirical formula with a linear approximation of its dependence on L. Writing the time constant as T_L [s] and the slope as a, we obtain Equation (3).

$$T_L(L, P, R) = a(P, R) * L. \tag{3}$$

From Fig. 5, the slope a of the approximate expression for the time constant is mainly determined by the size of the radius of the tube. Moreover, for each interior radius of the tube, slope a varies with the value of the input pressure. Thus, we believe slope a to be a function of the radius in the tube and the input pressure. Therefore, the dependence of slope a on the radius and input pressure must be determined. As the time constant can be classified mainly by the change in the radius of the tube from Fig. 5, the tendency of slope a for this change is found.

Figure 6(a) shows the tendency of slope a under changes of the radius at input pressures of 0.08, 0.12, and 0.16 [MPa]. From Fig. 6(a), we believe that the relation between slope a and the radius in the tube may be expressed by a power approximation. In this case, when variables b, c are substituted in Equation (3), we obtain an equation for slope a as follows:

$$a(P, R) = b(P) * R^{c(P)}. \tag{4}$$

From Fig. 6(a), it is found that variables b and c change with the input pressure. Therefore, relations between variables b and c and the input pressure are demanded. Figure 6(b) shows the tendencies of variables b and c under changes in the input pressure. The equations for variables b and c are as follows.

$$b(P) = 0.0350P + 0.0280, \tag{5}$$

$$c(P) = 5.13P - 2.79. \tag{6}$$

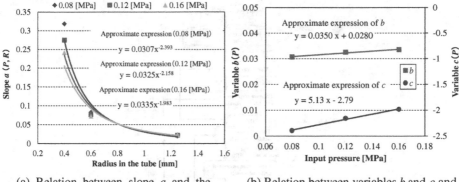

(a) Relation between slope a and the radius in the tube

(b) Relation between variables b and c and the input pressure.

Fig. 6. Tendency of slope a and variables b, c for each parameter

4.4 Determined Empirical Formula for Time Constant

From Equations (4)–(6), we obtain the equation for the time constant T_L, where H is the experimental correction value. As the range of the empirical formula, the radius of the tube are 0.4 [mm] $\leqq R \leqq 1.25$ [mm], input pressures are 0.08 [MPa] $\leqq P \leqq 0.16$ [MPa], and lengths of the tube are 10 [m] $\leqq L \leqq 30$ [m].

$$T_L (L, P, R) = (\alpha P + \beta)\, R^{(\gamma P + \delta)} * L * H$$

$$(\alpha = 0.0350,\, \beta = 0.0280,\, \gamma = 5.13,\, \delta = -2.79).$$

(7)

5 Dynamic Characteristic Model for Pneumatic Artificial Muscle that Considers the Length of Air Tube

5.1 Flow Chart of Dynamic Characteristic Model

By combining the determined dynamic characteristics model of the pneumatic transmission line and the dynamic characteristic model of the pneumatic artificial muscle [2, 3], a dynamic characteristic model for PAMs that considers the length of the air tube is obtained. Figure 7 shows the flow chart of the obtained dynamic characteristic model.

The dynamic characteristic model of the artificial muscle is fabricated by referring to [2] and [3].

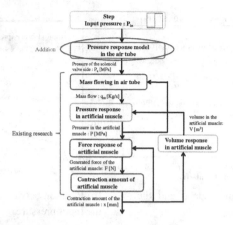

Fig. 7. Flow chart of the obtained dynamic characteristic model

5.2   Description of the Obtained Dynamic Characteristic Model

The dynamic characteristic model for PAMs that considers the length of the air tube is described here.

At first, using the newly-added reduced model (Equation 2) of Kagawa et al. [5], the pressure response of the pneumatic transmission line is considered. Thereafter, the characteristics of the pneumatic line for transmitting the pressure from the proportional solenoid valve side of the tube to the artificial muscle are considered. The mass flow (q_m) of air into the artificial muscle is determined. In Equations (8) and (9), q_m is expressed in terms of the pressures on the solenoid valve and artificial muscle sides.

$$0.528 < P/P_u \leqq 1, \qquad q_m(P_u, P) = CP_u\rho_0\sqrt{\frac{T_0}{T_u}}\sqrt{1 - \left(\frac{\dfrac{P}{P_u} - b}{1 - b}\right)}, \qquad (8)$$

$$0 < P/P_u \leqq 0.528, \qquad q_m(P_u, P) = CP_u\rho_0\sqrt{\frac{T_0}{T_u}}. \qquad (9)$$

T_0: Temperature of the standard state [K]
ρ_0: Density of the standard state [Kg/m^3]
C: Acoustic velocity conductance [dm^3/(s·bar)]
T_u: Air temperature [K]
P_u: Pressure on the solenoid valve side [MPa]
P: Pressure on the artificial muscle side [MPa]
b: Critical pressure ratio

Then, pressure in the artificial muscle is considered to vary by the mass flow from the air tube obtained from Equations (8) and (9). Now, given adiabatic changes in air in the artificial muscle, Equation (10) is derived using the law of energy conservation and the gas state equation. In Equation (10), the derivative of pressure with respect to time in the artificial muscle can be expressed in terms of the volume of the pressure of the artificial muscle and the mass flow into it.

$$\frac{dP}{dt} = RT\left(\frac{V}{\kappa}\right)^{-1} q_m - P\left(\frac{V}{\kappa}\right)^{-1}\frac{dV}{dt} \tag{10}$$

dP/dt: Change in pressure in the artificial muscle, κ: Ratio of specific heat
V: Volume of the artificial muscle [m^3], R: Gas constant [Pa·m^3·Kg^{-1}·K^{-1}]

The generation of force by pressure in the artificial muscle must be determined. A mechanical equilibrium model [6] that is expressed as a relation among the amount of the contraction of the artificial muscle, its contraction force, and the pressure in it is used. Equations (11)–(15) show relational expressions of force generated by the artificial muscle.

$$F(x,P) = \frac{PG_3(\phi_0) - G_1(\phi_0)}{G_2(\phi_0)}, \tag{11}$$

$$G_1(\phi_0) = \frac{2k_1 t}{d_0}\left[\frac{l_0}{d_0}\right]^2\left[\frac{\sin\phi_0 - \phi_0\cos\phi_0}{\phi_0^2}\right], \tag{12}$$

$$G_2(\phi_0) = \frac{M\tan\phi_0}{d_0 n b}, \tag{13}$$

$$G_3(\phi_0) = \left[\frac{l_0}{d_0}\right]^2\left[\frac{\phi_0 - \sin\phi_0\cos\phi_0}{\phi_0^2}\right] + \frac{2l_0}{d_0}\frac{\sin\phi_0}{\phi_0} - \frac{\pi d_0 M}{4nb}\tan\phi_0, \tag{14}$$

$$\phi_0(x) = \frac{2\alpha l_0^{1.5} x^{0.5}}{(l_0 - x)^2 + \alpha^2 l_0 x}. \tag{15}$$

F: Generated force of the artificial muscle [N]
x: Contraction amount of the artificial muscle [m]
k_1: Elastic modulus of rubber [MPa], t: Thickness of the artificial muscle [m]
d_0: Diameter of the artificial muscle [m], M: Fiber density coefficient
l_0: Length of the artificial muscle [m], n: Fiber number coefficient
α: Relation of expansion and contraction rates

The amount of the contraction of the artificial muscle for generated force is demanded. Because the movement of the rubber portion of the artificial muscle is very complicated, a simplified equation is considered in the model. In addition, because the artificial muscle is made of rubber, it is believed that the muscle has certain damper characteristics. Therefore, regarding the muscle as a mass damper system, the contraction amount of the muscle is expressed by Equation (16). Because the nonlinear spring characteristics are considered in the mechanical equilibrium model, they are not considered in Equation (16).

$$m\frac{d^2x}{dt^2}+c\frac{dx}{dt}=F$$ (16)

m: Mass of the artificial muscle [Kg], c: Damper coefficient [N·s/m]

Finally, the change in the volume of the artificial muscle is determined. As shown in Figure 8, regarding the shape displacement by the contraction of the artificial muscle as the change in the arc shape, a theoretical expression for the volume of the muscle can be determined as follows. From Fig. 8, the observed cross-sectional area when the muscle is cut in a plane perpendicular to the X axis is expressed as S, and the equations that are determined are as follows.

$$S = \pi d^2/4,$$ (17)

$$d = d_0 + 2r(\cos\phi - \cos\phi_0).$$ (18)

Equation (19) is the demanded geometric relation.

$$r = l_0/2\phi_0.$$ (19)

Thus, the volume of the artificial muscles can be determined by Equation (20).

$$\begin{aligned}
V &= 2\int_0^{r\sin\phi_0} S dx \\
&= 2\int_0^{\phi_0} S r\cos\phi_0 d\phi \\
&= \frac{l_0\pi}{4\phi_0^3}\left[d_0\phi_0^2 l_0 + \left(d_0^2\phi_0^2 + \frac{3}{4}l_0^2\right)\sin\phi_0 - \phi_0 l_0\cos\phi_0(l_0 + d_0\sin\phi_0) + \frac{l_0^2}{12}\sin3\phi_0\right].
\end{aligned}$$ (20)

From the above equations, the obtained model can display the contraction characteristics (contraction amount) of the artificial muscle under applied input pressures.

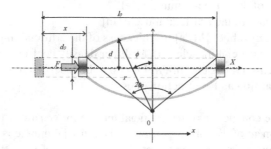

Fig. 8. Shape model of the artificial muscle

6   Comparison Between the Simulation Results and the Experimental Values

The simulation results of the obtained model and experimental results for the drive unit of the artificial muscle are compared. The dynamic characteristics of the muscle are seen as its contraction properties.

6.1   Determination of Contraction Properties in Drive Unit

Figure 9 shows the experimental environment. For the experimental method, the contraction of the artificial muscle is measured by a laser displacement meter when pressure is applied to the muscle. As the experimental conditions, the radius of the tube is 0.6 [mm], input pressures are 0.08 and 0.10 [MPa], and lengths of the tube are 10, 20, and 30 [m]. Moreover, Fig. 10 shows the specifications of the unit with the muscle used for the experiment.

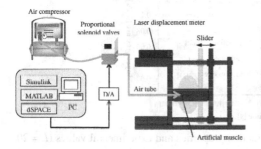
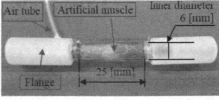

Fig. 9. Determination of the contraction properties

Fig. 10. Specifications of the unit with the artificial muscle

6.2   Comparison Between Simulated Contraction Properties and Experimental Values

As an example, Figure 11 shows a comparison between the simulated contraction properties and their experimental values during the step response when L = 20 [m], P = 0.10 [MPa], and R = 0.6 [mm]. Now, the experimental correction value for the empirical formula of the time constant used for the obtained model is 0.6, and dead time substituted in the reduced model (Equation 2) used the experimental values.

From Fig. 11, the steady-state value is confirmed to be almost the same for the experimental values and simulation results. However, the rising part (the part that the amount of contraction in Fig. 11 increases sharply) appears different for the simulated and experimental values.

Now, a comparison of the steady-state values between the simulation and experiment under all experimental conditions is performed. In addition, comparisons of the time constant that affects the rising part of the simulation results and experimental values under all experimental conditions are performed. Figure 12(a) shows the comparison between the steady-state values by simulation and experiment under all experimental conditions, and Fig. 12(b) shows the comparison of the time constant by simulation and experiment under the conditions.

From Figs. 12(a) and (b), the steady-state values and time constants from the simulation and experimental results are confirmed to be almost the same under all experimental conditions. Moreover, the time constant of the simulation is confirmed to exhibit the same tendency as the experimental values, whereby the time constant increases with the length of the tube and decreases when the input pressure increases.

Thus, we believe that this model can predict, to a reasonable degree of accuracy, the contraction properties of the artificial muscle in the 15A pipe inspection robot. Therefore, we believe that the obtained model can determine the optimal input pressure and operation time of the muscle.

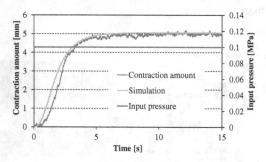

Fig. 11. Comparison between simulated contraction properties and experimental values ($L = 20$ [m], $P = 0.10$ [MPa], $R = 0.6$ [mm])

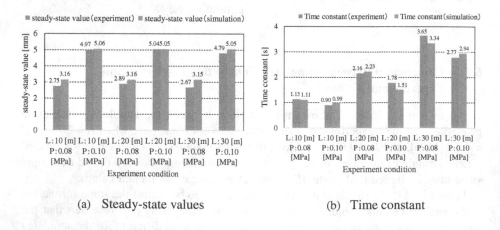

(a) Steady-state values

(b) Time constant

Fig. 12. Comparison between the simulated and experimental steady-state values

7   Conclusions

Summary

1) An empirical formula for the time constant using the reduced model of Kagawa et al. [5] was determined.
2) A dynamic characteristic model for the PAMs that considers the length of the air tube was obtained.
3) The simulated and experimental steady-state values and time constants were virtually the same under all experimental conditions.
4) We believe that this model can predict, to reasonable accuracy, the contraction properties of the artificial muscle in the 15A pipe inspection robot.

Future prospects

1) Check whether this model can be used for different sizes of the artificial muscle.
2) Determine an empirical formula for dead time used in the reduced model of Kagawa et al. [5].

References

1. Ikeuchi, M., Nakamura, T., Matsubara, D.: Development of an in-pipe inspection robot for narrow pipes and Elbows using Pneumatic artificial muscles. In: 2012 IEEE/RSJ International Conference on Intelligent Robots and Systems, Vilamoura, Algarve, Portugal, pp. 7–12, October 2012
2. Mori, M., Suzumori, K., Wakimoto, S., Kanda, T., Takahashi, M., Hosoya, T., Takematu, E.: Development of power robot hand with shape adaptability using hydraulic McKibben muscles. In: ICRA, pp. 1162–1168 (2010)
3. Hiroki, T., Hiroyuki, M., Taro, N.: Orbit Tracking Control of 6-DOF Lubber Artificial Muscle Manipulator Considering Nonlinear Dynamics Model. Transactions of the Japan Society of Mechanical Engineers (Trans. Jpn. Soc. Mech. Eng.)(part C) 77(779) (2011-7)
4. Hitomi, S., Nakamura, T.: Application of feed-forward linearization and compliance control for straight fibers type artificial muscles. In: 11th Robotic–symposia, pp. 222–227 (2006)
5. Toshiharu, K., Ato, K., Kazushi, S., Toshio, T.: A Study a Reduced Model for a Dynamics of a Pneumatic Transmission Line. Transactions of the Society of Instrument and Control Engineers (Transactions of SICE) 21(9), 53–58 (1985). (Japanese)
6. Nakamura, T., Saga, N., Yaegashi, K.: Development of pneumatic artificial muscle based on biomechanical characteristics. In: Proc. IEEE International Conference on Industrial Technology (ICIT 2003), pp. 729–734 (2003)

Three-Dimensional Finite Element Analysis of a Novel Silicon Based Tactile Sensor with Elastic Cover

Chunxin Gu, Weiting Liu[✉], and Xin Fu

The State Key Lab of Fluid Power Transmission and Control,
Zhejiang University, Hangzhou 310027, China
{cxgu,liuwt,xfu}@zju.edu.cn

Abstract. Tactile sensors are indispensable in robotic or prosthetic hands, which are normally covered with or embedded in soft materials. A novel tactile sensor with the structure of combining an elastic steel sheet and a piezoresistive gauge has been developed in the previous work. To better understand the mechanical effects of soft cover on this sensor, a three-dimensional finite element model (FEM), which is based on linear elastic behavior, is established. As usual, polydimethylsiloxane (PDMS) is adopted as the soft cover material. The results indicate that though soft cover diffusion of mechanical signals still exists, the steel sheet strengthens the measuring ability and at the same time lowers the density of sensing units to identify the single indentation location.

Keywords: Tactile sensor · Steel sheet · Soft cover · FEM · PDMS

1 Introduction

Like human skin, the soft cover for artificial tactile sensing plays a significant role in protecting the subsurface sensors from damage or for better hand-to-object contact. However, the skin-like soft cover blurs the signals that are transmitted to the embedded sensor due to its mechanical properties, which makes the measurement of the whole tactile system less reliable.

The existing researches have mainly focused on the transducers; only in a few works, the cover effects have been studied. R. S. Fearing etc. in [1] used a simple linear elastic model to predict strains beneath a compliant skin for a finger touching a knife edge, a corner, and a flat surface; M. Shimojo in [2] analyzed the mechanical spatial filtering effect of an elastic cover for different types of cover materials; P. Tiezzi etc. in [3] performed a series of experimental tests to analyze the influence of the thickness of the compliant layer on the resultant fingertip stiffness; J. Z. Wu etc. in [4] analyzed the contact interactions between the fingertips and objects with different curvatures via a 2D finite element fingertip model and in [5] predicted the time-dependent force responses of the fingertip via a 3D model; J.-J. Cabibihan etc. in [6] used a FEM based on viscoelastic and hyperelastic behavior to determine the effects of various thickness of the synthetic skin on the pressure distribution. All these works are based on the assumption that the embedded sensors are small and dependent which will not interfere with the mechanical behavior of the soft cover. However, because the tactile sensor we previously

H. Liu et al. (Eds.): ICIRA 2015, Part I, LNAI 9244, pp. 402–409, 2015.
DOI: 10.1007/978-3-319-22879-2_37

developed in [7] has the novel structure as the combination of the steel sheet and silicon gauge which makes the detecting value largely depend on the deflection of the steel sheet, the coupling effects between the steel sheet and the soft cover should be taken into consideration when analyzing the mechanical properties of the whole sensor.

In this paper, a three-dimensional FEM of a tactile sensor covered with a PDMS layer is proposed, which treats the PDMS material as linear elastic to simplify the modelling. And the function of the steel sheet is analyzed in aspects of valid mechanical signals transmission and single indentation location identification.

2   Sensor Design and Fabrication

The sensor consists of four parts as illustrated in Fig. 1, the PDMS layer, the steel sheet, the silicon gauges and the flexible printed circuit board (FPCB). The origin of coordinate is set on the center of the upper surface of the steel sheet. A tactile sensor with one silicon gauge but without PDMS layer has been developed, and more details about the structure of the sensor are available in the previous work [7]. To fit the finger of the artificial hand, the dimensions of the developed sensor are set as: L1, 15mm; L2, 15mm; H1, 0.3mm. The planar PDMS layer is glued to the steel sheet, and its bottom surface has the same dimensions as the upper surface of the steel sheet. The thickness H2 of the PDMS layer is set as 2mm which is close to that of human skin (about 521~1977μm[8]). On the bottom surface, three silicon gauges numbered as I, II, III are fixed along the symmetry axis parallel to the x-axis with exact coordinates as (-4, 0, -0.3), (0, 0, -0.3), and (4, 0, -0.3), respectively. At the same time, the valid stretch or compression direction of the silicon gauges should be parallel to the y-axis.

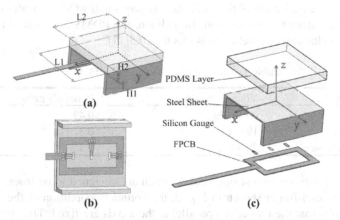

Fig. 1. Sensor structure. (a) A schematic of the sensor structure. (b) A view of the rear surface of the sensor. (c) An exploded view of the sensor (the golden leads are ignored).

The whole tactile system is fabricated easily combining the traditional machining process and the MEMS technology. The steel sheet is polished to guarantee the designed structure after being grinded from a steel block while the silicon gauges are fabricated with an SOI wafer based on the dry etching technology. After that, the silicon gauges are fixed on the rear surface of the steel sheet by the technology of glass sinter-ing. Around the gauges, the FPCB which is available in the market is glued on the sur-face. Using the ultrasonic bonding machine, the golden leads are connected to the FPCB and the silicon gauges respectively. Before being glued on the steel sheet, the PDMS layer is made at the mixing ratio of the base polymer to curing agent as 15:1.

3   Methods

By means of the finite element analysis (FEA) simulation tool (ANSYS Workbench 14.5), the mechanical behaviors of the sensor under various indenting conditions are analyzed.

3.1   Finite Element Model

The mechanical properties of the whole tactile system are considered as linear elastic. There are three kinds of materials in the system: PDMS for the soft layer, stainless steel 17-4PH for the steel sheet and structure steel for the indenter whose major mechanical parameters are shown in Table. 1. As the Young's modulus of human finger skin differ greatly due to different races, ages, sexes and measurement methods, a rough range of the Young's modulus is 0.01~20MPa[9] which means that the one we set for the PDMS is about at the mid-level of the range. Considering that the force sensing range for hu-man hand perception is about 0.1N-10N[10], the indenting force is set not larger than 10N. Because the Young's modulus of the steel sheet is large enough compared with the indenting force, the deflection of the steel sheet is very small which indicates only linear elastic properties among consideration. In addition, the PDMS layer could also be as-sumed as linear elastic when the strain is less than 0.5[11].

Table 1. Material properties used for FEA

Material	Young's modulus	Poisson's ratio
PDMS(15:1)	0.85MPa	0.49
17-4PH	197GPa	0.3
Structure steel	200GPa	0.3

As the steel sheet has two strong braces which are designed to be inserted into the slots of the artificial finger shown in Fig. 2, the boundary condition of the steel sheet can be simplified that the two edges parallel to the x-axis are fixed. The connection of the contact interface between the indenter and the upper surface of the PDMS layer is set as the no separation state, which means there is no gap in the contact interface along the z-direction when the indenting force is applied. The PDMS layer and the steel sheet are modeled as two parts with different materials but belong to the same body, which brings about the continuous state of the geometry structure and the ele-ments division. The elements are divided evenly with the dimension as 0.3mm. The model with elements division is shown in Fig. 2(c).

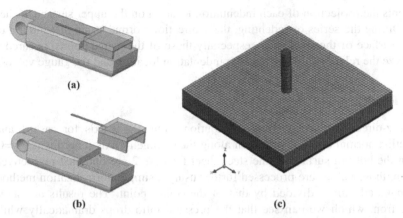

Fig. 2. Schematic of the sensor fixed in the artificial finger & simplified model of the sensor. (a) Assembly of a tactile sensor and an artificial finger. (b) An exploded view of the whole system. (c) The three-dimensional model simplified by removing the two supporting braces and fixing the two opposite edges of the steel sheet (the brown part is the indenter).

3.2   Indenting Conditions

Considering a simple contact condition that a rod indenter is applied on the surface of the soft cover (the PDMS layer here), the mechanical responses are compared with those of the usual tactile systems and the indenting on different locations is also implemented. The diameter of the rod indenter is set as 1mm while the indenting pressure is set as 0.5MPa causing the resultant force as about 0.4N.

1) Effects of the steel sheet deflection: As mentioned before, the steel sheet is the medium layer between the soft cover and the silicon gauges so that the combination of the soft cover and the steel sheet transmits the pressure values on the cover surface to the y-direction normal strain values on the bottom surface for the gauges. Nevertheless, in most tactile systems, the soft cover transmits the normal pressure directly to the sensing units. In other words, we care normal strain for this system but pressure or z-direction normal stress for other usual systems. It is meaningful to compare those two mechanical conditions under the same pressure circumstance. We use the same model introduced before but different support conditions for each cases: the whole bottom surface of the sheet fixed for the usual system (case 1) and two opposite edges fixed for this system (case 2). The indenter is applied at the center of the upper surface of the PDMS layer for both cases.

2) Identification of the indentation location: The rod indenter is applied at different points along the x-axis or lines parallel to the x-axis. The center of the bottom surface of the indenter is located respectively at 0~6mm (space interval as 1mm) from the origin of coordinate along the negative x-axis. After that, the indenting processes are repeated only by changing the y coordinate of the locations from 0 to 2, 4, and 6 respectively. The indentation locations are illustrated in Fig. 5(a) where the green "*"

represents the projection of each indentation location on the upper surface of the steel sheet. During the series of indenting, the y-direction normal strain distribution on the bottom surface of the steel sheet, especially those of the gauge points, is cared about to analyze the relationship between the indentation location and the gauge values.

3.3   Results

1) The z-direction normal stress distribution along the x-axis for case 1 and the y-direction normal strain distribution along the symmetry axis (coordinates: y=0, z=-0.3) on the bottom surface of the steel sheet for case 2 are obtained respectively. In addition, those values are processed further using a simple normalization method that is to make all values divided by that of the center point. The results are shown in Fig. 3, from which we can see that the pressure ratio drops dramatically while the strain ratio changes gently both relative to the distance from the center point. For example, at x=-2 or 2, the pressure ratio has declined to almost zero while the strain ratio still remains at about 0.8. The results indicate that the steel sheet strengthens the diffusion ability of the valid mechanical signal. In the aspect of force location identification, this may not be good for normal tactile sensor array with high density as it adds the complexity of signal coupling. But it will be useful for this kind of sensor because it will lower the demand of high density to determine the single indentation location due to the definite relationship between the values from different gauges.

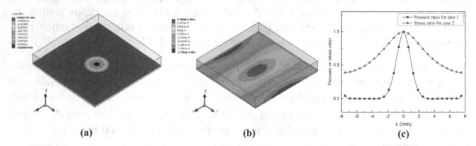

(a)   (b)   (c)

Fig. 3. Pressure or strain distribution for the two cases. (a) Normal pressure distribution on the bottom surface of the PDMS layer for case 1. (b) Y-direction normal strain distribution on the bottom surface of the steel sheet for case 2. (c) Comparison of the trends of the valid mechanical values change along the x-axis between the two cases.

2) For the identification of the indentation location, the y-direction normal strain distribution along the negative x-axis is shown in Fig. 4. The text "xi" in the legend means x coordinate value for each indentation location is "i". As the sensor structure is symmetrical about the center line, the distribution curve will be also symmetrical about the center line if the indentation is added along the positive x-axis. The strain distributions along the other lines whose y coordinate values are 2, 4, and 6 respectively have similar curves, which are not plotted. According to Fig. 4, we can see that when the indentation location changes, the pattern of the distribution curve also changes with the peak point following the indentation location. Besides, each curve inclines slowly when the point is

away from the indentation location, which means that the structure strengthens the expansion of the pressure signal distribution but makes it more definite due to the steel sheet's good linear elasticity and deflection continuity.

Taking advantages of these properties, it is suggested that the single indentation location could be uncovered using the outputs of the three silicon gauges (assuming that the indenter diameter is definite such as 1mm).Therefore, we calculate respectively the ratio of the strain value at x=-4 or 4 (where gauge I and II are) to that at x=0 (where gauge III is) and plot the ratio values relative to the indentation locations in Fig. 5(b). The text "I/II" or "III/II" in the legend represent the ratio of y-direction normal strain value between gauge I and II or between gauge III and II while "yj" means that the trajectory of the indentation points is (y=j, z=0) which is drawn in Fig. 5(a). The right half of the graph is plotted according to the symmetrical property. The combination of the strain ratio values from gauge I/II and III/II are unique at each point of the indentation location regardless of the normal indenting pressure values, making the reverse problem easily solved.

Specifically, the location identification process can be conducted as follows: firstly, compare the two ratio values from gauge I/II and III/II to determine which half space the indentation is in; secondly, for example if in the left half, draw a straight line parallel to the horizontal axis with the value obtained from gauge III/II and find out each intersection point with the calibrated curves; thirdly, compare the value from gauge I/II with the value at the calibrated curve for gauge I/II corresponding to each intersection point to finally determine the exact location. Above all, it is believed that the space revolution will be improved by reducing the space interval in the indenting calibration process.

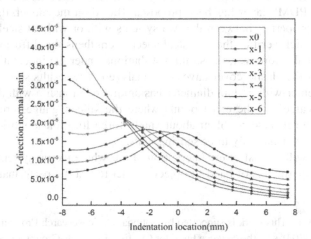

Fig. 4. Y-direction normal strain distribution along the line (y=0, z=-0.3) on the steel sheet under different indentation location conditions.

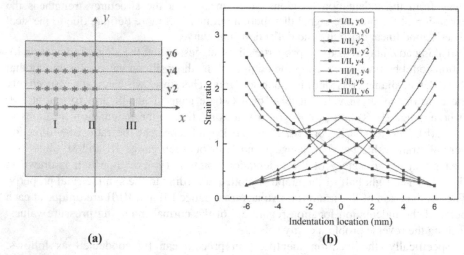

(a)   (b)

Fig. 5. (a) View of the upper surface of the steel sheet in the z-direction, where the green "*" represents the projection of each indentation location and the yellow rectangle represents the projection of each silicon gauge. (b) Ratio of y-direction normal strain of gauge I or III to that of gauge II along each red line in (a) relative to the indentation location.

4   Conclusions and Future Work

A three-dimensional finite element model for a silicon based tactile sensor covered with an elastic PDMS layer has been proposed. Based on the model, the comparison of mechanical properties between the two systems with or without steel sheet deflection is made, which verifies that the steel sheet strengthens the diffusion of the valid mechanical signal. Moreover, the strain distributions under the indentations on different locations of the elastic cover have been analyzed. The results show that the location of the indenter with certain diameter has unique relationship with the y-direction normal strain values on the exact points where the silicon gauges are fixed, which indicates that the reverse problem about single force location identification will be easily solved even using only three sensing units.

Future work will consist of the integration of the whole tactile system into artificial hand and the experiments of hand-to-object contact to test the performance in the real circumstance.

Acknowledgements. This work is supported by National Basic Research Program (973) of China (No.: 2011CB013303) and the Science Fund for Creative Research Groups of National Natural Science Foundation of China (No.: 51221004).

References

1. Fearing, R.S., Hollerbach, J.M.: Basic solid mechanics for tactile sensing. The International Journal of Robotics Research **4**, 40–54 (1985)
2. Shimojo, M.: Mechanical filtering effect of elastic cover for tactile sensor. IEEE Transactions on Robotics and Automation **13**, 128–132 (1997)
3. Tiezzi, P., Vassura, G.: Experimental analysis of soft fingertips with internal rigid core. In: Proceedings of the 12th International Conference on Advanced Robotics, ICAR 2005, pp. 109–114. IEEE (Year)
4. Dong, R.G., Wu, J.Z.: Analysis of the contact interactions between fingertips and objects with different surface curvatures. Proceedings of the Institution of Mechanical Engineers, Part H: Journal of Engineering in Medicine **219**, 89–103 (2005)
5. Wu, J.Z., Welcome, D.E., Dong, R.G.: Three-dimensional finite element simulations of the mechanical response of the fingertip to static and dynamic compressions. Computer Methods in Biomechanics and Biomedical Engineering **9**, 55–63 (2006)
6. Cabibihan, J.-J., Carrozza, M.C.: Influence of the skin thickness on tactile shape discrimination. In: 2012 4th IEEE RAS & EMBS International Conference on Biomedical Robotics and Biomechatronics (BioRob), pp. 1681–1685. IEEE (Year)
7. Gu, C., Liu, W., Fu, X.: A novel silicon based tactile sensor on elastic steel sheet for prosthetic hand. In: Zhang, X., Liu, H., Chen, Z., Wang, N. (eds.) ICIRA 2014, Part II. LNCS, vol. 8918, pp. 475–483. Springer, Heidelberg (2014)
8. Lee, Y., Hwang, K.: Skin thickness of Korean adults. Surgical and Radiologic Anatomy **24**, 183–189 (2002)
9. Pailler-Mattei, C., Bec, S., Zahouani, H.: In vivo measurements of the elastic mechanical properties of human skin by indentation tests. Medical Engineering & Physics **30**, 599–606 (2008)
10. Jones, L.A., Lederman, S.J.: Human hand function. Oxford University Press (2006)
11. Schneider, F., Draheim, J., Kamberger, R., Wallrabe, U.: Process and material properties of polydimethylsiloxane (PDMS) for Optical MEMS. Sensors and Actuators A: Physical **151**, 95–99 (2009)

Development of Delta-Type Parallel-Link Robot Using Pneumatic Artificial Muscles and MR Clutches for Force Feedback Device

Masatoshi Kobayashi[✉], Junya Hirano, and Taro Nakamura

Department of Precision Mechanics, Faculty of Science and Engineering,
Chuo University, 1-13-27 Kasuga, Bunkyo-ku, Tokyo 112-8551, Japan
m_kobayashi@bio.mech.chuo-u.ac.jp

Abstract. Force feedback devices have been recently adopted in virtual reality, rehabilitation, and medical training systems. Many of these devices convey the human forces sensed by the impedance control using motors; however, motorized actuators are disadvantaged by low output and backdrivability. To resolve these disadvantages, we developed a delta-type parallel-link robot using pneumatic artificial muscles and magnetorheological (MR) clutches. The artificial muscles deliver high output and backdrivability, while the MR clutches enable fast responses. Moreover, because the stiffness of its pneumatic artificial muscles directly responds to forces, our robot detects human forces without feedback from force sensors. This study introduces the prototype of our delta-type parallel-link robot, and evaluates its performance in elastic movement experiment. Finally, we confirmed that our prototype robot can obtain the stiffness and unloaded condition through its MR clutches without feedback from force sensors.

Keywords: Force feedback device · Pneumatic artificial muscle · Delta-type parallel-link mechanism · Magnetorheological (MR) clutch

1 Introduction

Recently, force feedback devices have been actively investigated in virtual reality, rehabilitation, and medical training systems [1-2]. Force feedback devices require a wide range of stiffness for handling soft and hard objects. They also require smooth, low-inertia motion, high backdrivability, and lightweight moving elements. Furthermore, they must safely operate within a large workspace.

Many of these devices generally detect human forces by impedance control using electric motors [3-5]. However, motor-driven force feedback devices generate insufficient output and undesired cogging torque. Furthermore, the backdrivability is lowered and backlash is caused via the reduction gears. Impedance control also requires feedback from force sensors, which delays the response to a sudden external force. Such delay is potentially dangerous to the operator.

© Springer International Publishing Switzerland 2015
H. Liu et al. (Eds.): ICIRA 2015, Part I, LNAI 9244, pp. 410–420, 2015.
DOI: 10.1007/978-3-319-22879-2_38

The mechanisms of force feedback devices are classifiable into serial- and parallel-link mechanisms. Serial-link mechanisms such as Phantom [6] realize a large workspace. However, the position accuracy and stiffness output are insufficient for their use in a force feedback device.

Previously, we proposed a delta-type parallel-link mechanism actuated by pneumatic artificial muscles for a soft manipulator [7]. The delta-type parallel-link mechanism is suitable for force feedback devices because it has the following characteristics [8]:

- Higher position accuracy and stiffness output than the serial-link mechanism.
- Larger workspace than the general parallel-link mechanism.
- The lightweight end effector of this mechanism realizes low-inertia motions.
- The mechanism has three translational degrees of freedom (DOF) and thus realizes three-dimensional movements.

We then focused on pneumatic artificial muscles as actuators. The stiffness of pneumatic artificial muscle can be structurally altered by applying air pressure. Thus, we can control the stiffness of force feedback devices without feedback of the force sensors [9]. Moreover, because pneumatic artificial muscles are flexible, pneumatic-driven actuators exhibit high backdrivability, no backlash and high safety. The main disadvantages are the difficulty of obtaining the unloaded condition.

To address these disadvantages of pneumatic artificial muscles, we newly install MR clutches between the transmission axes. The MR clutch uses an MR fluid with viscosity changes under an applied magnetic field. The advantages of the MR clutch are high output density and fast response.

Inspired by the above considerations, we propose a delta-type parallel-link mechanism using pneumatic artificial muscles and MR clutches as a prototype force feedback device. We named this prototype the *delta robot*. This paper introduces the development of the delta robot and its proposed control method. Next, we perform elastic movement experiment that demonstrate the basic properties of the device.

2   Pneumatic Artificial Muscle

Figure 1 is a schematic of our developed straight-fiber-type artificial muscle. The artificial muscle comprises natural latex rubber encapsulating a carbon fiber sheet. Both ends of the artificial muscles are fixed by the terminals. Therefore, under air pressure, the artificial muscle expands and contracts in the radial and axial directions, respectively.

The artificial muscles are antagonistically coordinated through a pulley system, as shown in Figure 2. This configuration enables independent control of the joint angle and joint stiffness. To control the artificial muscles, we must specify the air pressure, amount of contraction, and the contraction force. However, the properties of the artificial muscle are highly nonlinear and thus hard to control. For this reason, we linearize the muscle by applying the mechanical equilibrium model (EML) [10]. The model equations are expressed as follows:

$$P_1(\theta_d, \tau) = \left[G_{11}(\phi_{01})G_{22}(\phi_{02}) - G_{12}(\phi_{02})G_{12}(\phi_{01}) \right.$$
$$+ \frac{K_j}{K_{a2}} G_{21}(\phi_{01})G_{32}(\phi_{02}) + \frac{\tau}{r} G_{21}(\phi_{01})G_{22}(\phi_{02}) \right]$$
$$\left/ \left[G_{22}(\phi_{02})G_{31}(\phi_{01}) + \frac{K_{a1}}{K_{a2}} G_{21}(\phi_{01})G_{32}(\phi_{02}) \right] \right. \tag{1}$$

$$P_2(\theta_d, \tau) = \frac{K_j}{K_{a2}r_p^{\,2}} - \frac{K_{a1}}{K_{a2}} P_1 \tag{2}$$

$$\phi_{0i}(x_{di}') = \frac{2\alpha_i l_{0i}^{1.5} x_{di}'^{0.5}}{(l_{0i} - x_{di}')^2 + \alpha_i^{\,2} x_{di}' l_{0i}} \tag{3}$$

$$G_{1i}(\phi_{0i}) = \frac{4K_{ai}t_i}{d_{0i}} \left[\frac{l_{0i}}{d_{0i}} \right]^2 \left[\frac{\sin\phi_{0i} - \phi_{0i}\cos\phi_{0i}}{\phi_{0i}^{\,2}} \right] \tag{4}$$

$$G_{2i}(\phi_{0i}) = \frac{M\tan\phi_{0i}}{d_{0i}nb_i} \tag{5}$$

$$G_{3i}(\phi_{0i}) = 2\left[\frac{l_{0i}}{d_{0i}} \right]\left[\frac{\phi_0 - \sin\phi_{0i}\cos\phi_{0i}}{\phi_{0i}^{\,2}} \right]$$
$$+ 4\frac{l_{0i}}{d_{0i}} \frac{\sin\phi_{0i}}{\phi_{0i}} - \frac{M\pi d_{0i}}{nb_i}\tan\phi_{0i} \tag{6}$$

$$x_{d1}' = \frac{r\psi_1 - r\theta_d}{3} \tag{7}$$

$$x_{d2}' = \frac{r\psi_2 + r\theta_d}{3} \tag{8}$$

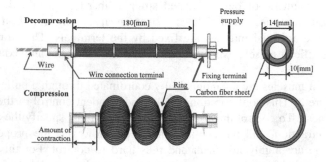

Fig. 1. Straight-fiber-type artificial muscles

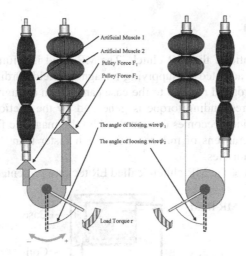

Fig. 2. Schematic of manipulator joint

Table 1. EML parameters of the artificial muscle

θ_d	[rad]	Desirable Angle	n		Number of Antinodes
τ	[Nm]	Load torque	K_j	[Nm/rad]	Joint Stiffness
P_1 P_2	[Pa]	Pressure	K_1 K_2		Stiffness Fixed number
F_1 F_2	[N]	Pull Force	l_{01} l_{02}	[m]	Length between cap and ring
x_{d1} x_{d2}	[m]	Desirable contraction	d_{01} d_{02}	[m]	Diameter of Artificial Muscle
ψ_1 ψ_2	[rad]	The angle of loosing wire	t_{01} t_{02}	[m]	Thickness of Artificial Muscle
b_1 b_2	[mm]	The width of the glass fiber	M		Fiber constant number
α_1 α_2		Approximation constant number			

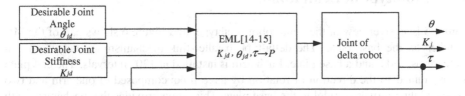

Fig. 3. Block diagram of artificial muscles

3   MR Clutch

Figure 4 is a schematic of the MR clutch. The MR fluid is a functional fluid whose apparent viscosity is adjusted by supplying a magnetic field. Ordinarily, the core part of the MR clutch is rotated relative to the case part. When a magnetic field is applied to the voltage, a corresponding torque is generated by the friction between the disk and the MR fluid (which becomes concentrated by the magnetic field). The MR fluid responds within several tens of milliseconds, much faster than the response time of pneumatic artificial muscles.

The specifications of the MR clutch, called ER tec, are presented in Table 2.

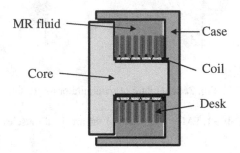

Fig. 4. Schematic of MR clutch

Table 2. Specifications of MR clutch (ER tec)

Parameter	Specification
Diameter [mm]	52.5
Thickness [mm]	30.6
Weight [g]	274
Maximum torque [Nm]	4.1
Minimum torque [Nm]	0.06
Maximum current [A]	0.9

4   Prototype of Delta Robot

Figure 5 is an overview of the delta robot prototype, and Figure 6 shows part of the MR clutches on the base plate. The delta-type parallel-link mechanism comprises an end plate, arms, rods, and a base plate. Each arm is installed at 120° intervals. The end plate is constrained to the direction of rotation by three links composed of one arm and two rods, and thus moves parallel to the base plate. This setup provides the mechanism with translational 3-DOF. The delta robot is driven by artificial muscles antagonistically configured through the pulleys and wires at their ends. Furthermore, the MR clutches are carried the delta robot at transfer axis on base plate, and their core and case parts are affixed to the pulley and to one side of the arm, respectively. Therefore, the rotation and torque of the pulley is transferred to the arm only during operation of the MR clutch.

In other words, the delta robot can switch between stiffness or the unloaded condition by operating the MR clutches, as shown in Figure 7.

The coordinate system and link parameters of the delta robot during the various experiments are presented in Figure 8 and Table 3, respectively. The origin O is the center of the base plate, and the negative y-coordinate is the direction of the A_i joint. Here, the subscript i is the link number, and θ_{1i} is the angle between the base plate and the arm.

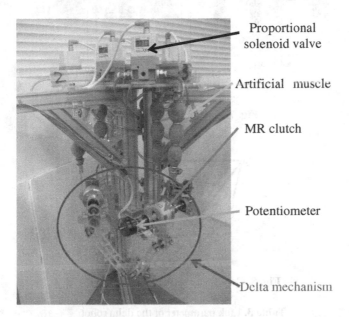

Fig. 5. Prototype of delta robot

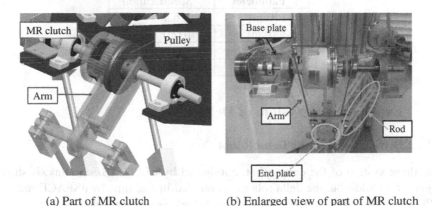

(a) Part of MR clutch   (b) Enlarged view of part of MR clutch

Fig. 6. Delta robot and part of the MR clutch

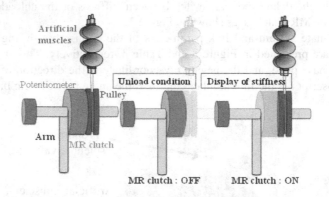

Fig. 7. Operating of MR clutch

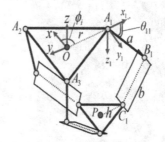

Fig. 8. Coordinate system of delta robot

Table 3. Link parameter of the delta robot

Parameter	Specification
r [mm]	126.5
a [mm]	100.0
b [mm]	173.0
h [mm]	43.0

5   Experimental System of Delta Robot

5.1   Drive System of the Delta Robot

The drive system of the delta robot, configured by MATLAB/Simulink, is shown in Figure 9. In addition, the delta robot is controlled in real time by dSPACE running on a PC. Each directive from dSPACE controls the proportional solenoid valves and MR drivers through a D/A converter. Under the directives, the proportional solenoid valves apply air pressure to the artificial muscles, causing them to contract, while the MR drivers apply a current to the MR clutches, switching them ON or OFF. Together, the artificial muscles and MR clutches realize stiffness of the delta robot's end plate.

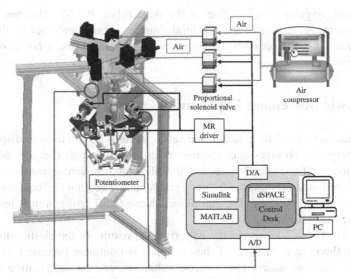

Fig. 9. Drive system of delta robot

5.2   Drive System of the Delta Robot

Figure 10 shows the control systems of the delta robot. The desirable end plate position is converted to the desirable joint angle by inverse kinematic calculations [11]. Each joint controller calculates the air pressure applied to the artificial muscles from the desirable joint angle and the torque calculated by inverse dynamics. The joint controller in Figure 10 includes the EML. On the other hand, the MR controller calculates the MR torque based on the desirable joint angle. The delta robot is then operated, and the joint angle is measured by the potentiometer.

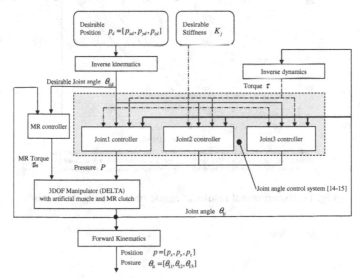

Fig. 10. Control system of the delta robot

To prevent singular configuration of the delta robot, the MR clutches are driven and air pressure is applied to the artificial muscles immediately before the singular configuration. This is necessary because a very stiff end plate is not moved by the artificial muscles.

6   Elastic Movement Experiment

In this experiment, the MR clutches and artificial muscles are used to display virtual elastic movements. To investigate the responses to forces applied at the end plate, we attached a load cell to the end plate, and the MR clutches were operated with the end plate positioned at x = 0 mm, y = 0 mm, z = 200 mm. The joint stiffness was varied as 0.10, 0.14, and 0.18 Nm/deg. The end plate achieved stiff artificial muscles by means of switching the MR clutches.

Figures 11, 12, and 13 shows the experimental results of the elastic motion experiment. The theoretical stiffness of the end plate is nonlinear because it is computed through the Jacobian matrix [12], whose values depend on the position of the end plate.

The experimental results reveal arbitrary stiffness of the delta robot without feedback from the force sensors. And the MR clutches enable the proper unloaded condition of the delta robot. However, the experimental and theoretical stiffness results deviate under low stiffness driving. We presume that the antagonism of the artificial muscles is insufficient under low air pressure. To solve this problem, we need to consider the wire elongation in the mechanism design.

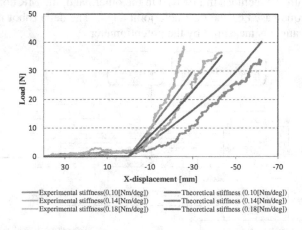

Fig. 11. Experimental results of elastic objects along the x-axis

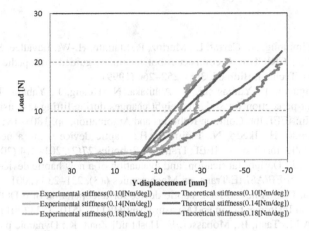

Fig. 12. Experimental results of elastic objects along the y-axis

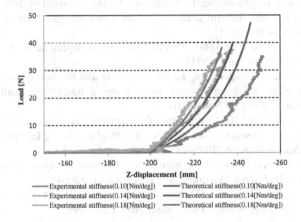

Fig. 13. Experimental results of elastic objects along the z-axis

7 Conclusion

In this paper, we proposed a delta-type parallel-link mechanism using pneumatic artificial muscles, and MR clutches, and we developed a prototype of the mechanism for a force feedback device. Through elastic movement experiment, we confirmed that our prototype robot can obtain the stiffness and unloaded condition through its MR clutches without feedback from force sensors.

In the future, we will perform an experiment to display arbitrary virtual three – dimensional elastic movements by using the delta robot. Then, we will also generate and evaluate smooth and rough virtual surfaces.

References

1. Brandt, G., Zimolong, A., Carrat, L., Merloz, P., Staudte, H.-W., Lavallee, S., Radermacher, K., Rau, G.: CRIGOS: A compact robot for image guided orthopedic surgery. IEEE Trans. Inform. Technol. Biomed. 3(4), 252–260 (1999)
2. Hara, M., Higuchi, T., Yamagishi, T., Ashitaka, N., Huang, J., Yabuta, T.: Analysis of human weight perception for sudden weight changes during lifting task using a force display device. In: IEEE Int. Conf. on Robotics and Automation, pp. 1808–1813 (2007)
3. Arata, J., Kondo, H., Ikedo, N., Fujimoto, H.: Haptic device using a newly developed redundant parallel mechanism. IEEE Trans. on Robotics 27(2), 201–214 (2011)
4. Yoon, J., Ryu, J.: Design, fabrication, and evaluation of a new haptic device using a parallel mechanism. IEEE/ASME Trans. on Mechatronics 6(3), 221–233 (2001)
5. Vu, M.H., Na, U.J.: A new 6-DOF haptic device for teleoperation of 6-DOF serial robots. IEEE Trans. on Instrumentation and Measurement 60(11), 3510–3523 (2011)
6. Tahmasebi, A.M., Taati, B., Mobasser, F., Hashtrudi-Zaad, K.: Dynamic parameter identification and analysis of a PHANToM haptic device. In: IEEE Conf. on Control Robots Applications, pp. 1251–1256 (2005)
7. Hirano, J., Tanaka, D., Watanabe, T., Nakamura, T.: Development of delta robot driven by pneumatic artificial muscles. In: IEEE/ASME Int. Conf. on Advanced Intelligent Mechatronics, pp. 1400–1405 (2013)
8. Clavel, R.: Conception d'un robot parallèle rapide à 4 degrés de liberté. Ph.D. Thesis, EPFL, Lausanne, Switzerland (1991)
9. Tanaka, D., Maeda, H., Nakamura, T.: Joint stiffness and position control of an artificial muscle manipulator for instantaneous loads using a mechanical equilibrium model. Advanced Robotics 25(3), 387–406 (2011)
10. Shinohara, H., Nakamura, T.: Position and force control based on mathematical models of pneumatic artificial muscles reinforced by straight glass fibers. In: IEEE Int. Conf. on Robotics and Automation, pp. 2809–2814 (2007)
11. Zsombor-Murray, P.J.: Descriptive geometric kinematic analysis of clavel's "Delta" Robot. Centre of Intelligent Machines. McGill University (2004)
12. Tsai, L.-W.: Robot Analysis: The mechanics of serial and parallel manipulators, pp. 223–240. Wiley-Interscience Publication, New York (1993)

Development of Semi-Active-Type Haptic Device Using Variable Viscoelastic Elements

Masakazu Egawa[✉], Takumi Watanabe, and Taro Nakamura

Department of Precision Mechanics, Faculty of Science and Engineering,
Chuo University, 1-13-27 Kasuga, Bunkyo-ku, Tokyo 112-8551, Japan
{m_egawa,t_watanabe}@bio.mech.chuo-u.ac.jp,
nakamura@mech.chuo-u.ac.jp

Abstract. Recently, the use of haptic technology has been expected in various fields, such as medicine and entertainment. In general, haptic devices are actuated by DC motors. However, DC motors have active rendering and can cause serious accidents when the system runs out of control. On the other hand, systems that employ a brake are stable and intrinsically safe. However, such systems have several haptic rendering limitations. Therefore, we have developed a semi-active-type haptic device using variable viscoelastic elements: an electrorheological (ER) devices and pneumatic artificial muscles. In this paper, we proposed control modes according to the haptic application, and we demonstrate several haptic experiments. Finally, we confirm that the novel semi-active system achieves haptic rendering of friction, viscous friction, virtual wall, and stiffness.

Keywords: Haptic device · Electrorheological Fluid · Pneumatic Artificial Muscle

1 Introduction

Recently, the use of haptic technology has been expected in various fields, such as medicine and entertainment. Haptic devices are classified into active haptic systems (e.g., using a motor) and passive haptic systems (e.g., using a brake) [1].

In general, a DC motor is used as an actuator of an active haptic device such as a commercial Phantom. An active haptic system can render various kinematic senses, but has the risk of harming the user [2]. Moreover, the system has low backdrivability because it uses reduction gears. Therefore, it is hard to render a free space (i.e., without resistance) by an active system. In terms of a control method, the system is based on impedance control with virtual stiffness and virtual viscosity. Therefore, feedback delay occurs and the control system becomes complex [3].

On the other hand, a passive haptic system is intrinsically stable and safe. Therefore, it is easy for such a system to safely render a high resistance force. However, because the system can render torque only in the direction opposite that of the motion, it has several haptic rendering limitations. For example, it cannot perfectly simulate a virtual spring [4]. Although a brake can engage when a relaxed spring

© Springer International Publishing Switzerland 2015
H. Liu et al. (Eds.): ICIRA 2015, Part I, LNAI 9244, pp. 421–432, 2015.
DOI: 10.1007/978-3-319-22879-2_39

is compressed, it cannot display the restoring force when a compressed spring returns to its natural length.

A feature of the haptic device is that it is moved directly by the operator using a joystick attached to the body. Therefore, safety is an important requirement. Furthermore, a virtual object is often modeled as an elastic body; therefore, haptic devices are strongly desired to render elasticity.

We paid attention to an ER device and a pneumatic rubber artificial muscle. The ER devices change its generation torque in milliseconds with the application of an electric field [5]. In general, an ER clutch transmits the torque generated by a DC motor. Smooth action at the no load can be achieved when the clutch is switched off. Moreover, the clutch limits the maximum torque delivered to the operator. Therefore, ER clutch is widely used in haptic devices [6]. On the other hand, pneumatic rubber artificial muscles are generally used as human-friendly actuators [7]. Because pneumatic rubber artificial muscles are light and soft, they are relatively safe, even if they accidently collide with humans. Artificial muscles render variable stiffness with the application of air pressure [8].

In this study, we developed a novel semi-active-type haptic device with an ER clutch, ER brake, and pneumatic artificial muscles. The merits of the proposal mechanism are as follows:

- The clutch limits the maximum torque delivered to the operator, and the system is fail-safe. Pneumatic artificial muscles are light, flexible and safe. Therefore, the system can produce larger forces without threatening the user's safety.
- The brake changes its friction torque structurally, and artificial muscle changes its stiffness structurally. Therefore, complicated feedback control is not needed but open-loop control can be applied. As a result, a quick response can be achieved.
- Combining the elasticity of artificial muscles and the friction of an ER brake, a wide variety of haptic renderings can be performed, such as virtual viscoelastic body, collision and instantaneous force.
- Without reduction gears, the system is backdrivable with no backlash.
- ER devices have high power density and it is easily miniaturized. Therefore, the proposed haptic system can be applied to wearable haptic device.
- ER devices and pneumatic artificial muscles are non-magnetic actuators. Therefore, the proposed system is an MRI-compatible haptic device.

In this paper, we developed a semi-active-type haptic device using the ER devices and pneumatic artificial muscles, and proposed a method to control this system. Next, several experiments rendering the haptic sensations were conducted to assess performance.

2   Mechanical Components

2.1   ER Devices

Figure 1 shows the configuration of an ER brake. The ER brake structure encloses ER fluid between the multiple disks. ER fluid is a type of smart fluid that varies its apparent viscosity with the application of an electric field. The system comprises both the

ER brake and an ER clutch. The ER brake is a device by which the input shaft is fixed by a shaft holder, and produces braking torque. The ER clutch varies transmission torque from a shaft to the other.

A particle type ER fluid is employed because its velocity dependence is low. The torque characteristics of the ER devices can be approximated by a quadratic equation as follows:

$$\tau_{ER} = K_{ER1}E^2 + K_{ER2}E + \tau_f \qquad (1)$$

where τ_{ER} is the transmission torque [Nm], K_{ER1} is the coefficient for torque increase quadratically due to the electric field [Nm/(V/mm)^2], K_{ER2} is the coefficient for torque increase linearly due to the electric field [Nm/(V/mm)], τ_f is the off-state friction torque [Nm] and E is the applied electric field [V/mm]. Table 1 shows the value of each coefficient as determined by an experiment previously conducted.

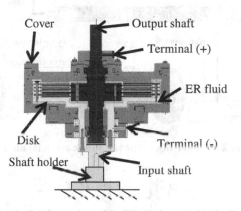

Fig. 1. Configuration of an ER device used as brake.

Table 1. Specifications of the ER Devices.

	K_{ER1} [Nm/(V/mm)^2]	K_{ER2} [Nm/(V/mm)]	τ_f [Nm]	Max torque [Nm]
ER clutch	2.4×10^{-7}	9.3×10^{-4}	0.2	3.0
ER brake	1.2×10^{-6}	1.6×10^{-3}	0.4	8.4

2.2   Pneumatic Artificial Muscle

Figure 2 shows a schematic of the straight-fiber-type artificial muscle we developed. This artificial muscle consists of natural latex rubber with a carbon fiber sheet, which is inserted along the long-axis direction. The two ends of the tube are fixed by terminals. Therefore, the artificial muscle expands in the radial direction and contracts in the axial direction when air pressure is applied.

As shown in the figure 3, a pair of artificial muscles is antagonized by a pulley fixed to the joint rotation axis. We can calculate the air pressure corresponding to the amount of contraction and the contraction force by using a mechanical equilibrium model

(EML) [9-10]. Therefore, we introduce EML as a feed-forward compensation and calculate the applied air pressure to achieve the desired joint stiffness K_d [Nm/rad] and the equilibrium joint angle θ_d [rad]. Figure 4 shows a block diagram of the feed-forward control method. The equations of the model are expressed as follows:

$$P_1 = P_1(\theta_d, \tau, K_d)$$
$$P_2 = P_2(\theta_d, \tau, K_d)$$
(2)

In the calculations, P is the pressures applied to artificial muscles [Pa], τ is the joint load torque [Nm]. The subscript number is used to discriminate between artificial muscles 1 and 2.

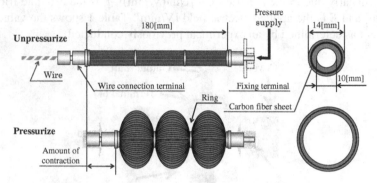

Fig. 2. Straight-fiber-type pneumatic artificial muscle.

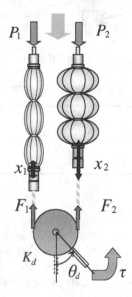

Fig. 3. Schematic diagram of manipulator joint.

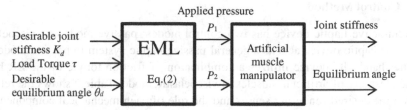

Fig. 4. Block diagram of feed-forward control using the EML.

3   Semi-Active-Type Haptic Device with Variable Viscoelastic Elements

3.1   Overall Design

Figure 5 shows a design overview of the 1-DOF semi-active haptic device we developed. This device is driven by an ER clutch, an ER brake, and antagonistic pneumatic artificial muscles. The input shaft of the ER clutch is connected to the artificial muscles and the ER brake. The output shaft of the ER clutch is connected to end effector. This mechanism has two potentiometers: the first potentiometer is primarily used to measure an angle of the end effector, whereas the second potentiometer confirms whether the input shaft of the ER clutch is rotated. If the input shaft is rotated, the antagonized artificial muscles render stiffness.

The ER clutch and the ER brake are dynamically controlled. The ER brake changes the braking torque, and the ER clutch changes the transmission torque to the output shaft. On the other hand, constant air pressure is applied beforehand to the artificial muscles. These artificial muscles are controlled under the EML for controlling the stiffness of a rotation shaft.

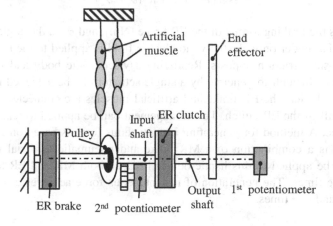

Fig. 5. Design overview of the 1-DOF semi-active haptic device.

3.2   Control Method

The semi-active haptic device has two control modes; passive mode and viscoelastic mode. The haptic device acts as a general passive haptic system in passive mode. On the other hand, the device utilizes a combination of friction torque of the ER brake and the elasticity of artificial muscles in viscoelastic mode. Table 2 shows the relation between the desired rendering senses and the role of each mechanical component. As common in the two modes, when the electric field of the ER clutch is switched off, the operator feels no resistance in free space (i.e., the ER brake and artificial muscles exert no influence on the end effector).

1) Passive mode: Maximum electric field is applied to the ER brake. Because the input shaft of the ER clutch is not be rotated, the artificial muscles are not engaged. Therefore, the ER clutch transmits only braking torque. The output torque τ_{out} in the passive mode is expressed as follows:

$$\tau_{out} = -\text{sgn}(\dot{\theta}_{out})\tau_{ERc} \tag{3}$$

where θ_{out} is the angle of the output shaft [rad], and τ_{ERc} is the transmission torque of the ER clutch [Nm]. If a constant electric field is applied to the ER clutch, the device generates Coulomb's friction. In addition, if the electric field of the ER clutch is controlled according to angular velocity, the device generates viscosity.

2) Viscoelastic mode: The ER brake is weakened to set a state that the input shaft of ER clutch can be rotated by an external force applied by the operator. Therefore, the ER clutch transmits the both friction torque of ER brake and elasticity of artificial muscle. The ER clutch has a roll of torque limiter. The output torque in the viscoelastic mode is expressed as follows:

$$\tau_{out} = -\text{sgn}(\dot{\theta}_{out})\tau_{ERb} + K_d(\theta_d - \theta_{in}) \tag{4}$$

where τ_{ERb} is the braking torque of the ER brake [Nm], and θ_{in} is the angle of the input shaft [rad]. To render only elasticity, no electric field is applied to the ER brake whereas the off-state friction appears. Rendering a viscoelastic body and instantaneous force which is difficult to generate by a single actuator can be achieved in this mode. In the haptic device, the ER brake and artificial muscles are connected in parallel to the input shaft of the ER clutch. This arrangement can be applied to rendering instantaneous force. A method for generating instantaneous force is based on a MR method [11], which is a combination of a MR brake and pneumatic artificial muscle. This method can be applied to this haptic device because both MR and ER brake display fast response times. The generation of instantaneous force achieves a force with minimal dead and rise times.

To clarify a condition which the input shaft is rotated, it needs to formulate a transmission torque from external torque to the input shaft; τ_t [Nm] is expressed as follows:

$$\tau_t = \begin{cases} \tau_h & \tau_{ERc} \geq |\tau_h| \\ \text{sgn}(\tau_h)\tau_{ERc} & \tau_{ERc} < |\tau_h| \end{cases} \tag{5}$$

where τ_h is the external torque by the operator. Therefore, the conditions which each of two modes satisfy are expressed as follows:

Passive mode:

$$\tau_{ERb} \geq | K_d(\theta_d - \theta_{in}) + \tau_t | \tag{6}$$

Viscoelastic mode:

$$\tau_{ERb} < | K_d(\theta_d - \theta_{in}) + \tau_t | \tag{7}$$

Table 2. Actuation Summary.

Rendering sense	Mode	ER clutch	ER brake	Artificial muscles
No resistance	-	OFF	-	-
Friction	Passive	ON	ON	-
Viscosity	Passive	Control	ON	-
Elasticity	Viscoelastic	ON	OFF	ON
Viscoelasticity	Viscoelastic	ON	Control	ON
Instantaneous force	Viscoelastic	Control	Control	ON

4   Experiment

4.1   Experimental Setup

The purpose of the experiments is to confirm whether the three mechanical components each functions a separate role and the system achieves representative haptic renderings. The experimental system is shown in Figure 6. The control system consists of a MATLAB interface and Simulink interface with dSPACE. The upward vertical direction of the arm was defined as 90 [deg]. Initial position of the arm was 60 [deg]. The end effector was moved by an operator, and reflective force was measured by a strain gauge. To ignore an influence of a gravity of the end effector, the output torque of the ER devices was gravity compensated.

428   M. Egawa et al.

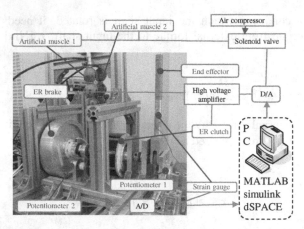

Fig. 6. Experimental system.

4.2   Friction

The aim of the experiment was to clarify the frictional property in passive mode. Free space (no resistance) was at less than 90 [deg] and the clutch was let in at more than 90 [deg]. The maximum electric field was applied to the ER brake and the ER clutch generates three kind of friction torque: the target output torque τ_d is 0.5 [Nm], 1.5 [Nm], and 2.5 [Nm].

Figure 7 shows the experimental results for friction. Little off-state friction is measured until 90 [deg]. This off-state friction was due to the output shaft of the ER clutch. At the friction space, measured torque was almost equaled to the target torque. In the case of τ_d =2.5 [Nm], output torque rose sharply because of the difference between static friction and dynamic friction. This phenomenon was often appeared when the applied electric field was high. Therefore, it is need to investigate both static and dynamic frictional property in various applied electric field.

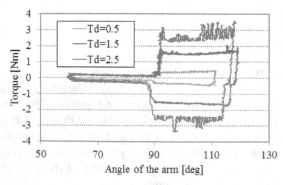

Fig. 7. Experimental result for friction.

4.3   Viscosity

In passive mode, friction torque was controlled so as to generate viscosity. The opera-tor moved the end effector backward and forward many times receiving viscosity. Virtual coefficient of viscosity c_d [Nms/deg] was set to three different values and the target output torque is expressed as follows:

$$\tau_{out} = -c_d \dot{\theta}_{out} - \text{sgn}(\dot{\theta}_{out})\tau_{fout} \qquad (8)$$

Because the off-state friction cannot be eliminated, target torque was a resultant tor-que of viscous friction generated by the ER clutch with the application of electric field and off-state friction. Figure 8 shows a block diagram of rendering viscosity.

Figure 9 shows the experimental results for viscosity. The gradients of the graph were changed according to the cd and the clutch generated variable viscosity. Since the ER fluid has really quick response, it is so high controllability that experimental values represented repeatability near theoretical ones. In the case of c_d=0.03 [Nms/deg], output torque was limited within 2.5 [Nm] because this torque was almost equaled to maximum transmission torque of the ER clutch.

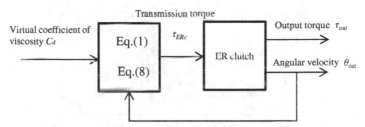

Fig. 8. Block diagram of rendering viscosity.

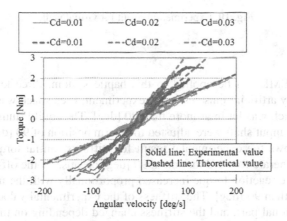

Fig. 9. Experimental result for viscosity.

4.4   Virtual Wall

This experiment was conducted in passive mode and a virtual wall was represented with the maximum electric field to the ER clutch. A virtual wall was set to 90 [deg] simulating that the operator cannot push any-more forward, and can leave freely backward. However, the ER clutch provides resistance force equally to the detaching movement of the end effector, representing sticky wall [12]. To prevent this pheno-menon, we used a case analysis whether the operator is touching or de-touching with the wall. The ER clutch was switched off when the detaching movement was de-tected; measured torque is smaller than $-\tau_{fc}$. τ_{fc} is the off-state friction torque of the output shaft of the ER clutch and the value is 0.2 [Nm].

Figure 10 shows the experimental result for virtual wall. The reaction torque dras-tically increased above 90 [deg]. The end effector detached by small torque of -0.45 [Nm] at virtual wall, the stick phenomenon did not appear. Although the end effector was stuck in the virtual wall exceeding maximum friction torque, the operator could easily leave the wall.

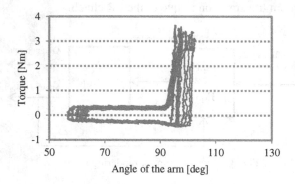

Fig. 10. Experimental result for virtual wall.

4.5   Stiffness

To confirm the EML can be applied to this haptic system, three kinds of stiffness were generated by artificial muscles in the experiment. Free space was at less than 90 [deg] and the clutch was let in at more than 90 [deg]. The equilibrium joint angle and initial position of input shaft were adjusted to an arm position of 90 [deg].

Figure 11 shows the experimental result for stiffness. Initial torque of 0.4 [Nm] was needed to generate forward motion at 90 [deg] because of the off-state friction of the ER brake. The reaction torque increased proportionally with the increasing of the position at more than 90 [deg]. The stiffness of the experimental value was calculated from the proportional part, and the stiffness changed depending on the application of air pressure. As shown in Table 3, the experimental and theoretical stiffness were almost equal. Since the error of the experimental and theoretical values was little when rendering force sense to the operator, we confirmed the applicability of EML.

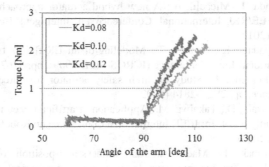

Fig. 11. Experimental result for friction.

Table 3. Desired stiffness and experimental stiffness.

Desired stiffness [Nm/deg]	0.08	0.10	0.12
Experimental stiffness [Nm/deg]	0.076	0.100	0.114

5   Conclusions and Future Work

5.1   Conclusions

In this paper, we proposed a semi-active-type haptic device using an ER devices and pneumatic artificial muscles. We developed a 1-DOF prototype and structured two control modes. Experimental results confirmed that the prototype can render various force senses.

5.2   Future Work

In the future, we will conduct experiments in viscoelastic mode to confirm the utility of variable viscoelastic components. Toward that end, we will conduct experiments to render viscoelastic body, instantaneous force, and inertia. Furthermore, we consider feasibility of multi degree of freedom into practical application.

References

1. Kim, B., Park, M., Hwang, C., KIm, M.: Passivity control of a passive haptic device. In: IEEE/RSJ Conference on Intelligent Robots and Systems, pp. 2905–2910 (2004)
2. Matsuoka, Y., Townsend, W.: Design of life-size haptic environments. Experimental Robotics. Experimental Robotics, vol. 7 (2001)
3. Galambos, P., Baranyi, P., Korondi, P.: Control design for impedance model with feedback delay. In: 19th International Workshop on Robotics, pp. 475–480 (2010)

4. Rossa, C., Lozada, J., Micaelli, A.: A new hybrid actuator approach for force-feedback devices. In: IEEE/RSJ International Conference on Intelligent Robots and Systems, pp. 4054–4059 (2012)
5. Terada, T., Koyanagi, K., Oshima, T.: Modeling of an ER fluid's time delay for servo systems. International Joint Conference (ICROS-SICE 2009), pp. 4767–4772 (2009)
6. Koyanagi, K., Furusho, J.: Study on high safety actuator for force display. In: SICE Annual Conference, pp. 2809–2814 (2002)
7. Noritsugu, T., Sasaki, D., Takaiwa, M.: Application of artificial pneumatic rubber muscles to a human friendly robot. In: IEEE International Conference on Robotics and Automation, pp. 2188–2193 (2003)
8. Nakamura, T., Tanaka, D., Maeda, H.: Joint stiffness and position control of an artificial muscle manipulator for instantaneous loads using a mechanical equilibrium model. Advanced Robotics 25(3), 387–406 (2011)
9. Nakamura, T., Shinohara, H.: Position and force control based on mathematical models of pneumatic artificial muscles reinforced by straight glass fibers. IEEE International Conference on Robotics and Automation, pp. 4361–4366 (2007)
10. Nakamura, T., Maeda, H.: Position and compliance control of an artificial muscle manipulator using a mechanical equilibrium model. In: IEEE International Conference on Robotics and Automation, pp. 3431–3436 (2007)
11. Tomori, H., Nagai, S., Majima, T., Nakamura, T.: Variable impedance control with an artificial muscle manipulator using instantaneous force and MR brake. In: IEEE International Conference on Robotics and Automation, pp. 4361–4366 (2007)
12. Sakaguchi, M., Furusho, J., Takesue, N.: Passive force display using ER brakes and its control experiments. In: Proceedings of the Virtual Reality, pp. 7–12 (2001)

Intelligent Visual Systems

Comparison of Grid-Based Dense Representations for Action Recognition

Yangyang Wang[✉], Yibo Li, Xiaofei Ji, and Yang Liu

College of Automation, Shenyang Aerospace University, Shenyang 110136, China
wyy200410l@163.com

Abstract. Dense descriptors have recently been widely used for human action recognition. Especially several methods of grid-based dense representations have been proposed in the literature. In order to evaluate which method performs better in the field of action recognition in unconstrained environments, four kinds of grid-based dense descriptions and recognition methods are tested on UCF sports dataset in this paper. And the recognition results of histogram of optical flow (HOF), histogram of oriented gradient (HOG), pyramid histogram of oriented gradients (PHOG) and Gist descriptor are compared. Furthermore, this paper also compares and analyzes recognition effects using 1-Nearest Neighbor (1NN) and support vector machine (SVM) combining these descriptors. It shows that the combination of Gist descriptor and SVM to reach the best recognition accuracy on UCF sports dataset.

Keywords: Action recognition · Grid-based dense representations · HOG descriptor · Gist descriptor · Nearest neighbor · Support vector machine

1 Introduction

Human action recognition in video sequences is an active topic of research. It covers a wide range of applications, such as video retrieval and annotation, sport videos analysis, safety monitoring, etc. Diverse approaches have been proposed in recognition of human actions based on videos [1,2]. And one of the main factors which directly affect action recognition accuracy is to obtain distinctive action descriptor.

Interest points-based representation is one of the most popular representations on human actions. It is also called as local descriptor. An action video is described as a collection of local spatio-temporal features [3]. Firstly, spatio-temporal interest points are detected from a video using detectors [4–6]. And then characteristics of the local patches around the points are described, for example, optical flow [7], HOG and its variant [8,9], and ESURF (extended speed-up robust features) [10]. These local representations are insensitive to partial occlusion and subtle viewpoint change. They have been widely employed with promising success in simple human action scenario, such as Weizmann dataset and KTH dataset [11].

© Springer International Publishing Switzerland 2015
H. Liu et al. (Eds.): ICIRA 2015, Part I, LNAI 9244, pp. 435–444, 2015.
DOI: 10.1007/978-3-319-22879-2_40

However, the descriptors based on interest points are sparse relative to the whole video. Especially to complex scenario, different manners of performing actions by different persons happened in diverse backgrounds. The sparse representations are not enough to exploit the underlying characteristics of actions. Therefor, one of the current trends is to adopt dense representations for action recognition [12]. For example, in the literature [13] the spatio-temporal interest points are densely sampled and combined with bag-of-word technique. Wang et al. densely tracked sampled points by optical flow fields and obtained dense trajectories [14]. Dense representations have shown excellent results [15]. And Wang et al. [16] has also compared and proved that dense grid-based representations outperform state-of-the-art interest point-based representations. But they only focus on the feature detectors and descriptors based on local spatio-temporal interest points from dense grids.

In this paper, a novel comparison framework for dense grid-based descriptors is developed. Different to [16], our emphases is to directly describe features of dense grids in the whole human action region. And then these descriptors are tested using 1NN and SVM to compare the performances of the recognition algorithms.

In our work, the comparison is carried out for four different kinds of dense descriptors, that is, HOF, HOG, PHOG, and Gist descriptors. The reason to choose these descriptors is that they are respectively reflect motion, appearance and texture characteristics of a video sequence. The four kinds of descriptors are respectively created through describing global features based on optical flow, gradient and filter response. Not only the performances of the descriptors are compared, but also the influence of 1NN and SVM on the recognition accuracy are reported.

Section 2 introduces how to extract and describe the HOF, HOG, PHOG, and Gist descriptors. Section 3 demonstrates 1NN and SVM recognition algorithms based on these descriptors. Section 4 presents the experimental results and analysis. Section 5 sums up this paper, and put forward the further work.

2   Grid-Based Dense Descriptors for Human Action

Grid-based dense descriptors are composed of a number of feature vectors which extracted from the spatial grids in a bounding box. To obtain the grid-based descriptors, firstly given a video with k frames, the human region in each frame is separated from the background, which is called as bounding box. Secondly, each box is normalized to equal size. Furthermore, the normalized box is divided into $m \times m$ spatial grids. The properties of each grid are represented using a feature vector. Finally, all feature vectors are concatenated to a global long descriptor. Four kinds of descriptors, HOG, HOF, PHOG and Gist descriptors are adopted to test the performances of action recognition in our work.

2.1   HOG and HOF Descriptor

HOG (histograms of oriented gradient) descriptors were adopted by Laptev et al. [8]. In their work the descriptors characterize the local gradient information of the spatio-temporal neighborhoods of detected interest points. In this paper, our grid-based HOG descriptors are obtained based on dense grids in each box. As shown in Fig. 1, the global gradient of a box is computed. And then the global features are divided into $m \times m$ grids. In each grid, a histogram of K gradient orientation bins is extracted. Finally, all histograms are concatenated to one vector. The final descriptor of a box has $m \times m \times K$ feature dimensions. The spatial locations of these grids are implied in HOG.

Similarly, HOF descriptor are produced by the same method. Followed by computing global optical flows of bounding boxes, each box is divided into $m \times m$ grids. In each grid, a histogram of K bins is extracted. Each bin covers a $360/K$ degree range. All histograms are concatenated to the final descriptor. It also has $m \times m \times K$ feature dimensions.

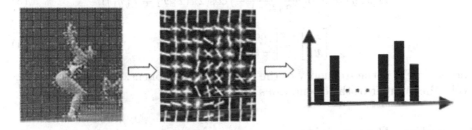

Fig. 1. The process of describing HOG descriptor.

2.2   PHOG Descriptor

PHOG (pyramid histogram of oriented gradients) descriptor is obtained based on HOG. The algorithm is to divide a grid into a number of sub-grids at several pyramid levels. For each sub-grid, a histogram of gradient orientation bins is extracted. And then all the histograms are concatenated to construct the final feature [17].

The detail process is described as flow: Firstly, the magnitude $T(x_i, y_i)$ and orientation θ of the gradient in each box of a video sequence is computed as:

$$T(x_i, y_i) = \sqrt{P_x(x_i, y_i)^2 + P_y(x_i, y_i)^2}, \tag{1}$$

$$\theta = \arctan \frac{P_x(x_i, y_i)}{P_y(x_i, y_i)}, \tag{2}$$

where $P_x(x_i, y_i)$ and $P_y(x_i, y_i)$ are respectively horizontal and vertical gradient of pixel (x_i, y_i). Secondly, the gradient distribution is divided into different sub-grids ($S_l = 2^l \times 2^l$) at every pyramid level l ($l = 1, 2, 3$). Thirdly, for each sub-grid, a histogram of K major gradient orientations is represented. The feature dimensions of each level is $2^l \times 2^l \times K$. Lastly, the feature vectors produced by the levels are concatenated to the final descriptor for each bounding box.

2.3 Gist Descriptor

Gist descriptor was proposed by Oliva and Torralba [18] and has been widely used in objection classification and retrieval [19,20]. The descriptor can directly represent the spatial configuration of an image without additional sub-grids parsing.

Having obtained the box of each frame, Gist descriptor is computed using a cluster of Gabor filters. Firstly, a box is divided into $m \times m$ grids, and representing each grid with its average filter response by using Gabor filter transfer functions with different orientations and spatial resolution. Secondly, all the averages are concatenated to a vector as the final descriptor of the box. The Gabor function is defined as:

$$G(x, y) = \exp(\frac{-(x_1^2 + y_1^2)}{2\sigma^{2(l-1)}}) \cos(2\pi(F_x x_1 + F_y y_1)), \tag{3}$$

$$\begin{bmatrix} x_1 \\ y_1 \end{bmatrix} = \begin{bmatrix} \cos\theta_l & \sin\theta_l \\ -\sin\theta_l & \cos\theta_l \end{bmatrix} \begin{bmatrix} x \\ y \end{bmatrix}, \tag{4}$$

where(F_x, F_y) is the frequency of the sinusoidal component. θ_l and l are respectively the number of the orientations and scales.

3   Action Recognition

In general, for the dense representations based on histogram, the approaches of template matching and discriminative models are popularly used in the recognition process. In order to compare and analyze the effects of the two kinds of approaches, 1NN and SVM recognition algorithms are chosen in this paper.

3.1 Nearest Neighbor Recognition Algorithm

According to the descriptors adopted in this work, each bounding box of a frame is described as a histogram representation. Therefore, the typical 1NN classifier is used for our frame-based recognition algorithm.

In our experiments, the Euclidean distance of histogram representation between a test frame and each training frame is calculated. The frame with minimum distance in the training frames will vote for the action type which it belongs to. All the frames in a test video are in turns carried out using the distance calculation and voting choices above. Finally, the action type of the test video will be the action which has the highest votes.

3.2   SVM Recognition Algorithm

A collection of feature vectors constitute the dense grid-based descriptor of human action. These feature vectors are extracted from the different spatial grids in a bounding box.

In order to make full use of the spatial distribution relations implied in grid-based descriptors, local patching coding based on bag-of-words method is respectively applied to our four dense descriptors. The detailed coding process were introduced by [19]. Dense descriptors are partitioned into four patches according to their location distribution. Each patch is respectively encoded and transformed to the form of visual words. In our case, the respective codebook sizes of four patches are 100, 100, 200 and 400. Thus an action video can be represented with the histogram concatenation of all the visual words. Finally SVM (support vector machine) with radius basis function kernel is used to recognize actions. This method effectively reduces the feature dimension and at the same time keeps the discrimination of the descriptors.

4   Experimental Results and Analysis

4.1   Experimental Results

Experimental data are from UCF sports dataset [21]. The dataset consists of 150 sports action videos. These videos are recorded in unconstrained environments with a wide range of viewpoints. The types of the actions include: diving, golfing, kicking, lifting, riding, running, skating, swinging 1 (gymnastics, on the pommel horse and floor), swinging 2 (gymnastics, on the high and uneven bars) and walking.

The experiments are tested on MATLAB R2011a. And leave-one-actor-out framework is adopted in the recognition process. An action video is respectively described using HOF, HOG, PHOG and Gist dense descriptor.

Using nearest neighbor recognition algorithm, recognition accuracy with respective optimal parameters is shown in Table 1. HOF is computed with parameters $m=4$ and $K=32$. HOG is computed with parameters $m=9$ and $K=12$. PHOG is computed with parameters $l=3$ and $K=12$. Gist descriptor is computed with parameters $m=\theta=8$ and $l=4$. The computational time for computing descriptors of a box is also listed in Table 1. And the recognition accuracy of each kind of actions for four descriptors in 1NN recognition algorithm are shown in Fig. 2.

Furthermore, to keep the equal descriptor length, the descriptor length of Gist in Table 1 is regarded as the benchmark. A comparison of the results with different representation is shown in Table 2. The description based on gradient is the fastest. HOF descriptor requires most time among all descriptors, since histograms have to be built on the basis of optical flow. And the recognition results for four descriptors with 4×4 grids are given in Table 3, their histogram bins of each grid are all set to 32.

Combined with SVM recognition algorithm mentioned in Section 3.2, the same descriptors in Table 2 are respectively transformed 800-dimensional histogram. The recognition results are shown in Table 4. The recognition accuracy of each kind of actions for four descriptors in SVM recognition algorithm are shown in Fig. 3.

Table 1. Comparison for four descriptors with optimal parameters and 1NN recognition algorithm

Descriptor	Descriptor length	Computational time (s)	Recognition accuracy (%)
HOF	512	0.5	58.67
HOG	972	0.005	78.4
PHOG	1008	0.01	76.8
Gist	2048	0.2	81.33

Table 2. Comparison for equal descriptor length with 1NN recognition algorithm

Descriptor	Descriptor parameters	Computational time (s)	Recognition accuracy (%)
HOF	$m=8$ and $K=32$	0.5	56.67
HOG	$m=8$ and $K=32$	0.007	76
PHOG	$l=3$ and $K=24$	0.014	76.7
Gist	$m=\theta=8$ and $l=4$	0.2	81.33

4.2 Experiment Analysis

According to the above experiment results, whether 1NN or SVM recognition algorithm, Gist descriptor achieves highest recognition rate than other descriptors. This is because that global structure information are captured in Gist descriptor by filtering an image with different orientations and scales. In 1NN recognition algorithm, under the special optimal parameters, HOG descriptor performs better than PHOG descriptor. While PHOG descriptor is superior descriptor for SVM recognition algorithm besides Gist descriptor. And HOF descriptor is inferior quality.

According to Table 1 and Table 2, the recognition accuracy for different descriptors is not always increasing as the change of descriptor length. When the feature dimension ranging about from 1000 to 2000, the performance of different dimension is quite close.

Table 3. Comparison for the same grid size with 1NN recognition algorithm

Descriptor	Recognition accuracy (%)
HOF	58.67
HOG	77.3
PHOG	76
Gist	79.3

Table 4. Comparison with SVM recognition algorithm

Descriptor	Descriptor length	Recognition accuracy (%)
HOF	800	32.67
HOG	800	71.3
PHOG	800	82
Gist	800	86

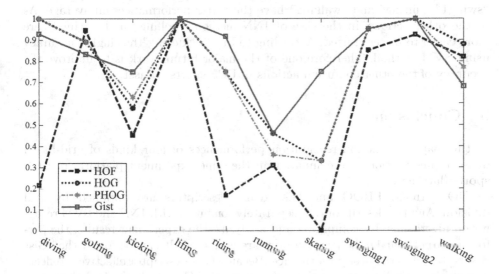

Fig. 2. Recognition accuracy for four descriptors with 1NN recognition algorithm.

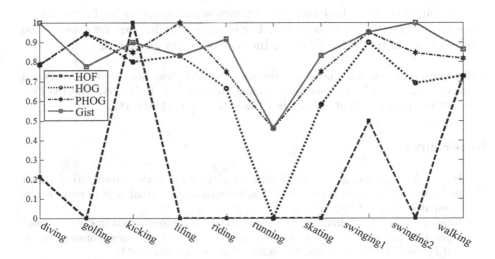

Fig. 3. Recognition accuracy for four descriptors with SVM recognition algorithm.

As we can also see, the performance of the same descriptor is differently with 1NN and SVM. For example, the recognition rates of PHOG and Gist descriptor have significant improvement with SVM method, while HOF and HOG descriptor give better results with 1NN method. This maybe SVM is more sensitive to intra-class discrimination. Furthermore in the classification process, for the feature dimension with about 1000, 1NN is about 2 seconds for processing each frame while SVM is about 0.01 seconds.

About the actions which be correctly recognized in the experiments, "swing1", "lifting" and "walking" have the better performances on average. As to the other actions, in the case of 1NN method, "golfing" and "swing2" are more easier to be recognized. According to Fig. 3, "kicking" has higher accuracy using SVM method. Therefore one of the major future work is to improve the accuracy of the other confusion actions in UCF sports dataset.

5   Conclusion

In this paper the action recognition performances of four kinds of grid-based dense representations are compared on the same experiment platform of UCF sports database.

HOF, HOG, PHOG and Gist dense descriptors are obtained by grid division. And the descriptors are separately combined with 1NN and SVM recognition algorithms. By comparing and analyzing the experiment results, the performances of four kinds of descriptors are evaluated. It can be found that Gist descriptor gives the best performance. Because Gist descriptor effectively reflects global structure information through a cluster of Gabor filters. And the representation only based on optical flow is the worst descriptor for UCF sports database.

The objective is to find the best representation methods for future use in more challenging action recognition. For example, similar experiments can be conducted for in the fields of human interaction action recognition.

Acknowledgments. The project supported by the Scientific Research General Project of Education Department of Liaoning Province (No. L2014066), and the Program for Liaoning Excellent Talents in University (No. LJQ2014018).

References

1. Ji, X., Wu, Q., Ju, Z., Wang, Y.: Study of Human Action Recognition Based on Improved Spatio-temporal Features. International Journal of Automation and Computing **11**, 500–509 (2014)
2. Jones, S., Shao, L.: A multigraph representation for improved unsupervised/semi-supervised learning of human actions. In: IEEE Conference on Computer Vision and Pattern Recognition, pp. 820–826. IEEE Press, Ohio (2014)
3. Shao, L., Zhen, X., Tao, D., Li, X.: Spatio-temporal Laplacian Pyramid Coding for Action Recognition. IEEE Transactions on Cybernetics **44**, 817–827 (2014)

4. Dollar, P., Rabaud, V., Cottrell, G., Belongie, S.: Behavior recognition via sparse spatio-temporal features. In: 2nd Joint IEEE International Workshop on Visual Surveillance and Performance Evaluation of Tracking and Surveillance, pp. 65–72. IEEE Press, Beijing (2005)
5. Laptev, I., Lindeberg, T.: Space-time interest points. In: 9th IEEE International Conference on Computer Vision, pp. 432–439. IEEE Press, Nice (2003)
6. Bregonzio, M., Gong, S., Xiang, T.: Recognising action as clouds of space-time interest points. In: IEEE Conference on Computer Vision and Pattern Recognition, pp. 1948–1955. IEEE Press, Florida (2009)
7. Zhang, X., Miao, Z., Wan, L.: Human action categories using motion descriptors. In: 19th IEEE International Conference on Image Processing, pp. 1381–1384. IEEE Press, Florida (2012)
8. Laptev, I., Marszalek, M., Schmid, C., Rozenfeld, B.: Learning realistic human actions from movies. In: IEEE Conference on Computer Vision and Pattern Recognition, pp. 1–8. IEEE Press, Anchorage (2008)
9. Vondrick, C., Khosla, A., Malisiewicz, T., Torralba, A.: HOGgles: visualizing object detection features. In: 14th IEEE International Conference on Computer Vision, pp. 1–8. IEEE Press, Sydney (2013)
10. Willems, G., Tuytelaars, T., Van Gool, L.: An efficient dense and scale-invariant spatio-temporal interest point detector. In: Forsyth, D., Torr, P., Zisserman, A. (eds.) ECCV 2008, Part II. LNCS, vol. 5303, pp. 650–663. Springer, Heidelberg (2008)
11. Chakraborty, B., Holte, M.B., Moeslund, T.B., Gonzlez, J.: Selective Spatio-temporal Interest Points. Computer Vision and Image Understanding 116, 396–410 (2012)
12. Shi, F., Petriu, E., Laganiere, R.: Sampling strategies for real-time action recognition. In: IEEE Conference on Computer Vision and Pattern Recognition, pp. 2595–2602. IEEE Press, Oregon (2013)
13. Marin-Jimnez, M.J., Yeguas, E., de la Blanca, N.P.: Exploring STIP-based Models for Recognizing Human Interactions in TV Videos. Pattern Recognition Letters 34, 1819–1828 (2013)
14. Wang, H., Kläser, A., Schmid, C., Liu, C.L.: Dense Trajectories and Motion Boundary Descriptors for Action Recognition. International Journal of Computer Vision 103, 66–79 (2013)
15. Shi, F., Petriu, E.M., Cordeiro, A.: Human action recognition from local part model. In: IEEE Int Haptic Audio Visual Environments and Games Workshop, pp. 35–38. IEEE Press, Hebei (2011)
16. Wang, H., Ullah, M.M., Kläser, A., Laptev, I., Schmid, C.: Evaluation of local spatio-temporal features for action recognition. In: British Machine Vision Conference, pp. 127–137. BMVA Press, London (2009)
17. Wang, J., Liu, P., She, M.F., Kouzani, A., Nahavandi, S.: Human action recognition based on pyramid histogram of oriented gradients. In: IEEE International Conference on Systems, Man, and Cybernetics, pp. 2449–2454. IEEE Press, Anchorage (2011)
18. Oliva, A., Torralba, A.: Modeling the Shape of the Scene: A Holistic Representation of the Spatial Envelope. International Journal of Computer Vision 42, 145–175 (2001)

19. Wang, Y., Li, Y., Ji, X.: Human Action Recognition Based on Global Gist Feature and Local Patch Coding. International Journal of Signal Processing, Image Processing and Pattern Recognition **8**, 235–246 (2015)
20. Ikizler-Cinbis, N., Sclaroff, S.: Object, scene and actions: combining multiple features for human action recognition. In: Daniilidis, K., Maragos, P., Paragios, N. (eds.) ECCV 2010, Part I. LNCS, vol. 6311, pp. 494–507. Springer, Heidelberg (2010)
21. Rodriguez, M.D., Ahmed, J., Shah, M.: Action mach: a spatio-temporal maximum average correlation height filter for action recognition. In: IEEE Conference on Computer Vision and Pattern Recognition, pp. 1–8. IEEE Press, Anchorage (2008)

Three-Label Outdoor Scene Understanding Based on Convolutional Neural Networks

Yue Wang and Qijun Chen[✉]

School of Electronics and Information Engineering,
Tongji University, Shanghai 201804, China
{1wangyue,qjchen}@tongji.edu.cn

Abstract. Scene understanding is the task of giving each pixels in an image a label, which is the class of the pixel belongs to. Traditional scene understanding is object-based approach, which has lots of limitations as the descriptors cannot give the whole characteristics. In this paper, a convolutional neural network based method is proposed to extract the internal features of the whole image, then a softmax regression classifier is applied to generate the label. Scene understanding used in self-navigating vehicles only concentrate on the road, so the number of classes is reduced in order to get higher accuracy by lower computational cost. A pre-processing is implemented on Stanford Background Dataset to obtain three-label images including road, building, and others. As a result, the system yields high accuracy on the three-label dataset with great speed.

Keywords: Scene understanding · Convolutional neural networks · Softmax regression · Self-navigating vehicles · Outdoor scene

1 Introduction

In the computer vision field, scene understanding is the task of giving each pixels in an image a label, which is the class of the pixel belongs to. Scene Understanding is one of the hardest problems in computer vision field and important for making robots useful in both indoor and outdoor environments. With rapid developing of sensing technology, large amount of high quality images or point clouds can be obtained easily. These big data can provide more information about the scene and it is important to extract the internal features accurately and quickly. This task is very challenging, as it implies solving detection, segmentation and multi-label recognition problems jointly.

The scene labeling problem is most commonly addressed with some kinds of local classifier constrained in its predictions with a graphical model, such as conditional random fields [1] and markov random fields [2], in which global decisions are made. These approaches usually consist of segmenting the image into superpixels [3] or segment regions [4] to assure a visible context consistency of the labeling and also to take into account the similarities between neighbor segment regions [5], then giving a high level understanding of the overall structure of the

© Springer International Publishing Switzerland 2015
H. Liu et al. (Eds.): ICIRA 2015, Part I, LNAI 9244, pp. 445–454, 2015.
DOI: 10.1007/978-3-319-22879-2_41

image. These models are trained to maximize the likelihood of classification correctly based on the features. The main limitation of scene labeling approaches based on graphical models is the computational cost at test time, which limits the model to simple contextual features.

In most cases, it is unnecessary to consider every label in the scene, thus only those which can influence the movements of robots is interesting. For example, when the scene understanding used in self-navigating vehicles, only road and signs need to be recognised. In this paper, a convolutional neural network approach is considered which can take into account long range label dependencies in the scenes while controlling the capacity of the network. The method relies on a convolution architecture with a logistic classifier which produce three labels. As a result, the three-label network shows a good performance with a low computational cost, which can get over 90% accuracy in less than a second.

Compared to graphical model approaches relying on image segmentation, our system has several advantages: (1)it does not require any engineered features, since deep learning architectures train filters to extract internal features in an end-to-end manner, (2)the number of classes can be set according to actual demand, thus the system achieves high accuracies under acceptable computational cost.

The paper is organized as follows. In Section 2, the related work are reviewed. The proposed methods based on convolutional neural networks are described in Section 3. Section 4 introduces the results of experiments on the Stanford Background Dataset. Finally, a discussion followed by a conclusion is presented in Section 5.

2   Related Work

Scene understanding problem for mobile robots has been widely studied. Traditional scene recognition is object-based approach. For example, if the robot recognizes desks, chairs and a blackboard, then it can be inferred this is a classroom. For indoor scene recognition, a generative hierarchical probabilistic representation based on objects is often used [2,4,6]. Common objects such as doors or furniture are used as intermediate semantic representation [2]. Furthermore, a 3D range sensor is used to increase detection accuracy [4]. Vasudevan et al. [6] produced a global topological map of places. Learning object features after segmentation gain some better results [7], they present a method called PLISS (Place Labeling through Image Sequence Segmentation) using online change-point detection. These approaches do an inference by detecting objects in order to get some robust and invariant keypoints [8]. They are easy to understand but have lots of limitations as the descriptors cannot give the whole characteristics and most of them are hand-crafted.

Instead of object detection, Lamon et al. [9] developed a descriptor for panoramic images called fingerprint of place. Tapus et al. [10] applied the fingerprint approach of Lamon and built topological maps of multi-room indoor environments. Liu et al. [11] propose a lightweight descriptor called FACT (Fast Adaptive

Color Tags) for omnidirectional vision, then use Dirichlet Process Mixture Model to combine color and geometry features [12]. Sometimes it is also necessary to process data without label by unsupervised learning. Farabet et al. [1] use convolutional networks to learn features, then combine the hierarchical model with purity trees and optimal covers to produce the scene labeling. Lai et al. [13] use the hierarchical matching pursuit algorithm, a state-of-the-art sparse coding technique for both RGB-D images and 3D point clouds to do 3D scene labeling. Among these, deep learning methods achieve good results in scene understanding and in this paper convolution neural networks will be used train our feature extractor.

In 2006, a breakthrough in feature learning and deep learning was initiated by Hinton [14] and quickly followed up in the same year [15,16], and soon after by Lee et al. [17] and many more later. The beginnings of deep learning in 2006 have focused on the MNIST digit image classification problem [14,15], and the latest records are still held by deep networks. Deep learning can be also used in many other applications, such as speech recognition, language processing and motion recognition. In computer vision fields, deep learning algorithms like CNN (Convolutional Neural Networks) [18], DBM (Deep Boltzmann Machine) [19], and DBN (Deep Belief Networks) [17] are well used. In the last few years, deep learning has moved from digits to object recognition and scene labeling in natural images, and many breakthrough have been achieved by using the convolutional neural networks.

Kavukcuoglu et al. [5] propose an unsupervised method to learning multi-stage hierarchies of sparse convolutional features, the obtained filters can improve performance on a number of visual recognition and detection tasks. Socher et al. [3] combine the convolutional and recursive neural networks (CNN and RNN) for learning features and classifying RGB-D images, and obtain state of the art performance on the standard RGB-D Object Dataset. Masci et al. [20] use convolutional networks with max pooling to segment images. Max pooling is a classic down sampling method used in CNN. Pinheiro et al. [21] propose an approach that consists of a recurrent convolutional neural network which allows to consider a large input context while limiting the capacity of the model. The approach does not rely on any segmentation technique nor any task specific features. Convolutional neural networks gain big success in computer vision fields, especially in scene understanding. If combine some suitable segmentation and classification algorithms with convolutional neural networks, a better result will be achieved.

3   System Description

The system use a 2-stage convolutional neural network to extract the internal features of the whole image. This is achieved by using convolutional and pooling layers. Then a classifier is added to the end of the convolutional neural network in order to give each pixel a label. The model proposed in this paper is described in Fig. 1.

In more detail, once input an image, do convolution computation with suitable amount of kernels and window size and reserve the margin part of the image

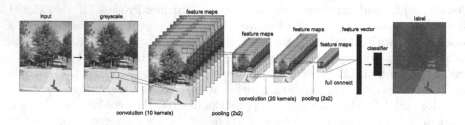

Fig. 1. Flow diagram of the three-label scene understanding system. The raw input image is transformed into grayscale image, then use a 3-satge convolutional neural network to extract the internal features. Both stages include convolutional layer and pooling layer. A fully connected layer is used to the feature maps in order to get the feature vector. Then a softmax regression classifier is added to produce the class label at last.

in order to keep the size of image unchanged after convolution. As images have the stationarity property, we can do sub-sampling after convolution to dramatically reduced computation. After some convolutional and pooling layers, a set of feature maps will be obtained, which need to be up-sampled to match the size of the raw image. Each pixel of the image thus has a corresponding feature vector. The final labeling is produced by putting the feature vectors into a classifier.

We formally introduce convolutional neural networks in Section 3.1 and we discuss how to capture internal feature maps while keeping a control over the capacity. In the main representation, convolutional neural networks are used to extract the internal features of the whole image. Then a classifier is added to the end of the convolutional neural networks in order to give each pixel a label. Section 3.2 introduces the optimization methods used to adjust parameters of our model. At last, in Section 3.3, we show how to infer the scene labeling by adding a multi-label classifier.

3.1   Feature Learning

Good internal features are hierarchical. In computer vision, pixels are assembled into edges, edges into motifs, motifs into parts, parts into objects, and objects into scenes. This suggests that scene recognition models should be hierarchical architecture, that each stage corresponds to each level of features. Convolutional neural networks provide a simple framework to learn such hierarchical features.

A typical convolutional network is composed of multiple stages. The output of each stage is made of a set of 2D arrays called feature maps. Each feature map is the outcome of a convolutional or a pooling filter applied over the full image. A non-linear activation function always follows a pooling layer.

In the context of scene labeling, given an image I which we are interested in labeling. It can be shown that the output image size s_l of the lth layer is computed as:

$$s_l = \frac{s_{l-1} - kW_l}{dW_l} + 1 \tag{1}$$

where s_0 is the input image size, kW_l is the size of the window of convolution or pooling kernels in the l_{th} layer, and dW_l is the pixel step size used to slide the convolution or pooling window over the input images. Given a network architecture and an input image, one can compute the output image size by applying (1) on each layer of the network.

For a network f with L stages and trainable parameters $(\mathbf{W}, \mathbf{b})$, we have:

$$f(I; (\mathbf{W}, \mathbf{b})) = \mathbf{W}_L \mathbf{H}_{L-1} \tag{2}$$

with the output of the l_{th} hidden layer computed as:

$$\mathbf{H}_l = \tanh(\text{pool}(\mathbf{W}_l \mathbf{H}_{l-1} + \mathbf{b}_l)) \tag{3}$$

for all $l \in \{1, \ldots, L-1\}$, $\mathbf{H}_0 = I$, $\mathbf{b}_l$ is the bias vector of layer l, and $\mathbf{W}_l$ is the Toeplitz matrix of connection between layer l and layer $l-1$. The pool function is the max-pooling operator and the tanh is the point-wise hyperbolic tangent function applied at each point of the feature map. Finally, the outputs of the neural networks are up-sampled to match the size of the raw image so as to produce a map of feature vectors which corresponding to each pixel of the raw image.

3.2 Optimization Algorithm

Gradient descent is a first-order optimization algorithm. To find a local minimum of a function using gradient descent, one takes steps proportional to the negative of the gradient of the function at the current point. During the last decade, the data sizes have grown faster than the speed, the computational complexity of learning algorithm becomes the critical limiting factor when one envisions very large datasets. Some improved optimization algorithm such as stochastic gradient descent and mini-batch gradient descent are proposed.

Batch gradient descent is the classical gradient descent algorithm. It needs to calculate all training date to adjust the parameters in each iteration step. Batch gradient descent can ensure finding the optimal solution, but when the training set is extremely large, this method is computational expensive or even doesn't work anymore.

The stochastic gradient descent algorithm is a drastic simplification. Instead of computing the gradient exactly, each iteration estimates this gradient on the basis of a single randomly picked example. It means that for each data in the training set, an optimal solution can be obtained, so there only one data needed in each iteration step. The limitation of this algorithm is the model may trap in local optimum.

In batch gradient descent we will use all examples in each generation, whereas in stochastic gradient descent we will use a single example in each generation. What mini-batch gradient descent does ia somewhere in between. One disadvantage of this is that there is an extra parameter, the mini-batch size which may take time to adjust. But if there is a good vectorized implementation this can run even faster than others.

In this paper, we use batch gradient descent when the number of classes is small, whereas using the other two optimization algorithms when learning the full labeled large dataset.

3.3   Scene Inference

Gradient descent is a first-order optimization algorithm. To find a local minimum of a function using gradient descent, one takes steps proportional to the negative of the gradient of the function at the current point. During the last decade, the data sizes have grown faster than the speed, the computational complexity of learning algorithm becomes the critical limiting factor when one envisions very large datasets. Some improved optimization algorithm such as stochastic gradient descent and mini-batch gradient descent are proposed.

Batch gradient descent is the classical gradient descent algorithm. It needs to calculate all training date to adjust the parameters in each iteration step. Batch gradient descent can ensure finding the optimal solution, but when the training set is extremely large, this method is computational expensive or even doesn't work anymore.

Gradient descent is a first-order optimization algorithm. To find a local minimum of a function using gradient descent, one takes steps proportional to the negative of the gradient of the function at the current point. Batch gradient descent is the classical gradient descent algorithm. It needs to calculate all training date to adjust the parameters in each iteration step. Batch gradient de-scent can ensure finding the optimal solution, but when the training set is extremely large, this method is computational expensive or even doesn't work anymore. In this paper, we only consider 3 labels, so we choose batch gradient descent in order to get the optimal solution.

The network is trained by adding a classifier using conditional probabilities. Logistic regression is the most basic classifier.It is a probabilistic, linear classifier. We had a training set $\{(x^{(1)}, y^{(1)}), \ldots, (x^{(m)}, y^{(m)})\}$ of m labeled examples, where the input features are $x^{(i)}$. With logistic regression, we were in the binary classification setting, so the labels were $y^{(i)} \in \{0, 1\}$. The inference is:

$$h_\theta(x) = \frac{1}{1 + \exp(-\theta^T x)} \tag{4}$$

When we deal with multi-label problems, the logistic regression must be generalized to a classifier called softmax regression. Thus, in the training set $\{(x^{(1)}, y^{(1)}), \ldots, (x^{(m)}, y^{(m)})\}$, we have $y^{(i)} \in \{1, 2, \ldots, k\}$. The inference takes the form:

$$h_\theta(x^{(i)}) = \frac{1}{\sum_{j=1}^k e^{\theta_j^T x^{(i)}}} \begin{bmatrix} e^{\theta_1^T x^{(i)}} \\ e^{\theta_2^T x^{(i)}} \\ \vdots \\ e^{\theta_k^T x^{(i)}} \end{bmatrix} \tag{5}$$

In this paper, the network is trained by applying a *softmax* function on the scores $f_c(I;(\mathbf{W}, \mathbf{b}))$, for each class of interest $c \in \{1, \ldots, k\}$. The conditional probabilities are:

$$p(c|I;(\mathbf{W}, \mathbf{b})) = \frac{e^{f_c(I;(\mathbf{W},\mathbf{b}))}}{\sum_{d \in \{1,\ldots,k\}} e^{f_d(I;(\mathbf{W},\mathbf{b}))}} \tag{6}$$

and maximizing the likelihood of the training data. More specifically, the parameters $(\mathbf{W}, \mathbf{b})$ of the network are learned in an end-to-end supervised way, by minimizing the negative log-likelihood over the training set:

$$L_f(\mathbf{W}, \mathbf{b}) = -\sum_I \ln p(l_{i,j}|I;(\mathbf{W}, \mathbf{b})) \tag{7}$$

where $l_{i,j}$ is the correct pixel label class at position (i, j) in image I. The minimization is achieved with the mentioned gradient descent optimization algorithms.

After adjusting the parameters of training set, the convolution kernels and pooling kernels are determined. Then once input a new image, the labels of each pixel will be output by applying this model. After that, we also add a denoising procedure. For each pixel, consider the 3×3 pixels around it, if the other 8 pixels have the same label, then change this pixel to that same label. This can help to improve the labeling accuracy.

4   Experiments

The Stanford Background Dataset contains 534 images of outdoor scenes composed of 9 classes. We do a pre-processing on Stanford Background Dataset in order to obtain three-label grayscale images including road, building, and others. Then we test the system on the new three-label dataset, which the training set contains 400 images and the validation set contains 134 images.

The system uses a 2-stage convolutional neural network, in which the first stage has 10 convolutional kernels and the second one has 20 convolutional kernels. At last, we use a softmax regression classifier to produce the three labels, and then give different labels different colors to make the label result visualization. We also add a denoising procedure, for each pixel, consider the 3×3 pixels around it, if the other 8 pixels have the same label, then change this pixel to that same label. Some examples of the labeling results on Stanford Background Dataset are shown in Fig. 2.

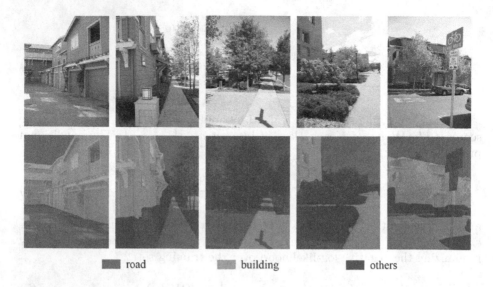

road   building   others

Fig. 2. Example of results using our 2-stage convolutional neural networks and softmax regression classifier on Stanford Background Dataset

The neural network is trained on 3.20 GHz Intel Core i5 CPU and NVIDIA Quadro 600 GPU, the cuda is used to speed up the training procedure. As a result, the speed of scene labeling for each image is less than a second. The accuracies of each label and the whole image are listed as follows.

Table 1. The accuracies of each label and the whole image

LABEL	ROAD	BUILDING	OTHERS	WHOLE
ACCURACY	93.7%	86.9%	90.8%	90.2%

Among the 3 labels, *road* gets the highest accuracy, which yields 93.7%. and the system gets 90.2% accuracy on the three-label dataset. The result shows reducing the number of labels according to actual demand is useful, it allows the system to get higher accuracy in less time. It shows that this system is suitable in self-navigating vehicles as it yields over 90% accuracy, and the road which we concerned most gets the highest accuracy.

5   Conclusions

Scene understanding problem for mobile robots has been widely studied. This pattern facilitates human cognition and recognition of their surroundings as well. These intuitive observations can be extended to similar tasks for mobile robots.

This paper propose a convolutional neural network based method to extract the internal features of the whole image, then apply into a logistic regression classifier to generate the label. The system has a good performance and the result shows reducing the number of labels according to actual demand allows the system to get higher accuracy in less time.

In future work, we prepare to do such things: (1) consider full labels using softmax classifier; (2) improve the algorithm of feature learning, try to extract better features; (3) use colour and depth information in order to increase the accuracy.

Acknowledgments. This work was supported in part by the National Natural Science Foundation of China (No. 91120308), the Key Program for Basic Research of Science and Technology Commission Foundation of Shanghai (No. 12JC1408800), and the International Technology Cooperation Project of Science and Technology Commission Foundation of Shanghai (No. 13510711100).

References

1. Farabet, C., Couprie, C., Najman, L., LeCun, Y.: Scene Parsing with Multiscale Feature Learning, Purity Trees, and Optimal Covers. arXiv preprint arXiv: 1202.2160 (2012)
2. Espinace, P., Kollar, T., Soto, A., Roy, N.: Indoor scene recognition through object detection. In: 2010 IEEE International Conference on Robotics and Automation (ICRA), pp. 1406–1413. IEEE Press (2010)
3. Socher, R., Huval, B., Bath, B., Manning, C.D., Hg, A.: Convolutional-recursive deep learning for 3D object classification. In: Proceedings of Advances in Neural Information Processing Systems (NIPS), pp. 665–673 (2012)
4. Espinace, P., Kollar, T., Roy, N., Soto, A.: Indoor Scene Recognition by a Mobile Robot through Adaptive Object Detection. Robotics and Autonomous Systems **61**(9), 932–947 (2013)
5. Kavukcuoglu, K., Sermanet, P., Boureau, Y., Gregor, K., Mathieu, M., LeCun, Y.: Learning convolutional feature hierarchies for visual recognition. In: Proceedings of Advances in Neural Information Processing Systems (NIPS), pp. 1090–1098 (2010)
6. Vasudevan, S., Gachter, S., Nguyen, V., Siegwart, R.: Cognitive Maps for Mobile Robot–An Object Based Approach. Robotics and Autonomous Systems **55**(5), 359–371 (2007)
7. Ranganathan, A.: PLISS: detecting and labeling places using online change-point detection. In: Proceedings of Robotics: Science and Systems (2010)
8. Leutenegger, S., Chli, M., Siegwart, R.: BRISK: binary robust invariant scalable keypoints. In: 2011 IEEE International Conference on Computer Vision (ICCV), pp. 2548–2555. IEEE Press (2011)
9. Lamon, P., Tapus, A., Glauser, E., Tomatis, N., Siegwart, R.: Environmental modeling with fingerprint sequences for topological global localization. In: 2003 IEEE/RSJ International Conference on Intelligent Robots and Systems (IROS), vol. 4, pp. 3781–3786. IEEE Press (2003)
10. Tapus, A., Siegwart, R.: Incremental robot mapping with fingerprints of places. In: 2005 IEEE/RSJ International Conference on Intelligent Robots and Systems (IROS), pp. 2429–2434. IEEE Press (2005)

11. Liu, M., Scaramuzza, D., Pradalier, C., Siegwart, R., Chen, Q.: Scene recognition with omnidirectional vision for topological map using lightweight adaptive descriptor. In: 2009 IEEE/RSJ International Conference on Intelligent Robots and Systems (IROS), pp. 116–121. IEEE Press (2009)
12. Liu, M., Siegwart, R.: DP-FACT: towards topological mapping and scene recognition with color for omnidirectional camera. In: Proceedings of the IEEE International Conference on Robotics and Automation (ICRA). IEEE Press (2012)
13. Lai, K., Bo, L., Fox, D.: Unsupervised feature learning for 3D scene labeling. In: 2014 IEEE International Conference on Robotics and Automation (ICRA). IEEE Press (2014)
14. Hinton, G.E., Salakhutdinov, R.R.: Reducing the Dimensionality of Data with Neural Networks. Science 313(5786), 504–507 (2006)
15. Bengio, Y., Lamblin, P., Popovici, D., Larochelle, H.: Greedy layer-wise training of deep networks. In: Proceedings of Advances in Neural Information Processing Systems (NIPS), pp. 153–160 (2007)
16. Ranzato, M., Poultney, C., Chopra, S., LeCun, Y.: Efficient learning of sparse representations with an energy-based model. In: Proceedings of Advances in Neural Information Processing Systems (NIPS), pp. 1137–1144 (2007)
17. Lee, H., Ekanadham, C., Ng, A.: Sparse deep belief net model for visual area V2. In: Proceedings of Advances in Neural Information Processing Systems (NIPS), pp. 873–880 (2008)
18. LeCun, Y., Bottou, L., Bengio, Y., Haffner, P.: Gradient-Based Learning Applied to Document Recognition. Proceedings of the IEEE 86(11), 2278–2324 (1998)
19. Salakhutdinov, R.R., Hinton, G.E.: Deep Boltzmann machines. In: Proceedings of International Conference on Artificial Intelligence and Statistics, pp. 448–455 (2009)
20. Masci, J., Giusti, A., Ciresan, D., Fricout, G., Schmidhuber, J.: A fast learning algorithm for image segmentation with max-pooling convolutional networks. In: Proceedings of the International Conference on Image Processing (ICIP), pp. 2713–2717 (2013)
21. Pinheiro, P., Collobert, R.: Recurrent convolutional neural networks for scene labeling. In: Proceedings of the 31st International Conference on Machine Learning (ICML), pp. 82–90 (2014)

Emphysema Classification
Using Convolutional Neural Networks

Xiaomin Pei(✉)

College of Information and Control Engineering, Liaoning Shihua University, Fushun, China
pxm_neu@126.com

Abstract. There has been paid more and more attention in diagnosing emphysema using High-resolution Computed Tomography. This may lead to improve both understanding and computer-aided diagnosis. We propose a novel classification framework using convolutional neural network(CNN). This model automatically extracts features from the raw image and generates classification. Experiments have been conducted on the database from clinical. Results a recognition rate of 92.54% for classification two kinds of emphysema with normal. The designed convolutional neural networks can get better results for classifying one kind of emphysema with normal.

Keywords: High-resolution computed tomography · Emphysema · Convolutional neural network

1 Introduction

Emphysema is among the top five diseases in the western world today in terms of rehabilitation and health care costs [1]. It is a main type of chronic obstructive pulmonary disease (COPD). It is a common chronic respiratory disorder characterized by the destruction of lung tissue and often reflected as areas of low attenuation in CT images. A proper classification of emphysematous and normal lung tissue is essential for further analysis of the disease. Emphysema diagnosis by radiologists is often based on visual recognition of imaging patterns augmented by anatomical knowledge. This may lead to improve both understanding and computer-aided diagnosis. High-resolution computer tomography(HRCT) is a minimally invasive imaging technique useful for patients with chronic symptoms. It has shown its potential for identifying changes in lung parenchyma and abnormalities associated with emphysema [2,3].

Computerised techniques for classifying emphysema have been explored using density values or texture [4-6].A technique employed is called densitometric analysis, by choosing a Hounsfiled unit threshold in the lung mask to discriminate emphysema from non emphysematous tissue [7,8]. Hayhurst et al [9] showed that calculating the percentage of voxels lesser than a threshold can measure the amount of emphysematous lung in HRCT images, which represent macroscopic and microscopic change due to emphysema. Texture-based classification had been carried out by authors in [10-12]. In [10] they try to extract textural features using simple descriptors. In [11],

© Springer International Publishing Switzerland 2015
H. Liu et al. (Eds.): ICIRA 2015, Part I, LNAI 9244, pp. 455–461, 2015.
DOI: 10.1007/978-3-319-22879-2_42

Riesz transform was used to obtain textural features. In [13], global and local descriptors were used to provide robust features and it achieves an improvement in the classification rate.

Feature extraction is one key factor in the success of a classification system. Since many classifiers cannot directly process raw images or data, feature extraction is a preprocessing step that aims at removing redundant information through reducing the dimension of the data. Features for classification should have the most distinguishable characteristics among different classes while retaining invariant characteristics within the same class. There are also researches on choosing features for classification [14]. A convolutional neural network is a feed-forward network that can learn high order and invariance properties from raw inputs without prior knowledge on the data. In this paper we propose an automatic feature extractor based on convolutional neural network for Emphysema Classification. Features were learned from multilayers of convolutional neural network. Classification was done by the classifier in deep convolutional network.

2   Datasets

We use datasets labeled by experienced pulmonologists: (1) The Bruijne and Sorensen Dataset (BS)[15]: Computed Tomography Emphysema Database which consists of 168 non-overlapping annotated ROIs of size 61×61 pixels from three different classes: normal tissue(NT), Centrilobular emphysema(CLE), and Paraseptal emphysema(PSE). (NT=59,CL=50,PS=59). (2) CT scans from CMU china. 25000 ROIs from 4560 CT scans are selected. Using the same parameters as CT scans in Bruijne and Sorensen Dataset: inplane resolution 0.78×0.78 mm, slice thickness 1.25mm,tube voltage 140kV and tube current 200mAs. 25000 61×61 pixel patches are also from three different classes NT,CLE and PSE.(NT=8331,CLE=8340,PSE=8329).

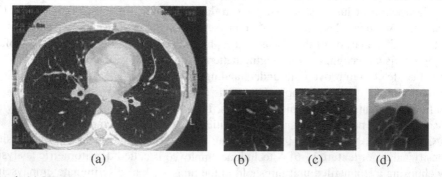

(a)   (b)   (c)   (d)

Fig. 1. (a) HRCT image for lung (b) normal tissue (c) Centrilobular emphysema (d) Paraseptal emphysema

3   Methods

Convolutional neural network is trained with the back-propagation algorithm like other neural networks. But architecture is different compared with other neural networks. Convolutional neural networks are designed to recognize visual patterns directly from pixel images with minimal preprocessing and without feature extraction[16-21]. They can recognize patterns with variability, distortions and geometric transformations. Compared with other standard feedforward neural networks with similarly-sized layers. CNNs are easier to train because they have much fewer connections and parameters.

In this paper, we design a convolutional neural network for emphysema classification. This model automatically retrieves features and recognizes unknown patterns based on CNN architecture. It contains feature map layers and retrieves discriminate features from the input raw images with two operations: convolutional filtering (1) and down sampling (2).

$$C^i_{k,l} = g(I^i_{k,l} \otimes W_{k,l} + B_{k,l})$$ (1)

$$S^i_{k,l} = g(I \downarrow^i_{k,l} w_{k,l} + Eb_{k,l})$$ (2)

In Eq.(1) $g(x) = \tanh(x)$ is a sigmoidal activation function. B respectively b are the biases, W and ω are the weights, $I^i_{k,l}$ is the i th input and $I \downarrow^i_{k,l}$ is the down-sampled i'th input of the neuron group k of layer l . E is a matrix whose elements are all one and $\otimes$ denoted a 2-dimensional convolution.

There are 8 layers. Each of the layers of the network is made up of one or several feature maps. The number of these feature maps is determined by the type of the problem and the degree of complexity of the data.

The input layer called Layer #1 is a matrix of normalized image with size 61by 61. Layer #2 is a convolutional layer with 12 feature maps, the convolutional filtering kernels on feature maps have the size of 5 by 5 pixels, reduces the feature size 61 to 57. Layer #3 is the down sampling layer have a ratio of 2. There are 12 input maps from previous layer, then there are 12 output maps after processed by layer #3, which would be one half of the input maps by a sub-sampling function and reduces the feature map size from input 57 to 29. Layer #4   is a convolutional layer with 18 feature maps, the convolutional filtering kernels on feature maps have the size of 5 by 5 pixels, reduce   feature map size from input 29 to 25. Layer #5 is   down sampling layer have a ratio of 5. Each feature map   reduce its feature size from   previous feature size 25 to 5. Layer #6 is the last convolutional layer which differ from Layer #4 as follows. Each one of its 100 feature maps is connected to a receptive field on all feature maps of Layer #5.And since the feature maps of Layer #5 were 5×5 size, the feature maps of Layer #6 were 1×1. Thus Layer #6 is equivalent to a fully connected layer. It is still labeled as a convolutional layer because if the input image is larger, the dimension of the feature maps would be larger. The fully connected layer Layer

#7 contained 60 units, connected to the 100 units of layer #6.All the units of the layers up to layer #7 had a sigmoidal activation function φ of the type.

$$y_i = \varphi(v_j) = A\tanh(Sv_j) \tag{3}$$

Where v_j is the activity level of the unit. A and S are two constant parameters for sigmoid function. Layer #8 ,the output layer is an Euclidean RBF layer of 3 units(for 3 classes) whose outputs y_j are computed by Eq. 4.

$$y_j = \sum_{i=1}^{60} (y_i - w_{ij})^2, j = 0,...,2 \tag{4}$$

Where y_i is the output of the i th unit of the layer #7.For each RBF neuron, y_j is a penalty term measuring the fitness of its inputs y_i to its parameters $w_{i,j}$.Layer #7 and Layer #8 form a trainable classifier which is a fully connected multi-Layer perceptron with a hidden layer(Layer #7) and an output layer(Layer #8).The training of the network is done using a simple backpropagation algorithm.

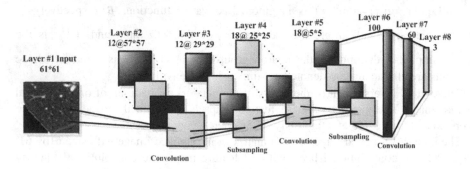

Fig. 2. Convolutional Neural network architecture

4   Experiments

Training convolutional neural network need so many datasets from clinical. We collecte HRCT images for Emphysema from CMU china.We distinguish between three types of ROIs: normal, centrilobular emphysema and paraseptal emphysema. Centrilobular emphysema was characterized by destroyed centrilobular alveolar walls, enlargement of respiratory bronchioles and associated alveoli. CT finds centrilobular areas are decreased attenuation, usually without visible walls, nonuiform distribution and predominantly located in upper lung zones. Paraseptal emphysema are characterized by a sub-pleural distribution of the lesions along the interlobular septa, localized near fissures and pleura,   frequently associated with bullae formation(area of emphysema larger than 1cm in diameter)[22-25].We select ROIs from HRCT images which have typical characteristics for centrilobular and paraseptal emphysema, we also select normal tissues from same location as lesions for training and testing convolutional neural network. The datasets are splitting to training and testing, 22000 images were used for training and 3000 images for testing.

Training data set is used to train two networks with different number of feature maps at layer #2 and layer #4. Table 1 gives the classifier performance for networks trained on our Datasets. From Table 1 we can see that with more Feature maps in Layer #2 and Layer#4 can get better results for recognition.

Table 1. Performance of convolutional neural networks for emphysema classification

Network	No.FM in Layer #2	No.FM in Layer #4	Data used for Training	Data used for Testing	Correct recognition percentage (%)
1	12	18	20000	3000	92.54%
2	6	12	20000	3000	89.78%

(FM   Feature map)

We split the datasets for different disease classification. First, we used part of the datasets for classification of normal and centrilobular emphysema,we call it datasets-NC. Datasets-NC contains 6000 normal and 6000 centrilobular emphysema ROIs. Second, we used part of the datasets for classification of normal and paraseptal emphysema, we call it datasets-NP. Datasets-NP contains 6000 normal and 6000 paraseptal emphysema ROIs. The architecture and parameter of convolutional neural networks for training these datasets are the same as network in Fig. 2, only different at the number of output. We also construct datasets-NPC which consists of 6000 normal ROIs,6000 paraseptal emphysema ROIs and centrilobular emphysema ROIs.We compare recognition results with CNNs for classification between normal and one kind of disease, normal and two kinds of disease.Table 2 gives the classier performance for networks trained on three Datasets. Compared the results for different datasets, we can see that distinguish between normal tissues and one kind of disease tissues can get higher accuracy. That's because centrilobular,paraseptal emphysema with little difference between each other.

Table 2. Performance of convolutional neural networks for different datasets

Datasets	Data used for training	Correct recognition percentage
Datasets-NC	12000	90.82%
Datasets-NP	12000	91.28%
Datasets-NPC	18000	90.70%

5   Conclusions

We propose a novel approach to classify emphysema patterns based on Convolutional Neural networks. The performance of the system was evaluated on datasets from clinical. It was achieved by employing convolutional neural networks without feature extraction. Results showed that our proposed model can obtain higher classification rates. Emphysema classification using convolutional neural networks is a try for the convolution neural networks applied in the field of computer aided diagnosis for emphysema. In the future, we will try to explore hybrid model of integrating superior classifier for recognizing more different types of emphysema.

References

1. Kinsella, M., Muller, N.L., Aboud, R.T., et al.: Quantification of emphysema by computed tomography using a "density mask" program and correlation with pulmonary function tests. Chest **97**, 315–321 (1990)
2. Friman, O., Borga, M., Lundberg, M., Tylen, U., Knutsson, H.: Recognizing emphysema-a neural network approach. In: Proceedings of Sixteenth International Conference on Pattern Recognition, Quebec, Canda, pp. 1–4 (2002)
3. Prasad, M., Sowmya, A., Wilson, P.: Multi-level classification of emphysema in HRCT lung images. Pattern Anal. Applic. **12**, 9–20 (2009)
4. Uppaluri, R., Mitsa, T., Sonka, M., Hoffman, E.A., McLennan, G.: Quantification of pulmonary emphysema from lung computed tomography images. American Journal of Respiratory and Critical Care Medicine **156**(1), 248–254 (1997). [pubMed:923756]
5. Depeursinge, A., Sage, D., Hidki, A., Platon, A., Poletti, PA., Unser, M., Mller, H.: Lung tissue classification using wavelet frames. In: 29th Annual International Conference of the IEEE Engineering in Medicine and Biology Society, EMBS 2007, pp. 6259–6262 (2007)
6. Park, Y.S., Seo, J.B., Kim, N., Chae, E.J., Oh, Y.M., Lee, S.D., Lee, Y., Kang, S.H.: Texture-based quantification of pulmonary emphysema on high-resolution computed tomography: Comparison with density-based quantification and correlation with pulmonary function test. Investigative Radiology **43**(6), 395–402 (2008). [PubMed:18496044]
7. van Ginneken, B., Hogeweg, L., Prokop, M.: Computer-aided diagnosis in chest radiography: Beyond nodules. European Journal of Radiology **72**(2), 226–230. [PubMed:19604661]
8. Cavigli, E., Camiciottoli, G., Diciotti, S., Orlandi, I., Spinelli, C., Meoni, E., Grassi, L., Farfalla, C., et al.: Whole-lung densitometry versus visual assessment of emphysema. European Radiology **19**(7), 1686–1692 (2009). [PubMed:19224221]
9. Hayurst, M., Flenley, D., Mclean, A., Wightman, A., Macnee, W., Wright, D., Lamb, D., Best, J.: Diagnosis of pulmonary emphysema by computerized tomography. The Lancet **324**, 320–322 (1984)
10. Sørensen, L., Nielsen, M., Lo, P., Ashraf, H., Pedersen, J., de Bruijne, M.: Texture-based analysis of COPD: A data-driven approach. IEEE Trans. Med. Imag. **31**(1), 70–78 (2012)
11. Depeursinge, A., Foncubierta–Rodriguez, A., Van de Ville, D., Müller, H.: Multiscale lung texture signature learning using the Riesz transform. In: Ayache, N., Delingette, H., Golland, P., Mori, K. (eds.) MICCAI 2012, Part III. LNCS, vol. 7512, pp. 517–524. Springer, Heidelberg (2012)

12. Nava, R., Marcos, J., Escalante-Ramírez, B., Cristóbal, G., Perrinet, L.U., Estépar, R.S.J.: Advances in texture analysis for emphysema classification. In: Ruiz-Shulcloper, J., Sanniti di Baja, G. (eds.) CIARP 2013, Part II. LNCS, vol. 8259, pp. 214–221. Springer, Heidelberg (2013)
13. Prasad, M., Sowmya, A., Wilson, P.: Multi-level classification of emphysema in HRCT lung images. Pattern Analysis and Applications 12(1), 9–20 (2009)
14. Mendoza, C.S., Washko, G.R.,Ross, J.C., Diaz, A.A., et al.: Emphysema quantification in a multi-scanner HRCT cohort using local intensity distributions. In: Proc IEEE Int. Symp. Biomed. Imaging, pp. 474–477 (2012)
15. Sørensen, L., Shaker, S., de Bruijne, M.: Quantitative analysis of pulmonary emphysema using Local Binary Patterns. IEEE Trans. Med. Imag. 29(2), 559–569 (2010)
16. Niu, X.-X., Suen, C.Y.: A novel hybrid CNN-SVM classifier for recognizing handwritten digits. Pattern Recognition 45, 1318–1325 (2012)
17. Bishop, C.M.: Neural Networks for Pattern Recognition. Oxford University Press, New York (1995)
18. Serre, T., Oliva, A., Poggio, T.: A Feedforward Architecture Accounts for Rapid Categorization. Proc. Natl. Acad. Sci. USA 104(15), 6424–6429 (2007)
19. Huang, F.J., LeCun, Y.: Large-scale learning with SVM and convolutional for generic object categorization. In: Proc. 2006 IEEE Compuer Society Conference on Computer Vision and Pattern Recognition, vol. 1, pp. 284–291 (2006)
20. Ji, S., Xu, W., Yang, M., Yu, K.: 3D convolutional neural networks for human action recognition. IEEE Trans. Pattern Anal. Mach. Intell. 35(1), 221–231
21. Jain, V., Murray, J.F., Roth, F., Turaga, S., Zhigulin, V., et al.: Supervised learning of image restoration with convolutional networks. In: IEEE 11th International Conference on Computer Vision, ICCV 2007, pp. 1–8. IEEE (2007)
22. Galban, C., Han, M., Boes, J., Chughtai, K., et al.: Computed tomography-based biomarker provides unique signature for diagnosis of COPD phenotypes and disease progression. Nat. Med. 18(11), 1711–1715 (2012)
23. Nava, R., Escalante-Ramírez, B., Cristóbal, G.: Texture image retrieval based on log-Gabor features. In: Alvarez, L., Mejail, M., Gomez, L., Jacobo, J. (eds.) CIARP 2012. LNCS, vol. 7441, pp. 414–421. Springer, Heidelberg (2012)
24. Prasad, M., Sowmya, A., Wilson, P.: Multi-level classification of emphysema in HRCT lung images. Pattern Anal. Applic. 12, 9–20 (2009)
25. Prasad, M., Sowmya, A.: Multi-level emphysema diagnosis in HRCT lung images through robust multi-view and meta-learning. In: Asia Conference on computer vision, Jeju, S. Korea, pp. 937–942 (2004)

Dynamic 3D Reconstruction of Tongue Surface Based on Photometric Stereo Technique

Yiheng Cai[✉], Linlin Zhang, Nan Sheng, Lina Wang, and Xinfeng Zhang

Signal and Information Processing Lab, College of Electronic Information
and Control Engineering, Beijing University of Technology, NO.100,
Pingleyuan, Chaoyang District, Beijing 100124, China
caiyiheng@bjut.edu.cn

Abstract. Tongue inspection is an important part of the diagnostic methods in traditional Chinese medicine. In the process of the objective research for tongue diagnosis, the tongue information we obtained from the analysis of two-dimensional (2D) tongue image is limited. The aim of this study is to reconstruct dynamic three-dimensional (3D) surface of tongue based on photometric stereo technique. We utilized a highlight excluding algorithm in order to reduce the effects of the highlight areas on the reconstruction results. The overall outline and detailed 3D surface of the tongue could be reconstructed in this way with an average relative error of 7.24%. The motion simulation of tongue was achieved via 3D morphing for better diagnosis. Our experience with this work demonstrated that the reconstruction method for the tongue was accurate and reliable for both clinic and research purposes. Further works still need to be done to improve the accuracy of the reconstruction result.

Keywords: Photometric stereo · Three-dimensional reconstruction · Highlight · Morphing · Tongue characterization

1   Introduction

Tongue inspection is an important part of the diagnostic methods in traditional Chinese medicine [1]. Traditional tongue inspection methods mainly rely on visual observation of the doctor. These methods always lack objective evaluation for judgment and analysis, which greatly restricts the inheritance and development of tongue inspection.

In recent years, with the development of image processing and pattern recognition techniques, objective and modern research of tongue inspection has received much attention. However, the feature extraction and analysis of tongue with existing objective research methods are based on 2D tongue image, which always cannot visually express detailed and dynamic information of tongue [2]. The 3D shape and dynamic characteristics of tongue are crucial indicator of physiological functions and pathological changes of human body. Therefore, the 3D dynamic reconstruction of tongue surface has important significance for the objective study of tongue inspection.

Currently, several methods are available for the recovery of 3D surface, such as pulsed laser based system which can measure range directly [3]. RGB-D sensor is

© Springer International Publishing Switzerland 2015
H. Liu et al. (Eds.): ICIRA 2015, Part I, LNAI 9244, pp. 462–472, 2015.
DOI: 10.1007/978-3-319-22879-2_43

used for 3D reconstruction based on structured-light measurement theory [4]. These methods above need special equipment and lack of convenience. So a method for 3D reconstruction from images of the illuminated object appears [5]. These researches focus on movement recovery act, stereo vision [6] and photometric stereo method. Photometric stereo method has wide application due to its easy implementation and better performance in the accuracy of recovery. Multiple images captured from the same view point under different illumination directions can be used by photometric stereo method to derive orientation and reflectance values for each pixel [3]. It is this method which is the subject of this paper.

We have done some researches on static tongue model to confirm the feasibility of photometric stereo method for 3D reconstruction of tongue surface [7]. Photometric stereo method is implemented in certain circumstances with a fixed relative position between the camera and the object. Images are acquired only by changing the direction of light. These conditions above can be satisfied in the case of the reconstruction of static tongue model. However, the real tongue cannot keep static. The tongue on the images captured under different lights cannot be guaranteed the same position. This will have a great impact on the reconstructive result. To circumvent these problems, high-speed camera is in need to obtain image sequences. In this paper, a 3D dynamic information collection system based on photometric stereo method is designed for 3D information acquisition. Then the surface normal and the depth information of tongue are obtained. We also utilize a highlight excluding algorithm in order to reduce the effects of the highlight areas on the reconstruction results. Finally, motion simulation of tongue is achieved via 3D morphing for intuitive diagnosis.

2   3D Reconstruction of Tongue Surface Based on Photometric Stereo Technique

2.1   3D Dynamic Information Collection System Based on Photometric Stereo Method

In order to apply photometric stereo method to the 3D reconstruction of the real tongue, the design of a 3D dynamic information collection system was in need firstly. The goal of this system was to achieve the synchronism by a timing control circuit.

Setup of the System
A spherical body with two square holes was fixed on a lifting platform. The hole on the left was for camera and the right for tongue. Two closed hoods were in need to prevent the ambient light incident. Since photometric stereo method has high demands of lights, the spherical body and two hoods provide a closed environment. The distance between the lens and the tongue was about 12 inches (Fig.1).

All images were captured by a high-speed camera. The frame rate, exposure time and trigger mode of the camera could be set manually. External trigger mode was used here with a trigger signal which produced by the timing control circuit. According to the dynamic features of tongue, we operate the camera at 40fps [8].

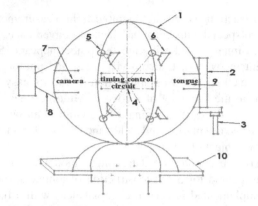

Fig. 1. 3D dynamic information collection system.

1.spherical body 2.hood for face 3.jaw bracket 4.timing control circuit 5.light holder 6.light source 7. hole for camera 8. hood for camera 9. hole for tongue 10.lifting platform

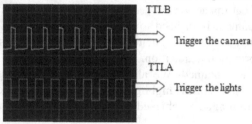

Fig. 2. The temporal relation between the two sets of signals

TTLA and TTLB (Fig.2) were two sets of trigger signals produced by the timing control circuit. The falling edges on TTLA were used to trigger the light sources to be lit up circularly, while the camera shutter was triggered by the falling edges of TTLB accordingly. There was a phase delay between them.

White LEDs were chosen as the light sources of the system. They were placed on a vertical ring in the spherical body. The inner surface of the spherical body was painted black to avoid the effect of reflectance.

Distribution of Light Sources
At least three images under different illumination directions are needed to recover the surface normal in photometric stereo method [9]. We have done some researches about the numbers and distribution of the light sources for the better reconstruction result.

Eight sources were evenly distributed on a vertical circle in our previous reconstruction of a static tongue model [7]. When the real tongue was imaged under this system, there were many highlight areas in the pictures obtained. Highlights, or specular reflections appear on a surface when the angle between the viewer and source of illumination is bisected by the surface normal [5]. Although attempts have been made to quantify the specular component in reflectance functions and reflectance maps, these functions are contrived for certain cases and fail to solve the general problem [10,11]. Since the tongue surface can be well approximated by a Lambertian component plus a specular component. Photometric stereo method can be used if three non-highlight intensity values are available for each point on the surface. To reduce the complexity, a minimal number of four light sources are chosen with the assurance of the accuracy.

If the highlight areas were not overlapped with each other in four images, three intensity values were available for each point. The position and orientation of the four sources were adjusted to insure the highlights to be non-overlapped. We can detect highlights by an appropriate threshold of Y component in YCbCr color model. The highlight area in each image (Fig. 3a) was marked with different colors and displayed in a uniform coordinate system (Fig. 3b). Through multiple sets of test results, we could draw a conclusion that the 4-sources distribution was a feasible method for highlight excluding.

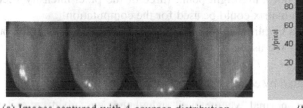 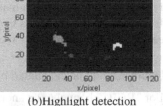

(a) Images captured with 4-sources distribution.   (b)Highlight detection

Fig. 3. Image sequences and highlight detection results.

2.2   Calculating of Normal Vector and Reflectance

Assuming a Lambertian surface illuminated by a light source of parallel rays, the observed image intensity I at each pixel is given by the product of the reflectance ρ and the cosine of the incidence angle θ (the angle between the direction of the incident light and the surface normal) [12]. The above incidence angle can be expressed as the dot product of two unit vectors, the light direction L and the surface normal n. For color images, the intensity of each pixel is composed of R, G and B component. Here we use R channel as an example to solve the problem.

Let us consider a Lambertian surface patch illuminated by three illumination sources with directions L^1, L^2, and L^3. In this case, we can express the intensities of the obtained pixels as:

$$I^k_{\ R} = \rho_R L^k \cdot n_R, k = 1, 2, 3 \tag{1}$$

n_R is the unit normal vector of the pixel in the surface. ρ_R is the reflectance of current pixel. And k represents the kth image among the three images. We stack the pixel intensities to obtain the pixel intensity vector $I_R = (I^1{}_R, I^2{}_R, I^3{}_R)^T$. We also stack the illumination vectors row-wise to form the illumination matrix $L = (L^1, L^2, L^3)^T$, Then, equation (1) could now be rewritten in a matrix form [12]:

$$I_R = \rho_R L \cdot n_R \qquad (2)$$

If the three illumination vectors L^1, L^2, and L^3 do not lie in the same plane, then matrix L is nonsingular and can be inverted. We multiplied by L^{-1} at both sides of equation (2), then,

$$L^{-1} I_R = \rho_R \cdot n_R \qquad (3)$$

Since n_R is the unit vector, the surface normal is the direction of the left vector of equation (3), the reflectance is the length of the same vector.

For the image sequences we obtained, highlight detection was done firstly per pixel.

If the pixel detected was not a highlight point, three of the pixel intensity vectors and the relevant illumination matrix could be used for the computation.

If the pixel detected was a highlight point, we used the intensity values of the other three images for calculation. We assumed that there was a highlight point in the first image, meanwhile, $I_R = (I^2{}_R, I^3{}_R, I^4{}_R)^T$, $L = (L^2, L^3, L^4)^T$. The normal vector and the reflectance could be calculated by equation (3).

For RGB channel, the normal vector was calculated per pixel, $n_R = (n_{Rx},$ $n_{Ry}, n_{Rz})^T$, $n_G = (n_{Gx}, n_{Gy}, n_{Gz})^T$, $n_B = (n_{Bx}, n_{By}, n_{Bz})^T$. The calculated normal vector could be used for generating the normal map. We could judge the correctness of the obtained normals according to the normal map.

2.3   Depth Generation

The next step in generating the actual tongue surface was the conversion from surface normals to depth information. There are some methods for depth generation such as variational method and pyramid method. Algebraic method is used according to the characteristics of tongue. Depth information was a data-orienting expression of 3D surface. Surface normals were strictly vertical with the two axes in the direction of the tangent plane [13].

For a point $(x, y, z_{x,y})$ on the surface, we can formulate two constraint equations. Assuming surface tangential at the horizontal direction could be expressed by a vector constituted with the point and its adjacent point $(x+1, y, z_{x+1,y})$ as follows:

$$V_1 = (x+1, y, z_{x+1,y}) - (x, y, z_{x,y}) = (1, 0, z_{x+1,y} - z_{x,y}) \qquad (4)$$

Similarly, surface tangential at the vertical direction could be indicated as following:

$$V_2 = (x, y+1, z_{x,y+1}) - (x, y, z_{x,y}) = (0, 1, z_{x,y+1} - z_{x,y})$$ (5)

Surface normals were perpendicular to surface tangential. Equation (6) could be obtained:

$$n \cdot V_1 = 0, n \cdot V_2 = 0$$ (6)

For the tongue image sequences obtained, there were two constraint equations for each pixel. When the size of an image was $m \times n$, the number of creating constraint equation was $2 \times m \times n$. Since the normal vector n has been obtained already, the depth value z could be solved by a series of matrix equations. According to the depth values, we could get the depth meshes. With the techniques of rendering and texture mapping in DirectX, the reconstruction result could be obtained.

2.4   Assessment of the Reconstruction Result

Analysis of the Reconstruction Result
We have chosen ten subjects with different tongue features for the experiment in order to demonstrate the feasibility of this system. Before image acquisition, every subject should stick out the tongue repeatedly for training, letting his tongue stick out of his mouth with the tip downward. Relaxation was needed in order to the surface of tongue could keep stable [14]. Two sets of images and reconstruction results of subject A and B with typical features were shown here.

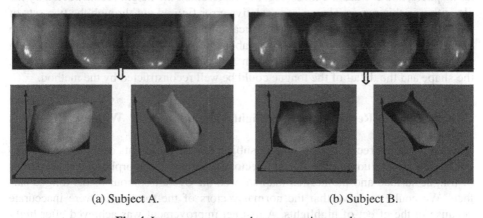

(a) Subject A.   (b) Subject B.

Fig. 4. Image sequences and reconstruction results.

In Fig. 4a, the surface of tongue was smooth, while in Fig. 4b, there were some surface dents could be seen. The 3D reconstruction results indicated that the overall outline and detailed information such as dents on the surface could be reconstructed well in this way. For the areas with obvious dens on the surface of subject B, the depth values of three rows on horizontal section (i=20,23,25) could be displayed in a 3D coordinates (Fig. 5a). It showed that with the increase of column j, there were obvious downward trends in depth value in the sunken area. The dents on the surface could be reconstructed accurately. The experimental data provided an important reference for the doctors to identify the tooth marks and crack of tongue, which represented some diseases of human body. The reconstruction results also had significant value in clinic and research areas.

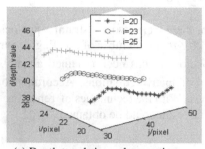 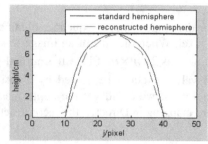

(a) Depth trends in sunken sections   (b) Comparison between the real height
         of subject B.   and the reconstructed height of the hemisphere

Fig. 5. Result analyses.

A standard Lambertian hemisphere with a radius of 8cm was reconstructed to evaluate the algorithm's performance. The image acquired was 50x50 pixels in dimensions. We performed the error analysis by the information of one row (i =25) on the horizontal section. The solid lines (Fig. 5b) represented the real height of the standard hemisphere, and the dashed lines (Fig. 5b) represented the height reconstructed by the algorithm in this paper. The average relative error figured out through the two sets of data was 7.24%. For the tongue, the vertical dimension between the back and the tip was about 5cm. So the average error was about 3.6mm. The size of the dents and teeth marks on the surface of tongue as known was between 5mm and 10mm. Therefore, the shape and the detail of the tongue could be well reconstructed by the method.

Comparison of Results with Highlights Excluding and Without Highlights Excluding

We compared the reconstruct results of subject A with highlights excluding and without highlights excluding. The normal vectors contained the morphological properties of tongue surface and the normal map showed the overall contour of the tongue surface. We could perceive that the normal vectors of the tongue tip were inaccurate because of the effect of highlights. A distinct improvement was achieved after highlights excluding (Fig. 6a).

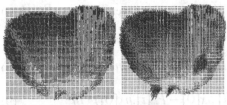

(a) Comparison of normal map.   (b) Comparison of depth map.

Fig. 6. Comparison of results without highlights excluding and with highlights excluding.

For the obvious highlight areas on the tip of the tongue, the reconstructive surface was rough as the dotted circle shown. There was an obvious improvement with smooth surface after highlights excluding (Fig. 6b).

3   Motion Simulation of Tongue

3.1   Motion Simulation Based on Morphing

The dynamic characteristics of tongue are crucial indicator for the doctors to diagnose. If the tongue sticking out shivers unconsciously, it is called shivering tongue in traditional Chinese medicine. Shivering tongue always represents the liver and kidney or hyperthyroidism. We focus on the identification and analysis of shivering tongue to illustrate the significance of the method in this paper.

Since we have got the meshes of a tongue in different states, how can we simulate the motion of the tongue in three-dimensional space?

3D morphing is a technique widely used in computer graphics to simulate the transformation between two different objects or to create new shapes by a combination of other existing shapes [15]. The shapes are meshes for us. Two kinds of meshes may be used in 3D morphing, surface (triangle) meshes [16,17] and volumetric (tetrahedron) meshes [18]. The process of morphing a mesh involves slowly changing the coordinates of the mesh vertices, starting at the so-called source mesh to the target mesh. The morphing operation merely moves vertices from the source mesh positions to match the target mesh positions. So the rules must be followed. First, each mesh must share the same number of vertices. Second, each vertex in the source mesh must have a matching vertex in the target mesh [19].

So we established a relationship of each vertex by sharing the same index number between the two meshes firstly. The next step was to choose a suitable path for each vertex from the source position to the target position. Since the meshes had the same topology, linear interpolation was used as the interpolation method.

Every four pictures were used to reconstruct a mesh. Two tongue meshes in adjacent states were chosen as the source mesh named S={Si}and the target mesh named T={Ti}, i=1,2…n. n was the numbers of vertices. The relationship between S and T was established as follows,

$$T_i = \phi(S_i)$$ (7)

470   Y. Cai et al.

Then linear interpolation was used for each position of vertex from $P_S(i)$ to $P_T(i)$:

$$P[i].x = P_S[i].x + (P_T[i].x - P_S[i].x) * m \tag{8}$$

$$P[i].nx[0] = P_S[i].nx[0] + (P_T[i].nx[0] - P_S[i].nx[0]) * m \tag{9}$$

m was a value between 0 and 1, which was used to control the deformation [20]. We had to interpolate the normal of vertex similarly to insure the smooth transition by lighting.

Since the motion simulation between two arbitrary meshes had been achieved, next we need to do was to realize the continuous range between multiple meshes by defining the MorphKeySet template. There were two variables in the MorphKeySet template, the first variable, NumKeys, was the number of meshes we used in the morphing. The second variable was the Keys array, which holded the name of the mesh and the time it was placed. The time of every mesh placed could be calculated by the frame rate of high-speed camera.

The MorphKeySet template was organized as follows:

```
MorphKeySet  MySet
{  10;
    0;      "MyMesh1";,
    100;    "MyMesh2";,
    200;    "MyMesh3";,        ...
    1000;   "MyMesh10";; }
```

Then the motion simulation could be achieved by linear interpolation between any two adjacent meshes.

3.2   Motion Simulation Result and Evaluation

The motion simulation result was shown in Fig. 7, which could reflect the shivering condition of the tongue. Three states in motion were chosen to display here.

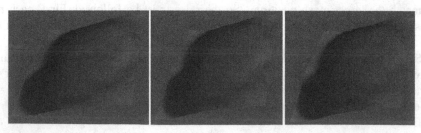

Fig. 7. Motion simulation result

The result provides an intuitive 3D dynamic tongue for the doctors. In order to evaluate the effect of the motion simulation, a large number of subjects were chosen for this experiment. Some feature points were selected to estimate the shivering frequency of the tongue by the time the feature points reached the same location. The experiments show that there are the same diagnostic results basically between observing the motion simulation result and the real tongue. The dynamic tongue reconstructed has important accessory diagnostic value for the doctors to observe the tongue objectively from multi-angles and multi-levels. It also benefits the medical records keeping, even long-distance diagnose.

4   Conclusion and Future Work

In this paper, a method based on photometric stereo technique was presented to reconstruct 3D surface of tongue. A 3D dynamic information collection system was designed firstly for the image acquisition of the real tongue. The tongue surface was reconstructed by calculating surface normal, reflectance and depth information of tongue. The experimental results show that the overall outline and detailed 3D surface of the tongue could be reconstructed in this way with an average relative error of 7.24% and also provide an important reference for the doctors to identify the tooth marks and crack of tongue. Finally, motion simulation was used to generate a realistic 3D dynamic tongue. Our results show that the dynamic features of the tongue can be simulated such as the shivering, which has important implications for doctors to observe the tongue of patient objectively, even long-distance diagnose.

In the reconstruction of tongue surface, though we have obtained the overall shape of tongue, the contour edge of tongue is not ideal. More efforts are still in need to improve the accuracy of the reconstructed result.

Acknowledgements. Thanks for the support of the National Natural Science Foundation of China (No. 61201360).

References

1. Xuejiao, C., Yuchen, W., Decai, W.: The overview of the development of traditional Chinese medicine tongue diagnosis. Jiangxi Journal of Traditional Chinese Medicine **43**(1), 72–75 (2012)
2. Changming, Y., Jianguo, J., Shu, Z.: Face recognition in complex domain based on 3D facial imaging system. Journal of Electronic Measurement and Instrument **25**(5), 420–426 (2011)
3. Lewis, R.A., Johnston, A.R.: A scanning laser rangefinder for a robotic vehicle. In: Proc 5th IJCAI, p. 762 (1977)
4. Engelhard, N., Endres, F., Hess J., et al.: Real-time 3D visual SLAM with a hand-held RGB-D camera. In: RSS Workshop on RGB-D Cameras (2010)
5. Coleman Jr., E., Jain, R.: Obtaining 3-dimensional shape of textured and specular surfaces using four-source photometry. Computer Graphics and Image Processing **18**(4), 309–328 (1982)
6. Haoran, L., Wenming, Z., Bin, L.: Three Dimensional Measurement Bsaed on the Binocular Vision. Acta Photonicasinica **38**(7), 1830–1834 (2009)

7. Lv, H., Cai, Y., Guo, S.: 3D reconstruction of tongue surface based on photometric stereo. In: 2012 IEEE 11th International Conference on Signal Processing (ICSP), vol. 3, pp. 1668–1671. IEEE (2012)

8. Park, J.I., Lee, M.H., Grossberg, M.D., et al.: Multispectral imaging using multiplexed illumination. In: IEEE 11th International Conference on Computer Vision, ICCV 2007, pp. 1–8. IEEE (2007)

9. Argyriou, V., Petrou, M., Barsky, S.: Photometric stereo with an arbitrary number of illuminants. Computer Vision and Image Understanding 114(8), 887–900 (2010)

10. Ikeuchi, K., Horn, B.K.P.: An application of the photometric stereo method (1979)

11. Silver, W.M.: Determining shape and reflectance using multiple images. Massachusetts Institute of Technology (1980)

12. Barsky, S., Petrou, M.: The 4-source photometric stereo technique for three-dimensional surfaces in the presence of highlights and shadows. IEEE Transactions on Pattern Analysis and Machine Intelligence 25(10), 1239–1252 (2003)

13. Yiling, Z., Hai, L.: Design and implementation of Image-based Modeling system based on Photometric Stereo. Zhejiang University, Zhejiang (2006)

14. Jiatuo, X., Zhifeng, Z., Ren, H.F.: Analysis of fat/thin tongue based on image processing method. In: The Second National Integrated Traditional Chinese and Western Medicine Diagnosis of Academic Symposium (2008)

15. Mocanu, B., Zaharia, T.: A complete framework for 3D mesh morphing. In: Proceedings of the 11th ACM SIGGRAPH International Conference on Virtual-Reality Continuum and its Applications in Industry, pp. 161–170. ACM (2012)

16. Kanai, T., Suzuki, H., Kimura, F.: Three-dimensional geometric metamorphosis based on harmonic maps. The Visual Computer 14(4), 166–176 (1998)

17. Lee, A.W.F., Dobkin, D., Sweldens, W., et al.: Multiresolution mesh morphing. In: Proceedings of the 26th Annual Conference on Computer Graphics and Interactive Techniques, pp. 343–350. ACM Press/Addison-Wesley Publishing Co. (1999)

18. Alexa, M., Cohen-Or, D., Levin, D.: As-rigid-as-possible shape interpolation. In: Proceedings of the 27th Annual Conference on Computer Graphics and Interactive Techniques, pp. 157–164. ACM Press/Addison-Wesley Publishing Co. (2000)

19. Yan, H.B., Hu, S.M., Martin, R.R.: 3D morphing using strain field interpolation. Journal of Computer Science and Technology 22(1), 147–155 (2007)

20. Dong, W., Su, H.: Research on the 3D Morphing Animation with Non-distorted Texture Mapping and Its Implementation. Computer Engineering 16, 66 (2004)

A Systematic Approach for the Parameterisation of the Kernel-Based Hough Transform Using a Human-Generated Ground Truth

Jonas Lang$^{(\boxtimes)}$, Mark Becke, and Thomas Schlegl

Regensburg Robotics Research Unit, Faculty of Mechanical Engineering,
Ostbayerische Technische Hochschule Regensburg, Regensburg, Germany
{jonas2.lang,mark.becke,thomas.schlegl}@oth-regensburg.de

Abstract. Lines are one of the basic features that are used to characterise the content of an image and to detect objects. Unlike edges or segmented blobs, lines are not only an accumulation of certain feature pixels but can also be described in an easy and exact mathematical way. Besides a lot of different detection methods, the Hough transform has gained much attention in recent years. With increasing processing power and continuous development, computer vision algorithms get more powerful with respect to speed, robustness and accuracy. But there still arise problems when searching for the best parameters for an algorithm or when characterising and evaluating the results of feature detection tasks. It is often difficult to estimate the accuracy of an algorithm and the influences of the parameter selection. Highly interdependent parameters and preprocessing steps continually lead to only hardly comprehensible results. Therefore, instead of pure trial and error and subjective ratings, a systematic assessment with a hard, numerical evaluation criterion is suggested. The paper at hand deals with the latter ones by using a human-generated ground truth to approach the problem. Thereby, the accuracy of the surveyed Kernel-based Hough transform algorithm was improved by a factor of three. These results are used for the tracking of cylindrical markers and to reconstruct their spatial arrangement for a biomedical research application.

Keywords: Feature detection · Human-generated ground truth · Hough transform · Image processing · Line detection · Systematic parameterisation

1 Introduction

There is still a challenge in automatically detecting shapes for object recognition. It remains a computationally expensive and non-trivial task, even for simple geometric patterns like lines, circles or ellipses. Where humans use additional heuristics, experiences from the past and a deeper image understanding, computers are restricted to more basic algorithms. Noise, missing or extraneous and imperfect data from the initial camera image, preprocessing steps or edge

© Springer International Publishing Switzerland 2015
H. Liu et al. (Eds.): ICIRA 2015, Part I, LNAI 9244, pp. 473–486, 2015.
DOI: 10.1007/978-3-319-22879-2_44

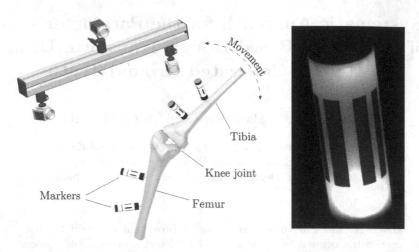

Fig. 1. Application of the marker tracking system and one of the corresponding marker images

detection further complicates the problem. Nevertheless, a large number of simple as well as sophisticated algorithms have been developed. One of the most famous line detection methods during the last decades is the Hough transform (HT). This algorithm transforms a curve searching problem into a much simpler peak searching problem. The effort for further development mainly focuses on computation time, robustness or accuracy. But these advancements often lack the numerical evaluation of results and the investigation of parameter influences. Some in-house or database images and the corresponding, extracted features are shown and sometimes only optically assessed.

To solve computer vision tasks, a whole chain of algorithms and processing steps is commonly used. Individual elements cannot be evaluated separately and without considering their interdependence. However, it is often difficult to examine this processing chain with all its parameters and influences in its entirety. The same is true for a hard, numerical key figure or classification coefficient to rate the end results of the image processing task. To solve this problem, a known ground-truth is required as a basis of comparison for the subsequent results. There are various possibilities, like simulation, measurement of the real scene or human-generated comparison data. All of them have their own strengths and weaknesses and it is highly dependent on the case of application which one to choose.

In the present paper, line and corner detection algorithms are used to reconstruct the spatial arrangement of cylindrical markers. These markers are tracked by a trinocular camera system which is pointed towards the examination object, in our case the bones of an experimental setup for biomechanical testing of knee joints moved manually or automatically, e.g. by a six axis robot, see Fig. 1. Those experiments try to imitate the human gait cycle by moving along a certain path or trajectory with the insertion of a specific force and torque [1]. By these means, long term stress tests can be simulated. In addition to the robot's internal position sensors, a computer vision system monitors the movements of the test object and thereby provides a correction factor for the robot's control loop.

Concerning line detection, a very recent survey analyses more than 200 papers dealing with the Hough transform over the last decades [2]. There is a myriad of derivatives and further developments with the Standard Hough Transform (SHT) as a starting point, like the Fast Hough Transform (FHT) [3], the Randomized Hough Transform (RHT) [4], the Probabilistic Hough Transform (PHT) [5] or the Kernel-based Hough Transform (KHT) [6], just to name a few. The latter will gain particular attention in this paper as it is the investigation subject in Sect. 3. Apart from the Hough transform to detect straight lines, there are some additional algorithms analysing eigenvalues or linking pixel clusters [7–9]. A test framework to assess the accuracy of the line detection process by the Hough transform is proposed in [10]. A local Hough transform to detect line candidates and further determination of their parameters by a global estimation are combined in [11].

The Hough transform generally only detects infinite lines. Nevertheless, there are several algorithms which overcome this restriction and provide information about the length and position or the start and end point of a line, respectively. This can either be the analysis of the neigbourhood of peaks in the parameter space or the application of additional independent processing steps in the image space. Each feature pixel is mapped into the parameter space as a sinusoid curve. A line as a succession of collinear pixels leads to a split up sinusoidal construct, also called butterfly. Depending on the location and length of the line, this butterfly takes a distinctly shaped appearance, see Fig. 2). Using these characteristics and suitable algorithms to interpret it, a finite line can be described completely [12–14].

Many computer vision applications for feature detection lack the existence of absolute, numerically exact comparison data or the real solution or dataset (i.e. true values) respectively. Especially edge detection or segmentation algorithms are often only assessed by a mere subjective verdict. Whereas a human-generated ground truth provides the possibility to get hard, mathematical classification numbers when evaluating the performance of an algorithm with its specific parameter set. Perhaps the most famous example for the use of human-generated ground truth is the *Segmentation Database of Berkley's Computer Vision Group* [15], which is frequently applied to evaluate and compare different segmentation algorithms [16,17]. An extension towards 3D segmentation tasks is realised in [18] and [19], while [20] compares 3D interest point detection algorithms with human selections. Edge and boundary detections were assessed in comparison with human-generated data in [21] whereas [22] and [23] try to introduce a more formal, mathematical comparison model.

The paper at hand is using human-generated comparison data to evaluate the influence of preprocessing steps and parameter choice on line detection through the Kernel-based Hough Transform (KHT) introduced by Fernandes [6]. In contrast to other works, the whole process is considered. The underlying

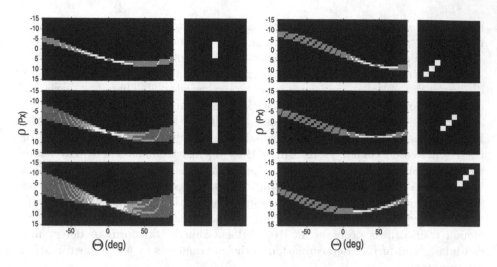

Fig. 2. The corresponding Hough space for different line lengths and positions

image processing sequence is described in Subsect. 2.1, a methodology for a paramaeter influence study using human-generated ground truth is proposed in Subsect. 2.2, mathematical quality factors are introduced in Subsect. 2.3 and exemplary results are given in Sect. 3.

2   Methodology

Because of the highly interdependent parameters of a image processing task, it is almost impossible to find a good or even the best parameter set by just guessing or trying out different combinations non-systematically. Apart from this, as described in Sect. 1, it is problematic to obtain numerical comparison data to assess the results of an algorithm. Both problems are addressed in the subsequent chapters. To solve the first, a systematic parameter study is conducted, to tackle the second, a mathematically describable ground-truth is developed.

2.1   Image Processing Sequence

A flowchart of the image processing chain used for this work and parts of its corresponding results are shown in Fig. 3. Images are acquired from monochrome industrial area cameras through Camera Link connection and PCI-E framegrabbers. In these initial images, the markers only occupy a small fraction of the total image area. Therefore, a region of interest (ROI) creation by using binary thresholding and a blob analysis is necessary to limit the subsequent computational effort. Furthermore, the ROIs are undistorted by the use of calibration data from a previous camera calibration process according to [24]. Afterwards, the contrast is adjusted, noise is reduced by Gaussian filtering and edges are

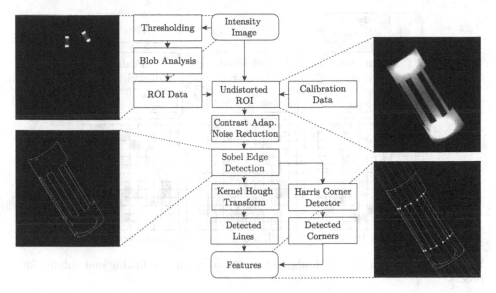

Fig. 3. Flowchart of the proposed image processing algorithm and the corresponding results

detected by the use of Sobel or Canny algorithms. The whole image processing chain is realised in Matlab. For time-efficiency several *OpenCV Library* functions are utilised.

To detect lines, the Standard Hough transform individually transforms each binary edge feature pixel into the parameter space. In contrast to that, the KHT links connecting feature pixels, subdivides them into collinear clusters and votes for these with a Gaussian voting scheme (see Fig. 4 for details). This leads to several advantages in comparison to the SHT. The computational effort is, depending on the image, reduced by a factor up to ten [6] and the parameter space is much clearer, with solitude real peaks and less noisy sub-peaks. As a result, the robustness and accuracy are increased and false positive line detection candidates are reduced. For most HT implementations, lines are used in their Hesse normal form, i.e., the line for a fixed ρ and Θ is

$$\rho = u \cos \Theta + v \sin \Theta \, \forall \, u, v \in \mathbb{R}. \tag{1}$$

Instead of evaluating the neighbourhood of peaks as described in Sect. 1, the results of a Harris corner detector are utilised to determine the end points of a line. This is less critical with respect to nearby lines or peaks and uses the corner characteristics of the rectangular areas on the marker design, see Fig. 3. The Harris feature points provide further possibilities to validate and optimise (i.e., error minimise) the reconstruction results. All the corner points at the top and at the lower end lie on an elliptical curve each. This property creates additional evaluation and cross checking data. The same is true for one upper and one lower (i.e., the start and end point) Harris feature for one line. Those two

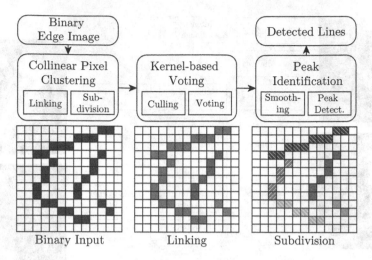

Fig. 4. Flowchart of the KHT algorithm with a exemplary linking and subdivision process

points span an additional line with certain θ and ρ values that can be compared to the corresponding θ and ρ values of the KHT line.

2.2 Human-Generated Ground Truth

There are various possibilities to achieve a hard, numerical ground-truth as a comparison base. First of all, a computational simulation and mapping of the real scene can be utilised. However, it is hard to take all aspects into consideration with influencing factors like discretization of pixel space, optical distortion or marker lighting and the ambient light situation. Second, a measurement of the exact position and rotation of both, all markers and all cameras, can be conducted. But this approach is very time-consuming and only feasible with special measurement equipment like a coordinate measurement machine or a measurement arm. Additionally, the procedure is always error-prone. Finally, a human-generated ground truth can be used. For this goal, different test subjects manually mark all the lines in all the images. Certainly, there is no exact true location for a line. It is a partly subjective task to tag the start and end point or the profile of a line when there is no hard black to white contrast but a grayscale gradient. Thus, averaging for each line marked by all persons leads to favourable results. This takes into consideration that every human being has a different perceptual view and minimizes those differences. Furthermore, outliers (e.g., wrongly marked lines by accident) are eliminated.

2.3 Mathematical Description

Several options for an evaluation criterion when comparing the distance or correspondence of two lines are possible. As the line data are available in their normal form, see (1), the most basic approach would be the distance in Hough space or the geometrical distance of two point pairs ρ and Θ for both lines, respectively. However, a difference of one pixel ρ would lead to the same distance rating result as one degree Θ, which does not represent the true circumstances. Therefore, the distance of the perpendicular foot (d_p) of two lines and the angular difference of its normal vectors (d_a) (see Fig. 5) are used. Depending on the position of the lines in the image or on the sensor, the latter leads to slightly different results. This is caused by the fact that when the perpendicular foot is displaced by one pixel, the influence on d_a is higher, the further away it lies from the principal point. Both will be used subsequently, the first for its descriptiveness with easy to grasp pixel units in image space, the second for its exact nature in the real three-dimensional scene. Alternatively to d_a the cosine distance of the two vectors v_{l1} and v_{l2} can be used, which leads to a reduced computational effort.

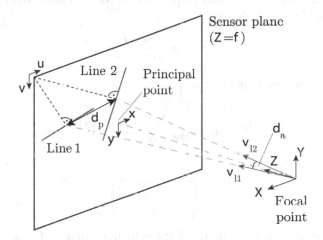

Fig. 5. Sensor plane with two lines and their corresponding perpendicular foot distance (dp) and perpendicular foot angle (da)

As an evaluation benchmark, the results of the KHT with its related parameter and preprocessing sets are compared to the human-generated ground truth. Therefore, all lines are allocated to each other and their distances are calculated. To assess the parameter set, the mean accuracy value of all lines in all 60 images is computed. Parameter sets which lead to incomplete detection results (i.e., in at least one image, at least one line has not been detected) are rejected. The 20 most significant lines (i.e., those with the highest bin or peak in the parameter space) are allocated and compared to the, depending on the view of the marker, eight to ten human-generated master lines.

The indices l_1 and l_2 represent the human-generated master line and its corresponding auto-detected KHT line which should be compared. n_{img} is the total number of all evaluated images, n_{line} the number of lines in each image. The perpendicular foot distance can be computed by

$$d_p = \sqrt{(u_{l_1} - u_{l_2})^2 + (v_{l_1} - v_{l_2})^2} \tag{2}$$

on the image plane with

$$u = \rho \cos \Theta \tag{3}$$

and

$$v = \rho \sin \Theta \tag{4}$$

and its mean value with

$$\bar{d}_p = \frac{1}{n_{\text{line}} n_{\text{img}}} \sum_{i=1}^{n_{\text{img}}} \sum_{j=1}^{n_{\text{line}}} d_p(i,j). \tag{5}$$

In addition, the mean angular difference of the normal vectors with v as the vector pointing from the principal point towards the perpendicular foot of a line (see Fig. 5) can be calculated by

$$d_a = \arccos \frac{v_{l_1} \cdot v_{l_2}}{\|v_{l_1}\| \, \|v_{l_2}\|} \tag{6}$$

in the three dimensional space using

$$\bar{d}_a = \frac{1}{n_{\text{line}} n_{\text{img}}} \sum_{i=1}^{n_{\text{img}}} \sum_{j=1}^{n_{\text{line}}} d_a(i,j) \tag{7}$$

together with

$$v = \begin{pmatrix} u_0 - \rho \cos \Theta \\ v_0 - \rho \sin \Theta \\ f \end{pmatrix} \tag{8}$$

where u_0 and v_0 are the coordinates of the principal point and f the focal length of the camera. Those parameters are individually evaluated for each camera using intrinsic camera calibration with a standard chequered pattern.

3   Parameter Study Results

The marking procedure for the realisation of the human-generated ground truth was implemented as a MATLAB script. The users always had the possibility to zoom in and out during the whole marking process to assess details of the line just as the whole line profile at once. The intensity images and not the already binarised edge images are used to generate the ground-truth, thus taking into account the influence of the edge detection algorithm and threshold, too. The data basis for all evaluations are 20 spatial poses of a cylindrical marker seen from three cameras which leads to a total of 60 images. Fifteen people have participated in the marking procedure. The task at hand was explained by a short demonstration and illustrated help files. Fig. 6 shows a flowchart for the parameter study process.

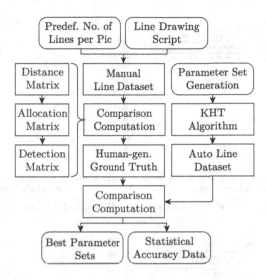

Fig. 6. Flowchart of the KHT parameter study

All of the five KHT parameters (see Table 1 for details) are varied in 10 steps and arranged in all possible combinations, leading to a total of $1 \cdot 10^5$ permutations. Additionally, the preprocessing is distinguished in six different alternatives: Sobel and Canny edge detection with three thresholds each. For the Sobel algorithm a Gaussian filter with the same parameters as the one for the Canny algorithm was applied. Although both algorithms actually use the same filter mask, there is a difference in thresholding and edge thinning. Therefore, both lead to different results. The standard MATLAB implementation of both algorithms with the thresholds 0.07, 0.10 and 0.13 is used.

The best five results of the parameter study for each edge detection algorithm and threshold combination can be seen in Table 3. Preferable results are achieved with Sobel edge detection and a relatively low threshold. In general, the Sobel results are about 50 percent more accurate than the Canny ones. In comparison to the original default parameter values (cf. Table 1) with a resulting $\overline{d_p}$ of 3.326 Pixels, the best parameter set 21137 is more than three times more accurate ($\overline{d_p} = 1.062$ Pixels) and almost in a sub-pixel range. Using the angle d_a as a distance criterion instead of the perpendicular foot distance d_p, the order of the best parameter sets stays the same. Table 2 shows a comparison for both variations. Their ratio varies only slightly depending on the relative position of a line in the image. Towards the border of the image sensor, one pixel difference leads to a larger angular difference then towards the image center (i.e., close to the principal point). A comparison between the detected default lines and the improved lines is shown in Fig. 8.

Figure 7 shows the effects on the detection accuracy when a single parameter is varied around its initial value. Parameter set 21137 is a narrow choice (see Fig. 7a), where small variations of the parameters lead to a rapid decrease of

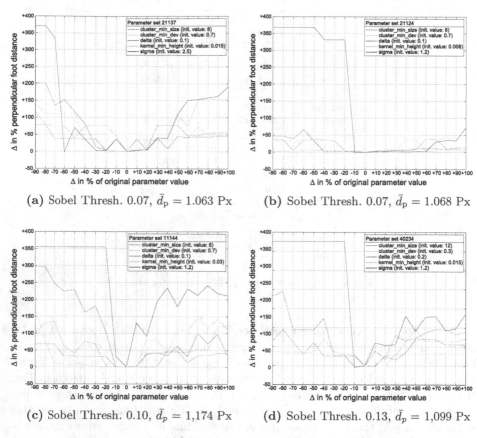

(a) Sobel Thresh. 0.07, $\bar{d}_{\mathrm{p}} = 1.063$ Px

(b) Sobel Thresh. 0.07, $\bar{d}_{\mathrm{p}} = 1.068$ Px

(c) Sobel Thresh. 0.10, $\bar{d}_p = 1,174$ Px

(d) Sobel Thresh. 0.13, $\bar{d}_p = 1,099$ Px

Fig. 7. Sensitivity of the detection accuracy to variations of single parameters

Fig. 8. Comparison between lines with the default parameter set (red, dashed) and lines with the improved parameter set (green, solid)

Table 1. Overview of the KHT parameters, its default values and the range for the parameter study

Parameter	Symbol	Def. value	Exam. range	Description
Cluster_Min_Size	c_σ	10	[2, ..., 25]	Minimal size of a cluster to be still considered as such
Cluster_Min_Dev	c_δ	2	[0.3,...,5]	Minimal distance for a feature pixel to be assigned to a cluster
Delta	δ	0.5	[0.03,...,1.5]	Discretisation of the parameter space
Kernel_Min_Height	κ	0.002	[0.001,...,0.4]	Minimum height of a kernel to still pass culling
Sigma	σ	2	[0.3,...,5]	Standard deviation for the Gaussian kernel

Table 2. Comparison between dp and da as a distance criterion with Sobel edge detection (Threshold = 0.07)

$\bar{d}_p$ (Px)	SD(d_p) (Px)	$\bar{d}_a$ (°)	SD(d_a) (°)
1.062	0.274	$2.016 \cdot 10^{-2}$	$0.521 \cdot 10^{-2}$
1.067	0.293	$2.026 \cdot 10^{-2}$	$0.556 \cdot 10^{-2}$
1.076	0.301	$2.043 \cdot 10^{-2}$	$0.571 \cdot 10^{-2}$
1.078	0.309	$2.046 \cdot 10^{-2}$	$0.586 \cdot 10^{-2}$
1.081	0.309	$2.052 \cdot 10^{-2}$	$0.587 \cdot 10^{-2}$

the accuracy (steep response curve), whereas parameter set 21124 (see Fig. 7b) lies in a much broader area with favourable parameters (flat response curve, except σ), where parameter changes lead to only marginally worse results. The same can be seen in Table 3d, where, in contrast to row one, the rows two to five represent similar parameter choices. The lowering of $\sigma = 2.5$ in parameter set 21137 (Fig. 7a) of 60 percent almost equals the $\sigma = 1.2$ of parameter set 21124 (Fig. 7b). Variations and especially a lowering of σ generally lead to the steepest rise in inaccuracy. Smaller δ values, i.e. a finer rasterisation of the parameter space, mainly lead to worse results. Bins of a certain size are necessary to avoid oversampling with too many sub peaks. Small σ values cause that peaks of different cluster do not melt into combined peaks anymore, which leads to worse results because of unintended sub-peaks of separate clusters only representing a part of the actual, entire marker line.

Table 3. Best five parameter sets and their corresponding perpendicular foot values with statistical values in (Px)

(a) Canny edge detection (Threshold 0.07)

$\bar{d}_p$	SD(d_p)	Min(d_p)	Max(d_p)	KHT Parameter Vector
1.475	0.481	0.863	2.232	12124: (6 1.1 0.10 0.008 1.2)
1.476	0.485	0.863	2.265	11124: (6 0.7 0.10 0.008 1.2)
1.479	0.455	0.915	2.222	1114: (4 0.7 0.10 0.004 1.2)
1.481	0.465	0.915	2.289	1104: (4 0.7 0.10 0.002 1.2)
1.483	0.491	0.928	2.248	1004: (4 0.7 0.05 0.002 1.2)

(b) Canny edge detection (Threshold 0.10)

$\bar{d}_p$	SD(d_p)	Min(d_p)	Max(d_p)	KHT Parameter Vector
1.467	0.476	0.877	2.208	11124: (6 0.7 0.10 0.008 1.2)
1.473	0.495	0.877	2.301	21124: (8 0.7 0.10 0.008 1.2)
1.474	0.483	0.883	2.222	1124: (4 0.7 0.10 0.008 1.2)
1.476	0.489	0.916	2.261	1004: (4 0.7 0.05 0.002 1.2)
1.480	0.495	0.877	2.301	20124: (8 0.3 0.10 0.008 1.2)

(c) Canny edge detection (Threshold 0.13)

$\bar{d}_p$	SD(d_p)	Min(d_p)	Max(d_p)	KHT Parameter Vector
1.494	0.475	0.915	2.359	1114: (4 0.7 0.10 0.004 1.2)
1.505	0.488	0.900	2.313	11105: (6 0.7 0.10 0.002 1.6)
1.509	0.498	0.915	2.359	11114: (6 0.7 0.10 0.004 1.2)
1.510	0.500	0.892	2.359	10114: (6 0.3 0.10 0.004 1.2)
1.514	0.440	0.889	2.169	11216: (6 0.7 0.20 0.004 2.0)

(d) Sobel edge detection (Threshold 0.07)

$\bar{d}_p$	SD(d_p)	Min(d_p)	Max(d_p)	KHT Parameter Vector
1.062	0.274	0.672	1.506	21137: (8 0.7 0.1 0.015 2.5)
1.067	0.293	0.772	1.619	21124: (8 0.7 0.1 0.008 1.2)
1.076	0.301	0.741	1.598	21125: (8 0.7 0.1 0.008 1.6)
1.078	0.309	0.730	1.598	21134: (8 0.7 0.1 0.015 1.2)
1.081	0.309	0.710	1.695	21136: (8 0.7 0.1 0.015 2.0)

(e) Sobel edge detection (Threshold 0.10)

$\bar{d}_p$	SD(d_p)	Min(d_p)	Max(d_p)	KHT Parameter Vector
1.173	0.336	0.435	1.526	40234: (12 0.3 0.20 0.015 1.2)
1.219	0.334	0.895	1.830	2004: (4 1.1 0.05 0.002 1.2)
1.225	0.414	0.464	1.969	31224: (10 0.7 0.20 0.008 1.2)
1.232	0.344	0.896	1.829	13004: (6 1.5 0.05 0.002 1.2)
1.241	0.379	0.867	1.859	23004: (8 1.5 0.05 0.002 1.2)

(f) Sobel edge detection (Threshold 0.13)

$\bar{d}_p$	SD(d_p)	Min(d_p)	Max(d_p)	KHT Parameter Vector
1.098	0.385	0.456	1.706	11144: (6 0.7 0.1 0.03 1.2)
1.105	0.404	0.469	1.768	21144: (8 0.7 0.1 0.03 1.2)
1.187	0.561	0.513	2.521	10144: (6 0.3 0.1 0.03 1.2)
1.242	0.380	0.717	1.882	32244: (10 1.1 0.2 0.03 1.2)
1.252	0.648	0.466	2.804	31144: (10 0.7 0.1 0.03 1.2)

4   Conclusion

An image processing chain to detect features for the calculation of the spatial arrangement of cylindrical markers has been presented. To improve its results, a parameter study with the use of a human-generated ground truth was conducted and statistically evaluated. This increased the accuracy of line detection results by a factor of three. Using this data, the movements of objects during the biomedical testing of knee prosthesis with a six axes robot can be assessed and used as a correction factor for the robot's control loop.

The proposed method can be adapted to other computer vision problems where complex parameter sets lead to manifold influencing factors which are only hard to survey. Instead of a merely subjective verdict, numerical comparison data can be achieved.

As future work, the results will be further assessed and cross checked with simulation and a measurement ground-truth. Furthermore, the implication of feature detection accuracy on the precision of the reconstruction algorithm will be evaluated.

References

1. Becke, M., Schlegl, T.: Toward an experimental method for evaluation of biomechanical joint behavior under high variable load conditions. In: IEEE International Conference on Robotics and Automation (ICRA), pp. 3370–3375 (2011)
2. Mukhopadhyay, P., Chaudhuri, B.B.: A Survey of Hough Transform. Pattern Recognition 48(3), 993–1010 (2015)
3. Li, H., Lavin, M.A., Master, R.J.L.: Fast Hough Transform: A Hierarchical Approach. Computer Vision, Graphics, and Image Processing 36(2–3), 136–161 (1986)
4. Li, Q., Xie, Y.: Randomised Hough Transform With Error Propagation for Line and Circle Detection. Pattern Analysis and Applications 6(1), 55–64 (2003)
5. Kiryati, N., Eldar, Y., Bruckstein, A.: A Probabilistic Hough Transform. Pattern Recognition 24(4), 303–316 (1991)
6. Fernandes, L.A.F., Oliveira, M.M.: Real-time Line Detection Through an Improved Hough Transform Voting Scheme. Pattern Recognition 41(1), 299–314 (2008)
7. Akinlar, C., Topal C.: Real-time line segment detection by edge drawing. In: 18th IEEE International Conference on Image Processing (ICIP), pp. 2837–2840, September 2011
8. Montero A., Nayak A., Stojmenovic M., Zaguia N.: Robust line extraction based on repeated segment directions on image contours. In: Computational Intelligence for Security and Defense Applications (CSIDA), pp. 1–7 (2009)
9. von Gioi, R., Jakubowicz, J., Morel, J.-M., Randall, G.: LSD: A Fast Line Segment Detector with a False Detection Control. IEEE Transactions on Pattern Analysis and Machine Intelligence 32(4), 722–732 (2010)
10. Nguyen, T.T., Pham X.D., Kim, D., Jeon, J.W.: A test framework for the accuracy of line detection by hough transforms. In: 6th IEEE International Conference on Industrial Informatics (INDIN), pp. 1528–1533 (2008)
11. Guerreiro, R., Aguiar, P.: Incremental local hough transform for line segment extraction. In: 18th IEEE International Conference on Image Processing (ICIP), pp. 2841–2844 (2011)

12. Furukawa, Y., Shinagawa, Y.: Accurate and Robust Line Segment Extraction by Analyzing Distribution Around Peaks in Hough Space. Computer Vision and Image Understanding **92**(1), 1–25 (2003)
13. Du, S., Tu, C., van Wyk, B., Chen, Z.: Collinear segment detection using HT neighborhoods. IEEE Transactions on Image Processing, 3612–3620 (2011)
14. Dai, B., Pan, Y., Liu, H., Shi, D., Sun, S.: An Improved RHT algorithm to detect line segments. In: 2010 International Conference on Image Analysis and Signal Processing (IASP), pp. 407–410 (2010)
15. Martin, D., Fowlkes, C., Tal, D., Malik, J.: A Database of human segmented natural images and its application to evaluating segmentation algorithms and measuring ecological statistics. In: Proc. 8th International Conference on Computer Vision, pp. 416–423 (2001)
16. Martin, D., Fowlkes, C., Malik, J.: Learning to detect natural image boundaries using local brightness, color, and texture cues. IEEE Transactions on Pattern Analysis and Machine Intelligence, 530–549 (2004)
17. Crevier, D.: Image Segmentation Algorithm Development Using Ground Truth Image Data Sets. Computer Vision and Image Understanding **112**(2), 143–159 (2008)
18. Attene, M., Katz, S., Mortara, M., Patane, G., Spagnuolo, M., Tal, A.: Mesh segmentation - a comparative study. In: IEEE International Conference on Shape Modeling and Applications (2006)
19. Benhabiles, H., Vandeborre, J.-P., Lavoue, G., Daoudi, M.: A Framework for the objective evaluation of segmentation algorithms using a ground-truth of human segmented 3D-models. In: IEEE International Conference on Shape Modeling and Applications (SMI), pp. 36–43 (2009)
20. Dutagaci, H., Cheung, C.P., Godil, A.: Evaluation of 3D Interest Point Detection Techniques via Human-generated Ground Truth. The Visual Computer **28**(9), 901–917 (2012)
21. Wang, S., Ge, F., Liu, T.: Evaluating edge detection through boundary detection. EURASIP Journal on Advances in Signal Processing, 213–227 (2006)
22. Lopez-Molina, C., De Baets, B., Bustince, H.: Quantitative Error Measures for Edge Detection. Pattern Recognition **46**(4), 1125–1139 (2013)
23. Yitzhaky, Y., Peli, E.: A Method for Objective Edge Detection Evaluation and Detector Parameter Selection. IEEE Transactions on Pattern Analysis and Machine Intelligence **25**(8), 1027–1033 (2003)
24. Zhang, Z.: A Flexible New Technique for Camera Calibration. IEEE Transactions on Machine Intelligence **22**(11), 1330–1334 (2000)

A Simple Human Interaction Recognition Based on Global GIST Feature Model

Xiaofei Ji[✉], Xinmeng Zuo, Changhui Wang, and Yangyang Wang

School of Automation, Shenyang Aerospace University, Shenyang, China
jixiaofei7804@126.com

Abstract. The precision of human interaction recognition mainly relies on the discrimination of action feature descriptors. The descriptors contained global and local information usually can be applied to classify interaction actions. A novel approach is proposed by using the Gist feature model to recognize human interaction actions, which has an advantage of simple feature, easy to operate, good real-time and flexible applications. Taking advantage of the theories with Gaussian pyramid and center-surround mechanism, the gist features from three channels are extracted to represent the human interaction motion, then the classification result is obtained by using frame to frame nearest neighbor classifier and weighted fusion them. The method is tested on UT-Interaction dataset. The experiments show that the method obtained stable performance with simple implementation.

Keywords: Interaction recognition · Gist feature · Nearest neighbor classifier · UT-interaction dataset

1 Introduction

Human interaction recognition and understanding have attracted growing interests in the computer vision community which can be used in intelligent surveillance, video analysis, human-computer interface, *etc.* Many approaches have been proposed to deal with interaction recognition [1-4]. However, due to the large intra-variations, viewpoint changes [5], clutter and occlusion and the other fundamental factors, which cause the interaction recognition is still a challenging research topic.

Many scholars deal with interaction recognition as a general action. This kind of method usually represents the interaction as an integral descriptor including all the people involved in the interaction. Then a classier is utilized to classify interactions Yuan *et al.* [6] proposed spatio-temporal context to describe local spatio-temporal features and the spatio-temporal relationships between them. Then a spatio-temporal context kernel (STCK) was introduced to recognize human interactions. The feature extraction of this method is simple, however the accuracy of recognition is not good enough. Burghouts *et al.* [7] improve the performance of this method by exploiting a novel spatio-temporal layout description of interaction, which can improve the discriminative ability of inter-class. Peng *et al* [8] combined four different features including DT shape, HOG, HOF and MBH to extract low-level feature of multi-scale

H. Liu et al. (Eds.): ICIRA 2015, Part I, LNAI 9244, pp. 487–498, 2015.
DOI: 10.1007/978-3-319-22879-2_45

dense trajectories, and explored four advanced feature encoding methods with a bag-of-features framework as the motion description of the whole interaction actions. Finally, SVM classifier was trained to recognize testing videos. Li *et al* [9] incorporated the advantages of both global feature (Motion Context, MC) and Spatio-Temporal (S-T) correlation of local Spatio-Temporal Interest Points (STIPs) to describe the human-human interaction actions, and proposed GA search based random forest and S-T correlation based match to achieve better performance of interaction recognition and understanding. This kind of method treats people as a single entity and do not extract the motion of each person from the group. So they do not need segment the feature of individual in the interaction. However the better performance always needs comprehensive motion features and matching method.

How to extract discriminative and simple features to describe interactions and design effective recognition methods to fuse different types of features have become two important solutions for interaction recognition. A novel feature model based on Gist is proposed by combining the orientation, intensity and color information to describe the human interactions, which usually applied to the classification of complicated scenes, and in human interaction recognition field, some scholars[12]only use the orientation channels to extract feature, which maybe loss some important information for recognition. Gist feature is chosen to the proposed method for three significant reasons: Firstly, Gist feature accords with the basic theory which human observe object and analyze what they have seen. Secondly, Gist feature captures global structure information by filtering an image with different orientations and scales. In case of realistic scenarios, it can be extracted more reliably than silhouettes feature. Thirdly, the computation time of Gist is much less than optical flow features. After the extraction of the three channels feature, the recognition result was obtained by fusing the probability of these three kinds of feature. Experiment tested on the UT-Interaction Dataset demonstrates the simplicity and practicability of the proposed method.

The rest of the paper is organized as follows. In Section 2, the Gist features extraction and representation are introduced. Section 3 provides a simple explanation of nearest neighbor classifier. And Section 4 gives experimental results and analysis. Finally, Section 5 concludes the paper.

2   The Global Feature Extraction

Gist model [10] first put forward by Olive in 2001, and it was applied to the classification of complicated scenes by later scholars. The Gist model is mainly a process model which simulates human eyes to receive some instant information and identify it. In this paper, the model is applied in human interaction description and recognition [11]. Gist model is a kind of the global feature model, which employs biologically plausible center-surround features from orientation, intensity, and color channels. We divide an image into a number of sub-regions. Then the gist features of the image from three channels are extracted in each sub-region. In the whole process of establishing feature model can be divided into four steps, *i.e.*, pretreatment process of videos, orientation channel descriptor extraction, intensity channel descriptor extraction and color channel descriptor extraction.

2.1   Interaction Detection

In order to improve recognition accuracy and efficiency, interaction detection for video is performed before feature extraction. The foreground information of the interaction is obtained by using frame difference. Redundant information in the frame can be eliminated by employing the frame difference. The process is shown in Fig. 1.

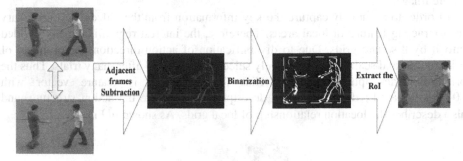

Fig. 1. Pretreatment process

2.2   The Orientation Channel Descriptor of Gist Model

The descriptor obtained from the orientation channels captures structural information by filtering an image with multiple scales and orientations [12]. In this paper, Gabor filter functions with various orientations and spatial resolution are adopted to filter the images. The algorithm is presented as Eq. 1 and Eq. 2.

$$G_{\theta_i}^s = k \exp(\frac{-(x_{\theta_i}^2 + y_{\theta_i}^2)}{2\sigma^{2(s-1)}}) \exp(2\pi j(u_0 x_{\theta_i} + v_0 y_{\theta_i})) \tag{1}$$

$$\begin{cases} x_{\theta_i} = x\cos\theta_i + y\sin\theta_i \\ y_{\theta_i} = -x\sin\theta_i + y\cos\theta_i \end{cases} \tag{2}$$

Where k is a positive constant, σ is standard deviation of Gauss function, s is the number of scales, θ_i is orientations of the scale s, where

$$\theta_i = \pi(i-1)/\theta_s, i = 1, 2, \cdots, \theta_s$$

First of all, the action interest region is divided into $n \times n$ square grids, and then the average filter response is calculated from each grid as Eq. 3, which is called as Gist vector. Finally these values are combined into a vector as the feature description of the frame. It is as Eq. 4

$$\zeta_e = \frac{1}{\frac{R}{n} \cdot \frac{C}{n}} \cdot \sum_{y=1}^{\frac{R}{n}} \sum_{x=1}^{\frac{C}{n}} f(x, y) \tag{3}$$

$$v = \{\zeta_1, \zeta_2, \cdots, \zeta_{n^2}\} \tag{4}$$

Where, R, C are the numbers of row and column respectively. $f(x, y)$ means the gray value function at 2D Coordinates (x, y). ζ_e is the element of feature description of orientation, $e = 1, 2, \cdots n^2$, v is the vector as the feature description of the whole image.

In order to accurately capture the key information from the videos, it is necessary for extracting feature of local areas. Therefore, the interest region is equally divided into 8 by 8 square grids. Due to diversification of action direction, the numbers of orientations and scales are respectively selected for 8 and 4 after many trials. Thus the orientation descriptor of global Gist feature model is feature vector with $2048(8 \times 8 \times 8 \times 4)$ dimension. The descriptor can capture the action structure, and also describes the location relationship of local grids. As shown in Fig. 2.

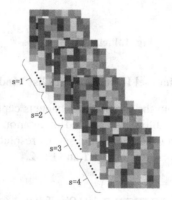

Fig. 2. The feature of orientation channel (4 scales, 8 orientations)

2.3   The Intensity Channel Descriptor of Gist Model

It is a necessary to simply introduce some methods or mechanism in building intensity channel descriptor of Gist model, which are also applied to extract the colour channel descriptor of Gist.

Gaussian pyramid: Gaussian pyramid is a representation method of multiscale images. The actual manipulations are followed: firstly an image is multiple processed by using Gaussian Blur (the size of the template is 3 by 3) and down-sampling to obtain the images with different scales.   In the Gaussian pyramid, the images with large scale have abundant information of image, and the images with small scale reveal local image information better. In this paper, each sub-channel has a nine-scale pyramidal image by down-sampling processing. As shown in Fig. 3.

Center-surround mechanism: It is commonly found in biological-vision, which compares image values in a center-location to its neighboring surround-locations. Across-scale difference between two different scales maps is obtained by interpolation to the center scale and pointwise absolute difference. Finally, it produce a feature map. [11].

Fig. 3. Nine-scale pyramidal images

The expression of intensity feature maps is calculated as Eq. 5

$$I = \frac{R+G+B}{3} \tag{5}$$

Where, R, G, B are the image of three raw color channels (red, green blue) respectively. Based on pyramid principle, a nine-scale pyramidal image $I(\sigma)$ is obtained, and $\sigma \in \{0,1,2,3,4,5,6,7,8\}$. Furthermore, making use of center-surround mechanism, six maps can get by the Eq. 6, as shown in Fig. 4.

$$I(c,s) = |I(c) \Theta I(s)| \tag{6}$$

Where, $s = c+k$, $c \in \{2,3,4\}$, $k \in \{3,4\}$, $I(c)$ is the intensity of c level and $I(s)$ is the intensity of s level. Θ means that the image $I(s)$ is magnified to the size of the image $I(c)$, then pointwise absolute difference is performed between the two images, finally, intensity feature descriptors with 384 dimensions($6 \times 8 \times 8$) of Gist model are obtained.

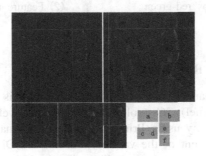

Fig. 4. Six intensity feature maps

2.4   The Color Channel Descriptor of Gist Model

According to Ewald Hering color opposition theories, there are two pairs of opposite color groups, which can be obtained according to the original image r, g, b outputs:

$$R = r - \frac{g+b}{2} \tag{7}$$

$$G = g - \frac{b+r}{2} \tag{8}$$

$$B = b - \frac{r+g}{2} \tag{9}$$

$$Y = r + g - 2(|r - g| + b) \tag{10}$$

The color opposition pairs are the two color channels' red-green and blue-yellow. As shown in Eq. 11 and Eq. 12. Each of the opponent pairs is used to construct six center-surround scale combinations. The process is similar to the intensity process, a total of 12 color feature maps, thus the color descriptor of Gist model is a feature vector with 768($12 \times 8 \times 8$) dimensions. As shown in Fig. 5.

$$RG(c,s) = |(R(c) - G(c)) \Theta (R(s) - G(s))| \tag{11}$$

$$BY(c,s) = |(B(c) - Y(c)) \Theta (B(s) - Y(s))| \tag{12}$$

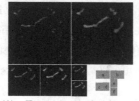   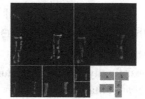

(1)   Feature maps' red-green   (2) Feature maps' blue-yellow

Fig. 5. 12 color feature maps

2.5   PCA Dimension Reduction

The number of three channels raw gist feature dimension is 1024 (orientation channel), 384 (intensity channel), 768 (color channel), respectively. The raw features are reduced the dimensions by using Principal Component Analysis (PCA), while still preserving up to 98 percent of the variance. After PCA Dimension Reduction, the number of three channels gist feature dimension is 134 (orientation channel), 80 (intensity channel), 91 (color channel).

3   Classifier Design

The nearest neighbor classify algorithm is not only a simple and effective identification method, but also has a fast recognition speed to recognize single human action. In order to realize the real-time detection, a kind of simple and effective recognition method should be chosen. The nearest neighbor classifier is chosen to recognize the extracted features respectively. The method is shown as follows:

Supposing that there are c classes as w_1, w_2, ... , w_c, each class has the number of N_i marked sample, then the discriminant function of class w_i is shown as Eq.13:

$$g_i(x) = \min \left\| x\text{-}x_i^k \right\|, k = 1, 2, ..., c \tag{13}$$

The subscript i of x_i^k means class w_i and k is the kth sample among total N_i in classes w_i. According to the function above, the decision rule can be defined as Eq. 14:
If

$$g_j(x) = \min g_i(x), i = 1, 2, ..., c \tag{14}$$

So

$$x \subseteq w_j \tag{15}$$

This decision method is called nearest neighbor method. The calculation formula of Euclidean distance between samples is shown as Eq. 16:

$$D = \sqrt{\sum_{i=1}^{N}(A_i\text{-}B_i)^2} \tag{16}$$

A and B are feature vectors and N is the number of the feature vectors.

The recognition method used in this paper is also called frame to frame nearest neighbor [13]. Firstly the training samples with known category have been chosen to form training set. The frames included the same interaction category have the same symbol in the training set. Then input the test sequence, the classifier try to forecast the symbol of test interaction sequences to be one action type by respectively calculating the Euclidean distance between feature of each test frame and each training frame feature according to the nearest neighbor decision rules. Then vote for the action category which the frame with minimum distance in the training samples belongs to. Finally the category of test sequence will be recognized as the action class symbol which has the highest votes.

4   Results Analysis

To test the effectiveness of our approach, the UT-Interaction set 1 benchmark dataset [14] is chosen to make a series of experiments. The dataset contains 6 classes of human-human interactions (handshake, hug, kick, point, punch, push) performed by 15 peoples under a realistic surveillance environment, where exists some interfering factors such as cluttered scenes, moving background, camera jitters/zooms and different clothes etc. Each class consists 10 video sequences, 60 video sequences in total. Illustrations of these interactions are presented in Fig. 6. The dataset is widely used as human-human interactions action analysis and recognition.

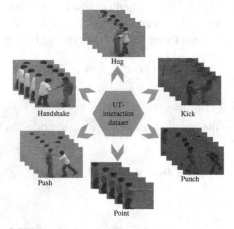

Fig. 6. UT-interaction dataset with pretreatment process

First of all, the interaction detection for video is performed before feature extraction, followed by using the frame difference to eliminate redundant information in the frame. In the experiments, Leave-one-out cross validation method is adopted throughout the process. In turns one action of each action class is chosen as test samples, and the rest of all actions as the training set, circulation continue until all actions are finished. Finally the experimental result is obtained.

4.1   Orientation Channel Feature Recognition Performance Test

UT-interaction dataset be used to test Gist feature model's performance. Choosing different scale grids to divide the feature maps, the Table 1 shows the recognition results of using the different scale grids.

Table 1. The recognition rate of using the different scale grids

Scale grids	The recognition rate /%
by 4 × 4	56.70
by 8 × 8	68.33

We can see that the recognition rate with 8 by 8 is better than that of 4 by 4. It demonstrates that dividing the feature maps with 4 by 4 cannot descript the local features. Instead of the former, dividing with 8 by 8 is able to obtain better description with the global and local information. However, the recognition rate doesn't embody the advantage of Gist. So adding the other two sub-channels is added to recognize the inter action as supplementary information.

4.2   Three Channels Features Recognition Performance Test

Gist feature model is usually applied to scenes classification, and the orientation channel is the main part. To some extent, the information from color channel and intensity channel can provide some supplementary information in the videos, that's why the feature information from the intensity channel and the color channel is taken into account in the proposed method.

Before fusing three channels feature information, the feasibleness is tested by extracting features from the color channel and intensity channel respectively and recognizing the interaction from the test videos. According to the analysis from context, a square grid sized 8 by 8 is used for dividing the feature maps. The recognition results under the three channels are shown in Table 2.

Table 2. Three channels' recognition rate (/%)

Orientation channel	Intensity channel	Color channel
68.33	51.67	58.33

Analyzed from the Table 2, although the recognition results from the intensity channel and the color channel are not ideal, it's a good signal that gets help from the two channels.

The proposed method chooses the frame by frame nearest neighbor to deal with the video, followed by voting for each frame with the video, and obtain the final results by weighted the voting results from three channels. The raw recognition result (75%) is as shown in Fig. 7 (1), and the recognition result (75%) after the PCA dimension reduction is as shown in Fig. 7 (2).

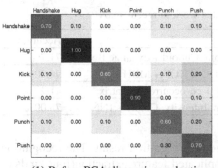
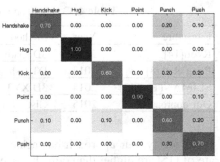

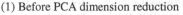

(1) Before PCA dimension reduction   (2) After PCA dimension reduction

Fig. 7. The Confusion Matrix of the proposed method Conclusion

Precision evaluation was carried out for the classified video by confusion matrix which confirmed the feasibility of this method, and the recognition accuracy is 75%. There is a high error rate between action "punch" and "push". Observed from two types of actions, it's easy to find the confusion reason is the higher similarity of the two types of actions, more specifically, the action variety of one person both comes from the upper limb (arm especially), and the whole body of the other person will backed away in the two types of actions. The confusion reason is also shown in Fig. 8.

Fig. 8. The two types of confusion actions

After the calculation and analysis, the computing time is as shown in table 3. We can see that after PCA dimension reduction, the average computing velocity with single frame and whole video is faster, which proves the advantage of Gist model.

Table 3. the average computing time comparisons

	Before PCA	After PCA
The single frame	0.343s	0.252s
The whole video	17.15s	12.60s

4.3   The Comparison of the Performance

The comparisons of performance between the proposed method and the recent related works based on UT-Interaction dataset are shown in Table 4.

Table 4. Comparison with related work in recent years

Literature	Year	Method	Accuracy
Ryoo[14]	2011	3D spatio-temporal Cuboid + dynamic BoWs	71.7%
Peng et al.[8]	2013	DT shape + HOG + HOF + MBH	94.50%
SLIMANI et al.[3]	2014	3D XYT + BoW + co-occurrence matrix	41%
Our approach	2015	Gist (Orientation, Intensity, Color) + 1NN	75.00%

Obviously, our approach has achieved a good recognition result. The recognition rate of [15] are higher than ours, but need complicated feature extraction(four kinds of features) and require a plurality of complex recognition model. Our approach need not segment the feature of interaction to individuals or create any complex discriminant model. So our method outperforms all of other state of methods.

5   Conclusion

In order to let the machine simulates human how to observe object and analyze the thing that they have seen, and extract feature from the global perspective, a novel approach is proposed by using the Gist feature model to recognize human-human interactive actions, which has an advantage of simple feature, easy to operate, good real-time and flexible applications. Gist feature model totally includes three channels, *i.e.*, the orientation channel, the intensity channel and the color channel. This is the novelty which using three channels to extract feature in human interaction recognition field. And taking advantage of the theories with Gaussian pyramid and center-surround mechanism, the gist features from three channels are extracted to represent the human interaction motion, then the classification result is obtained by using frame to frame nearest neighbor classifier and weighted fusion them. The experiments show that the method obtained stable performance. Future challenge is to reduce the feature dimension and improve the accuracy of recognition.

References

1. Cao, Y., Barrett, D., Barbu, A., et al.: Recognize Human Activities from Partially Observed Videos. IEEE Conf. on Computer Vision & Pattern Recognition **9**(4), 2658–2665 (2013)
2. Burghouts, G.J., Schutte, K.: Spatio-temporal layout of human actions for improved bag-of-words action detection. Pattern Recognition Letters **34**(15), 1861–1869 (2013)
3. El Houda Slimani, K.N., Benezeth, Y., Souami, F.: Human interaction recognition based on the co-occurrence of visual words. In: IEEE Conf. on Computer Vision & Pattern Recognition Workshops, Columbus, Ohio, USA, pp. 461–466 (2014)
4. Zhang, Y., Liu, X., Chang, M.-C., Ge, W., Chen, T.: Spatio-temporal phrases for activity recognition. In: Fitzgibbon, A., Lazebnik, S., Perona, P., Sato, Y., Schmid, C. (eds.) ECCV 2012, Part III. LNCS, vol. 7574, pp. 707–721. Springer, Heidelberg (2012)
5. Ji, X., Liu, H.: Advances in View-Invariant Human Motion Analysis: A Review. IEEE Transactions on Systems Man & Cybernetics Part C Applications & Reviews **40**(1), 13–24 (2010)
6. Yuan, F., Sahbi, H., Prinet, V.: Spatio-temporal context kernel for activity recognition. In: 1st Asian Conference on Pattern Recognition, Beijing, China, pp. 436–440 (2011)
7. Burghouts, G.J., Schutte, K.: Spatio-temporal layout of human actions for improved bag-of-words action detection. Pattern Recognition Letters **34**(15), 1861–1869 (2013)
8. Peng, X., Peng, Q., Qiao, Y.. Exploring dense trajectory feature and encoding methods for human interaction recognition. In: Proc. of the Fifth Int. Conf. on Internet Multimedia Computing and Service, Huangshan, China, pp. 23–27 (2013)
9. Li, N., Cheng, X., Guo, H., Wu, Z.: A hybrid method for human interaction recognition using spatio-temporal interest points. In: Proc of the 22nd Int. Conf. on Pattern Recognition, Stockholm, Sweden, pp. 2513–2518 (2014)
10. Oliva, A., Torralba, A.: Modeling the shape of the scene: A holistic representation of the spatial envelope. International Journal of Computer Vision **3**(42), 145–175 (2001)
11. Siagian, C., Itti, L.: Rapid biologically-inspired scene classification using features shared with visual attention. IEEE Transactions on Pattern Analysis and Machine Intelligence **2**(29), 300–312 (2007)

12. Wang, Y., Li, Y., Ji, X.: Recognizing human actions based on gist descriptor and word phrase. In: Proc. 2013 Int. Conf. Mechatronic Sciences, Electric Engineering and Computer, Shenyang, China, pp. 1104–1107 (2013)
13. Ji, X., Zhou, L., Li, Y.: Human action recognition based on AdaBoost algorithm for feature extraction. In: Proc. of the IEEE Int. Conf. on Computer and Information Technology, Xi'an, China, pp. 801–805 (2014)
14. Ryoo, M., Aggarwal, J.: Spatio-temporal relationship match: video structure comparison for recognition of complex human activities. In: Proc. of the IEEE Int. Conf. on Computer Vision, Kyoto, pp. 1593–1600 (2009)

SRDANet: An Efficient Deep Learning Algorithm for Face Analysis

Lei Tian$^{(\boxtimes)}$, Chunxiao Fan, Yue Ming, and Jiakun Shi

Beijing Key Laboratory of Work Safety Intelligent Monitoring, School of Electronic Engineering, Beijing University of Posts and Telecommunications, Beijing 100876, People's Republic of China
tianlei189@bupt.edu.cn, {fcxg100,sjk2012sjk}@163.com,
myname35875235@126.com

Abstract. In this work, we take advantage of the superiority of Spectral Graph Theory in classification application and propose a novel deep learning framework for face analysis which is called **Spectral Regression Discriminant Analysis Network (SRDANet)**. Our SRDANet model shares the same basic architecture of Convolutional Neural Network (CNN), which comprises three basic components: convolutional filter layer, nonlinear processing layer and feature pooling layer. While it is different from traditional deep learning network that in our convolutional layer, we extract the leading eigenvectors from patches in facial image which are used as filter kernels instead of randomly initializing kernels and update them by stochastic gradient descent (SGD). And the output of all cascaded convolutional filter layers is used as the input of nonlinear processing layer. In the following nonlinear processing layer, we use hashing method for nonlinear processing. In feature pooling layer, the block-based histograms are employed to pooling output features instead of max-pooling technique. At last, the output of feature pooling layer is considered as one final feature output of our model. Different from the previous single-task research for face analysis, our proposed approach demonstrates an excellent performance in face recognition and expression recognition with 2D/3D facial images simultaneously. Extensive experiments conducted on many different face analysis databases demonstrate the efficiency of our proposed SRDANet model. Databases such as Extended Yale B, PIE, ORL are used for 2D face recognition, FRGC v2 is used for 3D face recognition and BU-3DFE is used for 3D expression recognition.

Keywords: SRDA Network · Deep learning · Spectral Regression Discriminant Analysis · Face recognition · Expression recognition

1 Introduction

During the last decade, the single face analysis task has been extensively studied, such as single face recognition and single expression recognition. Numerous efforts have been made to design the hand-crafted features for 2D face analysis [1–5] and 3D face analysis [6–9]. However, most of the hand-crafted features

© Springer International Publishing Switzerland 2015
H. Liu et al. (Eds.): ICIRA 2015, Part I, LNAI 9244, pp. 499–510, 2015.
DOI: 10.1007/978-3-319-22879-2_46

are not competent for multiple face analysis tasks simultaneously, since their successes mainly depend on the successes of separate domains knowledge of face/expression recognition. Besides, the traditional single face analysis methods are not robustness to extreme intra-class variability. So, with the development of synergetic analysis for multiple visual tasks in the context of big data, the newly proposed algorithms need to provide an excellent performance for face recognition and expression recognition simultaneously.

In order to solve the problems mentioned above, learning higher-level features from the data is considered as a plausible way [10–12]. Researchers hope to discover a multi-level representation through deep neural networks, mainly because higher-level features can represent more abstract semantics of the data. There are two examples of such method: Convolutional Neural Network (CNN) [13–17] and Auto-Encoders (AEs) [18,19]. However, their successes depend not only on parameter tuning but also on depth of their architectures. In another word, when their architectures are not deep enough, the performances won't be as good as the hand-crafted features.

To overcome the problems mentioned above, we translate the idea of the Spectral Regression Discriminant Analysis (SRDA) into deep learning framework. The Spectral Regression Discriminant Analysis (SRDA) [20] is derived from Linear Discriminant Analysis (LDA) but solve the problems of LDA such as **singularity issue** and **small sample issue**. In previous work, SRDA, a supervised version of PCA, is used to extract the leading eigenvectors of input data. In this paper, SRDA is used to extract the principle component of the patches in face image which are used as filter kernels of the deep network. Our proposed SRDANet model has fewer deep network layers than traditional deep learning framework, without tuning parameter. **In the convolutional layer**, we extract the leading eigenvectors from patches in face image by SRDA which are used as filter kernels instead of randomly initializing kernels and update them by Stochastic Gradient Descent (SGD). **In the nonlinear processing layer**, there is no nonlinear processing layer until the end of **all** convolutional filter layer, and the binary hashing method is applied to nonlinearly process feature maps, instead of the *Sigmoid* or *ReLU* function [21], which is different from traditional deep learning network. It means that feature maps are extracted from the all cascaded convolutional layers, and they are used as the input of nonlinear processing layer. **In the feature pooling layer**, the block-based histograms method is applied to pooling feature. Since the output of nonlinear processing layer is a hashed decimal-valued image, conventional pooling methods such as Max-Pooling and Average-Pooling are not fit for our model. **In the output layer**, we vectorize the output of feature pooling layer as the final feature of our model. Figure 1 illustrates how our SRDANet model extracts multiple features from face images.

Organization of this Paper. In Section 2, we give a detailed illustration of SRDANet algorithm. In Section 3, we evaluate the computational complexity of our SRDANet. The extensive experimental results are presented in Section 4. Finally, we conclude our work in Section 5.

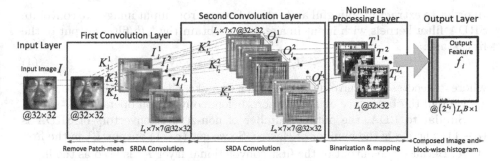

Fig. 1. The detailed layer architecture diagram of the two-stage SRDANet

2 SRDA Network Deep Learning Algorithm

SRDA Network takes both advantage of Spectral Graph Theory and the architecture of Convolutional Neural Network (CNN). Suppose that we are given a set of N samples $\{I_i\}_{i=1}^N$ which belongs to c classes, and its size is $d_1 \times d_2$ ($d_1 \times d_2 = d$). The patch size is $j_1 \times j_2$ at all stages. The number of filters in ith layer is L_i. In order to illustrate precisely, we set the number of stages to be 2.

2.1 The Convolutional Filter Layer

In order to take advantage of convolution, we reserve the basic convolutional processing of CNN. As for learning filter kernels, we just replace SGD with SRDA. For each input image, we take a $j_1 \times j_2$ patch around each pixel and collect all overlapping patches in the ith image I_i. Then, we vectorize all patches and subtract patch mean from each patch, then we obtain $\bar{P}_i = [\bar{p}_{i,1}, \bar{p}_{i,2}, \cdots, \bar{p}_{i,d}]$ for the ith image. We construct the same matrix for all input images and we have

$$P = [\bar{P}_1, \bar{P}_2, \cdots, \bar{P}_N] \in \mathbb{R}^{j_1 j_2 \times Nd} \tag{1}$$

Then, we learn the transformation vectors V^i from data matrix P. The SRDA is aimed to search for the project axis on which the data points of different classes are separate from each other while the data points of same classes are close to each other. Its objective function is that:

$$V^* = \arg\max \frac{V^T S_b V}{V^T S_t V} \tag{2}$$

where we call S_b the between-class scatter matrix and S_t the total scatter matrix. In order to clearly denote, $SR_{l_1}(a)$ means data a through SRDA algorithm and get the l_1th leading eigenvectors. Then we map each eigenvector to matrix of size $j_1 \times j_2$.

$$K_{l_1}^1 = vec2mat_{j_1, j_2}(SR_{l_1}(P)), l_1 = 1, 2, \cdots, c - 1 \tag{3}$$

In order to extract meaningful and robust feature from input image, we convolute SRDA filter kernels with input image I_i and obtain the $l_1 th$ filter output of the first stage:

$$I_i^{l_1} = I_i * K_{l_1}^1, i = 1, 2 \cdots, N \tag{4}$$

where * denotes 2D convolution. In order to make $I_i^{l_1}$ to have the same size of I_i, we pad the boundary of I_i with zero before convolution operation.

Similar to LDA, the output number of non-zero eigenvectors of SRDA is $(c-1)$, where c is the number of classes. So we get $(c-1)$ output $I_i^{l_1}$ in the first stage. Because the output of the first convolutional layer $I_i^{l_1}$ is used as the input of second convolutional layer, it is infeasible for SRDANet algorithm to run in a limited memory when c is large. So further dimensionality reduction is essential. We apply PCA algorithm to reduce dimension from $(c-1)$ to L_1. Therefore we replace $K_{l_1}^1$ in Eq. (3) with the following equation when c is too large to load into memory:

$$K_{l_1}^1 = vec2mat_{j_1,j_2} \left(PCA_{l_1} \left(SR \left(P \right) \right) \right), l_1 = 1, 2, \cdots, L_1 \tag{5}$$

where $PCA_{l_1} \left(X \right)$ denotes the $l_1 th$ principal eigenvector of $X X^T$.

Different from the architecture of conventional CNN, we only insert nonlinear processing layer when all convolutional layers have been processed. Since the operation that we insert nonlinear processing layer (such as *Sigmoid* or *ReLU* function) after each convolutional layer shows no chance of improving the performance of our SRDANet model through extensive experiments. So we simplified our network's architecture by considering the efficiency of training. The output of the first convolutional layer $I_i^{l_1}$ is used as the input of second convolutional layer instead of nonlinear processing layer. Almost repeating the same process as the first convolutional layer, we collect all patches of $I_i^{l_1}$ and obtain $Q = \left[\bar{Q}_1, \bar{Q}_1, \cdots, \bar{Q}_N \right]$. Then we compute the leading eigenvectors from data matrix Q by SRDA and perform dimensionality reduction by PCA. Similar to the first stage, we obtain the filter kernels of the second layer $K_{l_2}^2$. For each input $I_i^{l_1}$ of the second convolutional layer, we obtain L_2 outputs $O_i^{l_1}$:

$$O_i^{l_1} = \left\{ I_i^{l_1} * K_{l_2}^2 \right\}_{l_2=1}^{L_2}, l_2 = 1, 2, \cdots, L_2 \tag{6}$$

where $l_1 = 1, \cdots, L_1$ and $i = 1, 2, \cdots, N$.

2.2   The Nonlinear Processing Layer

We choose $O_i^{l_1}$ as the input of next nonlinear processing layer. It is obvious that the method of learning filter kernels in our model is different from conventional CNN, so the traditional nonlinear processing functions aren't suitable for our SRDANet model. For each $I_i^{l_1}$, there are L_2 outputs $O_i^{l_1}$ in the second convolutional layer. The Heaviside step function[1] is employed to binarize these outputs,

---

[1] $H\left(x \right) = \begin{cases} 1, x > 0 \\ 0, x \leq 0 \end{cases}$

and we obtain binary images

$$H_i^{l_1} = \left\{ \max\left(I_i^{l_1} * W_{l_2}^2, 0 \right) \right\}_{l_2=1}^{L_2}. \tag{7}$$

We consider the L_2 binary bits as a decimal number and convert the L_2 outputs in $O_i^{l_1}$ into a decimal image:

$$T_i^{l_1} = \sum_{l_2=1}^{L_2} 2^{l_2-1} H_i^{l_1}, \tag{8}$$

whose pixel is a decimal number in the range $\left[0, 2^{L_2} - 1 \right]$.

2.3   The Feature Pooling Layer

As far as we know, the max-pooling and average-pooling methods are not suitable to process the output of our nonlinear processing layer. We partition the image $T_i^{l_1}$ into B blocks and compute the histogram of the decimal values in each block.

We further concatenate all the B histograms into one vector as $Bhist\left(T_i^{l_1} \right)$. Finally, the feature of input image I_i is defined as a set of block-wise histograms vector:

$$f_i = \left[Bhist\left(T_i^1 \right), \cdots, Bhist\left(T_i^{L_1} \right) \right]^T \in \mathbb{R}^{(2^{L_2})L_1 B} \tag{9}$$

3   The Computational Complexity of the SRDANet Algorithm

In this section, we provide evidence to show how light the computational complexity of our SRDANet algorithm is. We take the two-stage SRDANet as an example for analysis. Forming the patch-mean-removed matrix P costs $(j_1 j_2 + j_1 j_2 d_1 d_2)$ *flops*[2] in each stage of SRDANet; And the computational costs of learning the SRDA filter kernels from whole training sets is $[\frac{1}{2} N d_1 d_2 (j_1 j_2)^2 + O(N d_1 d_2 j_1 j_2)]$; The convolution of SRDA filters take $L_i d_1 d_2 j_1 j_2$ *flops* for stage i; The conversion of L_2 binary bits to a decimal number costs $2 L_2 d_1 d_2$ *flops* in the output stage; The operation of concatenated block-wise histogram is of complexity $O\left(d_1 d_2 B L_2 \log 2 \right)$.

Assuming $d \gg \max\left(j_1, j_2, L_1, L_2, B \right)$, the overall complexity of SRDANet is simplify as

$$O\left(d j_1 j_2 \left(L_1 + L_2 \right) + d (j_1 j_2)^2 \right) \tag{10}$$

---

[2] *flops*, An acronym for **FL**oating-point **O**perations **P**er **S**econd.

4   Experimental Results

In this section, we evaluate the performance of our proposed SRDANet in different face analysis tasks, such as *face recognition* and *expression recognition* in 2D or 3D facial images. Figure 2 shows the cropped face images from ORL, PIE, FRGC v2 and BU-3DFE database.

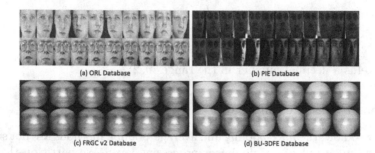

Fig. 2. Sample face images from face analysis database. (a) The ORL Database. (b) The PIE Database. (c) The FRGC v2 Database. (d) The BU-3DFE database.

4.1   Impact of Number of Filter Kernels

Before comparing SRDANet with existing LDANet [22] and other subspace methods, we first investigate the impact of the number of filter kernels of one-stage SRDANet and two-stage SRDANet. We use Yale database and down-sample the image to 32×32 pixels. We random choose 5 images per individual as training set and average the results over 10 random splits. For one-stage SRDANet, we vary the number of filter kernels in the first stage L_1 from 4 to 14. For two-stage SRDANet, We set the number of filter kernels in the second stage $L_2 = 8$ and vary L_1 from 4 to 14. The size of filter kernel is $j_1 = j_2 = 7$ and their non-overlapping block size is 7×7. The results are shown in Figure 3.

One can see that the accuracy of one-stage SRDANet and two-stage SRDANet increases for larger L_1. What's more, both one-stage SRDANet and two-stage SRDANet achieve best performance when L_1 approximately in the range $[8, 10]$.

4.2   2D Face Recognition on ORL Datasets

The ORL face database is used in this test. It consists of a total of 400 face images of 40 individuals (10 samples per individual). We random choose $n (= 2, 3, 4, 5)$ images per individual as training set. The rest of the database to form the testing set. For each given n, we average the results over 50 random splits. We compare the performance of SRDANet with LDANet [22], Discriminant Face Descriptor (DFD) [23], Sparse Representation Classifier (SRC) [24], Spectral Regression Discriminant Analysis (SRDA) [20], Spatially Smooth LDA (S-LDA) [25] and Completed LBP (C-LBP) [26]. We also use PCA as baseline algorithms.

The parameters of SRDANet are set to $L_1 = L_2 = 8$, $j_1 = j_2 = 7$, and the non-overlapping block size is 7×7. We use SVM classifier for SRDANet, LDANet, SRDA, C-LBP and use NN classifier with cosine distance for DFD and S-LDA. Different classifier is to secure the best performances of respective features.

The results are given in Table 1. One can see that our proposed algorithm again achieves the best accuracy. A prominent message drawn from the above experiments is that our proposed SRDANet can be very effective to extract the abstract representation of face images, and it is quite competitive to the current state-of-the-art face recognition algorithms.

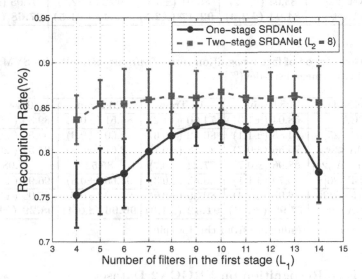

Fig. 3. Recognition accuracy of SRDANet on Yale for varying number of kernels in the first layer L_1.

4.3   2D Face Recognition on PIE Dataset

We investigate the performance of proposed SRDANet on PIE datasets for 2D face recognition. In PIE databases, original images were cropped into 32×32 pixels. The five near frontal poses is used in this test. There are around 170 images for each individual. We random choose $n(=5,10,20,30)$ images per individual as training set. The rest to form the testing set. For each given n, we average the results over 10 random splits. The parameters of SRDANet are same as the experiments of Section 4.2. Support Vector Machine (SVM) classifier is applied.

The results are given in Table 2. One can see that SRDANet algorithms achieves the best performance for all test sets. It is also observed that the standard deviation of SRDANet less than the standard deviation of LDANet. Compared to LDA, the SRDA provides more meaningful and stable eigen-solutions. Therefore, the SRDANet can extract more discriminative and stable features, so the SRDANet outperforms the LDANet and other subspace learning algorithms for different training samples.

Table 1. Comparison of face recognition rates of various methods with SVM classifier on ORL datasets (mean±std-dev(%))

Method	2 Train	3 Train	4 Train	5 Train
PCA	70.39 (±2.9)	77.17 (±2.4)	82.47 (±2.2)	85.09 (±2.4)
S-LDA [25]	81.71 (±2.5)	88.67 (±2.5)	92.58 (±1.7)	94.98 (±1.6)
SRDA [20]	80.93 (±3.0)	88.61 (±2.3)	92.08 (±1.9)	94.34 (±1.5)
SRC+PCA [24]	78.33 (±2.5)	86.46 (±2.2)	90.84 (±1.7)	93.54 (±1.6)
C-LBP [26]	83.06 (±2.6)	89.21 (±1.7)	94.21 (±1.6)	95.25 (±1.4)
DFD ($S = 5$) [23]	75.78 (±2.1)	82.57 (±2.5)	88.71 (±1.8)	92.35 (±1.3)
LDANet [22]	83.04 (±5.2)	89.91 (±5.5)	95.20 (±2.1)	97.05 (±1.9)
SRDANet	**84.94 (±2.6)**	**92.47 (±2.5)**	**96.42 (±1.5)**	**97.68 (±1.2)**

Table 2. Comparison of face recognition rates of various methods with SVM classifier on PIE datasets (mean±std-dev(%))

Method	5 Train	10 Train	20 Train	30 Train
PCA	58.04 (±1.3)	74.60 (±1.2)	85.51 (±0.5)	89.35 (±0.3)
S-LDA [25]	74.85 (±1.1)	86.52 (±0.6)	92.60 (±0.3)	94.55 (±0.3)
SRDA [20]	75.56 (±0.7)	87.18 (±0.7)	92.61 (±0.5)	94.34 (±0.2)
SRC+PCA [24]	68.96 (±0.7)	77.24 (±0.7)	85.52[*]	91.98[*]
C-LBP [26]	70.79 (±1.1)	86.85 (±0.3)	94.65 (±0.3)	96.98 (±0.2)
LDANet [22]	76.95 (±4.4)	89.39 (±2.9)	96.89 (±0.7)	97.99 (±0.4)
SRDANet	**77.99 (±2.6)**	**91.42 (±1.0)**	**96.96 (±0.4)**	**98.29 (±0.3)**

[*]Note that the results are from the 1st spilt.

4.4   3D Face Recognition on FRGC v2 Dataset

We choose FRGC v2 database for evaluating the performance of our SRDANet in 3D face recognition. Face images in FRGC v2 database were cropped into 128×128 pixels. We choose the subjects whose the number of samples more than 20, and leaves us a total of 1397 face images of 60 individuals. We random choose $n (= 2, 4, 6)$ images per individual as training set. The rest of the database to form the testing set. For each given n, we average the results over 10 random splits. In our SRDANet, the number of filters, the filter size and the non-overlapping block size are set to $L_1 = L_2 = 8$, $j_1 = j_2 = 7$ and 7×7, respectively.

The results are given in Figure 4. One can see that the our proposed SRDANet achieves the best performance for all test sets in 3D face recognition. Since the **small samples problem** of LDA also bring into LDANet model, and SRDA solve this problem by Spectral Regression Theory. Therefore, LDANet just achieves 70.12% accuracy, but our SRDANet achieves **92.99%** accuracy when the number of training samples is 2.

We can draw a conclusion that our proposed SRDANet can be very effective to extract the discriminative feature not only from 2D face images, but also from 3D face images.

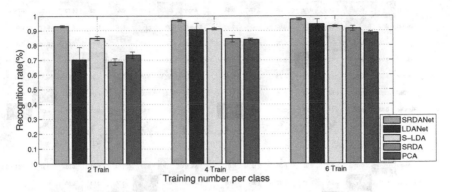

Fig. 4. Comparison of depth image face recognition rates(%) of various methods with SVM classifier on FRGC v2 datasets.

4.5 3D Expression Recognition on BU-3DFE Dataset

In this section, We evaluate the performance of SRDANet on BU-3DFE expression database for 3D expression recognition. The BU-3DFE database contains 100 subjects with 2500 facial expression, each individual has six different expressions (happiness, disgust, fear, angry, surprise and sadness) includes four levels of intensity and one neutral expression. Therefore, there are 25 (24 prototypic expressions and 1 neutral expression) expressions for each individual. We random choose $n_1 (= 50, 100, 150, 200)$ images from each prototypic expression and $n_2 (= 20, 25, 40, 50)$ images for the neutral expression as the training set. For each $(n_1 + n_2)$, we average the results over 10 random splits.

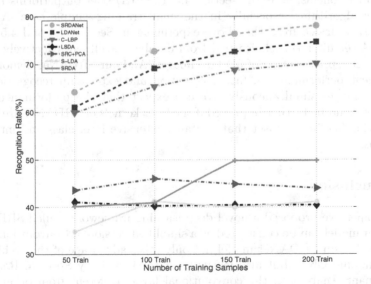

Fig. 5. Comparison of expression recognition rates(%) of various methods on BU-3DFE datasbase.

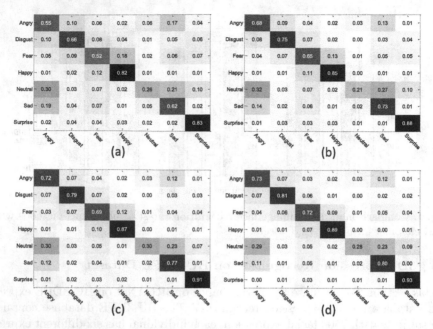

Fig. 6. Confusion matrices of different training number on the BU-3DFE dataset by our proposed SRDANet algorithm.

Figure 5 compares the our proposed SRDANet with the other existing algorithm, and Figure 6 presents the confusion matrices of different training number on the BU-3DFE dataset by our proposed SRDANet algorithm. We can get a similar conclusion as the section 4.4, the SRDANet outperforms LDANet and other algorithms, especially in the small training sample case. A prominent message drawn from the above experiment in Section 4.4 and 4.5 is that the SRDANet filter kernels learned from database itself can effectively extract the abstract representation of 3D face images. And our SRDANet demonstrates an excellent performance in face recognition and expression recognition with 3D facial images simultaneously. We can expect that the performance of our SRDANet could be further improved if the filter kernels of SRDANet are learned from a wide 3D face dataset that contains extensive inter-class and intra-class variations.

5   Conclusion

In this paper, we proposed a novel deep learning framework, called SRDA Network. Our model can be considered as a simplified version of Convolutional Neural Network. Our SRDANet model not only takes advantage of the architecture of conventional CNN, but also learns the filter kernels by Spectral Regression Discriminant Analysis in the convolutional layer. Different from other CNNs, there is no nonlinear processing layer until images are processed through all convolutional filter layer, which is a main characteristic of our SRDANet model. Then we perform nonlinear computation on the output of convolutional layer by hashing method and pooling the decimal-valued image using block-wise histogram. At last, the block-wise histogram is considered as the final features of input image. Extensive experiments demonstrate that our model can achieve excellent performance when faced with various face analysis tasks. So SRDANet model can effectively improve the performance and robustness of face analysis.

Acknowledgments. The work presented in this paper was supported by the National Natural Science Foundation of China (Grants No. NSFC-61402046, NSFC-61170176), Fund for Beijing University of Posts and Telecommunications (No.2013XZ10, 2013XD-04), Fund for the Doctoral Program of Higher Education of China (Grants No.20120005110002).

References

1. Tzimiropoulos, G., Zafeiriou, S., Pantic, M.: Subspace learning from image gradient orientations. IEEE Transactions on Pattern Analysis and Machine Intelligence **34**(12), 2454–2466 (2012)
2. Kang, C., Liao, S., Xiang, S., Pan, C.: Local sparse discriminant analysis for robust face recognition. In: 2013 IEEE Conference on Computer Vision and Pattern Recognition Workshops (CVPRW), pp. 846–853, June 2013
3. Ren, C.-X., Dai, D.-Q., Yan, H.: Coupled kernel embedding for low-resolution face image recognition. IEEE Transactions on Image Processing **21**(8), 3770–3783 (2012)
4. Ramirez Rivera, A., Castillo, R., Chae, O.: Local directional number pattern for face analysis: Face and expression recognition. IEEE Transactions on Image Processing **22**(5), 1740–1752 (2013)
5. Juefei-Xu, F., Savvides, M.: Subspace-based discrete transform encoded local binary patterns representations for robust periocular matching on nist's face recognition grand challenge. IEEE Transactions on Image Processing **23**(8), 3490–3505 (2014)
6. Ming, Y.: Robust regional bounding spherical descriptor for 3d face recognition and emotion analysis. Image and Vision Computing **35**, 14–22 (2015)
7. Chu, B., Romdhani, S., Chen, L.: 3d-aided face recognition robust to expression and pose variations. In: 2014 IEEE Conference on Computer Vision and Pattern Recognition (CVPR), pp. 1907–1914, June 2014
8. Ming, Y.: Rigid-area orthogonal spectral regression for efficient 3d face recognition. Neurocomputing **129**, 445–457 (2014)
9. Ming, Y., Ruan, Q.: Robust sparse bounding sphere for 3d face recognition. Image and Vision Computing **30**(8), 524–534 (2012). Special Section: Opinion Papers
10. Liong, V.E., Lu, J., Wang, G.: Face recognition using deep pca. In: 2013 9th International Conference on Information, Communications and Signal Processing (ICICS), pp. 1–5, December 2013
11. Lu, C., Zhao, D., Tang, X.: Face recognition using face patch networks. In: 2013 IEEE International Conference on Computer Vision (ICCV), pp. 3288–3295, December 2013

12. Bengio, Y., Courville, A., Vincent, P.: Representation learning: A review and new perspectives. IEEE Transactions on Pattern Analysis and Machine Intelligence **35**(8), 1798–1828 (2013)
13. Kavukcuoglu, K., Sermanet, P., Boureau, Y.-L., Gregor, K., Mathieu, M., Cun, Y.L.: Learning convolutional feature hierarchies for visual recognition. In: Advances in Neural Information Processing Systems, pp. 1090–1098 (2010)
14. Jarrett, K., Kavukcuoglu, K., Ranzato, M., LeCun, Y.: What is the best multi-stage architecture for object recognition? In: 2009 IEEE 12th International Conference on Computer Vision, pp. 2146–2153, September 2009
15. Huang, G.B., Lee, H., Learned-Miller, E.: Learning hierarchical representations for face verification with convolutional deep belief networks. In: 2012 IEEE Conference on Computer Vision and Pattern Recognition (CVPR), pp. 2518–2525, June 2012
16. Burges, C.J.C., Platt, J.C., Jana, S.: Distortion discriminant analysis for audio fingerprinting. IEEE Transactions on Speech and Audio Processing **11**(3), 165–174 (2003)
17. Kang, L., Kumar, J., Ye, P., Li, Y., Doermann, D.: Convolutional neural networks for document image classification. In: 2014 22nd International Conference on Pattern Recognition (ICPR), pp. 3168–3172, August 2014
18. Ngiam, J., Coates, A., Lahiri, A., Prochnow, B., Le, Q.V., Ng, A.Y.: On optimization methods for deep learning. In: Proceedings of the 28th International Conference on Machine Learning (ICML 2011), pp. 265–272 (2011)
19. Rifai, S., Mesnil, G., Vincent, P., Muller, X., Bengio, Y., Dauphin, Y., Glorot, X.: Higher order contractive auto-encoder. In: Gunopulos, D., Hofmann, T., Malerba, D., Vazirgiannis, M. (eds.) ECML PKDD 2011, Part II. LNCS, vol. 6912, pp. 645–660. Springer, Heidelberg (2011)
20. Cai, D., He, X., Han, J.: Srda: An efficient algorithm for large-scale discriminant analysis. IEEE Trans. on Knowl. and Data Eng. **20**(1), 1–12 (2008)
21. Krizhevsky, A., Sutskever, I., Hinton, G.E.: Imagenet classification with deep convolutional neural networks. In: Advances in Neural Information Processing Systems, pp. 1097–1105 (2012)
22. Chan, T.-H., Jia, K., Gao, S., Lu, J., Zeng, Z., Ma, Y.: Pcanet: A simple deep learning baseline for image classification? arXiv preprint arXiv:1404.3606 (2014)
23. Lei, Z., Pietikainen, M., Li, S.Z.: Learning discriminant face descriptor. IEEE Transactions on Pattern Analysis and Machine Intelligence **36**(2), 289–302 (2014)
24. Wright, J., Yang, A.Y., Ganesh, A., Sastry, S.S., Ma, Y.: Robust face recognition via sparse representation. IEEE Transactions on Pattern Analysis and Machine Intelligence **31**(2), 210–227 (2009)
25. Cai, D., He, X., Hu, Y., Han, J., Huang, T.: Learning a spatially smooth subspace for face recognition. In: Proc. IEEE Conf. Computer Vision and Pattern Recognition Machine Learning (CVPR 2007) (2007)
26. Guo, Z., Zhang, D.: A completed modeling of local binary pattern operator for texture classification. IEEE Transactions on Image Processing **19**(6), 1657–1663 (2010)

Pedestrian Classification and Detection in Far Infrared Images

Atmane Khellal[1], Hongbin Ma[1,2]($\boxtimes$), and Qing Fei[1,2]

[1] School of Automation, Beijing Institute of Technology,
Beijing 100081, People's Republic of China
{khe.atmane,mathmhb}@gmail.com
[2] State Key Lab of Intelligent Control and Decision of Complex Systems,
Beijing Institute of Technology, Beijing 100081, People's Republic of China

Abstract. In this paper, a new approach of learning features based on convolutional neural networks for pedestrian detection in far infrared images is presented. Unlike traditional recognition systems which use hand-designed features like SIFT or HOG, our convolutional networks architecture learns new features and representations more appropriate to the classification task in infrared images. Another pedestrian detector based on logistic regression is designed and compared to convolutional networks based classifier. Our system built over non-visible range sensor may have an important role in next generation robotics, especially in perception, advanced driver assistant systems (ADAS) and intelligent surveillance systems.

Keywords: Convolutional neural networks · Far infrared imagery · Learning features · Logistic regression · Pedestrian detection

1 Introduction

Vision based systems play an important role in the field of robotics, especially in perception, visual navigation and object recognition. However, computer vision community still limited to algorithms and techniques related to visible range sensors. This low interest in non-visible imaging systems can be interpreted by several reasons like the high cost of sensors and poor image quality. Recently, these traditional objections become less pertinent due to the fact that infrared imaging systems advance rapidly and its cost drops impressively. Exploring infrared thermal imaging technology and related algorithm design can complete and promote current robots systems. As a result, we believe that infrared systems can be considered as one of the important sensors and perception systems for next generation robots, to achieve more complicated tasks and to perform better than humans.

In this paper, we explore both classification and detection of pedestrians in infrared thermal images, which may be integrated with robot perception systems,

This work was partially supported by the National Natural Science Foundation in China (NSFC) under Grants 61473038.

H. Liu et al. (Eds.): ICIRA 2015, Part I, LNAI 9244, pp. 511–522, 2015.
DOI: 10.1007/978-3-319-22879-2_47

advanced driver assistant systems (ADAS) [12] or even intelligent surveillance systems. In traditional object recognition systems, we use some hand-designed features like SIFT (Scale-Invariant Feature Transform) or HOG (Histogram of Oriented Gradients) to build a classifier for some given objects. This approach is widely used by the computer vision community in different tasks even in infrared imagery [7–11]. However, some recent works have shown that to build a better recognition system, we need to design better features instead of building better classifiers. Moreover, mid-level features (or representations) are either difficult or impossible to hand-engineer [2]. In addition, we want to discover new features more appropriate and more specific to infrared images.

In this contribution, we investigate an approach of learning features based on Convolutional Neural Networks (CNN), which achieves state-of-the-art in different visual recognition challenges. The rest of the paper is organized as follows: first we introduce the classification problem and we propose a solution based on logistic regression, and another one based on convolutional networks; then the detection problem and the sliding window approach are presented with some simulation results analyzed in details; finally, we conclude this paper by summarizing our work and pointing out some forward perspectives.

2   Classification

We address the problem of building a pedestrian detector as classification problem. Therefore, given an input as a patch image, the model predicts the probability whether the patch contains a human or not. We first start by logistic regression, which is a simple learning algorithm for making such decision. Then, we introduce an algorithm for features learning based on convolutional neural networks.

Fig. 1. Pedestrian detector as a classifier.

As shown in Figure 1, the classifier's input (the ith example) represents the image patch, while the output represents the corresponding label, which in the ideal case indicates whether the image patch contains a human.

2.1   Logistic Regression Based Classification

We use the LSI Far Infrared Pedestrian Data set (LSIFIR)[13]. In the classification training set, each example is a 64×32 image patch. The first step is to unroll it into a vector representation $x \in R^{2048}$, then we add the intercept term (*bias*), so that we have $x \in R^{2049}$. Then we preprocess the data by applying feature normalization, to obtain data with 0 mean and 1 as standard deviation.

In logistic regression, we use the hypothesis defined in equation (1) to try to predict the probability that a given example belongs to the "1" class (presence of pedestrian) versus the probability that it belongs to the "0" class (background).

$$h_\theta(x) = \frac{1}{1 + \exp(-\theta^\top x)} \tag{1}$$

For a set of training examples with binary labels $\{(x^{(i)}, y^{(i)}) : i = 1, \ldots, m\}$, the following cost function measures how well a given hypothesis $h_\theta(x)$ fits our training data:

$$J(\theta) = -\sum_i \left(y^{(i)} \log(h_\theta(x^{(i)})) + (1 - y^{(i)}) \log(1 - h_\theta(x^{(i)})) \right) \tag{2}$$

The parameters of the model ($\theta \in R^{2049}$) can be obtained by minimizing the cost function $J(\theta)$, using different tools of optimization, as well as we provide the cost function and the gradient defined in equation (3).

$$\nabla_\theta J(\theta) = \sum_i x^{(i)} (h_\theta(x^{(i)}) - y^{(i)}) \tag{3}$$

After the training phase, we use the learned θ to predict an unseen input x by computing the hypothesis $h_\theta(x)$. The class label is predicted as 1 if $h_\theta(x) \geq 0.5$, and as 0 otherwise.

2.2 Training and Testing the Logistic Regression Model

The pedestrian infrared data set is divided into detection data set and classification data set, which contains 81 592 images. See Table 1 for details.

Table 1. LSIFIR classification data set.

Data type	Positive examples	Negative examples
Training set	10 208	43 390
Test set	5 944	22 050

In Figure 2, we show some positive examples (pedestrian) and negative examples (background) from the classification training data set.

To evaluate the learning algorithm, we use F score defined in equation (4), considering the issue that an evaluation based on accuracy is not significant since the difference between the number of positive and negative examples is very large.

$$F = \frac{2 \cdot prec \cdot rec}{prec + rec} \tag{4}$$

We compute $prec$ (precision) and rec (recall) as follow:

$$prec = \frac{tp}{tp + fp} \tag{5}$$

Fig. 2. Positive (top) and negative (bottom) training examples.

$$rec = \frac{tp}{tp + fn} \tag{6}$$

where

- tp is the number of true positives: the ground truth label says it is a pedestrian and our algorithm correctly classified it as a pedestrian.
- fp is the number of false positives: the ground truth label says it is not a pedestrian, but our algorithm incorrectly classified it as a pedestrian.
- fn is the number of false negatives: the ground truth label says it is a pedestrian, but our algorithm incorrectly classified it as non-pedestrian.

Table 2. Logistic regression results.

Data type	Accuracy	tp	fp	fn	prec	rec	F score
Training set	98.04	9540	380	668	0.9616	0.9345	0.9479
Test set	96.94	5236	146	708	0.9728	0.8808	0.9245

After training the model, we compute different evaluation parameters as shown in Table 2. We conclude that the model predicts well for both training and test sets.

We trained different models based on the maximum number of iterations, then the evaluation parameters are computed for training and test sets. The results are shown in Figure 3.

2.3 Convolutional Neural Network Based Classification

Convolutional Neural Networks (CNN) are very similar to ordinary neural networks with a specialized connectivity structure. Convolutional neural networks were inspired by the visual system structure. Moreover, CNN based pattern recognition systems are among the best performing systems [2]. Unlike standard multi-layer neural networks, CNN are much more easier to train [2]. In one significant work conducted by LeCun et al. [1], several convolutional networks using

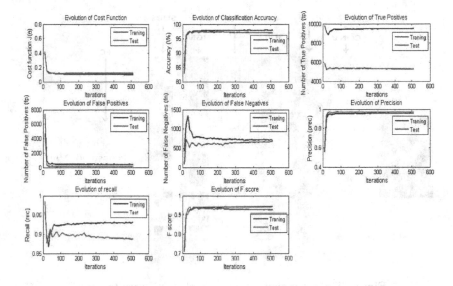

Fig. 3. Evaluation parameters for logistic regression based classification.

the error gradient designed and trained, obtaining state-of-the-art performance on several pattern recognition tasks. Recently, very deep convolutional neural networks are designed and trained successfully and won many computer vision challenges [3–5].

In logistic regression model, we use the raw pixels of the image patch as a feature vector to make decision (pedestrian or no). There exist many kinds of features, extracted from images, which can be used as feature vector for a variety of learning algorithms. Feature descriptors like SIFT (Scale-Invariant Feature Transform) or HOG (Histogram of Oriented Gradients) are widely used in many different computer vision tasks.

Unlike traditional object recognition systems which use hand-designed features, our CNN model learns new features, hence hopefully it may be suitable for infrared images. The reasons why we adopt a CNN based learning features approach are two-folds: first, we can learn more higher representations as well as we go deep through the network; second, infrared images are different from visible images, thus we need appropriate features. Now we ignore how to design such features, and instead, we try to learn them.

The input image will be transformed from raw pixels to new feature maps as long as it passes through different layers [6]. Figure 4 shows the neurons activations in each layer of our trained CNN model, starting from the input image patch, going through different feature maps (we plot only the first 6 feature maps of each layer), ending by the class score given by the output unit.

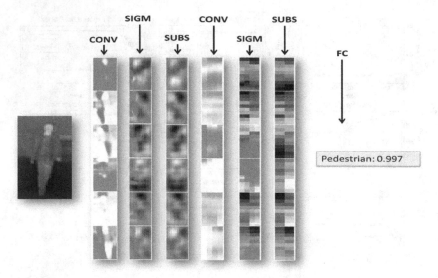

Fig. 4. The activations of an example CNN architecture.

2.4   CNN Layers

To build a CNN architecture, there are mainly three different types: convolutional layer (CONV), sub-sampling layer (SUBS) and fully connected layer (FC). Then, we stack these layers to form a full CNN architecture.

- INPUT $[64 \times 32 \times 1]$ holds the input image which is, in this case, a one channel 64×32 image.
- CONV layer convolves the input image with the learned convolutional filters. In our case, we have 6 kernels (9×9), so the resulting feature maps will have the dimension of $[56 \times 24 \times 6]$.
- SIGM layer will apply nonlinear function to the previous layer elements such as sigmoid function. Therefore, this layer has the same size as the CONV layer.
- SUBS layer will perform a down-sampling operation to the input feature maps. The resulting size is $[28 \times 12 \times 6]$.
- FC layer will compute the class score, like regular neural networks. In this layer, each neuron is connected to all the neurons of previous layer.

2.5   Architecture of Our CNN Model

We adopt a similar architecture to the one developed by LeCun for document recognition [1]. In our model shown in Figure 5, we use a [6c 2s 12c 2s] architecture. In the first convolution layer, we use 6 filters with 9×9 kernel size; while in the second convolution layer, we use 12 filters with same kernel size. For both sub-sampling layers, we down-sample the input one time. So that one feature map of size 20×4 becomes a 10×2 feature map. The following paragraphs describe in detail our architecture.

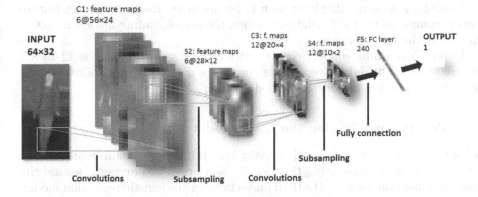

Fig. 5. Architecture of our Convolutional Neural Network model.

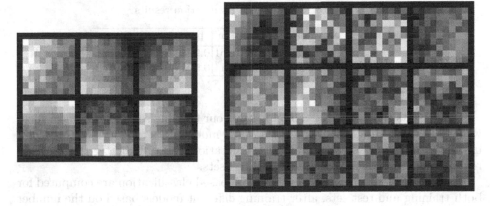

Fig. 6. Visualization of learned kernels example, first convolution filters (left), second convolution filters (right).

Layer C1 is a convolutional layer with 6 feature maps. Each unit in each feature map is connected to a 9×9 neighborhood in the input. The size of the feature maps is 56×24, which prevents connection from the input from falling off the boundary. An example of learned filters associated to this layer is illustrated in Figure 6, the left picture.

Layer S2 is a sub-sampling layer with 6 feature maps. Each unit in each feature map is connected to a 2×2 neighborhood in the corresponding feature map C1. The size of feature map is 28×12, and it has half the number of width and height as feature maps in C1(down-sampling operation).

Layer C3 is a convolutional layer with 12 feature maps. Each unit in each feature map is connected to several 9×9 neighborhoods. The size of the feature map is 20×4. Figure 6, the right picture shows an example of learned filters in this layer.

Layer S4 is sub-sampling layer with 12 feature maps. Each unit in each feature map is connected to 2 × 2 neighborhood in the corresponding feature map C3. the size of the feature map is 10 × 2.

Layer F5 is a fully connected layer with 240 units (in C4 there is 12 10 × 2 feature maps). The output represents the class score or the probability that an input image patch is considered as pedestrian.

2.6   Training and Testing the CNN Model

LeCun et al. in their paper [1] explain in detail how to train convolutional networks by back-propagating the classification error. In our work, we use the same classification data set (LSIFIR) used to train the logistic regression model, described in Section 2.2. The results for the training and the test sets are described in Table 3.

Table 3. CNN based classification results

Data type	Accuracy	tp	fp	fn	prec	rec	F score
Training set	99.60	10047	51	161	0.9949	0.9842	0.9895
Test set	99.13	5730	29	214	0.9949	0.9639	0.9792

The results show that the model fits well our training set and also generalizes well in the test set, which proves that the model does not over-fit the training data. Moreover, the CNN based classification performs better than logistic regression model for both training and test sets.

The evaluation parameters for the CNN based classification are computed for both training and test sets, after training different models based on the number of iterations. The results are shown in Figure 7.

3   Detection

Given an infrared thermal image with any size containing or not one or several pedestrians, the goal is to detect all the people presented in the image and localize them with bound boxes (Figure 8).

To get advantage of the pedestrian detector developed previously, we use a technique of sliding window. We fix the widow size to 64 × 32, the same size as the patch used for classification. Then, we classify the patch using CNN or logistic regression. The output of a classifier is stored in an image at the same location of the patch, an example is shown in Figure 9.

In the detection image, the white regions correspond to the pedestrian. So, more the region is white, more the probability that it contains a pedestrian is big, and more the region is black, more the pedestrian detection probability is low (the background probability is high).

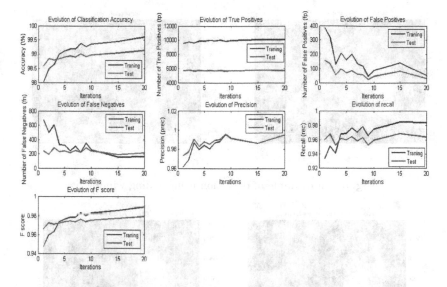

Fig. 7. Evaluation parameters for CNN based classification.

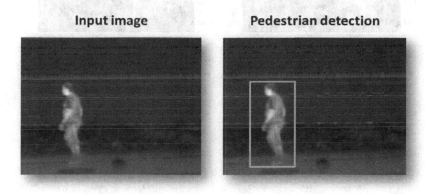

Fig. 8. Example of pedestrian detection, (left) the input image, (right) pedestrian detection ground truth.

To localize the pedestrian in the image, we estimate the contour of the white region in the detection image and we approximate it to a bound box as shown in Figure 10. Then the coordinates of this box will be projected in the original image to get the location of region of interest.

In general, for the images where the region of interest is bigger than the patch size, we first perform a down-sampling to the input image to get different images at different scales. Then, for each resulting image we repeat the procedure (sliding window and classification), so that we can detect pedestrian at different scales.

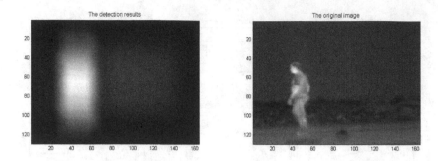

Fig. 9. An example of detection using sliding window.

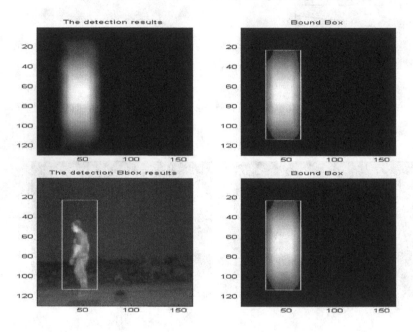

Fig. 10. An example of detected pedestrian localization.

Finally, we present a comparison between three classification algorithms; logistic regression (LR), convolutional neural networks (CNN) and another algorithm based on ordinary neural networks (NN). For the NN model, we use 2048 input units, 100 hidden units and one output unit. The comparison is conducted in terms of accuracy and F score, the results are shown in figure 11. Even though we note that the NN model slightly performs better than CNN model in the training set, but in the generalization aspect (test set), the CNN model shows the best performances.

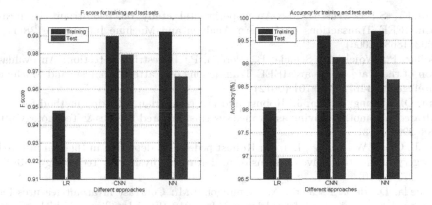

Fig. 11. Comparaison of different approaches.

4   Conclusion

In this paper, an approach of learning features based on convolutional neural networks is presented. We learn new feature representations by training a convolutional neural networks to detect pedestrians in far infrared images. These features are specific to infrared images and are very useful for the classification where the systems have shown high performances for both training and test sets. In this work, we demonstrate that the idea of learning features using convolutional neural networks works well as it works for visible images. And in the future we may explore whether and how the characteristics of infrared images can be fused into the design of CNN or CNN variants for the sake of improving the system performance further.

References

1. LeCun, Y., Bottou, L., Bengio, Y., Haffne, P.: Gradient-Based Learning Applied to Document Recognition. Proceedings of the IEEE **86**(11), 2278–2324 (1998)
2. Bengio, Y.: Learning deep architectures for AI. Foundations and Trends in Machine Learning **2**(1), 1–127 (2009)
3. Krizhevsky, A., Sutskever, I., Hinton, G.E.: Imagenet classification with deep convolutional neural networks. In: Advances in neural information processing systems, pp. 1097–1105 (2012)
4. Sermanet, P., Eigen, D., Zhang, X., Mathieu, M., Fergus, R., LeCun, Y.: Overfeat: Integrated recognition, localization and detection using convolutional networks. arXiv preprint arXiv:1312.6229 (2013)
5. Simonyan, K., Zisserman, A.: Very deep convolutional networks for large-scale image recognition. arXiv preprint arXiv:1409.1556 (2014)
6. Zeiler, M.D., Fergus, R.: Visualizing and understanding convolutional networks. In: Fleet, D., Pajdla, T., Schiele, B., Tuytelaars, T. (eds.) ECCV 2014, Part I. LNCS, vol. 8689, pp. 818–833. Springer, Heidelberg (2014)

7. Munder, S., Gavrila, D.M.: An experimental study on pedestrian classification. IEEE Transactions on Pattern Analysis and Machine Intelligence **28**(11), 1863–1868 (2006)
8. Dollar, P., Wojek, C., Schiele, B., Perona, P.: Pedestrian detection: An evaluation of the state of the art. IEEE Transactions on Pattern Analysis and Machine Intelligence **34**(4), 743–761 (2012)
9. Liu, Q., Zhuang, J., Ma, J.: Robust and fast pedestrian detection method for far-infrared automotive driving assistance systems. Infrared Physics & Technology **60**, 288–299 (2013)
10. Li, J., Gong, W., Li, W., Liu, X.: Robust pedestrian detection in thermal infrared imagery using the wavelet transform. Infrared Physics & Technology **53**(4), 267–273 (2010)
11. Olmeda, D., de la Escalera, A., Armingol, J.M.: Contrast invariant features for human detection in far infrared images. In: 2012 IEEE Intelligent Vehicles Symposium (IV2012), pp. 117–122 (2012)
12. Geronimo, D., Lopez, A.M., Sappa, A.D., Graf, T.: Survey of pedestrian detection for advanced driver assistance systems. IEEE Transactions on Pattern Analysis and Machine Intelligence **32**(7), 1239–1258 (2010)
13. Olmeda, D., Premebida, C., Nunes, U., Armingol, J., Escalera, A.: LSI far infrared pedestrian dataset. http://earchivo.uc3m.es/handle/10016/17370

One New Human-Robot Cooperation Method Based on Kinect Sensor and Visual-Servoing

Hongbin Ma[1,2]($\boxtimes$), Hao Wang[1], Mengyin Fu[1,2], and Chenguang Yang[3]

[1] School of Automation, Beijing Institute of Technology,
Beijing 100081, People's Republic of China
mathmhb@gmail.com
[2] State Key Lab of Intelligent Control and Decision of Complex Systems,
Beijing Institute of Technology, Beijing 100081, People's Republic of China
[3] School of Computing and Mathematics,
Plymouth University, Plymouth PL4 8AA, UK

Abstract. Human-robot interactions have received increased attention during the past decades for conveniently introducing robot into human daily life. In this paper, a novel Human-robot cooperation method is developed, which falls in between full human control and full robot autonomy. The human operator is in charge of the main operation and robot autonomy is gradually added to support the execution of the operator's intent. The proposed cooperation method allows the robot to accomplish tasks effectively with the help of human. The effectiveness of the method is verified by experiments, which is based on the Microsoft Kinect Sensor and the Virtual Robot Experimentation Platform.

1 Introduction

In recent years, the presence of robotic systems more or less evolved is common in our daily life. Since the start of this field of study from the last century, their evolution was incomparable. In the domains such as the automobile industry or the mobile phone factory, the uses of robots have become essential [1]. Humans and robots can perform tasks together and their relation became more important than a basic remote control to realize a task .The human/robot interaction is a large research area, which has attracted more and more interests in the research community.

The development of robots that incorporate teleoperation brings together two very different branches of human-robot interaction (HRI). One branch is social HRI, which focuses on studying psychological aspects of conversational interactions between people and robots. The human's role is as a service receiver in the interaction, in contrast to the robot's role as a service provider. The other branch is HRI for teleoperation, which typically focuses on issues like the workload of the operator (person remotely controlling the robot), situation awareness, and shared autonomy [2] for the remote operation of nonsocial robots.

This work is partially supported by National Natural Science Foundation (NSFC) under Grants 61473038 and 61473120. Also partially supported by Guangdong Provincial Natural Science Foundation of China 2014A030313266.

H. Liu et al. (Eds.): ICIRA 2015, Part I, LNAI 9244, pp. 523–534, 2015.
DOI: 10.1007/978-3-319-22879-2_48

Some researchers developed HRI systems using force sensor to control the motion of the robot. Mykoniatis developed a face recognizing robot to increase situation awareness and enhance HRI [3]. Wang designed a robot dancer control system to enhancing haptic HRI [4]. The system is based on force sensor feedbacks and laser range finders. However, the success rate is low and the controller needs further improvement. Mora developed a teleoperation approach for mobile social robots [5]. This approach incorporates automatic gaze control and three-dimensional spatial visualization. Wakita developed a motion control of an omnidirectional-type cane [6]. The system is based on a human-walking-intention model and force sensor. Spexard focused on achieve comprehending human-oriented interaction with an anthropomorphic robot [7]. This research brought together different interaction concepts and perception capabilities integrated on a humanoid robot. These systems cost lower than previous HRI systems, and enhanced the stability of the HRI, but limited in the success rate.

Among human-robot interaction techniques, visual interaction is dealt with as a major topic. In particular, Microsoft's Kinect sensor [7], which provides wealth of information like depth, which a normal video camera fails to provide. In addition, RGB-D data from the Kinect sensor can be used to generate a Skeleton model of humans with semantic matching of 15 body parts. Since human activities are a collection of how different body parts move across each time period, these information can be used to detect human activities.

The existing human-robot interaction systems enhance the stability and reduce the cost, but also have some limitations, such as comfort, accuracy and success rate. One solution to enhance the accuracy and success rate is to bridge the gap between full human control and full robot autonomy, the human operator is in charge of the main operation and robot autonomy is gradually added to support the execution of the operator's intent. Our objective is based on this idea. We aimed at enabling intuitive interaction between human and robot, in the context of an service scenario, where the two can collaborate to realize the task accurately. We will explore how a human-robot team can work together effectively.

The rest of this paper presents the cooperation system environment and details of the technique developed to accomplish the human-robot cooperation for the robotic application.

2   System Architecture

As an example of human-robot interaction, the objective of the human-robot cooperation system is to enable a robot to classify the objects of different shapes and colors on the table effectively with the direction of human. Note that in this experiment, the robot itself has no *priori* information of the position of the objects and the robot is not programmed in advance for achieving the task of classifying the objects. In other words, we aim to make the robot capable to achieve different unprogrammed jobs under the guidance of the operator, while the operator needs only to deliver his or her intent via rough motion intent. The required operations can be summarized as following:

- for the human: to direct the robot move arm to the nearby of the objects on the table.
- for the robot: to detect the objects and adjust the position of the gripper, then, grasp the objects and classify them.

Fig. 1. The structure of the human-robot cooperation system, which includes Kinect Sensor and Baxter® robot and the Virtual Robot Experimentation Platform

The proposed cooperation system is composed of a Laptop, one Microsoft Kinect Sensor and the Virtual Robot Experiment Platform software. The Microsoft Kinect Sensor is used to detect and track human movement, these video data gathered by the Kinect with 640 × 480 pixels at 30 Hz are streamed to a laptop using a USB 2.0 connection. These data as the raw control signal are analyzed and then conversed into control command of each joint of the robot by the remote OpenNI API, sent to the V-REP over serial port communication.

The service robot shown in Fig. 1 is Baxter® robot in V-REP. The humanoid Baxter® robot includes a torso based on a movable pedestal and two 7DOF (degree of freedom) arms installed on left/right arm mounts respectively. Each arm has 7 rotational joints and 8 links, as well as an interchangeable gripper (such as electric gripper or vacuum cup) which can be installed at the end of the arm. A head-pan with a screen, located on the top of torso, can rotate in the horizontal plane.

The control object is the Baxter robot's arms in V-REP. The robot simulator V-REP, with integrated development environment, is based on a distributed control architecture. Next to offering the traditional approaches also found in other simulators, V-REP adds several additional approaches which makes V-REP very versatile and ideal for multi-robot applications.

3   Tele-Operation

3.1   Control Signal Acquisition

The control signal of the teleoperation system is from skeleton data tracked by the Kinect, the first feature can be extracted is joint data. For each joint, we have three main information. The first information is the index of the joints. Each joint has a unique index value. The second information is the positions of each joint in x, y, and z coordinates. These three coordinates are expressed in meters. The x, y, and z axes are the body axes of the depth sensor. This is a right-handed coordinate system that places the sensor array at the origin point with the positive z axis extending in the direction in which the sensor array points. The positive y axis extends upward, and the positive x axis extends to the left (with respect to the sensor array).

3.2 Communication

After acquisition of the control signal, the next step is to send signal to V-REP. The communication between the OpenNI and the Virtual Robot Experimentation Platform has to be real time. V-REP allows the user to choose among various programming techniques simultaneously and even symbiotically. The cooperation system chooses the remote API client method to make V-REP to communicate with OpenNI.

The remote OpenNI API plugin start as the 'Kinect Server' communicating via a free port set in List. 1 before. The temporary remote API server service was chosen to start the 'Kinect Server' from within a script at simulation start. After the simulation started, the child script will detect the free serial port set in the 'Kinect Server' and receive skeleton data from the 'Kinect Server'.

3.3 Robot Arm Tele-Control

To remotely control the robot, the first required development was the arm control. Offsets were applied to define the workspace of the virtual arm. They provided a suitable workspace environment where the remote control of the Baxter® Robot arm was possible. Then the choice concerning was decided. As a mirror, the arms movements were matched to the Baxter® Robot Simulator arms, as the user moved its own arms. The Cartesian systems used for the Baxter® Robot Simulator and the Human in Real World are shown in Fig. 2.

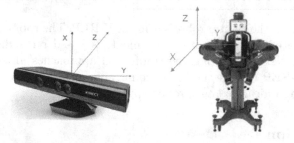

Fig. 2. The Cartesian world of Kinect Sensor and simulator

The control of the arms was performed by sending positions of shoulders, elbows and hands with a command via a free serial port to the Baxter® Robot Simulator. Then it moved its joints according to its inverse kinematics in order to reach the desired positions. In order to get the target positions in simulator, the coordinate transformation is performed from the real world space coordinates to the task space coordinates.

More specifically, the OpenNI tracker detects the position of the following set of joints in the 3D space $G = \{g_{i, \ i \in [1,I]}\}$. The position of joint g_i is implied by vector $P_{i0}(t) = [x_m \ y_m \ z_m]^T$, where t denotes the frame for which the joint position is located and the origin of the orthogonal X, Y and Z coordinates system is placed at the center of the Kinect sensor. The task space is also a 3-dimensional space, the position of joint g_i is implied by vector $P_{i1}(t) = [x_s \ y_s \ z_s]^T$ in the task space. The work space mapping is shown as below:

$$
\begin{bmatrix} x_s \\ y_s \\ z_s \end{bmatrix} = \begin{bmatrix} \cos\beta & -\sin\beta & 0 \\ \sin\beta & \cos\beta & 0 \\ 0 & 0 & 1 \end{bmatrix} \times \left(\begin{bmatrix} S_x & 0 & 0 \\ 0 & S_y & 0 \\ 0 & 0 & S_z \end{bmatrix} \begin{bmatrix} x_m \\ y_m \\ z_m \end{bmatrix} + \begin{bmatrix} T_x \\ T_x \\ T_z \end{bmatrix} \right) \tag{1}
$$

where δ is the revolution angle about the Y-axis of the manipulator base frame, $[S_x \ S_y \ S_z]^T$ and $[T_x \ T_y \ T_z]^T$ are the scaling factors and translations about the X, Y and Z axis. The mapping parameters of (1) are given by

$$
\delta = \frac{\pi}{2}, \quad \begin{bmatrix} S_x \\ S_y \\ S_z \end{bmatrix} = \begin{bmatrix} 0.7 \\ 0.7 \\ 0.7 \end{bmatrix}, \quad \begin{bmatrix} T_x \\ T_y \\ T_z \end{bmatrix} = \begin{bmatrix} 0.051 \\ 0.049 \\ 0.034 \end{bmatrix}. \tag{2}
$$

4   Visual Servoing System

In order to implement the function that the Baxter® robot can classify the objects of different shapes and colors on the table effectively while receives the trigger signal, the image-based visual servoing system [8] was applied. The visual servoing system consists of one fixed camera and two cameras on each wrist of the Baxter® robot that can image a scene containing the different objects on the table.

4.1   Image Jacobian

The image Jacobian matrix relates robot joint angle changes (or robots positions) to image feature changes. For an eye-in-hand system, if the object in the workspace is motionless, the image feature changes would be only related to the motion of the robot or camera, which can be expressed as

$$
y = f(\theta) \tag{3}
$$

Suppose $\theta = [\theta_1, \theta_2, \cdots, \theta_n]^T$ denotes the coordinate in robot joint space, $\mathbf{s} = [s_1, s_2, \cdots, s_p]^T$ is that of the robot end-effector in Cartesian space, and $\mathbf{y} = [y_1, y_2, \cdots, y_n]^T$ is image feature vector. Consequently $\dot{\mathbf{s}}$ is the velocity of robot end-effector including translation and rotation, $\dot{\theta}$ is the velocity in robot joint space, and $\dot{\mathbf{y}}$ is the velocity in image feature space. The relation between $\dot{\theta}$ and $\dot{\mathbf{s}}$ is

$$
\dot{\mathbf{s}} \approx \mathbf{J}_\alpha \cdot \dot{\theta} \tag{4}
$$

where

$$\mathbf{J}_\alpha = \frac{\partial \mathbf{s}}{\partial \theta} = \begin{bmatrix} \frac{\partial s_1}{\partial \theta_1} & \frac{\partial s_1}{\partial \theta_2} & \cdots & \frac{\partial s_1}{\partial \theta_n} \\ \cdots & \cdots & \cdots & \cdots \\ \frac{\partial s_p}{\partial \theta_1} & \frac{\partial s_p}{\partial \theta_2} & \cdots & \frac{\partial s_p}{\partial \theta_n} \end{bmatrix} \tag{5}$$

which is called robot Jacobian or feature Jacobian [9].

Similarly, the relation between $\dot{\mathbf{s}}$ and $\dot{\mathbf{y}}$ can be

$$\dot{\mathbf{y}} \approx \mathbf{J}_\beta \cdot \dot{\mathbf{s}} \tag{6}$$

where

$$\mathbf{J}_\beta = \frac{\partial \mathbf{s}}{\partial \theta} = \begin{bmatrix} \frac{\partial y_1}{\partial \theta_1} & \frac{\partial y_1}{\partial \theta_2} & \cdots & \frac{\partial y_1}{\partial \theta_n} \\ \cdots & \cdots & \cdots & \cdots \\ \frac{\partial y_p}{\partial \theta_1} & \frac{\partial y_p}{\partial \theta_2} & \cdots & \frac{\partial y_m}{\partial \theta_n} \end{bmatrix} \tag{7}$$

J_β is called image Jacobian [10], interaction matrix [11], or feature Jacobian. Thus we can get $\dot{\mathbf{y}} = \mathbf{J}_\beta \cdot \dot{\mathbf{s}} = \mathbf{J}_\beta \cdot \mathbf{J}_\alpha \dot{\theta}$. Let $\mathbf{J}_\theta = \mathbf{J}_\beta \cdot \mathbf{J}_\alpha$, then

$$\dot{\mathbf{y}} \approx \mathbf{J}_\theta \cdot \dot{\theta} \tag{8}$$

$\mathbf{J}_\theta$ is nominated as composite Jacobian or Visual-Motor Jacobian.

4.2   Online Estimation with Kalman Filtering

In an image-based robot visual servo system, the Jacobian matrix is a complex temporal nonlinear matrix. Its accuracy would affect directly to the entire system performance. Online estimation of of Jacobian matrix is to acquire the Jacobian matrix by dynamic computing based on real-time measurements of the robot jointangulars and observations of the image features [12]. The system model is not required in advance.

By a constructing state space model, the Jacobian estimation problem is transformed to that of system state estimation. The system state vector is composed of elements of Jacobian matrix. Furthermore, image features construct the measurement vector of the system. Kalman filtering algorithm applied to estimate the states. In order to estimate elements of total Jacobian matrix, measurement vector x is defined as $mn \times 1$vector, which is as follows

$$\mathbf{x} = [\mathbf{J_1}\ \mathbf{J_2}\ \cdots\ \mathbf{J_m}]^T \tag{9}$$

where $\mathbf{J_i} = [\frac{\partial y_i}{\partial \theta_1} \frac{\partial y_i}{\partial \theta_2} \cdots \frac{\partial y_i}{\partial \theta_n}]$is the ith row vector of Jacobian matrix $\mathbf{J}$.

Let $x(k)$ as system state, the difference of image feature as system's output, that is

$$z(k) = y(k+1) - y(k) \tag{10}$$

According to equation (10), we could get the state space model

$$\begin{cases} \mathbf{x}(k+1) = \mathbf{x}(k) + \omega(k) \\ z(k) = \mathbf{H}(k)x(k) + \upsilon(k) \end{cases} \tag{11}$$

where

$$\mathbf{H}(k) = \begin{bmatrix} \Delta\mathbf{h}(k)^T & & 0 \\ & \ddots & \\ 0 & & \Delta\mathbf{h}(k)^T \end{bmatrix}_{m \times mn} \tag{12}$$

$\omega(k)$ and $\upsilon(k)$ are process noise and measurement noise respectively.

If the distribution of system noises is Gaussian, the following recursive estimation can be obtained by using Kalman filtering algorithm:

$$\begin{cases} \hat{\mathbf{x}}^-(k+1) = \hat{\mathbf{x}}(k) \\ \mathbf{P}^-(k+1) = \mathbf{P}(k) + \mathbf{R}_\omega \\ \mathbf{K}(k+1) = \mathbf{P}^-(k+1)\mathbf{H}^T(k)[\mathbf{H}(k)\mathbf{P}^-(k+1)\mathbf{H}^T(k) + \mathbf{R}_\upsilon]^{-1} \\ \mathbf{P}(k+1) = [I - \mathbf{K}(k+1)\mathbf{H}(k)]\mathbf{P}^-(k+1) \\ \hat{\mathbf{x}}(k+1) = \hat{\mathbf{x}}^-(k+1) + \mathbf{K}(k+1)[\mathbf{z}(k+1) - \mathbf{H}^T(k)\hat{\mathbf{x}}(k)] \end{cases} \tag{13}$$

R_ω, R_υ are noise covariance matrixes, $\mathbf{P}(k)$ is state estimation error covariance matrix, whose initial value can be $\mathbf{P}(0) = 10^5\mathbf{I}_{mn}$.

The calculation process of Kalman filtering algorithm could be divided into two steps: time update and measurement update. Time update process is responsible for proceeding with the calculation of present state variable and estimating error covariance, in order to provide prior estimation for the next state. Measurement update is responsible for feedback, that is, it combines prior estimation and new measurement to construct updated posterior estimation. Time update equation is also called pre-estimation equation; measurement update equation is also called revised equation.

4.3   Control Algorithm

The task of visual servo is to control the motion of a robot via proper control law according to the measured image feature error, and eventually make the image feature error approximate to zero. The definition of image feature error is:

$$\mathbf{e} = \mathbf{y}_d - \mathbf{y} \tag{14}$$

The control is designed as:

$$\dot{\theta}_c = \hat{\mathbf{J}}_\theta^+ \mathbf{e} - \hat{\mathbf{J}}_\theta^+ \hat{\mathbf{J}}_t, \quad \hat{\mathbf{J}}_\theta^+ = (\hat{\mathbf{J}}_\theta^T \hat{\mathbf{J}}_\theta)^{-1} \hat{\mathbf{J}}_\theta^T \tag{15}$$

The second term of the control law formula (15) is to compensate for the image feature changes induced by the object motion. If the object is motionless, the

second term will be zero and the visual control task will become object positioning with visual feedback. If the object is moving, the visual control task will be
object tracking with visual feedback.

Suppose the sample interval is Δt , then the discrete control law is:

$$\Delta\theta(k) = \lambda\hat{\mathbf{J}}_\theta^+(k)\mathbf{e}(k) - \hat{\mathbf{J}}_\theta^+(k)\hat{\mathbf{J}}_t(k)\Delta t, \quad \lambda > 0 \tag{16}$$

The current image feature error is obtained through the contrast between
the current image feature and the desired image feature. The changes of image
feature errors and the robot joints are input into the corresponded filter and
the current Jacobian matrix is calculated. At last, the robot is controlled by
the control law according to Jacobian matrix and the image feature errors. The
above processes repeated until the image feature error approximates to zero.

4.4   Inverse Kinematics

For a serial manipulator for instance, the inverse kinematics problem would
be to find the value of all joints in the manipulator given the position (and/or
orientation) of the end effector. The inverse kinematics problem can be simplified
to solve the equation

$$e = J\Delta\theta \tag{17}$$

where $e = t - s$, the vector $t = (t_1, \ldots, t_k)^T$ are the target positions, the vector
s are the end effectors positions, and $\theta = (\theta_1, \ldots, \theta_n)^T$ are the joint angles.

There are many methods of modeling and solving inverse kinematics problems [13]. The inverse kinematics with end-effector posture constraints was proposed in order to implement the grasping motions accurately. There are kinds
of methods for the requirements of end-effector constraints. We proposed one
new inverse kinematics method combined the damped least squares method and
the geometric method. First step of the new method is to calculate using the
damped least squares method, the second step is to implement the end-effector
posture constraints using geometric method.

The damped least squares method avoids many of the pseudo-inverse
method's problems with singularities and can give a numerically stable method
of selecting $\Delta\theta$. It is also called the Levenberg-Marquardt method and was first
used for inverse kinematics by Wampler [14] and Nakamura and Hanafusa [15].
Rather than just finding the minimum vector $\Delta\theta$ that gives a best solution to
equation (5), we find the value of $\Delta\theta$ that minimizes the quantity

$$\| J\Delta\theta - e \|^2 + \lambda^2 \| \Delta\theta \|^2 \tag{18}$$

where $\lambda \in \mathbb{R}$ is a non-zero damping constant. This is equivalent to minimizing
the quantity

$$\left\| \begin{pmatrix} J \\ \lambda I \end{pmatrix} \Delta\theta - \begin{pmatrix} e \\ 0 \end{pmatrix} \right\| \tag{19}$$

The corresponding normal equation is

$$\begin{pmatrix} J \\ \lambda I \end{pmatrix}^T \begin{pmatrix} J \\ \lambda I \end{pmatrix} \triangle\theta = \begin{pmatrix} J \\ \lambda I \end{pmatrix}^T \begin{pmatrix} e \\ 0 \end{pmatrix} \qquad (20)$$

This can be equivalently rewritten as

$$(J^T J + \lambda^2 I)\triangle\theta = J^T e \qquad (21)$$

It can be shown that $J^T J + \lambda^2 I$ is non-singular. Thus, the damped least squares solution is equal to

$$\triangle\theta = (J^T J + \lambda^2 I)^{-1} J^T e \qquad (22)$$

Now $J^T J$ is a $n \times n$ matrix, where n is the number of degrees of freedom. It is easy to show that $(J^T J + \lambda^2 I)^{-1} J^T = J^T (JJ^T + \lambda^2 I)^{-1}$. Thus,

$$\triangle\theta = J^T (JJ^T + \lambda^2 I)^{-1} e \qquad (23)$$

The advantage of equation (23) over (22) is that the matrix being inverted is only $m \times m$ where $m = 3k$ is the dimension of the space of target positions, and m is often much less than n.

Additionally, (23) can be computed without needing to carry out the matrix inversion, instead row operations can find f such that $(JJ^T + \lambda^2 I)f = (e)$ and then $J^T f$ is the solution.

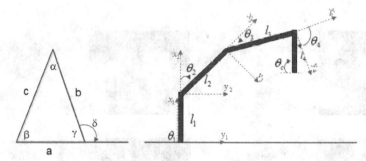

Fig. 3. The left is an arbitrary triangle and the right is the geometric based inverse kinematics

The geometric inverse kinematics method was applied to implement the end-effector posture constraints. The geometric method is based on exterior angle theorem and cosine rule. Given a triangle in the left of the Fig. 3, the exterior angle theorem says as in equations (24),a cosine rule says as in equations (25).

$$\delta = \alpha + \beta \qquad (24)$$
$$b^2 = a^2 + c^2 - 2ac\cos\beta \qquad (25)$$

Based on triangle in the left of the Fig. 3, an inverse kinematics based on geometric calculation is proposed. The right of the Fig. 3 redefine servomotor angle in the arm robot frame. Inverse kinematic is then calculated by solving the angle in robot configuration with end-effector posture constraints. The angle θ_c represents the angle between the end-effector and the horizontal line. In our experiment scene, θ_c is set to $\frac{\pi}{2}$. The following equation is got to implement the end-effector posture constraints,

$$\theta_4 = \frac{\pi}{2} - \theta_2 + \theta_3 \tag{26}$$

5   Experiment and Results

Our goal is to explore whether a human-robot team can accomplish the service task together effectively using the proposed cooperation method. Five different objects were used to test the detecting and classifying accuracy of the Baxter® visual servoing system. The first part of the experiment is to examine the validity of the control of the robot arm with motion. We design 4 motions to observe the motion of subject and 3D robot simulator. The first picture of the Fig. 4 show the procedure of the robots bending its elbow joint. The second picture of the Fig. 4 show the control of the robot simulator whose arms swing up and down. In each subfigure of Fig. 4, the left side shows the motion of the subject detected by the Kinect based on the skeleton tracking technology, while the right side displays the motion of the simulated robot Baxter®.From the figures, we find that the simulated robot Baxter® can track the motion of subject's arm effectively.

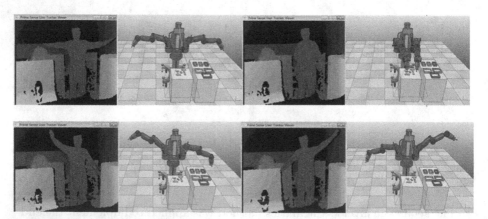

Fig. 4. The experiment results of teleopration simulation

One object tracking experiment is applied to test the result of the visual servoing system. The camera pinhole model is used and the image resolution is 480×480 pixels. The camera parameters are as follows. The focal length $f = 8mm$, the principal point is $(240, 240)$, and the scale factor in image horizontal axis and vertical axis are all $0.05mm/pixels$. Suppose the object is point-shaped, the selected image features are x-coordinate and y-coordinate in the image plane

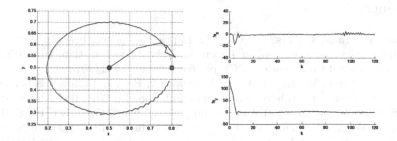

Fig. 5. The left is the robot tracking trajectory based on Kalman filter. The circle is the initial position of robot end-effector; The square is the initial position of object. Dashed: object's real motion trajectory; Crude real line: robot end-effector's motion trajectory. The right is the image feature error based on Kalman filter

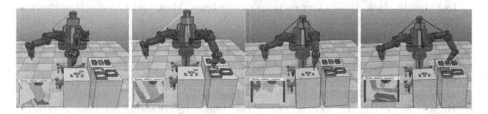

Fig. 6. The experiment results of the Baxter®: classifying function, Baxter® picks up them and classify them.

of the object. The controlling goal is to keep the robot end-effectors motion following the objects motion.

In order to verify the accuracy of the blob detection module, we put the objects on different positions. Fig. 6 illustrates that the Baxter® robot can pick up objects and classify them effectively.

6   Conclusion

In this paper, we have put forward one novel Human-robot cooperation method which falls in between full human control and full robot autonomy. In order to verify the effectiveness of the method by experiments, an experiment environment based on the Kinect sensor and V-REP was built up. In the experiments, the Kinect sensor was used to obtain the skeleton framework information of the human, then, the data is classified and transformed to the robot as the motion control signal. Moreover, the Blob detector was applied to detect the target and applied the closed loop control to implement autonomous adjustment in the visual servoing system. It can be found that the human-robot team can work effectively through the experiments.

References

1. Renon, P., Yang, C., Ma, H., Cui, R.: Haptic interaction between human and virtual icub robot using novint falcon with chai3d and matlab. In: 2013 32nd Chinese Control Conference (CCC), pp. 6045–6050. IEEE (2013)
2. Pitzer, B., Styer, M., Bersch, C., DuHadway, C., Becker, J.: Towards perceptual shared autonomy for robotic mobile manipulation. In: 2011 IEEE International Conference on Robotics and Automation (ICRA), pp. 6245–6251. IEEE (2011)
3. Sun, D., Wang, C., Shang, W., Feng, G.: A synchronization approach to trajectory tracking of multiple mobile robots while maintaining time-varying formations. IEEE Transactions on Robotics 25(5), 1074–1086 (2009)
4. Wang, H., Kosuge, K.: Control of a robot dancer for enhancing haptic human-robot interaction in waltz. IEEE Transactions on Haptics 5(3), 264–273 (2012)
5. Mora, A., Glas, D.F., Kanda, T., Hagita, N.: A teleoperation approach for mobile social robots incorporating automatic gaze control and three-dimensional spatial visualization. IEEE Transactions on Systems, Man, and Cybernetics: Systems 43(3), 630–642 (2013)
6. Wakita, K., Huang, J., Di, P., Sekiyama, K., Fukuda, T.: Human-walking-intention-based motion control of an omnidirectional-type cane robot. IEEE/ASME Transactions on Mechatronics 18(1), 285–296 (2013)
7. Spexard, T.P., Hanheide, M., Sagerer, G.: Human-oriented interaction with an anthropomorphic robot. IEEE Transactions on Robotics 23(5), 852–862 (2007)
8. Hojaij, A., Zelek, J., Asmar, D.: A two phase rgb-d visual servoing controller. In: 2014 IEEE/RSJ International Conference on Intelligent Robots and Systems (IROS 2014), pp. 785–790. IEEE (2014)
9. Jang, W., Bien, Z.: Feature-based visual servoing of an eye-in-hand robot with improved tracking performance. In: Proceedings of the 1991 IEEE International Conference on Robotics and Automation, pp. 2254–2260. IEEE (1991)
10. Asada, M., Tanaka, T., Hosoda, K.: Adaptive binocular visual servoing for independently moving target tracking. In: Proceedings of the IEEE International Conference on Robotics and Automation, ICRA 2000, vol. 3, pp. 2076–2081. IEEE (2000)
11. Espiau, B., Chaumette, F., Rives, P.: A new approach to visual servoing in robotics. IEEE Transactions on Robotics and Automation 8(3), 313–326 (1992)
12. Lv, X., Huang, X.: Fuzzy adaptive Kalman filtering based estimation of image Jacobian for uncalibrated visual servoing. In: 2006 IEEE/RSJ International Conference on Intelligent Robots and Systems, pp. 2167–2172. IEEE (2006)
13. Sun, Z., He, D., Zhang, W.: A systematic approach to inverse kinematics of hybrid actuation robots. In: 2012 IEEE/ASME International Conference on Advanced Intelligent Mechatronics (AIM), pp. 300–305. IEEE (2012)
14. Wampler, C.W.: Manipulator inverse kinematic solutions based on vector formulations and damped least-squares methods. IEEE Transactions on Systems, Man and Cybernetics 16(1), 93–101 (1986)
15. Nakamura, Y., Hanafusa, H.: Inverse kinematic solutions with singularity robustness for robot manipulator control. Journal of Dynamic Systems, Measurement, and Control 108(3), 163–171 (1986)

A Multiscale Method for HOG-Based Face Recognition

Xin Wei[2], Gongde Guo[2], Hui Wang[1(✉)], and Huan Wan[2]

[1] School of Computing and Mathematics,
University of Ulster at Jordanstown, Newtownabbey, UK
xinwei.mail@qq.com
[2] Key Lab of Network Security and Cryptology,
School of Mathematics and Computer Science, Fujian Normal University,
Fuzhou, People's Republic of China
ggd@fjnu.edu.cn, h.wang@ulster.ac.uk, huanwan.mail@qq.com

Abstract. Image representation is an important process in image classification, and there are many different methods for representing images. HOG (Histograms of Oriented Gradients) is a popular one which has been used in many applications including face recognition, pedestrian detection and palmprint recognition. In this paper, a novel method is presented to improve HOG-based image classification by using the multiscale features of images. For each image, multiple HOG feature vectors are extracted under different spatial dimensions (or 'scales'). These 'multiscale' feature vectors are then fused into a distance function to calculate the distance between two images. Experiments have been conducted on ORL face database, AR face database and FERET face database. Results show the use of multiscale HOG features has led to significant improvement in performance over the use of single scale HOG features. Results also show that the nearest neighbour classifier equipped with our distance function is comparable to the well-known and widely-used benchmark classifier.

Keywords: Image classification · Face recognition · HOG · Multiscale

1 Introduction

In the past twenty years, image classification technologies have advanced very quickly and have already found successful applications. At the centre of image classification is image representation, which usually represents an image as feature vectors. There are many image representation methods or *image descriptors*, including HOG (Histograms of Oriented Gradients) [5], LBP (Local Binary Pattern) [18], SURF (Speeded-Up Robust Features) [3], SIFT (Scale Invariant Feature Transform) [14], Gabor [13] and MSER (Maximally Stable Extremal Regions) [16]. These methods have been shown quite effective for different types of image analysis and have been successfully applied. Despite the success of image classification over the past decades, there is still a long way to go towards

© Springer International Publishing Switzerland 2015
H. Liu et al. (Eds.): ICIRA 2015, Part I, LNAI 9244, pp. 535–545, 2015.
DOI: 10.1007/978-3-319-22879-2_49

human level performance in accuracy, flexibility and adaptability. Researchers continue to search for better image representation methods or better ways of using existing image representation methods. Feature fusion is an example of the latter approaches.

Nikan et al. [17] proposed a local-based face recognition algorithm, which first filters the images based on the ratio of the gradient amplitude to the original image intensity. Then, each face image is divided into 4x6 regions. For each region, *local phase quantisation* (LPQ) and multi-scale *local binary pattern* (LBP) are used to extract the features. Finally, a local nearest neighbour classifier is constructed for each region and these classifiers are fused into a bigger one to make the final decision. In [9], a robust face recognition algorithm is proposed, which extracts the holistic feature and local feature by discrete cosine transform and Gabor wavelet transform, respectively. Then these two types of feature vectors are fused by weighted sum, resulting in a new feature vector. Finally the linear regression classifier is used for recognition. Huang et al. [11] proposed a method based on wavelet transform. In this method, each image is transformed by discrete wavelet, which generates several levels of sub-bands. Then, the top-level's three high frequency sub-bands will be fused in a way of linear combination (namely weighted sum). Thus, each face image is represented by the low-level sub-band and the composite high-level sub-band. Finally, these two sub-bands make up a big matrix which is then processed by PCA or LDA so as to obtain the final feature vector.

In [2], every image (face or palmprint) is divided into several regions and a Garbor feature vector will be extracted from each region. Then these feature vectors will be sorted according to their frequency information. Those feature vectors that have relatively low frequency information will be picked out, as low frequency information is believed to have higher discriminant features. After that, principle component analysis (PCA) and linear discriminant analysis (LDA) are used to reduce the feature dimensionality of the picked regions. Finally, these feature vectors are concatenated to a new one. In [20], every face image is down-sampled and up-sampled so as to obtain three face images at different resolutions. Then the Gabor features are extracted from these three images. However, the dimensionality is too high, so principal component analysis (PCA) is applied to each Gabor feature vectors. After that, the high-resolution feature vector and the original resolution feature vector will be projected to another feature space by two projective vectors (these two projective vectors can be found by maximizing the generalized canonical correlation discriminant criterion function), respectively. Then the two projected feature vectors are fused in a form of weighted sum. At last, the final feature vector will be generated by fusing this new feature vector and the low-resolution feature vector in the same way as the fusion of high-resolution and original resolution feature vectors.

Beyond these methods, in an earlier work [25], a multiscale strategy for exploiting existing image descriptors to improve image analysis performance is proposed, where different LBP feature vectors are constructed under different **spatial** scales and used together through a distance function. This strategy has been shown to be quite effective in LBP based face recognition. An important question we can ask about this approach is: Can we use this strategy in other feature descriptors? In this paper, we attempt to answer this question. We consider HOG image descriptor [5] as it also has a spatial information related parameter.

Since its inception HOG [5] has received a lot of attention, and many variants of HOG have been proposed [6,8,10,21]. HOG is the dominant image descriptor used in pedestrian detection and is also widely used in many other image analysis tasks such as face recognition [7] and palmprint recognition [12]. In the original HOG an image is divided into many cells (typically 8×8 pixels per cell) and some cells make up a block (typically 16×16 pixels per block). The HOG feature vector is composed of regionally normalized histogram of gradient orientations from each block. Some parameters in HOG are spatial related, such as block size, block stride, cell size and number of bins, which can be exploited in this strategy.

In this paper, we use the multiscale [1] strategy in HOG-based face recognition. For each image, multiple feature vectors are firstly extracted under different *numbers of bins* (a parameter of HOG). Then these feature vectors are weighted and fused into a distance function which can be used to calculate the distance between two images. At last, the nearest neighbour classifier is used to complete the classification. As the *number of bins* reflects the thickness of histogram (the level of detail), extracting more feature vectors under different numbers of bins will bring in more useful information. Hence, better recognition performance can be expected. Our method (the application of the multiscale strategy to HOG and use nearest neighbours for classification) is evaluated on the popular ORL face database [1], AR face database [15] and FERET face database [19] [2]. Results show that our method can indeed improve the performance of HOG-based face recognition, and it is also competitive to the popular baseline methods for image classification (PCA and 2DPCA).

The rest of the paper is organised as follows. Section 2 reviews HOG. Section 3 presents the proposed method. Section 4 shows experimental results. Section 5 summarises and concludes the paper.

2   Brief Review of HOG

HOG is a method introduced by Dalal *et al.* [5], originally for human detection. As object appearance and shape can often be described well by edges, and

---

[1] The term 'scale' has been used in different contexts for different meanings. In mathematics and computer science 'scale' may be a description of the spread or dispersion of a probability distribution (probability theory), or a linear transform that enlarges or shrinks objects (geometry). In the physical sciences, it may mean an informal system of general size categorizations (physics) or a measure in mass or volume of a chemical reaction or process (chemistry). In this paper we use the term to mean *the categorisation of levels of detail of an image that are used to construct an image feature vector.* This is more in line with the physics interpretation of the term.

[2] AR and FERET are among the current benchmark datasets for face recognition.

the distribution of local intensity gradients can well characterize edges. HOG is robust even in the face of large variations in appearances, cluttered backgrounds and different illumination. HOG has become very popular in the field of object recognition since it was proposed.

In brief, HOG takes the following steps to extract the features from one image:

(1) compute the gradients of each pixel in the image,
(2) divides an image into many cells and some cells make up a block,
(3) build histogram of orientation for each cell,
(4) normalize histograms within each block of cells,
(5) create the HOG feature vector by concatenating the histograms in each blocks.

Figure 1 shows the overview of HOG.

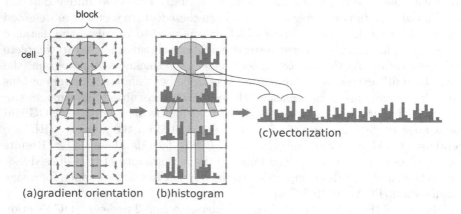

Fig. 1. Overview of HOG. [24]

3   A Multiscale Method for HOG-Based Image Classification

As mentioned above, HOG has some parameters, such as block size, block stride, cell size and number of bins. Since the *number of bins* reflects the thickness of histogram (the level of detail), it is reasonable to expect that the feature vectors extracted under different *numbers of bins* may carry different information. Therefore having several such feature vectors for one image together will bring in more information including useful information and useless information. The useful and discriminative information will be kept after dimension reduction is done, and the useless or misleading information will be dropped by dimension reduction (e.g. PCA). With more useful information considered, better recognition performance can be expected.

In our method we select the *number of bins* as the parameter to construct our multiscale HOG feature vectors. This parameter specifies the number of orientations to consider to construct the 'bins'. 4, 8 and 16 are all frequently used in the literature in multiples. It is possible to consider even 32, but we do not expect to gain too much additional information. Therefore we decide to set the *number of bins* parameter to three different values 4, 8 and 16, resulting in three feature vectors in different scales. The HOG's under these parameters are illustrated in Figure 2. Other parameters of HOG take their default values. Then these feature vectors are fused into a distance function which can be used to calculate the distance between two samples.

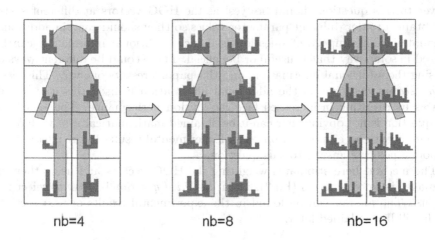

$$nb{=}4 \qquad\qquad nb{=}8 \qquad\qquad nb{=}16$$

Fig. 2. The HOG's having different numbers of bin

Let I and I' be two images. Let $V_{nb=4}, V_{nb=8}, V_{nb=16}$ be the multiscale HOG feature vectors for image I with *number of bins* being 4, 8 and 16 respectively; $V'_{nb=4}, V'_{nb=8}, V'_{nb=16}$ be the multiscale HOG feature vectors for image I'. Then the distance between the two images is defined as follows:

$$d(I, I') = f(V_{nb=4}, V'_{nb=4}) + f(V_{nb=8}, V'_{nb=8}) + f(V_{nb=16}, V'_{nb=16}) \qquad (1)$$

where $f(V_i, V'_i)$ is the Euclidean distance of vectors V_i and V'_i.

It is possible that different feature vectors at different scales may play different role in image classification, we modify distance between the two images as follows:

$$d(I, I') = w_1 * f(V_{nb=4}, V'_{nb=4}) + w_2 * f(V_{nb=8}, V'_{nb=8}) + w_3 * f(V_{nb=16}, V'_{nb=16}) \qquad (2)$$

where w_1, w_2, w_3 are the weights of the feature vectors.

Discussion

It is clear that if the weights w_1, w_2, w_3 take the values of $1, 0, 0$ respectively, Eq.(2) becomes

$$d(m, m') = f(V, V') \tag{3}$$

meaning that the image distance becomes the Euclidean distance in the HOG vector space. By adjusting the weights, the HOG vectors at different scales can contribute differently toward the image distance.

There are three questions to answer. *The first question* is: Is there additional information obtained by considering HOG vectors at different scales? The answer to this question should be 'yes' as the HOG vectors at different scales are obtained under different parameter values so they should contain additional information. The important question is if such additional information can be utilised in some way that is useful or beneficial. There could be different ways of utilising the additional information, and this paper presents one way. This leads to *the second question*: Is the additional information utilised this way useful, such as by resulting in increased face recognition performance? The answer to this question is not obvious nor can it be obtained mathematically, so we have to be content with an empirical approach. Experimental results in the next section support a positive answer to this question.

The method here stipulates weighting the HOG vectors and using them in an image distance function (Eq.2). Then, *the third question* is: How to select the weights? The answer can be found in the experimental results of next section, which will be explained later.

4   Experiments on ORL, AR and FERET Databases

4.1   Experiments Setting

In this section, the multiscale method for HOG-based image classification is evaluated through experiments on ORL face database [1], AR face database [15] and FERET face database [19].

In the first set of experiments we apply the HOG descriptor (as presented in Section 2) at a single scale. We calculate one HOG feature vector for one image, use Euclidean distance to measure image distance, and use nearest neighbour classifier for classification.

In the second set of experiments we apply the HOG descriptor at three scales, resulting in three HOG feature vectors for each image. We use the unweighted image distance function in Eq.(1) to measure image distance, and the nearest neighbour classifier for classification.

In the final set of experiments we also apply the HOG descriptor at three scales, but use the weighted image distance function in Eq.(2) to measure image distance, and choose the optimal weighting for each scale (i.e., consider all possible weight values in some range for different scales, and then select those weight values for the scales that give the best result).

PCA (Principal Component Analysis) [22, 23] is a popular dimension reduction method which can keep the most discriminative information and exclude the relatively less discriminative information. So PCA is very suitable for our multiscale method as multiscale feature vectors bring in more information including discriminative information and also relatively less discriminative information. LDA (Linear Discriminant Analysis) [4] is capable of transforming the original data into another feature space, maximizing the between-class scatter and minimizing the within-class scatter. So, a simple classifier (e.g. KNN) can also achieve good results after LDA is used. For the above reasons, PCA+LDA is chosen to be applied to all HOG vectors in all experiments to do dimension reduction. The nearest neighbour classifier is used in all experiments for classification.

As a benchmark method we also consider the well-known *Principal Component Analysis classifier* (PCA_c) [23] and its variant *2D Principal Component Analysis classifier* ($2DPCA_c$) [26] as face recognition algorithms. They represent images using a small set of basis images, and classification can be achieved by comparing the basis set based representations of images. These basis images are called *eigenfaces*, which are eigenvectors of the covariance matrix in the high-dimensional vector space of face images [23, 26].

4.2   ORL, AR and FERET Face Databases

The ORL database [1] contains 400 images of 40 individuals and each individual has 10 images. For some of the individuals, the pictures were taken at different time, under different illumination, facial expression and facial details (wearing glasses or not). In our experiments, five images of each individual are used for training and five images for testing.

The AR face database [15] consists 1400 images from 100 people with 14 images for each person. The face portion of each image is cropped out and normalized to the size of $100 * 80$. The first seven images of each person in the first session are selected for training, which have the following characteristics: (1) neutral expression, (2) smile, (3) anger, (4) scream, (5) left light on, (6) right light on, and (7) all side lights on. The first seven images of each person in the second session are used for testing.

The FERET face database [19] contains more than 10000 face images with different expressions and under different lighting conditions. It is one of the most widely used face databases. In our experiments, a subset of the FERET database is used, which is composed of images named for $01100**101\_0860521.tif$ to $01200 * *010\_860521.tif$("* *" can be "ba", "bj", "bk", "be", "bf", "bd", or "bg"). This subset contains 700 images from 100 people with 7 images per person. The first 4 images of each person are used for training, and the remaining 3 images are used for testing. The face portion of each image is cropped out and normalized to the size of $100 * 100$. Figure 3 shows some examples of ORL, AR and FERET face databases.

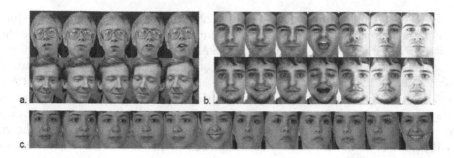

Fig. 3. some examples of (a) ORL database, (b) AR database, and (c) FERET database.

4.3   Experiments Results

Table 1 lists the recognition accuracy of different methods on ORL, AR and FERET face databases. On AR face database, it is clear that the recognition accuracy of HOG increased by up to 4.57% after multi-scaling is applied and increased by up to 4.87% after weighted multi-scaling is applied. Also on FERET face database, it can be seen that the recognition accuracy of HOG increased by up to 2.33% after multi-scaling is applied while increased by up to 3.33% after weighted multi-scaling is applied. Besides, our method also leads to accuracy increse on ORL database. However, the accuracy of optimal weighting strategy is the same as the accuracy of Uniform weighting as the optimal weights are exactly the same. In summary, the experimental results on ORL, AR and FERET face databases clearly shows that the multiscale method brings in more useful information as multiple feature vectors are extracted, resulting in higher recognition accuracy.

Table 1. Recognition accuracy of different methods on ORL, AR and FERET face database

Method	ORL	AR	FERET	Part of Parameters
PCA	88.5%	67.43%	75.67%	-
2DPCA	90.5%	72.14%	65.67%	-
HOG	93.5%	90.57%	84.00%	default
HOG/UW3S	95.5%	95.14%	86.33%	number of bins: 4, 8, 16; uniform weighting
HOG/OW3S	95.5%	95.43%	87.33%	weights: 1, 1, 1 (ORL) 2, 4, 3 (AR); 2, 3, 2 (FERET)

4.4   Discussion

Looking at the results from ORL, AR and FERET face databases we can also observe the following.

- When multiple HOGs are used in our way, the classification performance has increased on all three databases. This indicates that the effectiveness of our method is consistent.
- When multiple HOG vectors are used, the face recognition accuracy has increased from when only one HOG vector is used for each image. This observation strongly supports the expectation that utilising multiple HOG vectors does create value.
- There is a question in Section 3 that can be answered here: How to select the weights? One way is to use mathematical optimisation. Another way is empirical – varying the weights for different scales and selecting the set of weights that gives the best performance. As can be seen in the results, there is little difference between the simple unweighting (i.e. all scales are given equal weights) and the more involved optimal weighting. Therefore it is sufficient to adopt the unweighted (or uniform weighting) option.

5   Conclusion

In this paper a novel method for image classification, i.e., multiscale HOG-based image classification, is presented based on the multiscale strategy for exploiting existing image descriptors. The novelty lies in the fact that the multiple feature vectors used correspond to different values of the same HOG parameter, i.e., number of bins. This method is evaluated on ORL, AR and FERET face databases. Experimental results show this method is indeed better than the HOG-based image classifier using a single HOG vector, and this performance improvement pattern is consistent in both three face databases. The studies suggest that the multiscale strategy for exploiting existing image descriptors works for HOG as well as LBP.

However the method has a higher computation cost since multiple HOG vectors will be extracted. Fortunately, the increase in computation cost is only linear in the number of scales.

In future work we will consider another method of utilising fine-grained information: we use only one scale but many bins. For example, in the multiscale method we use 4, 8, 16 bins at 3 different scales. In this 'future' method we use 28 bins at a single scale. This will hopefully provide more insight into the effects of information granularity in representing images.

Acknowledgments. This work is partially supported by Hu Guozan Study-Abroad Grant for Graduates of Fujian Normal University.

References

1. http://www.cl.cam.ac.uk/research/dtg/attarchive/facedatabase.html
2. Ahmad, M.I., Ilyas, M.Z., Md Isa, M.N., Ngadiran, R., Darsono, A.M.: Information fusion of face and palmprint multimodal biometrics. In: 2014 IEEE Region 10 Symposium, pp. 635–639. IEEE (2014)
3. Bay, H., Tuytelaars, T., Van Gool, L.: SURF: Speeded Up Robust Features. In: Leonardis, A., Bischof, H., Pinz, A. (eds.) ECCV 2006, Part I. LNCS, vol. 3951, pp. 404–417. Springer, Heidelberg (2006)
4. Belhumeur, P.N., Hespanha, J.P., Kriegman, D.J.: Eigenfaces vs. fisherfaces: Recognition using class specific linear projection. IEEE Transactions on Pattern Analysis and Machine Intelligence 19(7), 711–720 (1997)
5. Dalal, N., Triggs, B.: Histograms of oriented gradients for human detection. In: IEEE Computer Society Conference on Computer Vision and Pattern Recognition, CVPR 2005, vol. 1, pp. 886–893. IEEE (2005)
6. Dang, L., Bui, B., Vo, P.D., Tran, T.N., Le, B.H.: Improved hog descriptors. In: 2011 Third International Conference on Knowledge and Systems Engineering (KSE), pp. 186–189. IEEE (2011)
7. Déniz, O., Bueno, G., Salido, J., De la Torre, F.: Face recognition using histograms of oriented gradients. Pattern Recognition Letters 32(12), 1598–1603 (2011)
8. Felzenszwalb, P.F., Girshick, R.B., McAllester, D., Ramanan, D.: Object detection with discriminatively trained part-based models. IEEE Transactions on Pattern Analysis and Machine Intelligence 32(9), 1627–1645 (2010)
9. Gao, Z., Ding, L., Xiong, C., Huang, B.: A robust face recognition method using multiple features fusion and linear regression. Wuhan University Journal of Natural Sciences 19(4), 323–327 (2014)
10. Hou, C., Ai, H., Lao, S.: Multiview Pedestrian Detection Based on Vector Boosting. In: Yagi, Y., Kang, S.B., Kweon, I.S., Zha, H. (eds.) ACCV 2007, Part I. LNCS, vol. 4843, pp. 210–219. Springer, Heidelberg (2007)
11. Huang, Z.-H., Li, W.-J., Wang, J., Zhang, T.: Face recognition based on pixel-level and feature-level fusion of the top-levels wavelet sub-bands. Information Fusion 22, 95–104 (2015)
12. Jia, W., Rong-Xiang, H., Lei, Y.-K., Zhao, Y., Gui, J.: Histogram of oriented lines for palmprint recognition. IEEE Transactions on Systems, Man, and Cybernetics: Systems 44(3), 385–395 (2014)
13. Liu, C., Wechsler, H.: Gabor feature based classification using the enhanced fisher linear discriminant model for face recognition. IEEE Transactions on Image Processing 11(4), 467–476 (2002)
14. David, G.: Lowe. Distinctive image features from scale-invariant keypoints. International Journal of Computer Vision 60(2), 91–110 (2004)
15. Martinez, A., Benavente, R.: The AR Face Database. CVC Tech. Report 24, Report 24, (1998)
16. Matas, J., Chum, O., Urban, M., Pajdla, T.: Robust wide-baseline stereo from maximally stable extremal regions. Image and Vision Computing 22(10), 761–767 (2004)
17. Nikan, S., Ahmadi, M.: Local gradient-based illumination invariant face recognition using local phase quantisation and multi-resolution local binary pattern fusion. IET Image Processing 9(1), 12–21 (2014)
18. Ojala, T., Pietikainen, M., Maenpaa, T.: Multiresolution gray-scale and rotation invariant texture classification with local binary patterns. IEEE Transactions on Pattern Analysis and Machine Intelligence 24(7), 971–987 (2002)

19. Jonathon Phillips, P., Moon, H., Rizvi, S.A., Rauss, P.J.: The feret evaluation methodology for face-recognition algorithms. IEEE Transactions on Pattern Analysis and Machine Intelligence **22**(10), 1090–1104 (2000)
20. Pong, K.-H., Lam, K.-M.: Multi-resolution feature fusion for face recognition. Pattern Recognition **47**(2), 556–567 (2014)
21. Satpathy, A., Jiang, X., Eng, H.-L.: Human detection by quadratic classification on subspace of extended histogram of gradients. IEEE Transactions on Image Processing **23**(1), 287–297 (2014)
22. Swets, D.L., Weng, J.J.: Using discriminant eigenfeatures for image retrieval. IEEE Transactions on Pattern Analysis & Machine Intelligence **8**, 831–836 (1996)
23. Turk, M.A., Pentland, A.P.: Face recognition using eigenfaces. In: Proceedings of the IEEE Computer Society Conference on Computer Vision and Pattern Recognition, CVPR 1991, pp. 586–591. IEEE (1991)
24. Watanabe, T., Ito, S., Yokoi, K.: Co-occurrence histograms of oriented gradients for human detection. IPSJ Transactions on Computer Vision and Applications **2**, 39–47 (2010)
25. Wei, X., Wang, H., Guo, G., Wan, H.: A General Weighted Multi-scale Method for Improving LBP for Face Recognition. In: Hervás, R., Lee, S., Nugent, C., Bravo, J. (eds.) UCAmI 2014. LNCS, vol. 8867, pp. 532–539. Springer, Heidelberg (2014)
26. Yang, J., Zhang, D., Frangi, A.F., Yang, J.-Y.: Two-dimensional pca: a new approach to appearance-based face representation and recognition. IEEE Transactions on Pattern Analysis and Machine Intelligence **26**(1), 131–137 (2004)

Estimation and Identification

Magnetic–Gravity Gradient Inversion for Underwater Object Detection

Meng Wu[⊠] and Jian Yao

School of Remote Sensing and Information Engineering, Wuhan University,
Wuhan 430079, People's Republic of China
wumenghust911@gmail.com

Abstract. A new underwater object detection method based on joint Gravity-Gradient and Magnetic-Gradient Inversion algorithms is proposed in this paper. Magnetic gradient and gravity gradient anomalies induced by an underwater object can be measured and inversed to estimate the relative changes in distance between underwater object and underwater vehicle. The weight least squares estimation is introduced to combine equations of magnetic gradient and gravity gradient inversion. Therefore, the joint inversion method gets an optimal relative position of underwater object using two kinds of data. Simulation results show that the proposed method is more efficient than the previous Joint gravity – gradient tensor inversion and magnetic gradient tensor inversion respectively.

Keywords: Magnetic gradient inversion · Gravity gradient inversion · Underwater object detection · Weight least squares estimation

1 Introduction

Underwater object detection is an area of research with broad commercial and military applications. How to detect underwater abnormal object is still a major challenge in underwater navigation field nowadays [1-2]. Over past decades, underwater object detection systems based on gravity gradient and magnetic anomaly have been researched widely [3-8]. In the field of underwater object detection based on gravity gradient tensor, Lin Wu used gravity gradient inversion method to detect abnormal objects in underwater environment [4]. Furthermore, in reference [5], joint gravity and gravity gradient inversion is presented for passive subsurface object detection. In reference [7], Zu Yan presented a novel method for underwater object detection based on the gravity gradient differential and the gravity gradient differential ratio caused by the relative motion between the AUV and the object. In the field of underwater object detection based on magnetic gradient tensor, YuHuang proposed a localization method combined with magnetic gradient tensor and draft depth [8]. Hao yan-ling put forward an available solution, where the geomagnetic anomaly is inversed a magnetic dipole target and the relative position of vehicle is determined by magnetic magnitude and gradients of targets [9].

© Springer International Publishing Switzerland 2015
H. Liu et al. (Eds.): ICIRA 2015, Part I, LNAI 9244, pp. 549–555, 2015.
DOI: 10.1007/978-3-319-22879-2_50

After considering the invisibility requirement of AUVs and the existing problems in using gravity gradient and magnetic gradient for detecting objects, this paper proposed a novel and practical solution to combine magnetic gradient Inversion and Gravity Gradient Inversion in an information fusion way, where the two kinds of data are complementary to each other and a optimal detection is calculated by the weighted least squares estimation.

This paper is organized as follows. In Section 2, the joint gravity gradient and magnetic gradient inversion algorithms are introduced and analyzed, In Section 3, experiment and stimulation results are discussed. Conclusions are summarized in Section 4.

2   Joint Gravity Gradient and Magnetic Gradient Inversion Method

2.1   Structure of Joint Geophysical Inversion Method

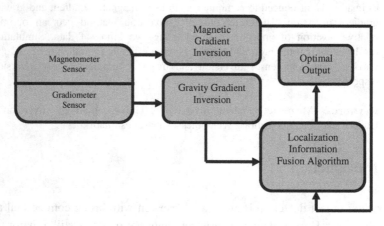

Fig. 1. Block diagram of Magnetic-Gravity Gradient Inversion for Underwater Object Detection

2.2   Introduction of Information Fusion Algorithm

In Fig1, after calculating the relative distances from gravity gradient inversion method and magnetic gradient inversion method, which are denoted by by P_{gg} and P_{mg} respectively, the weighted least squares estimation method in reference [4][5] is constructed to get an optimal relative distance between underwater vehicle and underwater object. The optimal equation is shown as follows:

$$
\begin{cases}
\hat{P}_{Joint} = W_{gg}P_{gg} + W_{mg}P_{mg} \\[4pt]
\delta_T = \sqrt{\sum w_{ig}^2 \delta_{ig}^2} \\[4pt]
\sum w_{ig} = 1 \\[4pt]
0 \le w_{ig} \le 1 \\[4pt]
i = g,\ m
\end{cases}
\tag{1}
$$

Where W_{gg} and W_{mg} are weights coming from gravity gradient inversion and magnetic gradient inversion. $\hat{P}_{Joint}$ is the optimal localization result from two Inversion methods, according to methods introduced in reference [8][9], the relative distance between an underwater object and underwater vehicle can be expressed as follows:

$$
\begin{cases}
\dfrac{\partial B_x}{\partial X} = a = \dfrac{3\mu}{4\pi}\dfrac{x\left(3r^2-5x^2\right)m_x + y\left(3r^2-5x^2\right)m_y + z\left(3r^2-5x^2\right)m_z}{r^7} \\[6pt]
\dfrac{\partial B_y}{\partial Y} = b = \dfrac{3\mu}{4\pi}\dfrac{x\left(3r^2-5y^2\right)m_x + y\left(3r^2-5y^2\right)m_y + z\left(3r^2-5y^2\right)m_z}{r^7} \\[6pt]
c = -(a+b) \\[6pt]
\dfrac{\partial B_y}{\partial X} = d = \dfrac{3\mu}{4\pi}\dfrac{y\left(3r^2-5x^2\right)m_x + x\left(3r^2-5x^2\right)m_y - 5xyzm_z}{r^7} \\[6pt]
\dfrac{\partial B_z}{\partial Y} = e = \dfrac{3\mu}{4\pi}\dfrac{z\left(r^2-5y^2\right)m_y + y\left(r^2-5z^2\right)m_z - 5xyzm_x}{r^7} \\[6pt]
\dfrac{\partial B_z}{\partial X} = f = \dfrac{3\mu}{4\pi}\dfrac{z\left(r^2-5x^2\right)m_x + x\left(r^2-5z^2\right)m_z - 5xyzm_y}{r^7} \\[6pt]
A_6 k^6 + A_5 k^5 + A_4 k^4 + A_3 k^3 + A_2 k^2 + A_1 k^1 + A_0 = 0 \\[4pt]
A_6 = d^2(a+2b) - e^2(a-b) + 2def \\[4pt]
A_5 = -2d\left[(a-b)(a+2b) + \left(d^2+e^2+f^2\right)\right] \\[4pt]
A_4 = (a-b)^2(a+2b) + d^2(4a-7b) + \left(f^2-2e^2\right)(a-b) + 6edf \\[4pt]
A_3 = -4d\left[(a-b)^2 + \left(d^2+e^2+f^2\right)\right] \\[4pt]
A_2 = (a-b)^2(a+2b) + d^2(4b-7a)\left(2f^2-e^2\right)(a-b) + 6edf \\[4pt]
A_1 = 2d\left[(a-b)(a+2b) - \left(d^2+e^2+f^2\right)\right] \\[4pt]
A_0 = d^2(a+2b) + f^2(a-b) + 2edf \\[6pt]
q = \dfrac{\left[d(k^2-1)-(a-b)k\right]}{(ek-f)(k^2+1)} \\[6pt]
z = \dfrac{\pm 3}{\sqrt{\left[(ak+d)q+f\right]^2 + \left[(dk+b)q+e\right]^2 + \left[(fk+e)q+c\right]^2}} \\[6pt]
x = kqz \\[4pt]
y = qz \\[4pt]
P_{mg} = \sqrt{x^2+y^2+z^2} \\[6pt]
P_{gg} = R(x,y,z) = \sqrt{\dfrac{GM}{\Gamma_{xx}(x,y,z)+\Gamma_{yy}(x,y,z)}\left(1 - \dfrac{3}{\left(\dfrac{\Gamma_{zz}(x,y,z)}{\Gamma_{xx}(x,y,z)}\right)^2 + \left(\dfrac{\Gamma_{zz}(x,y,z)}{\Gamma_{yy}(x,y,z)}\right)^2 + 1}\right)}
\end{cases}
\tag{2}
$$

Where $\Gamma_{ij}(x,y,z)$ denotes the gravity gradient value with coordinate x, y and z, respectively; $\theta(x,y,z)$ and $\varphi(x,y,z)$ stand for the relative orientation of an underwater object; $R(x, y, z)$ is a relative distance between the underwater vehicle and an underwater object that needs to be detected; M is the mass of the object. G is the gravitation constant [4-5]; In order to get the values A, it is necessary to measurement magnetic gradients (a, b, d, e, f), According to the method in reference [10], seven single-axis magnetometer configuration is as the simplest scheme of measuring magnetic gradient tensor. After obtaining parameters A, the six order equation in Eq. (2) can be calculated to get the value of k, then, k is to calculate q. Finally, the underwater vehicle position parameters (x, y, z) relative to the underwater object can be calculated by Eq. (2)

3   Stimulation Results

In order to do a comparison with the method in reference [4-5], a cube model was also designed for the estimation of the changes in distance between an underwater vehicle and an underwater object. The relevant parameters are listed in Table 1. Assuming position of underwater vehicle is fixed and underwater object is moving. Based on equation (2), after getting the values of gravity gradient and magnetic gradient tensors, the relative distance between underwater vehicle and underwater object can be calculated. Then equation (1) is to combine each gradient inversion information by weighted least squares estimation method in reference [4][5].

Table 1. Parameters of Underwater Object Model

Model	Size (m)
x direction	20
y direction	20
z direction	20
Density contrast (kg/m3)	25
Actual mass of object(kg)	$6.624 * 10^4$
m_x	$10^6 \ A \cdot m^2$
m_y	$2 \times 10^5 \ A \cdot m^2$
m_z	$10^5 \ A \cdot m^2$

In this part of the stimulation, a fixed underwater vehicle and a moving underwater object were assumed. The density distribution and the size of the object were ignored and the object was assumed to be a particle. While the object was moving, sequences of varying gravity gradient and magnetic gradient responses at the fixed position of underwater vehicle could be measured simultaneously. According to Eq. (2), relative positions between underwater vehicle and moving object can be calculated. Besides, a constant relative error of 12.33% was added to magnetic gradient inversion method, which means that δ_{mg} in (1) was equal to 12.33%. When the relative error of gravity gradient (δ_{gg}) in (1) is 22.65%, here, the weights are calculated with (1) so that W_{gg}=0.219 and W_{mg}=0.781. The estimation result is shown in Fig. 2.

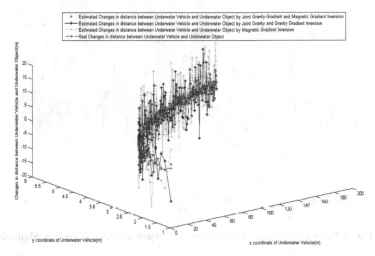

Fig. 2. A Comparison between actual changes in distance between underwater vehicle and underwater object and estimated changes in distance between underwater vehicle and underwater object based on Three Inversion methods

In Fig.2, the actual change in distance between underwater vehicle and underwater object is depicted with a blue line. The estimated change in distance between underwater vehicle and underwater object by the proposed joint gravity gradient and magnetic gradient inversion method is denoted by red stars. The black stars are to denote the estimated changes in distance between underwater vehicle and underwater object by Joint gravity and gravity gradient inversion. Finally, the green line is to illustrate the estimated changes in distance between underwater vehicle and underwater object by magnetic gradient inversion. It is clearly seen that the estimated changes in distance between underwater vehicle and underwater object by joint gravity gradient and magnetic gradient inversion method is in much better agreement with the actual changes in distance between underwater vehicle and underwater object. In fact, the changes in distance between underwater vehicle and underwater object is a good solution to reflect the relative position changes between underwater vehicle and underwater object, the bias in estimated changes in distance between underwater vehicle and underwater object come from measurement outliers from magnetometers and gradiometers. The accuracy of each geophysical sensor has a great impact on the accuracy of underwater object localization. In addition, the fluctuation of various terrains in underwater environment has a great impact on the accuracy of Joint gravity and gravity gradient inversion method. Furthermore, electromagnetic disturbance in underwater environment is another important factor which affects the accuracy of magnetic gradient inversion method. So, the joint gravity gradient and magnetic gradient inversion method is a good solution to eliminate such kinds of negative influences on underwater object detection. The Fig.2 shows its effectiveness in underwater object detection.

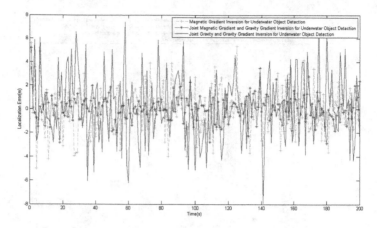

Fig. 3. Error Comparison among three Underwater Object Detection Methods

In Fig.3, it is clearly to see that Joint Magnetic Gradient and Gravity Gradient Inversion methods for underwater object detection show the best localization performance among three methods. In fact, gravity and gravity gradient is sensitive to the fluctuation of various terrain in the underwater environment, on the contrary, magnetic gradient is less affected by fluctuation of underwater terrain and time-vary disturbances. In this way, magnetic gradient is a good solution to combine with gravity gradient to realize underwater object detection. In the future, it is necessary to pay more attention to improve the accuracy of magnetometer and gradiometer and develop a better solution to realize data fusion method to fuse gravity gradient and magnetic gradient information.

4   Conclusions

Underwater Object detection method based on joint gravity gradient and magnetic gradient inversion has been proposed. The weighted least square is introduced to fuse inversion information from gravity gradient inversion and magnetic gradient inversion respectively. With such method, the changes in distance between underwater vehicle and an underwater object can be detected accurately. Simulation results show that the joint inversion method is better than mono-magnetic gradient inversion method and joint gravity and gravity gradient inversion method.

Acknowledgements. This work was partially supported by the Hubei Province Science and Technology Support Program, China (Project No. 2015BAA027), the Jiangsu Province Science and Technology Support Program, China (Project No. BE2014866), and the South Wisdom Valley Innovative Research Team Program.

References

1. Heald, G.J., Griiffiths, H.D.: A review of underwater detection techniques and their applicability to the landmine problem. In: Proc. 2nd Int. Conf. Detection Abandoned Land Mines, pp. 173–176 (1998)
2. Montanari, M., Edwards, J.R., Schmidt, H.: Autonomous underwater vehicle-based concurrent detection and classification of buried targets using higher order spectral analysis. IEEE. J. Ocean. Eng. 31(1), 188–199 (2006)
3. Zheng, H., Wang, H., Wu, L., Chai, H., Wang, Y.: Simulation Research on Gravity-Geomagnetism Combined Aided Underwater Navigation. Royal Institute of Navigation 66(1), 83–98 (2013)
4. Wu, L., Tian, X., Ma, J., Tian, J.W.: Underwater object detection based on gravity gradient. IEEE Geosci. Remote Sens. Lett. 7(2), 362–365 (2010)
5. Wu, L., Ke, X.P., Hsu, H., Fang, J., Xiong, C.Y., Wang, Y.: Joint gravity and gravity gradient inversion for subsurface object detection. IEEE Geosci. Remote Sens. Lett. 107(4), 865–869 (2013)
6. Wu, L., Tian, J.W.: Automated gravity gradient tensor inversion for underwater object detection. J. Geophys. Eng. 7(4), 410–416 (2010)
7. Yan, Z., Ma, J., Tian, J., Liu, H., Yu, J., Zhang, Y.: A Gravity Gradient Differential Ratio Method for Underwater Object Detection. IEEE Geosci. Remote Sens. Lett. 11(4), 833–837 (2014)
8. Huang, Y., Wu, L.-H., Sun, F.: Underwater Continuous Localization Based on Magnetic Dipole Target Using Magnetic Gradient Tensor and Draft Depth. IEEE Geosci. Remote Sens. Lett. 11(1), 178–180 (2014)
9. Yan-ling, H., Ya-feng, Z., Jun-feng, H.: Preliminary analysis on the application of geomagnetic field matching in underwater vehicle navigation. Progress in Geophysics 18, 64–67 (2008)
10. Huang, Y., Feng, S., Yan-ling, H.: Simplest magnetometer configuration scheme to measure magnetic field gradient tensor. In: Proceedings of the 2010 IEEE International Conference on Mechatronics and Automation, vol.5, pp. 1426–1430 (2010)
11. Huang, Y., Feng, S., Yan-ling, H.: Vehicle attitude detection in underwater magnetic anomaly localization experiment. In: 2010 3rd International Symposium on Systems and Control in Aeronautics and Astronautics (ISSCAA), pp. 788–792 (2010)

Real-Time Data Acquisition and Model Identification for Powered Parafoil UAV

Li Bingbing[1,2], Qi Juntong[1], Lin Tianyu[1], Mei Sen[1], Song Dalei[1(✉)], and Han Jianda[1]

[1] State Key Laboratory of Robotics, Shenyang Institute of Automation,
Chinese Academy of Sciences, Shenyang 110016, China
{libingbing,qijt,daleisong,jdhan}@sia.cn, razorwoods@126.com,
meisensia@163.com
[2] University of Chinese Academy of Sciences, Beijing 100049, China

Abstract. Powered Parafoil UAV (PPUAV) is a kind of different aircraft from common UAVs. It consists of parafoil canopy, payload and suspension lines and has the advantages of simple structure, low cost and high load weigh ratio, which is suitable for large-area and long-time surveillance and airdrop missions. This type of UAV has the problems of adding mass and flexible connection, and it is not easy to build the accurate model, which brings challenges to theoretical study and practical applications. This paper designs the structure of PPUAV and real-time data acquisition system, obtains flight data for modeling, identifies and validates the transfer functions using Matlab System Identification Toolbox. Finally the model of PPUAV is obtained, which gives a practicable method of PPUAV modeling and does a preliminary work for the following controller design and mission planning.

Keywords: Powered parafoil · UAV · Data acquisition · Transfer functions · System identification · Modeling

Nomenclature

la	left brake deflection
ra	right brake deflection
T	thrust
S	symmetric brake deflection
v_x	longitudinal velocity
v_z	vertical velocity
w	yaw rate

© Springer International Publishing Switzerland 2015
H. Liu et al. (Eds.): ICIRA 2015, Part I, LNAI 9244, pp. 556–567, 2015.
DOI: 10.1007/978-3-319-22879-2_51

1   Introduction

1.1   The Characteristics of PPUAV

With the development of society and economy, Unmanned Aircraft Vehicles (UAVs) have been more and more widely used in varied domains. Compared with common UAVs which have rigid contact surface, Powered Parafoil UAV (PPUAV) has a soft wing called parafoil, which is connected to the payload by suspension lines and provides lift for the system. This type of UAV can takeoff from the ground or be released from a plane and has the ability to perform long-time surveillance. When landing, it impacts the ground with low velocity. It is suitable for field investigations, search and rescue, and cargo delivery [1]. The system has the characteristics of complexity, uncertainty, nonlinearity, time variation, large delay and large inertia, and is easily affected by the atmospheric environment [2]. The system is usually treated as a rigid body, with the models built in 6 degree of freedom (DOF), 8DOF, etc. The researches make contributions to theoretical study, with approaches to data acquisition, flight tests designing, model analysis discussed inadequately, which causes challenges to practical applications. This paper discusses the structure, control mechanisms, fight test design, data acquisition and model analysis of the PPUAV, which provides a method of PPUAV modeling and flight test and does a preliminary work for PPUAV modeling and control design. The PPUAV system is shown in Fig. 1.

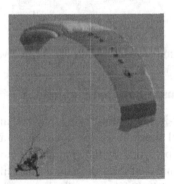

Fig. 1. PPUAV system

1.2   The Structure of Powered Parafoil UAV

Powered Parafoil UAV consists of parafoil canopy, payload, suspension lines and GN&C system. The GN&C system consists of winches, global position system (GPS), magnetic compass, inertial measurement unit(IMU), pitot tube, flight computer and data-transmit module to uplink commands and downlink data.

The structure of the system is shown in Figs. 2-3: 1 Powered Parafoil system, 2-1 GPS, 2-2 data acquisition board, 2-3 data-transmit module, 2-4 magnetic compass 3, 4 IMU, 5-1 canopy lines, 5-2 hanging lines, 5-3 manipulating lines, 6 GPS and magnetic compass, 7 oil tank, 8-1 pitot tube, 8-2 airspeed calculating module, 9 data-transmit module, 10 flight computer, 11 IMU, 12 cradle head, 13 winches, 14 wheels, 15 tachometer, 16 engine, 17 propeller, 18 protection ring, 20 fuselage.

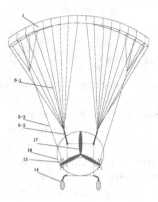

Fig. 2. Powered Parafoil UAV structure

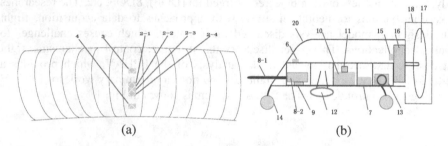

(a)   (b)

Fig. 3. (a) Parafoil canopy structure. (b) Payload vehicle structure

1.3   Control Mechanisms of Powered Parafoil UAV

PPUAV consists of ram-air-inflated fabric wing, payload and suspension lines.

The predominant control mechanism for powered parafoil UAV is left and right brake deflection and thrust provided by the engine. For most parafoils, deployment of the right brake causes a significant drag rise and a small lift increase on the right side of the canopy with slight right tilt. The overall effect causes the parafoil to turn right when a right brake is deployed. With a engine installed on the back of the payload, the system can adjust its longitudinal and vertical velocity [3-8].

To improve the accuracy of the parafoil system, several new flight control mechanisms have been created. One new control mechanism is called the dynamic incidence angle control. It is realized by changing the rigging length between the parafoil and the payload dynamically in flight, thus to realize direct glide slope control [3,9]. Another new control mechanism is realized by dynamically opening vent holes on the upper surface of the canopy to create a virtual aerodynamic spoiler. Symmetric activation of canopy spoilers yields longitudinal control while asymmetric activation creates lateral control [10].

Over the past few decades, a lot of models of different parafoil system were developed to address the varied issues, including 3DOF [11] (degrees-of-freedom), 6DOF [12], 8DOF [13], 9DOF [14] and 12DOF [15] models. In order to develop the GN&C successfully, reasonably accurate dynamic models that exhibit similar nonlinear behaviors with the actual airdrop systems are required. Thus several methods of system identification were developed [16-20].

In the paper, we consider the system as a rigid body, left brake deflection la, right brake deflection ra, symmetric brake deflection S and thrust T as inputs and longitudinal velocity v_x, vertical velocity v_z and Eular angles as outputs. The PPUAV model is obtained with lateral motion taken into account.

2   Real-Time Data Acquisition and Flight Tests Design

2.1   Real-Time Data Acquisition

Sensors and Structure of the System

The sensors installed on the aircraft consists of inertial measurement unit (IMU), global position system (GPS) , magnetic compass, tachometer and pitot tube.

The platform, on which a 3-axis gyro, a 3-axis acceleration sensor, a compass, and a GPS are installed, can store the data of velocities, angular rates, Euler angle accelerations, and positions into an SD card through an ARM processor. An Extended Kalman Fliter is used to estimate the sensors' values.

The sensors installed on the lower surface of the canopy are used to accurately measure the movement of the canopy, to reduce the error caused by the soft structure of the system, thus to control the system more accurately with the system's stability enhanced. The sensors installed on the vehicle measure the movement of the payload. Thus the total movement of the system is measured, which provides the basis for flight control.

Data Communication

The data acquisition module on the canopy is used to acquire the data from IMU, GPS and magnetic compass and transmit them to the flight computer.

The data acquisition module on the vehicle is used to acquire the data from IMU, GPS, magnetic compass and pitot tube and transmit them to the flight computer. As the communication is established, the onboard avionics transmits the system's flight status, including GPS position and 3axis compass information, aircraft attitude data, filtered and estimated velocities and acceleration data, control inputs and outputs and task management information to the ground control station (GCS) . After parsing incoming data packets, GCS displays the flight trajectory, waypoints, target location, as well as the battery and gas level, engine temperatures and revolutions. The GCS also uploads task information to the onboard controller, such as task waypoints, target location, and payload control commands.

2.2   Flight Test Design for the PPUAV Modeling

Three flight tests are needed. The objective of the first test is to gather aerodynamic data during manual flight by executing planned longitudinal and lateral maneuvers, and to analyze the main relationship between inputs and outputs, thus to obtain the aerodynamic characteristics of the system. In the second flight test, the powered para-foil is flown to a suitable altitude and the flight computer is switched to automatic mode to verify the correctness and the effectiveness of the model and the controller. The third flight test is autonomous. Fig. 4 shows the flight simulation in Matlab.

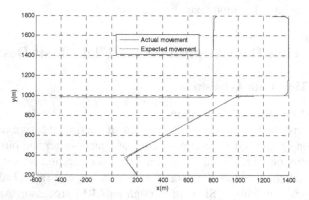

Fig. 4. Flight simulation in Matlab

2.3   Flight Test Arrangement for Mathematical Modeling

Manual flight test requires a windless environment. The powered parafoil was flown to the altitude of 200 m, then the maneuvers were executed after it flied stably. The maneuvers are presented by the percentage of the maximum value of left and right deflection and thrust. Each maneuver is held for a pre-determined time to ensure that the dynamics induced as the result of the maneuver have been damped out as shown in Table 1.

Table 1. Maneuvers of the flight test

Left deflection(%)	Right deflection(%)	Thrust(%)	Duration(s)
10	0	0	5
20	0	0	5
30	0	0	5
40	0	0	5
50	0	0	5
0	10	0	5
0	20	0	5
0	30	0	5
0	40	0	5
0	50	0	5

(Continued)

Table 1. (*Continued*)

Left deflection(%)	Right deflection(%)	Thrust(%)	Duration(s)
10	0	0	10
20	0	0	10
30	0	0	10
40	0	0	10
50	0	0	10
0	10	0	10
0	20	0	10
0	30	0	10
0	40	0	10
0	50	0	10
10	10	0	5
20	20	0	5
30	30	0	5
40	40	0	5
50	50	0	5
10	10	0	10
20	20	0	10
30	30	0	10
40	40	0	10
50	50	0	10
0	0	10	10
0	0	30	10
0	0	50	10
0	0	10	15
0	0	30	15
0	0	50	15

3 System Identification of Powered Parafoil UAV

3.1 Matlab System Identification Toolbox

Modern engineering of airdrop systems leans heavily on flight dynamic modeling and simulation to predict a multitude of drop events virtually so that guidance, navigation, and control (GN&C) software can be developed and tested in a cost efficient manner [21]. Matlab System Identification Toolbox is used to construct mathematical models of dynamic systems from measured input-output data. The toolbox also provides algorithms for embedded online parameter estimation. The tools simplifies the calculation and make the process concise and direct [22].

3.2 Identification and Verification of the Transfer Functions

The Powered Parafoil UAV is a kind of underactuated aircraft and the variation of a single input will lead to variations of multiple outputs.

Increasing or decreasing the thrust T provided by the engine power affects the longitudinal and vertical velocity. Deflecting the right or left trailing edge of the parafoil turns the aircraft right or left. If the trailing edge of the wing is pulled on both sides at the same time, the longitudinal velocity of the aircraft will decrease and the vertical velocity will increase. The data is acquired 50 times per second.

Symmetric Brake Deflection-Longitudinal Velocity
The flight data was imported to the Matlab System Identification Toolbox to obtain the transfer function from symmetric brake deflection to longitudinal velocity. Fig. 5 shows the curves of input and output and the transfer function was verified. The transfer function is as follow:

$$\frac{-0.0002958s - 0.002586}{s^2 + 0.775s + 1.357} \tag{1}$$

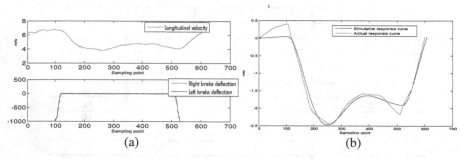

(a)   (b)

Fig. 5. (a) Curves of inputs and outputs. (b) Transfer function verification

Symmetric Brake Deflection-Vertical Velocity
Fig. 6 shows the curves of input and output and the transfer function was verified. The transfer function is as follow:

$$\frac{-0.002016s - 0.0002064}{s^2 + 0.8429s + 0.8064} \tag{2}$$

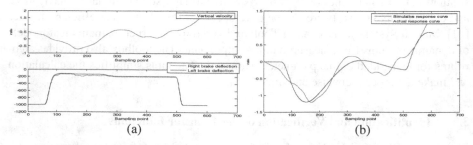

(a)   (b)

Fig. 6. (a) Curves of inputs and outputs. (b) Transfer function verification

Right Brake Deflection-Yaw Rate
Fig. 7 shows the curves of input and output and the transfer function was verified. The transfer function is as follow:

$$\frac{0.0001213s + 0.0002496}{s^2 + 0.9468s + 0.9751} \tag{3}$$

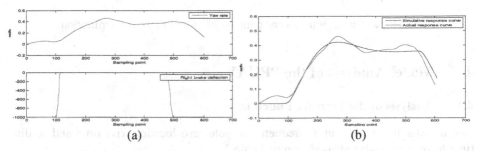

(a)   (b)

Fig. 7. (a) Curves of inputs and outputs. (b) Transfer function verification

Thrust-Longitudinal Velocity
Fig. 8 shows the curves of input and output and the transfer function was verified. The transfer function is as follow:

$$\frac{0.0002851s + 0.0002664}{s^2 + 2.159s + 0.533} \tag{4}$$

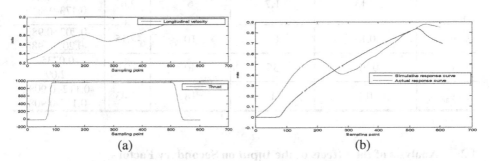

(a)   (b)

Fig. 8. (a) Curves of inputs and outputs. (b) Transfer function verification

Thrust-Vertical Velocity
Fig. 9 shows the curves of input and output and the transfer function was verified. The transfer function is as follow:

$$\frac{-0.0003046s - 0.001045}{s^2 + 0.2243s + 0.8359} \tag{5}$$

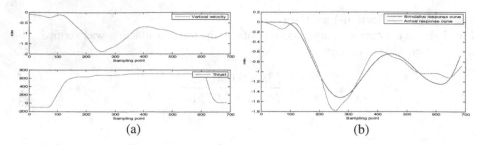

Fig. 9. (a) Curves of inputs and outputs. (b) Transfer function verification

4   Model Analysis of the PPUAV

4.1   Analysis of the Transfer Functions

The transfer functions' cutoff frequencies, pole-zero location, rise time and settling time have been analysed, as shown in Table 2.

Table 2. Analysis of the transfer functions

Channel	Cutoff frequency (Hz)	Rise time (second)	Settling time (second)	Zero	pole
$S - v_x$	0.28	1.5	7	-8.74	-0.338+1.11j; -0.338-1.11j
$S - v_z$	1.4	0.5	8	-0.102	-0.421+0.793j; -0.421-0.793j
ra -w	0.13	1.7	6	-2.06	-0.473+0.867j; -0.473-0.867j
la-w	0.13	1.5	10	-2.15	-0.202+0.98j; -0.202-0.98j
$T- v_x$	0.016	35	40	-0.934	-0.0735; -2.09
$T- v_z$	0.1	1	18	-3.43	-0.112+0.908j; -0.112-0.908j

4.2   Analysis of the Effects of the Input on Secondary Factors

The influences from brake deflection to pitch, asymmetric brake deflection to roll, thrust to pitch,and thrust to roll are analysed, as shown in Fig. 10. The figure shows that brake deflection and thrust have minor effects on pitch and roll.

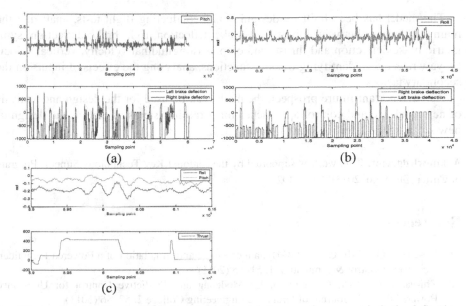

(a)   (b)

(c)

Fig. 10. (a) Curves of brake deflection and pitch. (b) Curves of asymmetric brake deflection and roll. (c) Curves of thrust, pitch and roll

5 Model of the PPUAV

Using the transfer functions, a mathematical model of Powered Parafoil UAV is obtained, which is used for trim control of the system. Fig. 11 shows the model and the specific expression of the model using transfer functions.

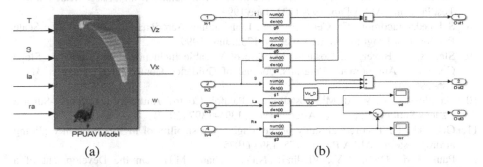

(a)   (b)

Fig. 11. (a) Inputs and outputs of PPUAV. (b) Transfer functions expression of the model

6 Conclusion

The PPUAV is an excellent platform for surveillance, search and rescue and cargo delivery. The platform is nonlinear, time-varing and strongly coupled. Thus modeling of the PPUAV system is of great significance.

The model has been built by designing and conducting flight tests, analyzing the main relationship between inputs (left brake deflection, right brake deflection, symmetric brake deflection and thrust) and outputs (longitudinal velocity, vertical velocity, yaw rate, etc.), identifying transfer functions, analysing the effects of inputs on the secondary factors.

As an important future prospect, the model can be used for the design and analysis of new guidance and control approaches for parafoil systems as well as the design of new software tools for mission planning.

Acknowledgment. This work is supported by the National Key Technology Support Program of China (Grant No. 2013BAK03B02).

References

1. Ke-chang, Q., Zi-li, C.: 8-DOF Dynamics Model and Simulation of a Powered Paraglider. Command Control & Simulation **1**, 51–55 (2011)
2. Zhi-gang, X., Zi-li, C., Yu-tian, L.: Modeling and Predictive Control for Unmanned Powered Parafoil. Journal of Ordnance Engineering College **2**, 52–56 (2011)
3. Michael, W., Mark, C.: Adaptive Glide Slope Control for Parafoil and Payload Aircraft. Journal of Guidance, Control, and Dynamics **36**(4), 1019–1034 (2013)
4. Oleg, A.Y., Nathan, J.S., Robyn A.T.: Development and Testing of the Miniature Aerial Delivery System Snowflake. AIAA Paper 2009–2980 (2009)
5. Zhu, Y., Moreau, M., Accorsi, M., Leonard, J., Smith, J.: Computer Simulation of Parafoil Dynamics. AIAA Paper 2001–2005, May 2001
6. Gupta, M., Zhenlong, X., Zhang, W., Accorsi, M., Leonard, J., Benney, R., Stein, K.: Recent Advances in Structural Modeling of Parachute Dynamics. AIAA Paper 2001–2030, May 2001
7. Iacomini, C., Cerimele, C.: Lateral-Directional Aerodynamics from a Large Scale Parafoil Test Program. AIAA Paper 99–1731, June 1999
8. Schroeder Iacomini, C., Cerimele, C.J.: Longitudinal Aerodynamics from a Large Scale Parafoil Test Program. AIAA Paper 99–1732, June 1999
9. Slegers, N., Beyer, E., Costello, M.: Use of Variable Incidence Angle for Glide Slope Control of Autonomous Parafoils. Journal of Guidance Control and Dynamics **31**(3), 585–596 (2008)
10. Gavrilovski, A., Ward, M., Costello, M.: Parafoil Control Authority with Upper-Surface Canopy Spoilers. Journal of Aircraft **49**(5), 1391–1397 (2012)
11. Goodrick, T.: Theoretical study of the longitudinal stability of high-performance gliding airdrop systems. AIAA Paper 1975–1394 (1975)
12. Paul, A.M., Oleg, A.Y., Vladimir, N.D., Richard, M.H.: On the Development of a Six-Degree-of-Freedom Model of a Low-Aspect-Ratio Parafoil Delivery System. AIAA Paper 2003–2105 (2003)
13. Yakimenko, O.A.: On the Development of a Scalable 8-DoF Model for a Generic Parafoil-Payload Delivery System. AIAA Paper 2005–1665 (2005)
14. Slegers, N., Costello, M.: Comparison of measured and simulated motion of a controllable parafoil and payload system. AIAA Paper 2003–5611 (2003)
15. Vishnyak, A.: Simulation of the payload-parachute-wing system flight dynamics. AIAA Paper 93–1250 (1993)

16. Michael, W., Costello, M., Slegers, N.: Specialized System Identification for Parafoil and Payload Systems. Journal of Guidance, Control, and Dynamics **35**(2), 588–597 (2012)
17. Majji, M., Juang, J.-N., Junkins, J.L.: Observer/Kalman-Filter Time-Varying System Identification. Journal of Guidance Control and Dynamics **33**(3), 887–900 (2010)
18. Manoranjan, M., Jer-Nan Junkins, J., John, L.: Time-Varying Eigensystem Realization Algorithm. Journal of Guidance, Control, and Dynamics. Journal of Guidance, Control, and Dynamics **33**(1), 13–28 (2010)
19. Valasek, J., Chen, W.: Observer/Kalman Filter Identification for Online System Identification of Aircraft. Journal of Guidance Control and Dynamics **26**(2), 347–353 (2003)
20. Yakimenko, O., Statnikov, R.: Multicriteria parametrical identification of the parafoil-load delivery system. In: 18th AIAA Aerodynamic Decelerator Systems Technology Conference and Seminar. AIAA Paper 2005–1664 (2005)
21. Yuhu, D., Jiancheng, F., Wei, S., Xusheng, L.: Identification of Small-scale Unmanned Helicopter Based on Least Squares and Adaptive Immune Genetic Algorithm. ROBOT **1**, 72–77 (2012)
22. Xiao-hui, Q., Qing-min, T., Hai-rui, D.: System Modeling Based on System Identification Toolbox in Matlab. Ordnance Industry Automation **10**, 88–90 (2006)

Visual Servoing Control of Baxter Robot Arms with Obstacle Avoidance Using Kinematic Redundancy

C. Yang[1,2]($\boxtimes$), S. Amarjyoti[1], X. Wang[1,3], Z. Li[2], H. Ma[3], and C.-Y. Su[2,4]

[1] Center for Robotics and Neural Systems, Plymouth University, Plymouth, UK
cyang@ieee.org
[2] College of Automation Science and Engineering,
South China University of Technology, Guangzhou, China
[3] School of Automation, Beijing Institute of Technology, Beijing, China
[4] Department of Mechanical and Industrial Engineering, Concordia University,
Montreal, Canada

Abstract. In this paper, a visual servoing control enhanced by an obstacle avoidance strategy using kinematics redundancy has been developed and tested on Baxter robot. A Point Grey Bumblebee2 stereo camera is used to obtain the 3D point cloud of a target object, which is then utilized to manipulate the closely coupled dual arms of Baxter robot. The object tracking task allocation between two arms has been developed by identifying workspaces of the dual arms and tracing the object location in a convex hull of the workspace. By employment of a simulated artificial robot as a parallel system as well as a task switching weight factor, the actual robot is able to restore back to the natural pose smoothly in the absence of the obstacle. Two sets of experiments carried out demonstrate the effectiveness of the developed servoing control method.

Keywords: Visual servoing · Point cloud · Obstacle avoidance

1 Introduction

Visual sense is one of the most essential markers of intelligence and contributes immensely to the interactions with our environment. Visual feedback plays an important role for an anthropomorphic robot to operate in human surroundings. Especially, the advancements in 3D Vision Guided Robots (VGRs) have led to accurate non-contact geometrical measurements and reduction of workspace ambiguity. Therefore, strides made in robotic and machine vision are of paramount importance for engineering, manufacturing and design processes [1].

This work is supported in part by EPSRC grants EP/L026856/1 and EP/J004561/1,Guangdong Provincial Natural Science Foundation of China under Grant 2014A030313266 and National Natural Science Foundation (NNSF) of China under Grants 61473120 and 61473038.

© Springer International Publishing Switzerland 2015
H. Liu et al. (Eds.): ICIRA 2015, Part I, LNAI 9244, pp. 568–580, 2015.
DOI: 10.1007/978-3-319-22879-2_52

Visual servoing(VS) is defined as the use of visual feedback mechanisms for the kinematic control of a robot. Based on the positioning of the camera on the link and control techniques, VS ramifies into several types. Eye-in-hand and eye-to-hand VS are represented by the position of the camera on the robotic manipulator. Being attached on the robot arm, eye-in-hand VS provides a narrower field of view as compared to eye-to-hand servoing. Many control schemes use either a direct visual servoing or a dual loop system [7]. A visual processing method for hand gesture recognition is developed in [9] for control a simulated iCub robot on YARP platform. In [8], a vision-based non-contacting method of robot teleoperation is presented, whereas a human operator teleoperate a six degrees of freedom (DOF) manipulator using the three-dimensional human hand-arm motion of the operator, in order to complete an object manipulation task naturally.

These schemes are robust but come at the price of reduced performance and speed variables. The drawbacks of time delay between image acquisition and its relation to latency in robot controller can be solved by the use of an open loop based VS for computing the reference input of the joint controller from the targets projection once [3]. It is practically useful to integrate visual sensing with other sensing technologies to develop a hybrid sensing system for better control performance. A combination of force sensor, CCD cameras and laser projector is used for accurate drilling on a 3D surface of unknown pose [2]. Eye-to-hand pose estimation and eye-in-hand object grasping in a hybrid control loop have been studied in [13][10][15]. In this paper, we use a similar approach that subsumes stereo image processing for pose estimation and the range data from IR sensors near the end effectors for precise gripping.

Similar to our human arm, each of Baxter robot manipulators is of 7 joints. When the DOF in joint space is larger than that required in the task space, there is redundancy. At most 6 DOF is required for manipulation in the Cartesian space, and the redundancy can be used to generate motion for obstacle avoidance. In [11,16], the solution is decomposed into a particular and a homogeneous component, which achieve an exact end-effector control and maximizing the distance to obstacles with redundant DOFs. In this work, the robot end effector is controlled to follow the reference trajectory generated according to the detected target object's position, while at the same time the joint space redundancy is utilized to avoid obstacle in the environment. Using our proposed method, the redundant Baxter robot arm is able to avoid obstacle without sacrificing the tracking performance of the end effector. In addition, the developed control method also enables the robot arm to restore its natural pose when the obstacle is absent, by employment of a simulated artificial robot as a parallel system based on kinematic model of the Baxter robot.

2   Baxter® Robot and Its Kinematics

2.1   Workspace Identification of the Baxter® Robot Arms

Baxter® robot is a semi-humanoid robot consisting of two arms and a rotational head on its torso, and it can be installed on a movable pedestal. Several attributes

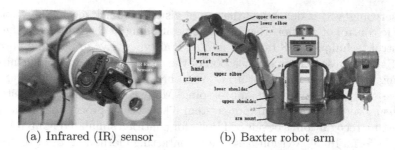

(a) Infrared (IR) sensor   (b) Baxter robot arm

Fig. 1. Left: IR Range sensor; Right: Illustration of the 7DOF Baxter robot arm (Modified from [4]).

such as the presence of a CCD camera and an infra-red (IR) range sensor at the end effector as shown in Fig. 1(a), make the Baxter robot more intelligent. As illustrated in Fig. 1(b), each arm of Baxter robot constitutes of 8 links and 7 rotational joints with additional interchangeable grippers, including electric gripper and a vacuum cup. The kinematics model of a Baxter robot arm was built in our previous work [6], where the details of DH parameters as well as joint rotation limits are provided. Robot manipulator workspace boundary estimation is essential for improvising of robotic algorithms and optimization of its overall design and analysis. In this paper, we first extend the Monte Carlo method used on a single arm [6] to dual arms of Baxter robot, in order to identify the reachable workspace of two arms. Homogenous radial distribution is used to generate 6000 points in the joint space for each arm separately. The joint angle values are chosen randomly and forward kinematics is implemented to evaluate the end effector points, creating a point cloud of the reachable workspace for dual arms. The generated point cloud of the workspace for both the manipulators is shown in Fig. 2(a). Next, Delaunay triangulation is applied to the points in the 3D space to generate a set of points with a circumcircle without any points in its interior. This facilitates the creation of a convex hull of the joint space. It is used to constrain the workspace and to distinguish the ranges of control between the left and right arms for efficient manoeuvrability.

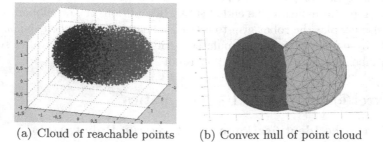

(a) Cloud of reachable points   (b) Convex hull of point cloud

Fig. 2. Workspace Identification

3   Visual Sensor and its Preprocessing

A client-server UDP network as shown in Fig. 3(a) is employed to integrates various system components and confers parallel processing. The machine vision processing is done in the client computer connected with a Point Grey® Bumblebee2 stereo camera with IEEE-1394 Firewire connection. The Point Grey Bumblebee2 stereo camera, as shown in Fig. 3(b), is a 2 sensor progressive scan CCD camera with fixed alignment between the sensors. Video is captured at a rate of 20 fps with a resolution of 640×480 to produce dense colored depth maps to assist in tracking and a viable pose estimation of the object. The resolution-speed trade-off has to be managed concisely as an increased frame speed gives a smooth robot trajectory whereas enhances the processing time. And an increased resolution provides a denser, more accurate point cloud for feature extraction but with increased latency.

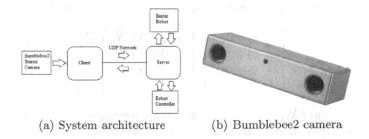

(a) System architecture   (b) Bumblebee2 camera

Fig. 3. Left: Communication network; Right: Point Grey @ Bumblebee 2 camera

3.1   Camera Calibration

Camera calibration is necessary as the use of lenses introduces nonlinearities and deviates from the simple pin-hole model such as lens distortion, namely radial and tangential distortion. The camera parameters, namely, the intrinsic, extrinsic and distortion are evaluated by the use of a 2D checker-board pattern. 3D reference models were avoided due to computational complexity and high cost of precise calibration objects. In this work, 20 checkerboard images were fed to the calibrator algorithm encompassing differential angles in the projection space. This provides enough values to estimate the camera geometry parameters, including 4 intrinsic parameters of f_x, f_y, c_x, c_y, 5 distortion parameters of k_1, k_2, k_3 (radial), p_1, p_2 (tangential), and the camera extrinsic parameters of Φ, Ψ, Θ (rotation) together with T_x, T_y, T_z (translation). The point "Q" on the object plane is related with the image plane point "q" by the following equations. A summary to the definition of these parameters is provided in Table 1.

$$\begin{bmatrix} x \\ y \\ 1 \end{bmatrix} = sA[r_1 \ r_2 \ r_3 \ T] \begin{bmatrix} X \\ Y \\ 0 \\ 1 \end{bmatrix}, \quad x = f\frac{X}{Z} + c_x, \ y = f\frac{Y}{Z} + c_y, \ q = sHQ \quad (1)$$

The camera parameters are finally obtained using the Closed Form solution and the Maximum Likelihood estimation [12]. Furthermore the distortion is nullified from the image points by introducing the following equations

$$x_{corrected} = x(1 + k_1 r^2 + k_2 r^4 + k_3 r^6), \quad y_{corrected} = y(1 + k_1 r^2 + k_2 r^4 + k_3 r^6)$$

Table 1. Definition of variables

X, Y	World co-ordinates of the point	c_x, c_y	Projection displacement parameter
x, y	Image plane co-ordinates of the point	H, A	Homography matrix, and Intrinsic matrix
f	Focal length of the camera lens	r_1, r_2, r_3	spatial rotational matrices
s	scaling factor	T	translational matrix

The co-ordinate transformation of the detected feature points from the Bumblebee2 co-ordinates to Baxter co-ordinates can be achieved by a homogeneous transformation matrix to be obtained as

$$T = \begin{bmatrix} x_1 & x_2 & x_3 & x_4 \\ y_1 & y_2 & y_3 & y_4 \\ z_1 & z_2 & z_3 & z_4 \\ 1 & 1 & 1 & 1 \end{bmatrix} \begin{bmatrix} X_1 & X_2 & X_3 & X_4 \\ Y_1 & Y_2 & Y_3 & Y_4 \\ Z_1 & Z_2 & Z_3 & Z_4 \\ 1 & 1 & 1 & 1 \end{bmatrix}^{-1} \tag{2}$$

where (x_i, y_i, z_i) and (X_i, Y_i, Z_i), $i = 1, 2, 3, 4$, are four non-collinear points measured from the robot coordinate and the Bumblebee2 coordinate, respectively.

4   Object Detection and Localization

4.1   3D Reconstruction

The images captured by the Bumblebee2 stereo camera in active ambient lighting are shown in Fig. 4. Both the images are calibrated using the camera intrinsics and are corrected for distortion. Subsequently, the undistorted images are stereo

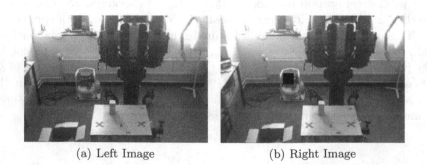

(a) Left Image   (b) Right Image

Fig. 4. Images captured from Bumblebee2

rectified in order to align the epipolar lines of both the projection planes and ensure the presence of similar pixels in a specified row of the image. The images obtained are then frontal parallel and are ready for correspondence estimate. The Essential and the Fundamental matrix are calculated by using Epipolar geometry. Essential matrix is a 3×3 matrix with 5 parameters; two for translation and three for the rotation values between the camera projection planes. On the other hand, the Fundamental matrix represents the pixel relations between the two images and has seven parameters, two for each epipole and three for homography that relates the two image planes. Bouguets algorithm is then implemented to align the epipolar lines and shift the epipoles to infinity. Fig. 5(a) depicts the results of stereo rectification where the red and cyan colors represent the left and right images with row aligned pixels.

(a) Rectified stereo Images   (b) Disparity Map   (c) 3D reconstruction of the image

Fig. 5. Stereo images and 3D reconstruction.

Table 2. Definition of variables

1	x^l	column value of left image pixel	5	f	focal length
2	x^r	column value of right image pixel	6	d	disparity
3	D	Depth	7	Q	Projection matrix
4	T	Baseline	8	X/W, Y/W, Z/W	3D world co-ordinates

The definition of variables used below is provided in Table 2. Stereo correspondence is a method of matching pixels with similar texture across two co-planar image planes. The distance between the columns of these perfectly matched pixels is defined as $d = x^l - x^r$.

Block matching is implemented for evaluating the correspondence between the images. Block sizes of 15 pixel window are used to find the matches by the use of SAD (sum of absolute differences). The disparity range is kept low as [0 40] in order to match the indoor low texture difference and taking into account computational speed. Semi Global method is used to force the disparity values to the neighboring pixels for a more comprehensive result [5]. The disparity output is shown in Fig. 5(b). Disparity is inversely proportional to the depth of the pixel and is related by the Triangulation equation $D = T\frac{f}{d}$. Triangulation refers to the estimation of depth of an object by visualizing its location from two different known points. The reconstruction of the image in the Cartesian co-ordinates is obtained by the use of projection matrix evaluated using Bouguets algorithm (3). Fig. 5(c) depicts the 3D reconstruction of the robots workspace.

$$Q\left[x, y, d, 1\right]^{T} = [X, Y, Z, W]^{T} \tag{3}$$

4.2   Object Detection

Color based segmentation is used in order to separate a single color object from the background. The image is converted into L*a*b* color space and the Euclidean distance between red-green and yellow-blue opponent components of the object and a and b matrices calculated. The minimum value gives the most accurate estimate of the object. Further, the corners of the object are calculated by Harris corner detector and the centroid calculated by intersection of the diagonals. The depth value of the centroid is then extracted from reconstructed point cloud of the task space. Fig. 6 shows the calculated robot co-ordinates of the centroid of the object after co-ordinate transformation.

Fig. 6. Object detection results

4.3   Switching Scheme for Dual Arms

The depth of the object to be manipulated is mapped to the convex hull to expedite the process of decision making, because a point cloud matching would be hinder the processing speed. This is done by checking the presence of the point's respective co-ordinates in the convex-hull projection on the three Cartesian planes. Hence, if (X_1, Y_1, Z_1) be the point representing the object, its presence in the 3D hull is detected by the five steps listed in Algorithm 1.

The above procedure ensures an efficient motion to reach the object in the robot workspace. It mimics the human intuition of using the nearest possible arm for grabbing in order to avoid the use of excessive body movements and minimize the use of energy.

Algorithm 1. Object detection in the 3D workspace hull
1: Check if the XY plane projection of the hull contains the point (X_1, Y_1), YZ plane contains point (Y_1, Z_1) and XZ plane contains point (X_1, Z_1) using Ray Casting Algorithm;
2: Obtain the presence decision in the workspace of both arms;
3: If the point is present in both of the manipulator workspace, give the control priority to the arm with the smallest Euclidean distance from manipulator;
4: Divide the control based on the detection of the point in left or right workspaces;
5: If the point lies outside both the workspaces, stop arm movement to avoid singularity.

4.4 Detection of Collision Points

The collision points, p_{cr} and p_{co}, are defined as the two points either on the robot or on the obstacle, which covers the nearest distance between the robot and the obstacle. The forward kinematics of the robot is used to estimate the collision points. Each link of the robot can be seen as a segment in 3-D space. The coordinates of the endpoints of the segments, i.e. the coordinates of the joints, in Cartesian space can be obtained by forward kinematics.

The problem of estimating the collision points can be seen as the problem of searching the nearest point between the obstacle segment $q_{l1} - q_{l2}$ and the segments that stand for the robot links. Firstly, the distance between segment $q_{l1} - q_{l2}$ and the ith segment of the robot links $\lfloor x_i, y_i, z_i \rfloor - \lfloor x_{i+1}, y_{i+1}, z_{i+1} \rfloor$ can be calculated by 3 D geometry, which is denoted by d_i; the nearest points on the robot link and the obstacle are denoted by p_{cri} and p_{coi}, respectively. Then we have the collision point $p_{cr} = p_{cri_{min}}$ and $p_{co} = p_{coi_{min}}$ and the distance $d = d_{i_{min}}$ where $i_{min} = \arg\min_{i=0,1,\dots,n} d_i$.

5 Obstacle Avoidance and Restoring Control

Once the target object's co-ordinates has been detected, end effector desired trajectory for the visual tracking task can be defined as the moving trajectory of the target. While for visual guided manipulation task, the end effector desired trajectory can be generated according to the location of the target object. Denote the joint velocities as $\dot{\theta}$ and end-effector velocity as $\dot{x}$, then they must satisfy $\dot{x} = J\dot{\theta}$ where J is the Jacobian matrix of each arm. When the collision points p_{cr} and p_{co} are found, the manipulator needs to move away from the obstacle with a desired velocity $\dot{x}_o$ given by (4), as shown in Fig. 7(a).

$$\dot{x}_o = \begin{cases} \mathbf{0} \ , d \geq d_o \\ \dfrac{d_o - d}{d_o - d_c} v_{max} \dfrac{p_{cr} - p_{co}}{d} \ , d_c < d < d_o \ , \\ v_{max} \dfrac{p_{cr} - p_{co}}{d} \ , d \leq d_c \end{cases} \tag{4}$$

where $d = \|\boldsymbol{p_{cr}} - \boldsymbol{p_{co}}\|$ is the distance between the obstacle and the manipulator; v_{max} is the maximum obstacle avoidance velocity; d_o is the distance threshold that the manipulator starts to avoid the obstacle; d_c is the minimum acceptable distance and the manipulator will avoid at the maximum speed.

After the obstacle is removed, the manipulator is expected to restore back its original state to eliminate the influence of the obstructing. To achieve that, an artificial parallel system of the manipulator is designed in the controller in real time to simulate its pose without the influence of the obstacle, as shown in Fig. 7(b), where the dashed black line indicates the parallel system. The restoring velocity $\dot{\boldsymbol{x}}_r$ is designed as a closed-loop system as $\dot{\boldsymbol{x}}_r = \boldsymbol{K_r e_r}$, where $\boldsymbol{K_r}$ is a symmetric positive definite matrix and $\boldsymbol{e_r} = [e_{r1}\ e_{r2}\ e_{r3}]^T$ is the position errors of the joints between the parallel system and the real system. Based on our previous work [14], the control strategy is given below

$$\dot{\boldsymbol{\theta}}_d = \boldsymbol{J}_e^\dagger (\dot{\boldsymbol{x}}_d + \boldsymbol{K}_e \boldsymbol{e}_x) + \left(\boldsymbol{I} - \boldsymbol{J}_e^\dagger \boldsymbol{J}_e\right) [\alpha \boldsymbol{z}_o + (1 - \alpha) \boldsymbol{z}_r] \tag{5}$$

where

$$\boldsymbol{z}_o = \left[\dot{\boldsymbol{x}}_o^T \boldsymbol{J}_o \left(\boldsymbol{I} - \boldsymbol{J}_e^\dagger \boldsymbol{J}_e\right)\right]^\dagger \left(\dot{\boldsymbol{x}}_o^T \dot{\boldsymbol{x}}_o - \dot{\boldsymbol{x}}_o^T \boldsymbol{J}_o \boldsymbol{J}_e^\dagger \dot{\boldsymbol{x}}_e\right) \tag{6}$$

$$\boldsymbol{z}_r = \left[\dot{\boldsymbol{x}}_r^T \boldsymbol{J}_r \left(\boldsymbol{I} - \boldsymbol{J}_e^\dagger \boldsymbol{J}_e\right)\right]^\dagger \left(\dot{\boldsymbol{x}}_r^T \dot{\boldsymbol{x}}_r - \dot{\boldsymbol{x}}_r^T \boldsymbol{J}_r \boldsymbol{J}_e^\dagger \dot{\boldsymbol{x}}_e\right) \tag{7}$$

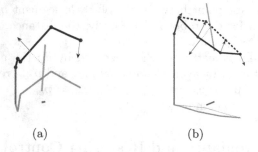

(a)   (b)

Fig. 7. The artificial parallel system built using Baxter® kinematics model [6]. The solid black line indicates the real manipulator. The dashed black line indicates the simulated manipulator in the artificial parallel system.

5.1 Object Gripping with Hybrid Servoing

Stereo camera provides a wider angle of view for the eye to hand servoing system, while for meticulous sensing of the object, an IR range sensor equipped at the end effector provides an eye-in-hand feedback, and helps to reduces the processing time and provides satisfactory accuracy. The IR range sensor used provides 16 bit digital data and checks for the location of object along the Z and Y axes. The camera and robot X axes are perfectly aligned and hence the real-time error is minimal. But the error in other axes needs to be neutralized, partially due to lack of visibility of some portions of the object and objects with ambiguous dimensions. A low pass filter is applied to the range data in order to remove the occasional anomalies.

6   Experiment Studies

In order to test the proposed visual servoing method, two groups of experiments of visual tracking task and visual guided manipulation task are performed, with obstacle and without obstacle for comparison.

6.1   Visual Tracking Task

In the visual tracking experiments, the robot is controlled to follow the object using two manipulators. The experiment video frames are shown in Figs. 8(a)-8(d) and Fig. 8(e)-8(h) for the experiments without and with obstacle respectively. The end-effector trajectory of this two comparative experiments are shown in Fig. 9.

As can be seen from Fig. 9, the robot automatically choose the appropriate manipulator to track the object based on the workspaces. When the obstacle is found near the manipulator, the corresponding manipulator change its pose to avoid collision and, at the same time, keeping its end-effector position tracking the object.

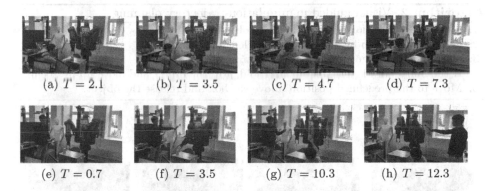

(a) $T = 2.1$ (b) $T = 3.5$ (c) $T = 4.7$ (d) $T = 7.3$

(e) $T = 0.7$ (f) $T = 3.5$ (g) $T = 10.3$ (h) $T = 12.3$

Fig. 8. Video frames of visual tracking with obstacle without and with obstacle.

6.2   Visual Guided Manipulation Task

In the visual guided manipulation experiments, the robot is controlled to pick up the object and put it into a box at a predefined position. The end-effector trajectory is planed using Algorithm 2. The experiment video frames are shown in Fig. 10(a)-10(d) and Fig. 10(e)-10(h) for the experiments without and with obstacle respectively. The end-effector trajectory of this two comparative experiments are shown in Fig. 11. As can be seen from the results, the robot can pick up the object precisely even if the position of object has been changed. When the obstacle is found near the manipulator, the corresponding manipulator change its pose to avoid collision but can still move the end-effector right above the box and drop the object into the box.

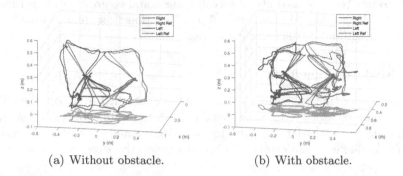

(a) Without obstacle.   (b) With obstacle.

Fig. 9. End-effector trajectory of visual tracking. The blue and red dashed lines indicates the reference end-effector trajectory of the left and right manipulators respectively; the blue and red solid lines indicates the actual trajectory of the end-effectors of the left and right manipulators respectively.

Algorithm 2. Visual guided manipulation trajectory planning

1: Move to the position 10cm behind the object;
2: Move forward step by step with a step length of 0.5cm;
3: Check the IR range sensor after each 0.5cm step;
4: If the range data is less than a threshold, grip the object and move up by 20cm;
5: Move to the predefined position above the box and release the object.

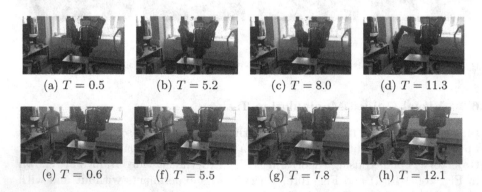

(a) $T = 0.5$ (b) $T = 5.2$ (c) $T = 8.0$ (d) $T = 11.3$

(e) $T = 0.6$ (f) $T = 5.5$ (g) $T = 7.8$ (h) $T = 12.1$

Fig. 10. Video frames of visual guided manipulation without and with obstacle.

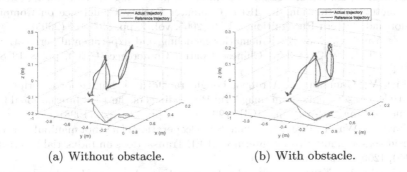

(a) Without obstacle.   (b) With obstacle.

Fig. 11. End-effector trajectory of the visual guided manipulation. The dashed lines indicates the reference end-effector trajectory; and the solid lines indicates the actual trajectory of the end-effectors.

7   Conclusion

This paper developed and implemented a visual servoing control method with automatic obstacle avoidance on the dual arm Baxter robot. Stereo imaging and range sensors are utilized to attain eye-to-hand and eye-in-hand servoing, such that both target object and obstacle are detected and localized. Kinematic redundancy has been exploited for obstacle avoidance. A simulated artificial robot of same kinematics as Baxter robot is employed as a parallel system to enable the robot restoring back to the natural pose when the obstacle is absent. Two switching mechanisms have been designed, one for task allocation between two arms for visual tracking task, while the other for smooth switching between obstacle avoiding task and restoring task. An object tracking and an object manipulation experiments have been performed to test the develop control method, and the results have demonstrated the effectiveness and robustness of the developed approach.

References

1. Association for safe international road travel statistics. http://www.asirt.org/initiatives/informing-road-users/road-safety-facts/road-crash-statistics.aspx
2. Chang, W.C., Shao, C.K.: Hybrid fuzzy control of an eye-to-hand robotic manipulator for autonomous assembly tasks. In: Proceedings of the SICE Annual Conference 2010, pp. 408–414 (2010)
3. Daigavane, P., Bajaj, P.: Road lane detection with improved canny edges using ant colony optimization. In: 2010 3rd International Conference on Emerging Trends in Engineering and Technology (ICETET), pp. 76–80 (2010)
4. Guizzo, E., Ackerman, E.: How rethink robotics built its new baxter robot worker. IEEE Spectrum (2012)

5. Hirschmuller, H.: Accurate and efficient stereo processing by semi-global matching and mutual information. In: IEEE Computer Society Conference on Computer Vision and Pattern Recognition, CVPR 2005, vol. 2, pp. 807–814 (2005)

6. Ju, Z., Yang, C., Ma, H.: Kinematics modeling and experimental verification of baxter robot. In: 2014 33rd Chinese Control Conference (CCC), pp. 8518–8523 (2014)

7. Kamat, V., Ganesan, S.: A robust hough transform technique for description of multiple line segments in an image. In: Proceedings of the 1998 International Conference on Image Processing, ICIP 1998, vol. 1, pp. 216–220 (1998)

8. Kofman, J., Wu, X., Luu, T., Verma, S.: Teleoperation of a robot manipulator using a vision-based human-robot interface. IEEE Transactions on Industrial Electronics **52**(5), 1206–1219 (2005)

9. Li, C., Ma, H., Yang, C., Fu, M.: Teleoperation of a virtual icub robot under framework of parallel system via hand gesture recognition. In: Proceedings of the 2014 IEEE World Congress on Computational Intelligence, WCCI, Beijing (2014)

10. Luo, R., Chou, S.C., Yang, X.Y., Peng, N.: Hybrid eye-to-hand and eye-in-hand visual servo system for parallel robot conveyor object tracking and fetching. In: 40th Annual Conference of the IEEE Industrial Electronics Society, IECON 2014, pp. 2558–2563 (2014)

11. Maciejewski, A.A., Klein, C.A.: Obstacle avoidance for kinematically redundant manipulators in dynamically varying environments. The International Journal of Robotics Research **4**(3), 109–117 (1985)

12. Su, H., He, B.: A simple rectification method of stereo image pairs with calibrated cameras. In: 2010 2nd International Conference on Information Engineering and Computer Science (ICIECS), pp. 1–4 (2010)

13. Urmson, C., Whittaker, W.: Self-driving cars and the urban challenge. IEEE Intelligent Systems **23**(2), 66–68 (2008). doi:10.1109/MIS.2008.34

14. Wang, X., Yang, C., Ma, H.: Automatic obstacle avoidance using redundancy for shared controlled telerobot manipulator. In: 2015 IEEE 5th Annual International Conference on Cyber Technology in Automation, Control, and Intelligent Systems (CYBER) (accepted, 2015)

15. Wijesoma, W., Kodagoda, K., Balasuriya, A., Teoh, E.: Road edge and lane boundary detection using laser and vision. In: Proceedings of the 2001 IEEE/RSJ International Conference on Intelligent Robots and Systems, vol. 3, pp. 1440–1445 (2001)

16. Zlajpah, L., Nemec, B.: Kinematic control algorithms for on-line obstacle avoidance for redundant manipulators. In: IEEE/RSJ International Conference on Intelligent Robots and Systems, vol. 2, pp. 1898–1903 (2002)

Hand Gesture Based Robot Control System Using Leap Motion

Sunjie Chen[1], Hongbin Ma[1,2(✉)], Chenguang Yang[3], and Mengyin Fu[1,2]

[1] School of Automation, Beijing Institute of Technology,
Beijing 100081, People's Republic of China
mathmhb@gmail.com
[2] State Key Lab of Intelligent Control and Decision of Complex Systems,
Beijing Institute of Technology, Beijing 100081, People's Republic of China
[3] School of Computing and Mathematics, Plymouth University,
Plymouth PL4 8AA, UK
cgyang82@gmail.com

Abstract. Gesture based human-robot interface is a highly efficient robot control strategy for its simple operation and high availability. This paper develops a hand gesture based robot control system using Leap Motion. The process that the robot responds to human's hand gesture contains noise suppression, coordinate transformation and inverse kinematics. A Client/Server structured robot control system is developed, which provides the function of controlling virtual universal robot UR10 with hand gesture. Finally, experimental results demonstrate that the system is effective and practical.

Keywords: Leap motion · Gesture · Robot · Human-robot Interface · Simulation

1 Introduction

Nowadays, robots are so popular that programming seems not to be efficient enough to control them. That is, we need to create a more practical easy-to-use robot control system. Voice recognition analyzes voice signal with models such as hidden Markovian models (HMMs) such that the voice can be interpreted as text information by the robots. However, there has a large amount of dialect so that a big database needs to be built. Analyzing meaning of voice must refer to specific context, so it will be difficult as well. EEG samples physiological electric signals when brain is active. However, the approach of using EEG to reflect the user's requirement and intention is not robust or reliable enough. Gesture recognition samples the characteristic of hand action to reflect the user's requirement or

This work was partially supported by the National Natural Science Foundation in China (NSFC) under Grants 61473038 and 61473120. Also supported in part by State Key Laboratory of Robotics and System (HIT) SKLRS-2014-MS-05.

© Springer International Publishing Switzerland 2015
H. Liu et al. (Eds.): ICIRA 2015, Part I, LNAI 9244, pp. 581–591, 2015.
DOI: 10.1007/978-3-319-22879-2_53

intention. Hence, gesture based control system is low-cost, highly flexible and efficient.

Gesture tracking is an important function of gesture based human-robot interaction. There are two methods for gesture tracking research, one relies on vision, and the other one bases on data glove. The former one can be affected by the light, complexion, and it is difficult to capture detail. The latter one may depend more on the device [1].

Low-cost sensors, the Kinect and the Xtion Pro, make it convenient to achieve gesture recognition. However, they concentrate more on body action. The Leap Motion is small in size, and has high precision and low power dissipation. It can detect details of hand action [2].

The Leap Motion has two cameras inside, which can take photos from different directions to obtain hand action information in 3D space. The detection range of Leap Motion is between 25mm and 600mm upon the sensor. The shape of the detection space is similar to an inverted cone [3]. Figure 1 depicts the coordinate system for the Leap Motion.

Fig. 1. Coordinate system of Leap Motion

Recently, Xu *et al.* [4] put forward a method of remote interaction through detecting the position of palm based on the Leap Motion, and mapping the physical space and information space. As one typical application of the Leap Motion, a MIDI controller using the Leap Motion was reported in the literature [5] and it is also well known that many games can be played with motion sensors such as Kinect. With the fast development of motion sensing technology, the Leap Motion can serve as an excellent replacement of Kinect for desktop applications. After extensive testing, one may easily draw a conclusion that the motion sensing technology is more practical and more attractive than traditional way. As to the dynamic hand gesture recognition, Wang *et al.* [6] proposed an effective method which can recognize dynamic hand gesture by analyzing the information of motion trajectory captured by the Leap Motion.

This paper develops a hand gesture based robot control system using the Leap Motion. The system first obtains gesture information using the Leap Motion. Then, with the aid of algorithms, it analyzes data and understands robot control signals from the gesture information. Finally, the system achieves the goal of controlling the robot with gesture in real time.

2   Hardware and Software

The software development with the Leap Motion should base on the official SDK and the driver. When it is powered on, the Leap Motion sends hand action information periodically. Every package information is called a frame [7]. The sensor will assign these information with an ID. With the ID, the user may call functions such as Frame::hand(), Frame::finger() to check any object's information.

With C++ compiler of Microsoft Visual Studio 2010 on Windows or Gnu C/C++ Compiler on Linux, we can implement functions of reading and analyzing data from Leap Motion.

V-REP is a robot simulator with integrated development environment. Each object/model can be individually controlled via an embedded script, a plugin, a ROS node, a remote API client, or a custom solution [8].

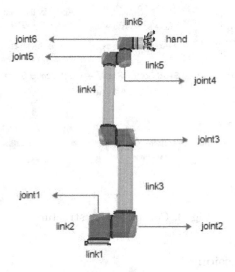

Fig. 2. UR10&BarrettHand

UR10 robot is produced by Universal Robots. Unlike traditional industrial robot, UR10 is light and cheap. It weights 28kg, and can load 10kg [9]. Its working range is 1300mm.(see Figure 2)

The BarrettHand is a multi-fingered programmable grasper with the dexterity to secure target objects of different sizes, shapes and orientations. Even with its low weight (980 g) and super compact (25mm) base, it is totally self-contained [10].

3   Control System

To achieve the goal of controlling robot with gesture, this paper develops a Client/Server structured remote human-robot interface control software [11].

First, the client sends a signal to the server to inform that the client is pre-pared to accept next action command. Then, the server gets the signal, and receives current frame information from the Leap Motion. The server finishes the work about noise suppression, coordinate transformation, inverse kinematics and packs the control signals sending to V-REP as well. Finally, V-REP runs the simulation. Figure 3 shows the process of the loop.

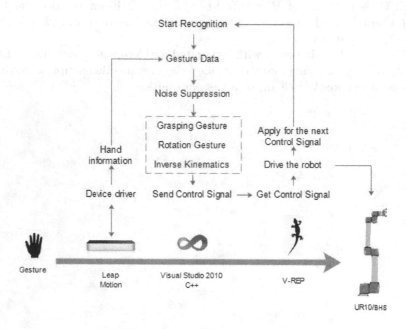

Fig. 3. Control system structure

3.1   Noise Suppression

The data, including the position of palm, direction vector of finger and normal vector of palm, will be handled by the server. The precision of the Leap Motion can be 0.01mm in theory. Actually, there are some destabilizing factors which will affect the precision, such as shaking of hand, magnetocaloric effect, calculating, etc. This paper adapts a speed-based low-pass filter [12] to eliminate noise. The point of this method is changing the cut-off frequency of low-pass filter with the velocity of palm. The filter can be mathematically described by

$$\hat{X}_i = \alpha X_i + (1 - \alpha)\hat{X}_{i-1} \tag{1}$$

where X_i is a vector containing the coordinates and direction information given by Leap Motion, $\hat{X}_i$ is the vector after filter, and α is a factor which can be calculated by

$$\alpha_i = \frac{1}{1 + \tau_i/T_i} \tag{2}$$

$$\tau_i = \frac{1}{2\pi f_{ci}} \tag{3}$$

where T_i is the period of updating the data, τ_i is a time constant, f_{ci} is cut-off frequency, which can be determined by

$$f_{ci} = f_{cmin} + \beta \mid \hat{V}_i \mid \tag{4}$$

where $\hat{V}_i$ is a derivative of $\hat{X}_i$, representing the velocity of palm. Based upon experience, make

$$f_{cmin} = 1HZ, \beta = 0.5 \tag{5}$$

3.2 Grasping Gesture

We add a hand on the terminal of robot, which can execute grasping task. We preset every joint of finger which can rotate between $0 - 120°$. When the hand is open, joint rotation angle is defined as 0. On the contrary, when the hand is closed, the angle is $120°$. We use a coefficient μ to describe the level of grasping. With the API of the Leap Motion SDK, the parameter μ can be given by hand::grabStrength(), and the finger joint angle is 120μ.

3.3 Rotation Gesture

To describe the rotation gesture mathematically, we build the coordinate system (Figure 4) for the hand, just the same as Leap Motion. Therefore, the problem of hand rotation is equivalent to the problem of coordinate rotation. For example, the coordinate system for Leap Motion is named as frame A, and for hand it is named as frame B. Starting with the frame B coincident with frame A. Rotate frame B first about $\hat{Y}_B$ by an angle $\alpha(\alpha \in [0, 360°])$, then about $\hat{X}_B$ with an angle $\beta(\beta \in [-90°, 90°])$, finally, about $\hat{Z}_B$ by an angle $\gamma(\gamma \in [0, 360°])$. We know the orientation of frame B relative to frame A. So if we can obtain the Euler angles of these rotation process, we know how to control joints 4, 5, 6 to reappear the rotation gesture.

Because all rotations occur about axes about frame B, the rotation matrix is

$$_B^A R = R_Y(\alpha)R_X(\beta)R_Z(\gamma) = \begin{pmatrix} c\alpha & -s\alpha & 0 \\ s\alpha & c\alpha & 0 \\ 0 & 0 & 1 \end{pmatrix} \begin{pmatrix} c\beta & 0 & s\beta \\ 0 & 1 & 0 \\ -s\beta & 0 & c\beta \end{pmatrix} \begin{pmatrix} 1 & 0 & 0 \\ 0 & c\gamma & -s\gamma \\ 0 & s\gamma & c\gamma \end{pmatrix} \tag{6}$$

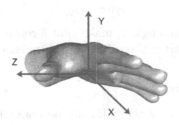

Fig. 4. Coordinate system for hand

$$_{B}^{A}R = \begin{pmatrix} sas\beta s\gamma + cac\gamma & sas\beta c\gamma - c\beta s\gamma & sac\beta \\ c\beta s\gamma & c\beta c\gamma & -s\beta \\ cas\beta s\gamma - sac\gamma & cas\beta s\gamma + sas\gamma & cac\beta \end{pmatrix} \tag{7}$$

where $_{B}^{A}R$ is a rotation matrix that specifies the relationship between coordinate system A and B. And $c\alpha$ means cosine of angle α and $s\alpha$ means sine of angle α.

According to the definition of rotation matrix, we have

$$_{B}^{A}R = \begin{pmatrix} ^{A}\hat{X}_{B} & ^{A}\hat{Y}_{B} & ^{A}\hat{Z}_{B} \end{pmatrix} \tag{8}$$

where the unit vectors giving the principal directions of coordinate system B, when written in term of coordinate system A, are called $^{A}\hat{X}_{B}, ^{A}\hat{Y}_{B}, ^{A}\hat{Z}_{B}$. We can obtain the normal vector of hand with the function Hand::palmNormal(), and $^{A}\hat{Y}_{B}$ is in the opposite direction with the normal vector. We also can get the vector from palm to finger with the function Hand::direction(), and $^{A}\hat{Z}_{B}$ is in the opposite direction with it as well. What's more, we can obtain that

$$^{A}\hat{X}_{B} = ^{A}\hat{Y}_{B} \times ^{A}\hat{Z}_{B} \tag{9}$$

when the coordinate system B employs a right-handed Cartesian coordinate system.

Assuming

$$_{B}^{A}R = \begin{pmatrix} r_{11} & r_{12} & r_{13} \\ r_{21} & r_{22} & r_{23} \\ r_{31} & r_{32} & r_{33} \end{pmatrix} \tag{10}$$

Now, all the things have been prepared. First, we can get angle β which satisfies

$$-s\beta = r_{23} \tag{11}$$

Then, we can get angle α which satisfies

$$sac\beta = r_{13} \tag{12}$$

$$cac\beta = r_{33} \tag{13}$$

Finally, we can get angle γ which satisfies

$$c\beta s\gamma = r_{21} \tag{14}$$

$$c\beta c\gamma = r_{22} \tag{15}$$

We control joint 4 to rotate angle β, make joint 5 rotate angle α, and make joint 6 rotate angle γ. Reappearing the rotation gesture is achieved.

3.4   Inverse Kinematics

Every frame of the position of the palm can be read when the hand is tracked. We use the palm position information to control joints 1, 2, 3 of robot with inverse kinematics. At the beginning, we build coordinate system for joints 1,

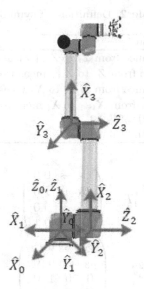

Fig. 5. Coordinate system for UR10

2, 3 (Figure 5). The position of black point, denoted by (x, y, z), is given by Hand::palmPosition().

The coordinate system for robot is not coincident with that for the Leap Motion, hence

$$x = palmPosition()[2]/150 \tag{16}$$

$$y = palmPosition()[0]/150 \tag{17}$$

$$z = palmPosition()[1]/150 - 0.4 \tag{18}$$

Table 1 shows the link parameters (Denavit-Hartenberg parameters), whose definitions are given in Table 2 [13], for UR10. In this paper, we can ignore the length between $\hat{X}_1$ and $\hat{X}_2$, as well as the length between $\hat{X}_2$ and $\hat{X}_3$.

Table 1. Link parameters for UR10

i	α_{i-1}	a_{i-1}	d_i	θ_i
1	0	0	0	θ_1
2	$-90°$	0	0	θ_2
3	0	L_1	0	θ_3

We compute each of the link transformations:

$$^0_1T = \begin{pmatrix} c_1 & -s_1 & 0 & 0 \\ s_1 & c_1 & 0 & 0 \\ 0 & 0 & 1 & 0 \\ 0 & 0 & 0 & 1 \end{pmatrix} \tag{19}$$

Table 2. Definitions of symbols

Symbol	Definition
a_i	the distance from $\hat{Z}_i$ to $\hat{Z}_{i+1}$ measured along $\hat{X}_i$
α_i	the angle from $\hat{Z}_i$ to $\hat{Z}_{i+1}$ measured along $\hat{X}_i$
d_i	the distance from $\hat{X}_{i-1}$ to $\hat{X}_i$ measured along $\hat{Z}_i$
θ_i	the angle from $\hat{X}_{i-1}$ to $\hat{X}_i$ measured along $\hat{Z}_i$
L_1	the length of link3
L_2	the length of link4

$$\begin{matrix} ^1_2T = \begin{pmatrix} c_2 & -s_2 & 0 & 0 \\ 0 & 0 & 1 & 0 \\ s_2 & c_2 & 0 & 0 \\ 0 & 0 & 0 & 1 \end{pmatrix} \end{matrix} \tag{20}$$

$$\begin{matrix} ^2_3T = \begin{pmatrix} c_3 & -s_3 & 0 & L_1 \\ s_3 & c_3 & 0 & 0 \\ 0 & 0 & 1 & 0 \\ 0 & 0 & 0 & 1 \end{pmatrix} \end{matrix} \tag{21}$$

where c_1 (or s_1) means cosine (or sine) of angle θ_1, and c_{12} means $cos_1 cos_2$.
Then,

$$^0_3T = {}^0_1T \ {}^1_2T \ {}^2_3T \begin{pmatrix} c_{123} - c_{13}s_2 & -c_{12}s_3 - c_{13}s_2 & -s_1 & L_1c_{12} \\ s_1c_{23} - s_{123} & -s_{13}c_2 - s_{123} & c_1 & L_1s_1c_2 \\ s_2c_3 + c_2s_3 & -s_{23} + c_{23} & 0 & L_1s_2 \\ 0 & 0 & 0 & 1 \end{pmatrix} \tag{22}$$

The position of black point relative to frame 3 is

$$^3P = \begin{pmatrix} L_2 \\ 0 \\ 0 \end{pmatrix} \tag{23}$$

The position of black point relative to frame 0 is

$$^0P = \begin{pmatrix} X \\ y \\ z \end{pmatrix} \tag{24}$$

Then,

$$\begin{pmatrix} ^0P \\ 1 \end{pmatrix} = {}^0_1T{}^1_2T{}^2_3T \begin{pmatrix} ^3P \\ 1 \end{pmatrix} \tag{25}$$

From equation (25), we can get ^0_3T, assuming

$$^2_3T = \begin{pmatrix} r_{11} & r_{12} & r_{13} & l_1 \\ r_{21} & r_{22} & r_{23} & l_2 \\ r_{31} & r_{32} & r_{33} & l_3 \\ 0 & 0 & 0 & 1 \end{pmatrix} \tag{26}$$

Now it is easy to obtain the value of θ_i ($i = 1, 2, 3$).

4   Experiment

An experiment was designed to test the system performance.The user first puts his hand upon the Leap Motion, and then can do the gestures such as translation, grasping and rotation. The results of simulation shown in Figure 6 demonstrate that the system can respond correctly and quickly to the gesture, which means that the system is efficient and practical.

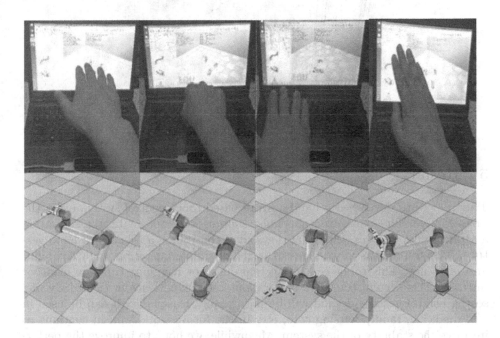

Fig. 6. Experiment result

Then, we test the accuracy of the system. The workspace of robot is a sphere with radius $1.4470m$. When the user conducts an action, the response time of system is limited in $0.1s$. The user can put his hand at any position upon Leap Motion. We get the position (29.7574mm, 175.155mm, 40.302mm) instantly. In theory, the terminal position of robot should be (0.2687m, 0.1983m, 0.7677m), and the real position is (0.2498m, 0.3629m, 0.7573m). The open loop error is 4.65%. Do more experiments, the average open loop error is limited in 5%. That is the system has high precision.

Finally, we add a table into the scene, and put a cup on the table. We select five people to experience the system by grasping the cup with the Leap Motion (see Figure 7). The testers' user experiences and feedbacks were recorded. Results demonstrate that people are satisfied the system. Users also suggest that grasping gesture should be adapted to different shape things.

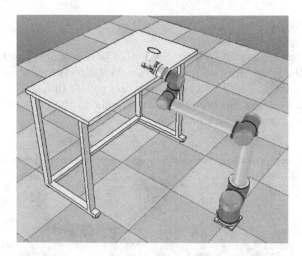

Fig. 7. Grasping the cup

5   Result

This paper has developed a hand gesture based robot control system using the Leap Motion. The system contains noise suppression, coordinate transformation and inverse kinematics, achieving the goal of controlling a robot with hand gesture. The system has advantages in terms of simple operation, high flexibility and efficiency. This robot control system does not have any complex menu or buttons. This system considers more about users experience in their daily life, so the control gesture desired will be natural and reasonable. It can be used in tele medicine, family nursing care, etc. In the future work, we can continue to improve the stability of the system. Meanwhile, we hope to improve the performance of the robot, increase the number of robots, and recognize more than one hands to control different robots, respectively.

References

1. Pan, J.J., Xu, K.: Leap motion based 3D. gesture. China Science Papaer **10**(2), 207–212 (2015)
2. Jiang, Y.C.: Menacing motion-sensing technology, different leap motion. PC. Fan **11**, 32–33 (2013)
3. The principle of leap motion. http://www.3dfocus.com.cn/news/show-440.html
4. Xu, C.B., Zhou, M.Q., Shen, J.C., Luo, Y.L., Wu, Z.K.: A interaction technique based on leap motion. Journal of Electronics & Information Technology **37**(2), 353–359 (2015)
5. Pan, S.Y.: Design and feature discussion of MIDI. controller based on leap motion. Science & Technology for Chinas Mass Media **10**, 128–129 (2014)

6. Wang, Q.Q., Xu, Y.R., Bai, X., Xu, D., Chen, Y.L., Wu, X.Y.: Dynamic gesture recognition using 3D trajectory. In: Proceedings of 2014 4th IEEE International Conference on Information Science and Technology, Shenzhen, Guanzhou, China, pp. 598–601 (April 2014)
7. Elons, AS., Ahmed, M., Shedid, H., Tolba, M.F.: Arabic sign language recognition using leap motion sensor. In: 2014 9th International Conference on Computer Engineering & Systems (ICCES), Cairos, pp. 368–373 (December 2014)
8. V-rep introduction. http://www.v-rep.eu/
9. UR10 introduction. http://news.cmol.com/2013/0530/33267.html
10. Barretthand introduction. http://wiki.ros.org/Robots/BarrettHand
11. Qian, K., Jie, N., Hong, Y.: Developing a gesture based remote human-robot interaction system using Kinect. International Journal of Smart Home 7(4), 203–208 (2013)
12. Casiez, G., Roussel, N., Vogel, D.: 1 filter: a simple speed-based low-pass filter for noisy input in interactive systems. In: Proceedings of the 2012 ACM Annual Conference on Human Factors in Computing Systems, Austin, TX, USA, pp. 2527–2530 (May 2012)
13. Craig, J.J.: Introduction to Rbotics: Mechanics and Control, 3rd edn. China Machine Press, Beijing (2006)

Particle Filter Based Simultaneous Localization and Mapping Using Landmarks with RPLidar

Mei Wu[1], Hongbin Ma[1,2]($\boxtimes$), Mengyin Fu[1,2], and Chenguang Yang[3]

[1] School of Automation, Beijing Institute of Technology,
Beijing 100081, People's Republic of China
mathmhb@gmail.com
[2] State Key Lab of Intelligent Control and Decision of Complex Systems,
Beijing Institute of Technology, Beijing 100081, People's Republic of China
[3] School of Computing and Mathematics, Plymouth University,
Plymouth PL4 8AA, UK

Abstract. Simultaneous localization and mapping (SLAM) is one active research area in robotics. SLAM using only landmarks is an efficient method without relying on dead reckoning (DR) or inertial navigation system (INS), hence informations such as position provided by inertial devices will simply abandoned. To optimize the use of available information, one novel approach of SLAM for indoor positioning with only RPLidar, a low cost laser lidar, is proposed in this paper. First, one improved structure of SLAM using landmarks with particle matching algorithm is introduced. Second, a novel landmark selection method is presented, which takes the quality of observation into consideration too besides the angles between the landmarks. Third, the number of the landmarks needed in the triangulation approach in localization is decreased by utilizing the range information provided by the RPLidar. Experimental results show that the new approach for SLAM with only RPLidar works well, which demonstrates that the low cost low precision laser lidar can also play significant role in robotics with the aid of particle matching and landmark selection algorithms.

Keywords: SLAM · Particle filter · RPLidar · Landmark selection · Particle machining

1 Introduction

Nowadays more and more robots appear in human life, and researches about robotics have been extended from traditional industry to new fields, like medical service, education, entertainment, exploration and surveillance, biological engineering, disaster rescue, etc. Among the rapidly growing robots, one class, e.g.

This work was partially supported by the National Natural Science Foundation in China (NSFC) under Grants 61473038 and 61473120. Also supported in part by Fundamental Research Funds for the Central Universities under Grant 2015ZM065.

© Springer International Publishing Switzerland 2015
H. Liu et al. (Eds.): ICIRA 2015, Part I, LNAI 9244, pp. 592–603, 2015.
DOI: 10.1007/978-3-319-22879-2_54

industrial robots, is designed to accomplish some specialized jobs and work in dangerous, dirty and dull environments where humans are inherently ill-suited; another class, e.g. service robots, aims to enter humans daily life, and these robots are becoming smaller, smarter and sensitive.

Both these robots have to tackle with the problem of localization and mapping. Due to the intrinsic complexity of the environment and difficulties in autonomous positioning, simultaneous localization and mapping (SLAM) problem has become one of the most important research area of robotics, and the key to realize autonomous navigation. Problems such as, if it is possible for a mobile robot in an unknown position in an unknown environment to build a consistent map of the environment by observations of sensors while simultaneously determining its position, are not fully answered. Generally speaking, localization needs the information of the environment map, while mapping relies on the robot's position and pose. Due to the accuracy and cost of sensors as well as the unknown environment, both localization and mapping problems are challenging, which call for the technique for localization and mapping at the same time.

SLAM has been applied in many systems and has resolved lots of practical problems. SLAM system may constitute by many kinds of sensors like radar, laser scanner, vision sensor and so on. There are two types of SLAM problems: the indoor SLAM problem and the outdoor SLAM problem. Situation are different in these problems. For example, in indoor SLAM, GPS sensor may be useless due to the poor or unavailable GPS signal, and the landmarks in outdoor SLAM are less regular than those in indoor environment. In this paper, we will investigate the indoor SLAM problem for home robots. To minimize the size and expense of the robot, we try to use the least number of the sensors, and introduce some low-cost sensors into the system. These sensors may have lower sample rate and fewer sample points, which may influence the result of position and mapping. Specially speaking, a laser scanner, RPLidar is brought to the indoor SLAM problem, with some new ideas of improving the usability of such low-cost range sensors for indoor navigation.

The SLAM problem was first proposed by Randall C. Smith and Peter Cheeseman in 1986 [1] and have attracted many researchers from then on. To address the nonlinear observation model of SLAM, various nonlinear estimation methods have been applied in SLAM systems [2–4], such as EKF(extend kalman filter)-SLAM [5], UKF(unscented kalman filter)-SLAM [6], PF(particle filter)-SLAM [7], UPF(unscented particle filter)-SLAM [8], FastSLAM [9,10] and distributed particle filter SLAM [11–15].

These methods mentioned above provide various solutions for many problems and each method has its own pros and cons. One can notice that the results of probabilistic estimation methods may depend on the accuracy of the observation model and motion model of robot. However, in practice, the accuracy of these models may be poor. For example, actions like collision and sliding will not be detected by the inertial sensor of robot which may introduce errors in the position of robot and these errors may accumulate in inertial navigation system. Different from motion model the error in observation model will not accumulate.

Noticing this issue, some methods of localization by landmarks are reported in the literature (e.g. [16]), which in fact abandons the inertial information. And Hao *et al* [17] advanced the selection method of landmarks in localization. These results were proved to be valid.

However, these mentioned work does not take full advantages of the information of the robot such as position information and observation information. To optimize the use of these information, a novel approach is suggested in this contribution. These information of the robot are utilized in three ways. First, we will introduce an improved structure of SLAM using landmarks with particle matching algorithm, where the position of the robot is utilized to generate the particles in particle filter. Second, a novel selection method will be presented and, besides the angles between the landmarks, this method can take the quality of observation into consideration too. Third, we are able to decrease the number of the landmarks needed in the triangulation approach in localization, by giving the range of the position from the position information of the robot. Experiment result shows that SLAM using landmarks can yield a better performance than the SLAM without landmarks.

The remainder of this paper is organized as follows. First the system model of particle filter (PF) based SLAM using landmarks is introduced in Section 2. Then in Section 3, the idea of particle matching algorithm applied in this paper will be briefly introduced which is the previous step of landmark selection, that is presented in Section 4. And triangulation for localization, the next step of landmark selection, is discussed in Section 5. The experimental results and the environment are presented in Section 6. Finally, in the last section, we shall briefly conclude this paper by summarizing our work and future research.

2   The Model of Particle Filter (PF) SLAM Using Landmarks

In this section the problem of position estimation in an unknown planar environment is defined. This paper introduces a SLAM system using only landmarks based on particle filter (PF), hence DR or INS are not required in this system. In this paper a robot with laser lidar moves in an unknown environment, and the robot can realize simultaneous localization and mapping using the data from a low-cost laser scanner RPLidar.

2.1   The Coordinate System of SLAM

The coordinate system of the global map is the global coordinate system (global map), and it is spanned by axes x^r and y^r. Different from the local coordinate system (robot-centered coordinate, local map), the global coordinate system is spanned by axes x^i and y^i. We can describe landmark z in both coordinate systems: Vector z_m^r describes the m-th landmark in the global coordinate system while vector z_m^i describes the m-th landmark in the local coordinate system.

In the global coordinate system, the position of the robot can be described by the location (x^r, y^r) and the orientation θ^r of the robot, where θ^r is the angle between the global map coordinate axis X^r and the local map coordinate axis X^i.

The global map is unknown at first, thus the first local map is utilized as the initial global map. The initial position of the robot is $(0, 0, 0)$ in global coordinate system. The x^r axis is the direction angle of the robot. After the first scan, the present local map is generated by the latest scan data. When the robot moves to a new position, the position of the robot is (x^r, y^r), and the x^r axis is the orientation of the robot. The relationship between local coordinate system and global coordinate system is shown in Fig. 1.

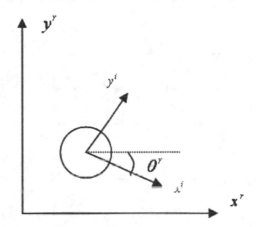

Fig. 1. The relationship between local coordinate system and global coordinate system

2.2   The Structure of PF-SLAM Using Landmarks

The structure of PF-SLAM using landmarks is shown in Fig. 2, which illustrates the following main steps of SLAM:

(1) Scanning and landmark extraction: Scanning and landmark extraction is an important step in SLAM. In this step, the sensor returns a serial of data from the environment. Although the position of the robot can not be acquired directly, we can estimate it from the data returned by the sensor. Before position estimation, landmark extracting is necessary. After landmark extracting, we can get the position of the landmarks in the local map. To obtain the position of the landmarks in global map, coordinate conversion is inevitable.

(2) Landmark matching: After extracting the landmarks, these landmarks can be divided into two categories: one class contains the landmarks which are already

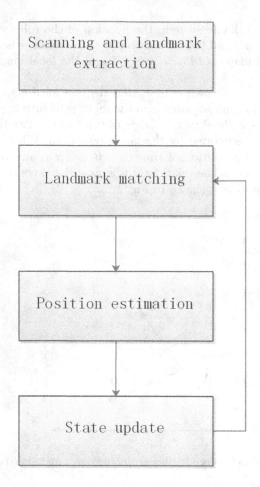

Fig. 2. The structure of the PF-SLAM using landmarks

stored in the global map, and the other class contains the landmarks which have not been observed previously. After categorization, the former landmarks will be utilized in position estimation, while the latter landmarks will be added to the global map. In this paper, a particle filter based landmark matching algorithm is introduced in Section 3.

(3) Position estimation: The specific algorithm of the position estimation will be described in Section 4 and Section 5 .

(4) State update: Update the state of the robot and the global map.

3   The Particle Filter Matching Algorithm

In this section a landmark matching algorithm using particle filter is introduced.

The most common matching algorithm in SLAM is nearest-neighbor (NN) matching [18]. The criterion of NN is based on the position deviation between the landmarks observed and the landmarks stored in the global map. If the position deviation is lower than a certain threshold, for example 0.03m, the landmark is matched with the landmark stored in the global map. We use the term *landmark set* to refer to the set of matched landmarks. However, the observation position of these landmarks rely on the estimation position of the robot. If the estimation position of the robot is wildly inaccurate, the NN algorithm will not work well.

To solve this problem, some algorithms [19] are introduced to find the landmark set by comparing the relative position of the landmarks observed and the landmarks stored in the map. These algorithms can work under the assumed conditions. However, we would like to highlight that both the deviation and the relative position of landmarks are useful in landmark matching. In order to use both of these criteria mentioned above, the particle filter matching algorithm is introduced.

Particle filter is introduced to find these landmark set in this paper. Particle filter is a filtering method based on Monte Carlo Bias and recursive estimation. The basic idea of particle filter algorithm is essentially a set of on-line posterior density estimation algorithms that estimate the posterior density of the state-space by directly implementing the Bayesian recursion equations with a set of particles to represent the posterior density.

In particle filter, the prior conditional probability is $p(\mathbf{x}_0)$, where $\{x_{0:k}^i, \omega_k^i\}_{i=1}^{N_s}$ is an approximation to the posterior probability distribution $\mathbf{x}_k$, $p(\mathbf{x}_{0:k}|\mathbf{z}_{1:k})$. Here $\{x_{0:k}^i, i = 0, 1, \cdots, N_s\}$ is the particle set, and $\{\omega_k^i, i = 0, 1, \cdots, N_s\}$ are the weights of particle sets. And $x_{0:k} = \{x_j, j = 0, 1, \cdots, k\}$ is the sequence of states from time 0 to time k.

For completeness, the eduction process of the posterior probability can be summarized as follows [8, 10].

- The importance weight can be recursively updated by

$$\bar{\omega}_k^i \propto \omega_{k-1}^i p(\mathbf{z}_k|\mathbf{x}_k^i) \tag{1}$$

- After normalizing the weight ω_k^i by equation (2), we can get the weight $\dot{\omega}_k^i$ for posterior probability calculation.

$$\omega_k^i = \bar{\omega}_k^i / \sum_{i=1}^{N_s} \bar{\omega}_k^i \tag{2}$$

- The posterior probability $p(\mathbf{x}_k|\mathbf{z}_{1:k})$ can be expressed as equation (3).

$$p(\mathbf{x}_k|\mathbf{z}_{1:k}) = \sum_{i=1}^{N_s} \omega_k^i \delta(\mathbf{x}_k - \mathbf{x}_k^i) \tag{3}$$

If $N_s \to \infty$, equation (3) is approximating to the true posterior probability $p(\mathbf{x}_k|\mathbf{z}_{1:k})$.

It is hypothesized that, the speed of the robot is slow enough and the position changes between each scan are under the range of the distribution of these particles. Particle filter is adopted in landmark matching, under this hypothesis. The position of the robot calculated in last scan is the prior information of the particle filter, which can generate the distribution $\{x_{0:k}^i, i = 0, 1, \cdots, N_s\}$ and weights $\{\dot{\omega}_{0:k}^i, i = 0, 1, \cdots, N_s\}$ of these particles. The posterior information $p(\mathbf{x}_k|\mathbf{z}_{1:k}) \approx \sum_{i=1}^{N_s} \dot{\omega}_k^i \delta(\mathbf{x}_k - \mathbf{x}_k^i)$ of these particles are calculated by these position deviations of landmark set, the less the value of the position deviation, the more the weight of the particle. After calculating the weights of these particles, we can choose the matching condition of the particle with the highest weight.

4   The Landmark Set Selection Method

After determining the match condition of landmarks, the next step is to select the landmark set for position calculation. Different from the selection algorithm proposed by [17] which takes the angle between observations as selection criterion, the improved method takes the quality of observation into selection criterion too.

In this paper, two key elements are determined in landmark set selection. One element is the angle between observations while the other is the quality of observation. In the associate method introduced in this section, these two elements mentioned above are added to the selection criterion together.

It is assumed that the universal set of all landmarks is denoted by $land(m)$, where m is the number of landmark set. This paper mainly focuses on the situation of $m \geq 2$, which is the common case in indoor environment. One subset of $land(m)$ utilized in localization is $sets(n)$. And the number of these landmark set needed in localization is 2. Thus the number of landmark set in $sets(n)$ is 2 and there are totally C_m^2 kinds of $sets(n)$. The most credible $sets(n)$ is chosen by the selection criterion for localization.

The angle between these two observations in $sets(n)$ is one issue in the selection criterion. When the angle between these landmarks is approaching 90 degree, the influence of the position error is smaller. And if the angle between these landmarks is around 180 degree, the distance between two possible positions of the robot will become very small, which may cause some difficulties in localization.

The quality of the observation is another issue in the selection criterion. It is estimated by the position deviation of the landmark set. The greater the value of the deviation is, the fewer the quality of the observation is.

Both these two issues are considered in the selection method proposed in this paper. The formula used in landmark set selection is introduced as follow.

Variable β is defined as the weight of the quality element, which satisfies equation (4).

$$0 < \beta < 1 \tag{4}$$

The credibility of the landmark set can be calculated by equation (5), where e stands for the quality of the observation, a stands for the angle coefficient of $sets(n)$, and r represents the credibility of $sets(n)$.

$$\beta e + (1 - \beta)a = r \tag{5}$$

After calculating the credibility of each landmark in the landmark set, the $sets(n)$ with the highest r is chosen for position calculation.

5   The Advanced Position Calculation Method

This section is mainly about how to apply advanced triangulation technique in position calculation.

From the conclusion of [20], if the visual angle between two landmarks Z_1^r and Z_2^r measured at an unknown position p and the distance between the landmarks are known, then position p lies on an arc of a circle spanned by landmarks Z_1^r and Z_2^r, shown in Fig. 3.

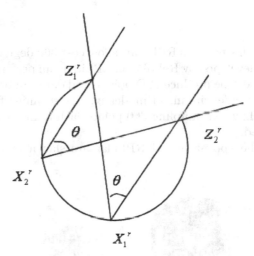

Fig. 3. Triangulation approach

As to the laser lidar, the distance between p to Z_1^r and p to Z_2^r are known, the point p should satisfy two range constraints, as shown in Fig. 4. Hence, the intersections of these two circles determine possible positions of p. Since two circles can usually have two intersection points, a third landmark is needed to determine the position p uniquely. With the idea introduced in [21], by making full use of temporal and spatial data provided by the laser lidar, the number of landmarks needed can be decreased to 2.

Based on the hypothesis given in Section 2, the position changing of the robot is under a certain range. And the selection method mentioned in Section 4 assured that the distance between these two intersections is large enough. From

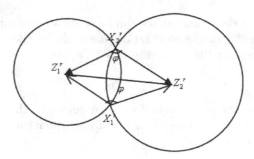

Fig. 4. Localization with range sensor

the position calculated in the last scan and the position rage of the robot, we can select the right position from these two intersections.

6   Experiment Study

6.1   RPLidar

The sensor used in this paper is RPLidar, a low cost 360 degree 2D laser scanner (LIDAR) solution developed by RoboPeak. RPLidar can perform 360 degree scan within 6 meter range. The produced 2D point cloud data can be used in mapping, localization and object/environment modeling. The scanning frequency of RPLidar can reach 5.5Hz when sampling 360 points each round. Hence the accuracy of its data is limited.

The size and the appearance of RPLidar are shown in Fig. 5 and Fig. 6, respectively.

Fig. 5. The appearance of RPLidar

6.2   Experiment

To verify the validness of the method proposed in this contribution, one simple yet non-trivial experiment is conducted. In this experiment, RPLidar moves around in a box and a serial of data is recorded. From the data recorded by

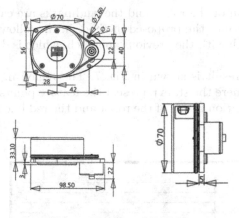

Fig. 6. The size of RPLidar

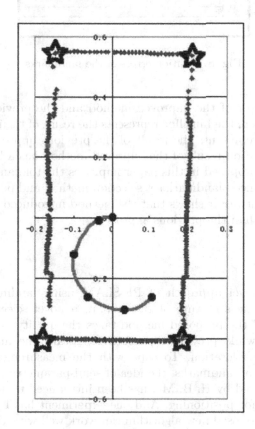

Fig. 7. Data from RPLidar

RPLidar, the position of the robot and the landmarks are calculated. To investigate the performance of the proposed method, in the following experiment we compare our approach with the previous approach, which is detailedly presented in the literature [17].

The experiment result is shown in Fig. 7. The region in this experiment is 0.4m × 1.0m large, where the stars represent the feature points. The points in the picture are the estimation result of the robot and the red line is the approximate path of the robot.

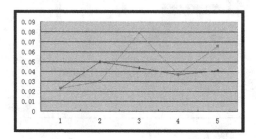

Fig. 8. Position errors of the landmarks

The position error of the improved method and the previous approach are shown in Fig. 8, where the blue line represents the result of the improved method and the pink line represents the result of the previous approach. It is easy to find that the estimation result of the improved method has a better tolerance.

The algorithm proposed in this paper improves the tolerance of the calculation result, by advanced landmark set selection method and position calculation method. Experiment result shows that the method introduced in this paper has a better tolerance than the previous approach.

7   Conclusion

In this paper, a novel approach of PF-SLAM using landmark is proposed. After introducing the structure of the system, a novel selection method has been described and the proposed method takes the quality of observation into consideration too while previous approach only take the angle between the landmarks into consideration. To cope with the measurement errors and to reduce the number of landmarks, the idea of semi-parametric adaptation under constraints, originated by H. B. Ma, has been introduced to advance the triangulation approach for positioning. And one experiment has been conducted to test whether the proposed new algorithm can work well with the low-cost radar sensor RPLidar, which has been proved to be useful in the robot navigation and the efficiency of the algorithm introduced in this contribution has been verified from the experiment result. In the future we may extend the proposed algorithm to 3D environment and apply RPLidar into our intelligent robot systems.

References

1. Smith, R.C., Cheeseman, P.: On the representation and estimation of spatial uncertainty. Robotics Research **5**(4), 231–238 (1986)
2. Durrant, H., Bailey, T.: Simultaneous localization and mapping: part i. Robotics & Automation Magazine **13**(2), 99–110 (2006)
3. Bailey, T., Durrant, H.: Simultaneous localization and mapping(slam): part ii. Robotics & Automation Magazine **13**(3), 108–117 (2006)
4. Dissanayake, M., Newman, P., Clark, S.: A solution to the simultaneous localization and map building (slam) problem. Robotics and Automation **17**(3), 229–241 (2001)
5. Shoudong, H., Gamini, D.: Convergence and consistency analysis for extended kalman filter based slam. Robotics **23**, 1036–1049 (2007)
6. Pan, Q., Yang, F., Ye, L., Cheng, Y.M.: Survey of a kind of nonlinear filters-ukf. Control and Decision **20**(5), 481–484 (2005)
7. Montemerlo, M., Thrun, S., Koller, D.: Fast slam 2.0: An improved particle filtering algorithm for simultaneous localization and mapping that provably converges. In: The 18th International Joint Conference on Artificial Intelligence vol.(3), pp. 1151–1156 (2003)
8. Julier, S.: Unscented filtering and nonlinear estimation. Proceedings of the IEEE **92**(3) (2004)
9. Garrido, S., Moreno, L., Blanco, D.: Exploration and mapping using the vfm motion planner. Instrumentation and Measurement **58**(8), 2880–2892 (2009)
10. Ma, Y., Ju, H., Cui, P.: Research on localization and mapping for lunar rover based on rbpf-slam. Intelligent Human Machine Systems and Cybernetics **2**, 2880–2892 (2009)
11. Sheng, X., Hu, Y.-H.: Distributed particle filters for wireless sensor network target tracking. IEEE International Conference on Acoustics, Speech, and Signal Processing **4**, 845–848 (2005)
12. Sheng, X., Hu, Y.-H.: Parameswaran ramanathan.distributed particle filter with gmm approximation for multiple targets localization and tracking in wireless sensor network. Information Processing in Sensor Networks, 181–188 (2005)
13. Zuo, L., Mehrotra, K., Varshney, P.K., Mohan, C.K.: Band width-efficient target tracking in distributed sensor networks using particle filters. In: International Conference on Information Fusion, pp. 1–4 (2006)
14. Gu, D., Sun, J., Hu, Z., Li, H: Consensus based distributed particle filter in sensor networks. In: International Conference on Information and Automation, pp. 302–307 (2008)
15. Cong, L., Qin, H.L., Xing, J.H.: Distributed genetic resampling particle filter. Advanced Computer Theory and Engineering **2**, 32–37 (2010)
16. Betke, M., Gurvits, L.: Mobile robot localization using landmarks. Transactions on Robotics and Automation **13**(2), 251–263 (1997)
17. Hao, Y.M., Dong, D.L., Zhu, F., Wei, F.: Landmarks optimal selecting for global location of mobile robot. High Technology Letters **11**(8), 82–85 (2001)
18. Bar-Shalom, Y., Fortmann, T.E.: Tracking and Data Association. Academic Press (1988)
19. Ding, S., Chen, X., Han, J.D.: A new solution to slam problem based on local map matching. Robots **31**(4), 296–303 (2009)
20. Gellert, W., Köstner, H., Hellwich, M., Kästner, H.: The VNR Concise Encyclopedia of Mathematics. Van Nostrand Reinhold, New York (1977)
21. Ma, H., Lum, K.Y.: Adaptive estimation and control for systems with parametric and nonparametric uncertainties. Adaptive Control, pp. 15–64 (2009)

Fast Transformations to Provide Simple Geometric Models of Moving Objects

David Adrian Sanders[1(✉)], Giles Tewkesbury[1], and Alexander Gegov[2]

[1] School of Engineering, University of Portsmouth, Portsmouth, UK
david.sanders@port.ac.uk
[2] School of Computing, University of Portsmouth, Portsmouth, UK

Abstract. Models are compared for use with a sensor system working in real time (*in this case a simple image processing system*). A static robot work-cell is modelled as several solid polyhedra. This model is updated as new objects enter or leave the work-place. Similar 2-D slices in joint space, and spheres and simple polhedra are used to model these objects. The three models are compared for their ability to be updated with new information and for the efficiency of the whole system in accessing data concerning new objects. The system supplies data to a "Path Planner" containing a geometric model of the static environment and a robot. The robot structure is modelled as connected cylinders and spheres and the range of motion is quantised.

Keywords: Manufacturing · Navigation obstacles · Robot · Model

1   Introduction

Industrial processes are being improved to meet the requirements of lean and agile manufacturing. Navigation in these dynamic industrial environments is challenging, especially when the motion of the obstacles populating the environment is unknown beforehand and is updated at runtime. Although computers are getting faster and faster, real time applications still require efficient modelling and programming techniques. Some traditional motion planning approaches can be relatively slow when applied in real-time, whereas reactive navigation methods often have too short a look-ahead horizon. This paper presents simple but fast transformations to improve existing manufacturing processes by providing simple geometric models of objects moving through the workspaces of industrial robots. This can improve path and trajectory planning in real time.

A complex industrial environment consists of moving machinery, objects to be manipulated and worked, and obstacles to be avoided [1,2,3,4,5]. Free space available to the moving machinery depends on the accuracy of the models used for this changing environment. Robot navigation in dynamic environments is a challenging task [6], especially when the motion of obstacles is unknown beforehand. Traditional motion planning approaches can be too slow to be applied in real-time, whereas reactive navigation methods have generally a too short look-ahead horizon. Van-der-Stappen [7] presented an efficient paradigm for computing the exact solution of the motion planning

© Springer International Publishing Switzerland 2015
H. Liu et al. (Eds.): ICIRA 2015, Part I, LNAI 9244, pp. 604–615, 2015.
DOI: 10.1007/978-3-319-22879-2_55

problem with very few obstacles but he accepted that motion planning algorithms often use long worst-case running times [8]. These running times are why exact algorithms are rarely used in real-time. Processing speed depends on the complexity of the problem, the complexity of the models used and system processing speed. Process speed is increasing but problems are also becoming more complex.

Assumptions about size and distribution of obstacles leads to a significant reduction in complexity. The complexity of the free space is known to be linear in the number of obstacles and De-Berg [9] studied the complexity of the motion planning problem for a bounded-reach robot with few obstacles and Tang [10] investigated how to topologically and geometrically characterize the intersection relations between movable polygon models often used for manufacturing environments. Large presented real-time motion planning approaches based on the concept of the Non-Linear Vobst (NLVO) [11]. Given a predicted environment, velocities which lead to collisions with static and moving obstacles, were modeled.

The complexity of motion planning algorithms depends on the complexity of the models used for moving obstacles to provide the set of collision-free placements left to the robot. Complexity can be high, resulting in relatively long computing times. Reducing complexity reduces the running time of motion-planning-algorithms.

In this paper, models in two different spaces are considered and they are tested with a simple image processing system working in real-time. A static geometric model of a robot work-cell is held in computer memory as solid polyhedra. This static model is updated as new objects enter or leave the work-place. 2-D slices of joint space, spheres and simple polyhedra are used to model these objects. Spheres are suggested as the simplest models of dynamic obstacles. Halperin [12] devised techniques to manipulate a collection of loosely connected spheres in three dimensional space. He analyzed sphere models and pointed to properties that make them easy to manipulate. He presented efficient algorithms for computing union boundaries.

The mapping from workspace to configuration space (or joint space) is important and then the avoidance of the obstacles in one or other space is then important. Sanders [1],[3,4,5] completed studies of geometric modelling techniques and regarded the following as meaningful criteria in depicting a robot and its work place: Fast intersection calculations, ease of use with path planning algorithms, fast model generation, low memory storage requirements, and efficiency. Sharma [13] for example recently studied minimum time needed to transfer vehicles from source to destination, avoiding conflicts with other vehicles. The other vehicles were effectively obstacles and a conflict occurred when the distance between any two vehicles was smaller than a velocity-dependent safety distance. Fiorini [14] presented a method for robot motion planning in dynamic environments in a velocity space. The models considered in this paper are compared for their ability to be updated with new information to provide data to planning systems like that. Efficiency in accessing data concerning new objects is also considered.

The robot machinery structure is modelled as connected cylinders and spheres and the range of motion is quantized. A fast sub-optimal path is to be derived using simplified models that avoids the modelled objects and seeks a direct path in terms of total actuator movement. The approach depends on inspecting a 3-D graph of quantised joint space.

The static model of the robot work-cell consists of solid polyhedra. The remaining free space model is updated as new objects penetrate or depart from the working volume. 2-D slices in joint space, multiple spheres, and six sided parallelepiped are used to model the dynamic objects in order to compare them.

2   Robot and Static Environment

Most computer representations of factory surroundings have flat surfaces and straight linear edges and this geometry resembles the objects often found in manufacturing work cells. These models are difficult to deal with in real-time. If both robot and dynamic objects are modelled by polyhedral shapes then the accuracy may be high but computation time is extended. The transformation of the static environment need only be made once though, so that computation time is not a problem. An accurate model was therefore selected and Polyhedra were used to model the static environment. The most influential factor in representing the robot was speed of intersection calculation (providing the model enclosed the whole robot). A large number of industrial robots have two major links, (an upper arm and a forearm) and three major joints (Base, Shoulder and Elbow). The simplest possible representation for this type of robot was two lines jointed at one end. Fixed distances from the lines were then defined as enclosing the outer casing of the robot. This gave two connected cylinders with hemispherical ends. The advantages of this representation were that the cylinders modelled the robot links efficiently and the intersection calculations between the robot arm and obstacles were simple. The end effector was then represented as a sphere with a radius adequate to surround the end effector. Work-pieces were included by increasing the radius of the sphere.

2.1   Dynamic Mapping

Speed of intersection calculation was compared for several models representing dynamic objects, that is objects that moved in the working volume of the robot. Three models compared favourably: 2-D slices in joint space, spheres and six sided parallelepiped in cartesian space. In all cases it was assumed that at least the 2-D cross section of the dynamic object in the X-Y plane and the height (Z) of the dynamic object was available from a sensor system (a simple vision system in this case). The dynamic objects were effectively only two and a half dimensional. That is, they had a two dimensional shape and a height. 3-D dynamic object shapes considered during the work described in this paper were cylinders and cubes. Parallelepiped models, spheres or similar 2-D planar slices in joint space modelled these 3-D shapes to a workable accuracy and in the case of the 2-D slices, more quickly in discretised 3-D space.

2-D slices are described here. Models were calculated by considering two pairs of boundaries: the angles of the base joint, $\theta 1$, which bounded the dynamic object ($\Theta 1min$ and $\Theta 1max$); and the maximum distance Dmax and minimum distance Dmin from the origin (maximum and minimum radii). The dynamic object was then modelled as a series of 2-D planar slices. The reference slice was calculated within a boundary of a line from the Origin bounded by Dmax and Dmin and the limits of the Z axis. The "blocked" configurations for the shoulder and elbow joints $\theta2$ and $\theta3$ were then calculated for this bounded plane and copied for all $\theta1$ within the two bounding angles, $\Theta1min$ and $\Theta1max$. For the global path planning methods described in the literature, this reduced the number of searches and tests for "blocked" points. The major part of the algorithm was reduced to copying values within a 3-D graph. The dynamic object was first modelled as a 2-D rectangle as this was the simplest model which could be derived from the row and column limits of an object under a camera.

3   Transformation into Joint Space

Data were processed to transform dynamic objects into a joint configuration space. A point in cartesian space is not transformed into a point in joint space. If the point is within the working volume of the robot then it is transformed into one or more complex three dimensional shapes. These complicated profiles may be depicted within a computer as geometric shapes, units of space or by approximating the profiles by mathematical curves. The method selected in this work represented the dynamic objects as regions within joint space consisting of small units. The technique was not limited to any specific design of machinery and may be used with any number of degrees of freedom. The work described here was based on the implementation for the three major axes of a KUKA KR125 robot at Ford Motor Company. A graph was created which consisted of a three dimensional structure of unit regions. The 3-D graph had each dimension corresponding to a principal degree of freedom of the robot arm, $\Theta1$, $\Theta2$ and $\Theta3$. The wrist configurations were not considered but these were included as being within a sphere. Each unit was initially set to `"clear"' status and the positions (in joint space) at which the robot intersected dynamic objects were then calculated. Each unit represented a range of configurations for the robot, in terms of, ($\Theta1cent$, $\Theta2cent$, $\Theta3cent$), plus a degree of movement away from these central joint values. All units together represented the whole robot work-space and the number of units in the graph, NodeTotal, was given by:

$$\{(\Theta1_{max}-\Theta1_{min})/2 \times \delta\Theta1\}*\{(\Theta2_{max}-\Theta2_{min})/2 \times \delta\Theta2\}*\{(\Theta3_{max}-\Theta3_{min})/2*\delta\Theta3\} \qquad (1)$$

where, $\Theta1max$, $\Theta1min$ = upper, lower limits of $\Theta1$.
$\Theta2max$, $\Theta2min$ = upper, lower limits of $\Theta2$
$\Theta3max$, $\Theta3min$ = upper, lower limits of $\Theta3$.

If at any configuration in a unit, the robot intersected a dynamic object, then the unit was set to "blocked". If at all configurations within a unit the robot did not intersect a dynamic object then the unit remained "clear". The path planning problem for the global approach was then reduced to finding a series of neighbouring units between the START and GOAL configurations that were still "clear". If free space is assumed to be larger than blocked space then a fast method was to consider each dynamic object and test for the nodes which could contain the transformed dynamic object. This was the method adopted and the algorithm was as follows:

For a node where the robot could intersect the dynamic object, recursively test all the neighbouring units to see if they are also within the reach of the robot.

Data structures were initialised to form a 3-D graph of joint space and trigonometric solutions were calculated. All units in the graph were set to `"clear"' status and flags were associated with each node. The static model of the work cell left a number of clear nodes that represented safe configurations that would not collide with the static environment. Dynamic object data was simulated or received from the vision system and the first task for the program was to read this data. The two and a half dimensional model was then created.

3.1   2-D Slices

Firstly the limits in x were increased by the radius of the upper-arm:

$$\text{StartRow\_clearance} = \text{StartRow} - \text{UpperRad\%} \tag{2}$$

$$\text{EnddRow\_clearance} = \text{EndRow} + \text{UpperRad\%} \tag{3}$$

The modulus of the ends and centre point on an edge StartCol were calculated. This is shown below for the furthest end from the Origin.

$$\text{Corner(TopLeft, Angle\%)} = \text{InvTan (StartCol / EndRow\_clearance)} \tag{4}$$

$$\text{Corner(TopLeft, Modulus\%)} = \sqrt{(\text{EndRow}^2 + \text{StartCol}^2)} \tag{5}$$

The following parameters of the model were found: the inside radius from the origin, (Dmin), outside radius from the origin, (Dmax), smallest base angle, ($\Theta 1$min), and largest base angle, ($\Theta 1$max).

If the dynamic object was matched to a template then the height of the dynamic object was extracted from the template, otherwise if the dynamic object height was unknown, the height (Z) was set to infinity. The segment was extrapolated to the Y axes so that calculation took place in the Y, Z plane. The modelled dynamic object was expanded by the radius of the robot's upper-arm in the Y and Z plane. $\Theta 1$ was set to its new lower limit and the inverse kinematic solution was found for all the points within the dynamic object, as shown in the following code:

```
FOR Yaxis = (Radius%(min%)UpperRad%) TO (Radius%(max%) +
UpperRad%)
  FOR Zaxis =  255 TO (Radius%(Z%) + UpperRad%)
       CALL InvKinematics
  NEXT Zaxis
NEXT Yaxis
```

The coordinates in Y and Z were converted to robotic joint angles using the inverse kinematic solution in the subroutine InvKinematics. Firstly the distance from the origin to the cartesian point (L3) and the angle to the point (CurvΘ) were calculated:

$$\text{Curv}\Theta = \text{InvTan Zaxis / Yaxis} \tag{6}$$

$$\text{sqL3} = \text{Yaxis}^2 + \text{Zaxis}^2 \tag{7}$$

$$\text{L3} = \sqrt{\text{sqL3}} \tag{8}$$

The upper-arm was checked against L3 to see if a collision was possible. If within the reach of the upper-arm and if $\Theta 2$ was within its limit, then $\Theta 2$ was set to CurvΘ and $\Theta 3$ was set to "blocked" between its limits. If L3 was less than the Forearm plus upper-arm then the Forearm collided with the point. $\Theta 2$ and $\Theta 3$ were calculated using the cosine rule and if $\Theta 2$ and $\Theta 3$ were within their limits a flag was set to "blocked".

The method is being successfully used with sensors [15,16] mobile robots [17,18,19,20,21] and wheelchairs [22,23].

3.2   Spheres

The graph data structure described earlier was initialised. Limits of the graph corresponded to the angular limits for the robot's joints within the range of the work cell and obstacles outside this work-space were ignored. As the graph carried out intersection checks at a limited number of positions, only a limited number of trigonometric solutions were required and these were calculated at the start. Before the obstacles were calculated all the units in the graph had a flag set to `CLEAR' status. Four other flags were used with each node, these were: `New obstacle', `Forearm tested', `Upper arm tested', and `On list'. Each unit code was stored as one byte of computer memory in an array and the flags used one bit each. The obstacle data was received from a file or from the vision system and the first task for the program was to read this data.

The task was then split into two sub-tasks, firstly to calculate the upper arm and then to calculate the forearm blocked space on the graph. A configuration was calculated at which the part of the arm under consideration was closest to the obstacle centre. If the forearm was being considered, then the configuration where the Foretip was at the centre of the sphere was calculated. For the upper arm, the configuration was calculated for which the centre line of the upper arm pointed at the sphere centre. If the obstacle was within the reach of the link being tested, then this configuration was the first unit for the transformed obstacle. The base angle was calculated from the X,Y coordinates of the sphere. Firstly the modulus (L3) and the angle (SphΘ) from the robot to the centre of the sphere was calculated and a test was conducted to see if the sphere was out of range, in which case no further processing was necessary.

$$\text{Waist Angle } \Theta 1 = \text{InvTan}(Y/X) \tag{9}$$

$$\text{Modulus XY} = \sqrt{(X^2+Y^2)} \tag{10}$$

$$\text{Sph}\Theta = \text{InvTan}(Z/\text{ModulusXY}) \tag{11}$$

$$L3$$
$$= \sqrt{(X^2 + Y^2 + Z^2)} \qquad (12)$$

The cosine rule was used to calculate the shoulder $\Theta 2$ and elbow $\Theta 3$ angles.
L1 = Upper-Arm = 220mm
L2 = ForeArm = 160mm

$$\Theta 3 = \text{InvCos} \left[\left(L1^2 + L2^2 - L3^2 \right) / \left(2 * L1 * L2 \right) \right] \qquad (13)$$

$$\Theta 2 = \text{InvCos} \left[\left(L1^2 + L3^2 - L2^2 \right) / \left(2 * L1 * L3 \right) \right] + \text{Sph}\Theta \qquad (14)$$

If the sphere centre was too close to the robot then $\Theta 3$ would exceed its lower limit ($\Theta 3 < 90°$). In this case $\Theta 3$ was set to 90° and $\Theta 2$ was calculated using InvTan:

```
If Θ3 < 90° THEN
     Θ3 = 90°
     Θ2 = InvTan ( L2 / L1) + SphΘ
END
```

This gave a starting configuration close to the centre. When the lower limit of $\Theta 2$ was exceeded, ($\Theta 2 < -30°$), the angle was set to minus 30° and the distance between the upper-arm and sphere centre was calculated (the modulus) using the subroutine FindModulus, from which the cosine could be used to find the new $\Theta 3$:

```
If -30° < Θ2 THEN
   Θ2 = -30°
   Θ3 = InvCos[(L1² + L2² - Modulus²) / (2* L1 * Modulus)]
END
```

The first configuration was set to blocked . Its neighbouring units were also tested and if they were set to blocked then their neighbours were checked. The position problem was solved using forward kinematic calculations and the minimum distance between the obstacle and the robot arm was calculated, (provided that it had not completed the calculation before). The method continued recursively until the whole obstacle transformation was found. All units were set to blocked, which had any two opposite neighbouring units which were also blocked. Any units which were on the edge of the now solid obstacle were recorded on a list. All the neighbours of the units on the list were tested, and the process repeated until the surface of the transformed sphere was completely defined.

Nodes which were blocked were stored on a list of units to be expanded later. When a unit was expanded it was retrieved from the list and new blocked points were added to the list. When all the nodes on the list were exhausted the obstacle transformation was complete. The most important consideration was processing speed. Times for calculating obstacles were recorded during the project and examples are presented in the results section.

3.3   Simple Polyhedral Shapes

Polyhedra are commonly used to model dynamic objects. The third modelling method described in this paper modelled the dynamic objects as simple six sided parallelepiped. This was the most elementary polyhedral model. The system found the position of the edges of the model in X and Y by calculating the limits of the rows and columns set by the vision program. If possible, the height of the object was then retrieved from an associated template. Edge positions were expanded with the model radius of the part of the robot under test (ie upper-arm or forearm), as demonstrated below for an expansion of the forearm in X.

$$\text{Expand\_XLow\%} = \text{EdgePosition\%(LowX\%)} - \text{ForRad} \qquad (15)$$

$$\text{Expand\_XHigh\%} = \text{EdgePosition\%(HighX\%)} + \text{ForRad} \qquad (16)$$

The cartesian coordinates of the arm were then tested against the expanded polyhedral edge limits.

4   Results

The most important consideration for the system was that it should be suitable for real time applications. Times for transforming dynamic objects were recorded during the project and as an example, the times for the three models to transform a large cube into joint space are shown in Table 1. The times were recorded with the Z axis of the cylinder at X = 0 mm and Y = 300 mm with respect to the origin.

Table 1. Transformation times for a cube.

Model	Time (Seconds)	Number of blocked nodes recorded.
One Sphere	8.8	2426
Two Spheres	14.1	2336
Simple Polyhedron	26.3	1983
2-D Slices	5.3	2489

The 2-D Slice Model: The advantage of modelling the dynamic object as a series of similar 2-D slices was that once the collision coordinates of $\Theta 2$ and $\Theta 3$ had been calculated for a particular $\Theta 1$ then these collisions could be repeated for the limits of $\Theta 1$ which collided with the dynamic object. This reduced the main processing task to copying data rather than calculating forward or reverse kinematic solutions. The representation of dynamic objects using similar 2-D slices was the fastest to transform into discrete 3-D joint configuration space.

Once a dynamic object increased above a certain size or was moved closer to the origin, part of the dynamic object intersected both the Upper Arm and ForeArm joint space. Thus the joint-space occupied by the dynamic object suddenly increased and calculation time increased. For the transformation methods a graph of calculation time vs discrete work-space volume can be expected to be linear, that is the calculation time for a dynamic object was approximately proportional to the number of units tested, the total number of nodes being the work-space volume.

The computer time required for dynamic object transformations was short. The initial conversion time to model the static environment was slow; Up to three minutes of computer time depending on the complexity of the model, but the transformation was only performed when the system was powered up.

The Sphere Model: An obstacle was modelled first as a single sphere of the smallest radius which would enclose the obstacle. Later, if time allowed it was modelled by two smaller spheres and then four spheres. Nodes set to blocked associated with the first sphere tested usually also collided with other spheres. The forward kinematic solutions did not need to be recalculated for these nodes but the total calculation time increased with the number of spheres because the overhead of calculation for each sphere was greater than the saving in time achieved as the spheres became smaller. This meant the single sphere calculation was faster than the calculations for multiple spheres although the single sphere model was less accurate and had a larger volume. The problem when using more than one sphere was that the centre of several spheres would be set to blocked (with some surrounding nodes) after the expansion of the first sphere. As these nodes were blocked, later spheres were sometimes not retested so that many nodes were not added to the list.

5 Discussion

Transforming a geometric algorithm into an effective computer program is a difficult task. The accuracy of the models affected the performance of the Path Planner. High accuracy models required more computation time and therefore longer solution times. Low accuracy models required links or dynamic objects to be oversized to eliminate the chance of undetected collisions. Lowering the accuracy led to the rejection of valid solutions.

For dynamic models, speed of calculation was important. The simplest possible intersection calculations for the local methods were made using the sphere model. Calculation was reduced to finding the distance from the robot to a point and subtracting the radius of the sphere to give the distance to the surface of the sphere. Modelling with more than one sphere was considered. As the real environment for a robot becomes more complex so more spheres are needed for the model. It was considered how increasing the number of spheres might increase the accuracy of the model. A cubic number of spheres was used, i.e. 1, 8, 27, 64 etc. The spheres formed a regular pattern and were equal in size. An infinite number of spheres was required to model the cube completely but modelling objects using the same sized spheres was inefficient. For example, in modelling a cube using sixty-four spheres of the same size, eight of the spheres are totally enclosed and might easily be replaced by a single larger sphere without increasing the model volume.

To compare the modelling of obstacles using single and multiple spheres, as an example, a model of a cylinder using one and two spheres is compared. The volume of two spheres of radii 35 mm was compared to that of one sphere of 70 mm as shown below.

Volume of Two Spheres: 2 x 4/3 x π x 353 = 359,188 mm^3

Volume of One Sphere: 4/3 x π x 703 = 1,436,755 mm^3

The area of the two spheres would be much smaller except that the model of the robot must then be considered to find the union volume,

$$\text{Robot} \cup \text{Model.} \tag{17}$$

The Upper-arm model radius= 80mm so: Union radius for a single sphere is 70+ 80 = 150. Union radius for two spheres is 35+80 = 115. Union volume of a single sphere is 4/3 x π x 1503 = 14,137,167 mm^3. Union volume of two spheres is 2 x 4/3 x π x 1153 = 12,741,211 mm^3.

There was a similar number of collisions for both models. When points within the second sphere were not tested to see if they had collided during the calculations for a previous sphere, this partially explained the lack of improvement in processing time for the model using two spheres.

Considering the simple six sided parallelepiped model, the volume of the model for the horizontal cylinder was less than that of the 2-D slice model.

Parallelepiped Volume = (60 + 160)^2 x (140 +80) = 10,648,000 mm3. This potentially reduced the number of blocked nodes, but shape and therefore the calculations were more complex so calculation time increased.

6   Conclusions

Using the two dimensional slice model of the cylinder, $\Theta 2$ and $\Theta 3$ were only determined for a single slice. This reduced the processing time as this slice of "blocked" nodes was copied for all $\Theta 1$ within the bounding base joint angles.

The number of "blocked" nodes produced was similar to other models, so that the intersection volume was approximately the same as for the sphere and polyhedral models. This suggested an equivalent accuracy.

The method of modelling dynamic objects by similar 2-D slices had the fastest intersection calculation times. Using the 2-D slices described, software models of the dynamic work-place were quickly passed to the main computer by the vision system. Similar 2-D slices were less complex than polyhedra, only requiring the two bounding angles of the base joint $\Theta 1$, the inner and outer radius and a height (five items of data).

2-D slices in a joint actuator space are the most efficient of the three models considered and they represent dynamic obstacles at least as effectively as the other two models.

References

1. Sanders, D.A.: Recognizing shipbuilding parts using artificial neural networks and Fourier descriptors. Proc. Institution of Mechanical Engineers Part B-Journal of Eng. Man. **223**(3), 337–342 (2009)
2. Rasol, Z., Sanders, D.A.: An automatic system for simple spot welding tasks. Total Vehicle Technology Conf., pp. 263–272 (2001)
3. Sanders, D.A., Harris, P.: Image modelling for real time manufacturing applications using 2-D slices in joint space and simple polyhedra. Journal of Design and Manufacturing **3**, 21–27 (1993)
4. Sanders, D.A.: Real time geometric modelling using models in an actuator space and cartessian space. Journal of Robotic Systems **12**(1), 19–28 (1995)
5. Sanders, D.A., Lambert, G., Pevy, L.: Pre-locating corners in images in order to improve the extraction of Fourier descriptors and subsequent recognition of shipbuilding parts. Proc. Institution of Mechanical Engineers Part B-Journal of Eng. Man. **223**(9), 1217–1223 (2009)
6. Carpin, S.: Randomized motion planning: A tutorial. Int. Jrnl. of Robotics & Automation **21**(3), 184–196 (2006)
7. Berretty, R.-P., Overmars, M.H., van der Stappen, A.F.: Dynamic motion planning in low obstacle density environments. Computational Geometry **11**(3–4), 157–173 (1998)
8. Sanders, D.A., Moore, A., Luk, B.L.: A Joint Space Technique for Real Time Robot Path Planning. Robots in Unstructured Environments, IEEE, 91TH376-4, pp. 1683–1689 (1991). ISBN 0-7803-0078-5
9. de Berga, M., Katzb, M.J., Overmarsa, M.H., van der Stappena, A.F., Vleugels, J.: Models and motion planning. Computational Geometry **23**(1), 53–68 (2002)
10. Tang, K.: A geometric method for determining intersection relations between a movable convex object and a set of planar polygons. IEEE Transactions on Robotics **20**(4), 636–650 (2004)
11. Large, F., Laugier, C., Shiller, Z.: Navigation Among Moving Obstacles Using the NLVO: Principles and Applications to Intelligent Vehicles. Autonomous Robots **19**(2), 159–171 (2005)
12. Halperin, D., Overmars, M.H.: Spheres, molecules, and hidden surface removal. In: Proc. 10th Annual. Symp. on Computational Geometry, pp. 113–122 (1994)
13. Sharma, V., Savchenko, M., Frazzoli, E., Voulgaris, P.G.: Transfer time complexity of conflict-free vehicle routing with no communications. International Journal of Robotics Research **26**, 255–271 (2007)
14. Fiorini, P.: Robot motion planning among moving obstacles, PhD disertation, University of California (1995)
15. Sanders, D.A.: Environmental sensors and networks of sensors. Sensor Review **28**(4), 273–274 (2008)
16. Sanders, D.A., Lambert, G., Graham-Jones, J., et al.: A robotic welding system using image processing techniques and a CAD model to provide information to a multi-intelligent decision module. Assembly Automation **30**(4), 323–332 (2010)
17. Sanders, D.A.: Comparing ability to complete simple tele-operated rescue or maintenance mobile-robot tasks with and without a sensor system. Sensor Review **30**(1), 40–50 (2010)
18. Sanders, D.A.: Comparing speed to complete progressively more difficult mobile robot paths between human tele-operators and humans with sensor-systems to assist. Assembly Automation **29**(3), 230–248 (2009)

19. Sanders, D.A., Graham-Jones, J., Gegov, A.: Improving ability of tele-operators to complete progressively more difficult mobile robot paths using simple expert systems and ultrasonic sensors. Industrial Robot **37**(5), 431–440 (2010)
20. Sanders, D.A., Stott, I.J., Robinson, D.C., et al.: Analysis of successes and failures with a tele-operated mobile robot in various modes of operation. Robotica **30**, 973–988 (2012)
21. Sanders, D.A., Tewkesbury, G.E., Stott, I.J., et al.: Simple expert systems to improve an ultrasonic sensor-system for a tele-operated mobile-robot. Sensor Review **31**(3), 246–260 (2011)
22. Sanders, D.A., Langner, M., Tewkesbury, G.E.: Improving wheelchair-driving using a sensor system to control wheelchair-veer and variable-switches as an alternative to digital-switches or joysticks. Industrial Robot **37**(2), 157–167 (2010)
23. Sanders, D.A., Stott, I.J., Graham-Jones, J.: Expert system to interpret hand tremor and provide joystick position signals for powered wheelchairs with ultrasonic sensor systems. Industrial Robot **38**(6), 585–598 (2011)

Adaptive Control System

Analysis and Correction of Ill-Conditioned Model in Multivariable Model Predictive Control

Hao Pan[1,2(✉)], Hai-Bin Yu[1], Tao Zou[1], and Dewei Du[1,2]

[1] Shenyang Institute of Automation, Chinese Acedemy of Science, Shenyang 110016, China
panhao@sia.cn
[2] University of Chinese Academy of Sciences, Beijing 100049, China

Abstract. The ill-conditioned model is a common problem in model predictive control. The model ill-conditioned can lead to control performance declining obviously from steady-state model of process in this paper. The direction of output movement is relevant to whether the model is ill-conditioned by simulation and analysis. Model mismatch also leads to model ill-conditioned becoming more serious. The geometry tools and SVD in linear algebra are used to analyze the essential reason of ill-conditioned model, and an offline strategy is proposed which can solve the ill-conditioned model problem together with existing online strategies. Finally, the simulations are used to prove the conclusions which presented in this paper are correct.

Keywords: Model predictive control · Model ill-conditioned problem · Singular value decomposition (SVD) · Model identification · Model mismatch

1   Introduction

Model predictive control (MPC) is a kind of constrained, multivariable and model-based control method and was first applied to control the oil refining process [1]. At present, MPC has been widely applied to many industries, for example, petroleum, chemical industry, paper making, food processing, aviation and so on, which can significantly reduce the variance of the control process output and have the strong robustness [2]. The classic algorithm of MPC mainly includes the Dynamic Matrix Control (DMC) [3], the Model Algorithm Control (MAC) [4], and the Generalized Predictive Control (GPC) [5]. Currently, the most industrial MPC software is based on DMC, and to extend widely.

MPC is a model-based control, so the model has a great influence on the control effect. The controller model was usually obtained through system identification in practice, but the model of identification is ill-conditioned or has error. This article focuses on the control effect of the ill-conditioned MPC controller model. The intuitive performance of ill-conditioned model is the large condition number of the process model, has strong correlation between input and output to lead to difficult control of the process. The ill-conditioned of controller model can be divided into the following two cases: 1, There is strong correlation in the process itself [6], which is a pathologi-

---

* This work is supported by National Nature Science Foundation under Grant No. 61374112, and General Research Project of Education Department of Liaoning Province, Grant NO.L1013158.

© Springer International Publishing Switzerland 2015
H. Liu et al. (Eds.): ICIRA 2015, Part I, LNAI 9244, pp. 619–631, 2015.
DOI: 10.1007/978-3-319-22879-2_56

cal process physically, such as the distillation tower in the chemical process; 2, As the model of the multivariate process is obtained through system identification, the sub-units of the model gained by fitting the input and output data may be sick.

The model ill-conditioned problem is more common in process control, and has great influence on the stability of the process [7]. The output set point of a certain direction movement probably brings about dramatic action of the controller, therefore it is neces-sary to eliminate or reduce the influence of the model ill-conditioned on the process.

There are several methods to eliminate model ill-conditioned and are introduced in following. For the square system(the number of inputs equal to outputs'), Grosdidier P [8] removed the outputs of linear correlation, and made them not participate in the calcu-lation of control function in rolling optimization processes, which leaded to the output offset. Another solution was to use output range control strategy by relaxing the control requirements of some outputs, that is, set points were extend to a set range, which adapted to the process of low output demand and was widely applied in industry. In fact, this solution was same as Grosdidier's method. In light of the standard regularization theory, J.Marroquin [9] came up with a new random method to handle model ill-conditioned problem in numerical calculation. In addition, Honeywell's advanced control software adopted SVT (Singular Value Thresholding) method to eliminate the influence of the model ill-conditioned on the control effect, which avoided the inputs possibly leading to the instability by neglecting the input which the singular value in the corres-ponding direction smaller than the threshold which was set by controller conFiguration [10]. AspenTech's advanced control software DMCplus adopted IMS(Input Move Sup-pression) control strategy to solve the model ill-conditioned problem, which decreased the inputs action by directly increasing the values of diagonal elements in LS problem and decreasing the condition number. Thanks to the two-layer predictive controller in DMCplus [2, 11-12, 19], this strategy was useful. However, the methods mentioned above were all online solution strategies. Therefore, this paper put up with an offline strategy as a complementary method, it can quite well handle the model ill-conditioned problem by cooperating with the online strategies above.

Starting with the steady state gain matrix of the multivariable process, this paper will use geometry and linear algebra tools to analysis the reason why ill-conditioned model influenced the control effect, then proposed the method to improve the model, and used the simulation to verify the effectiveness of the given method at last.

2   Brief Introduction of Two-Layer Model Predictive Control

Currently, predictive control technology has developed to the fourth generation, the main feature is the two-layer model predictive control(TMPC). TMPC adds a steady state optimization (SSO) layer above dynamic control layer of traditional MPC. Lite-rature [22] introduces a new generation state space controller of AspenTech, the struc-ture of TMPC is shown in Fig.1.

Compared with the traditional MPC, TMPC increases steady-state optimization func-tion modules including external targets, optimizing input parameters and output feedback items. Steady-state optimization has optimization function for economic performance which can be automatic optimization nearby the steady-state operating point and find the optimal process set value, this process is called the steady-state target calculation. Steady-state target calculation needs to use steady-state model of the process.

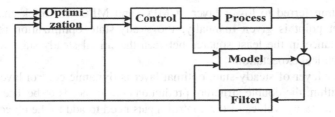

Fig. 1. Structure of two-layer model predictive control

Considered invariant open-loop stable linear system, its transfer function model of dynamic characteristics is

$$y(s) = G(s)u(s) \tag{1}$$

and steady-state model is

$$\Delta y_s = K \Delta u_s \tag{2}$$

where, K represent steady-state gain matrix. The constraint of inputs and outputs are

$$\begin{cases} u_{LL} \leq u_s(k) \leq u_{HL} \\ y_{LL} \leq y_s(k) \leq y_{HL} \end{cases} \quad k \geq 0 \tag{3}$$

The constraint conditions involve hard constraints and soft constraints, the soft constrains can relax in engineering permitted range when could not find a feasible solution, but the hard constraints can not relax.

Considered the inputs and outputs constraint conditions of controlled process, the steady-state optimization of TMPC is linear programming (LP) or quadratic programming (QP) problem. This paper solves LP problem as example, the problem descriptions are

$$\begin{cases} \min_{\Delta u_s(k)} J = c^T \Delta u_s, \\ \text{s.t. } \Delta y_s(s) = K \Delta u_s(k) + e_k, \\ u_{LL} \leq u_s(k-1) + \Delta u_s(k) \leq u_{HL} \\ y_{LL} \leq y_s(k-1) + \Delta y_s(k) \leq y_{HL} \end{cases} \tag{4}$$

where, e_k can be concluded by dynamic predictive errors of process,

$$e_k = y_k - \tilde{y}(k|k-1) \tag{5}$$

where, y_k represent output measured values of k time, $\tilde{y}(k|k-1)$ represent output predictive values of k time in $k-1$ time.

In order to solve steady-state optimization problem, analysis of feasibility is needed. If the viable solutions are nonexistent, feasibility determination and soft constraints adjustment will be done and constraint boundary of output variables will be relaxed appropriately according to the priority order.

Under feasible of steady-state optimization condition, the final optimal control input can be calculated by the steady-state target calculation, and the optimal process output values can be obtained further more. Optimal input and output values as the set values are transferred to lower layer of MPC, and MPC will track changing of set point. If set point is given manually, the steady-state optimization represents the shortest distance in the least squares between the initial steady-state working point and mathematical expectation.

The lower layer of steady-state optimal layer is dynamic control layer which uses DMC algorithm. the double structure prediction control needs to be The penalty term of steady-state target values of the control inputs need to add to the objective function of DMC in the framework of TMPC, and the formation of the following forms of control objective function

$$J(k) = \left\| w(k) - \tilde{y}_{PM}(k) \right\|_Q^2 + \left\| u(k) - u_{ss}(k) \right\|_V^2 \left\| \Delta u_M(k) \right\|_R^2 \tag{6}$$

There are not obvious differences in model prediction and feedback collection between TMPC and traditional MPC.

3   Affect of Ill-Conditioned of Steady-State Gain Model for Control Performance

The function of prediction model is based on the established mathematical model to predict outputs, including steady-state prediction and dynamic prediction. Steady-state optimization is a process of calculation of the optimal operating point by steady-state target calculation, the steady-state prediction needs steady-state model. By model identification, a step response coefficient of inputs and outputs model will be built. At present, most of the large-scale production process with continuous, stable production characteristics, identified model can characterize the input-output relationship of processes more accurately, so that the steady-state calculation has good application effect. Meanwhile, the steady-state analysis is relative simple, and can generally reflect the dynamic performance and controllability of the system. Therefore, steady-state gain matrix which including steady-state gain coefficient information of processes is to be an example,

$$K = \begin{bmatrix} k_{11} & \cdots & k_{1m} \\ \vdots & \ddots & \vdots \\ k_{p1} & \cdots & k_{pm} \end{bmatrix}$$

where, K represent steady-state gain matrix, m, p represent the number of inputs and outputs respectively. This paper proposes a strategy that tests the singularity of steady-state models and its subunits (2×2 matrix) before identified models are used as controller model [13-15]. If the models are ill-conditioned, the models will be modified and updated to avoid the problems of ill-conditioned model.

At first, the influence of ill-conditioned model to control is analyzed. It is assumed that a subunit model of process was as following. The heavy oil separation column is used to be object of study that is common in chemical field, its identification, process model is written as following by identifying.

$$G(s) = \begin{bmatrix} \dfrac{-12.62}{50s+1}e^{-27s} & \dfrac{9.84}{60s+1}e^{-28s} & \dfrac{5.88}{50s+1}e^{-27s} \\[2.5ex] \dfrac{9.84}{(50s+1)}e^{-18s} & \dfrac{-6.88}{(60s+1)}e^{-14s} & \dfrac{6.9}{(40s+1)}e^{-15s} \\[2.5ex] \dfrac{4.38}{33s+1}e^{-20s} & \dfrac{4.42}{44s+1}e^{-22s} & \dfrac{7.20}{19s+1}e^{-24s} \end{bmatrix}$$

Then, sub models of process model are used to be object of study,

$$G_{2\times2} = \begin{bmatrix} \dfrac{-12.62}{50s+1}e^{-27} & \dfrac{9.84}{60s+1}e^{-28} \\[2.5ex] \dfrac{9.84}{50s+1}e^{-18} & \dfrac{-6.88}{60s+1}e^{-14} \end{bmatrix}$$

and the steady-state gain matrixes are

$$K = \begin{bmatrix} -12.62 & 9.84 \\ 9.84 & -6.88 \end{bmatrix}$$

The condition numbers can be calculated by steady-state gain matrixes, and the equation is written as following.

$$cond(K) = \frac{\sigma_{max}}{\sigma_{min}} \tag{7}$$

Where, σ is singular value of matrix, σ_{max} is maximum singular value of matrix, and σ_{min} is minimum singular value of matrix. The condition number of this matrix is 40. If the condition number is big, this model will be considered as ill-conditioned by knowledge of linear algebra. In order to analyze the affect of control performance when the controller models are ill-conditioned, two different sets of data are chosen to simulate.

This paper used standard DMC algorithm to simulate. The set point 1 is $[-1.6, -1.2]^T$, set point 2 is $[-1.6, 1.2]^T$, and they will use in all of simulation. The weight coefficient matrix Q, R are unit matrixes, the input and output curves are shown in Fig. 2 as below.

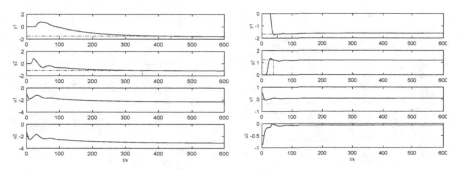

Fig. 2. Input and output curves of set point 1   **Fig. 3.** Input and output curves of set point 2

It is seen from Fig.2 that the outputs are finally stabile in set point 1 $[-1.6, -1.2]^T$, where point of intersection of dotted line represents the two output variables stabilize the range of $\pm 5\%$ error. In other words, setting time t_s are 438 seconds and 351 second respective, which indicates the controller is very slow response for process.

The weight coefficient matrix Q, R are still unit matrixes, and use set point 2 to simulate, the input and output curves are shown in Fig.3.

By the simulation shows that the outputs are finally stabile in $[-1.6, 1.2]^T$, and the process reaches the set point 2 quickly, meanwhile setting time t_s are only 51 seconds and 39 seconds. There are two very different controller's action and control performance for the same ill-conditioned model by two different set points. In particular, the response time have a lot of difference. The reason of this case is only on account of changing of moving direction of the outputs, and the detailed explanation will be introduced next.

The model mismatch is very common in practice. There are many reasons for the model mismatch such as the accuracy of the identification results, characteristic's changing in control processes [16, 20, 21], and lead to controller performance decline. Therefore, it is necessary to analyze model ill-conditioned problem under model mismatch condition.

The actual process model assumes the existence model mismatch of a $\pm 20\%$ range for k_{11}, the remaining parameters does not exist mismatch, $\pm 20\%$ mismatch are defined as mismatch mode 1 and mode 2 mismatch, the model gain matrixes are written as following.

$$K = \begin{bmatrix} -10.1 & 9.84 \\ 9.84 & -6.88 \end{bmatrix}, K_1 = \begin{bmatrix} -15.1 & 9.84 \\ 9.84 & -6.88 \end{bmatrix}$$

where, mismatch mode 1 represents forward mismatch, which means the condition number of matrix decrease after model mismatched. The final condition number of matrix K is 14.9, which smaller than 40 that has been condition number before model mismatched. The mode 2 is opposite to mode 1, the final condition number of matrix K_1 is 14.9

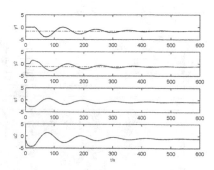

Fig. 4. Input and output curves of set point 1 under mismatch mode 1

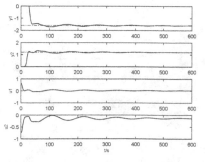

Fig. 5. Input and output curves of set point 2 under mismatch mode 1

In order to analyze model ill-conditioned problem under model mismatch condition, use set point 1 and set point 2 to simulate, Q, R are still unit matrixes, and, the input and output curves are shown in Fig.4 and Fig.5 as above.

From Fig. 4 and Fig.5, it is seen that the process has a turbulence which is still related to the direction of outputs movement. Setting times before model mismatched are larger than after model mismatched for two sets of set point. Meanwhile, the potential static state input point is $u_s = [-2.2816, -3.0888]^T$ which corresponding to the set point $[-1.6, -1.2]^T$, the output value of actual process steady-state model is

$$y_{process} = K_{process\_gain} \times u_s = [-7.3496, -1.2]^T$$

It is seen that 20% model mismatch in k_{11} is amplified 459.35% in output, which is caused by a large movement of control input. That is to say that the range of the controller action amplifies the mismatch of the model, which causes the controller become very sensitive for model mismatch. And then, changing the direction of model mismatch and use mismatch model 2, set point 1 and set point 2 to simulate, Q, R are still unit matrixes, and the input and output curves are shown in Fig.6 and Fig.7 as below.

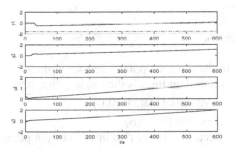

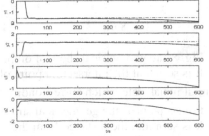

Fig. 6. Input and output curves of set point 1 2 under mismatch mode 2.

Fig. 7. Input and output curves of set point under mismatch mode 2.

It is seen from Fig.6 and Fig.7 that two groups of output do not reach set point under model mismatch, and model mismatch leads to controller action more drastic and causes the curves of input and output detergency. When movement direction of set point is changing, the condition number of matrix will change, and model ill-conditioned level will change, and then the diverging direction of curve will change too. All these phenomena show that the changing of output movement direction will affect control performance under model ill-conditioned.

So far, two conclusions can be obtained. (1) The ill-conditioned model can lead to extreme action of the controller, which can bring about poor control performance. (2) The changing of output movement direction will affect control performance under model ill-conditioned.

4   Analysis of Poor Control Performance Under Model Ill-Conditioned

For ill-conditioned model, there is a relationship between whether the controller produces a large action and movement direction of output. For the 2-input 2-output square system (sub process model is a steady-state process), when set point is given, the potential steady-state input point has been determined [22].To solve input set point actual is a process for solving linear equations group, and in geometrically speaking, the solution is the intersection point of the two lines.

For two groups of set point are given in section II, the geometrically solution is shown as following. For output initial state $[0,0]^T$, the geometrical show is shown in Fig.8(a) as below.

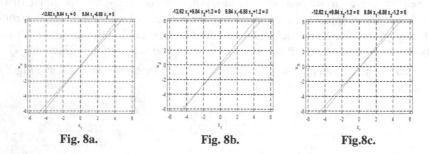

Fig. 8a.   Fig. 8b.   Fig.8c.

Fig. 8. are geometrical show of output initial state, set point 1 and set point 2.

It is seen that the intersection point of two straight lines is the origin point, it means output initial state which corresponding to input initial state is $[0,0]^T$.

For set point 1, the geometrical show is shown in Fig. 8(b).

It is seen that the distance from the intersection point of two lines to origin point is much further than initial state, which means the amplitude of the input initial state will be larger, so controller will generate large motion.

For set point $[-1.6,1.2]^T$, the geometrical show is shown in Fig. 8(c).

It is seen that the intersection point of the two lines is very close to origin point, which means the amplitude of the input initial state is small, so controller will generate little motion.

By the simulation results, a conclusion will be gotten that the movement direction of output will influence the control performance for ill-conditioned model. Because of the changing of output movement direction on geometric changes the distance between input steady state point and the original point, which changes the initial amplitude of input, and lead to the controller action range changed.

It is intuitive to judge whether a dynamic model is ill-conditioned just based on the condition number of model, the condition number is bigger, the model ill-conditioned will become more serious, the control performance will be poorer. This conclusion can be proved through simulation of two groups of different mismatch model in section 3. For mismatch $k_{11} = -10.1$, the condition number of steady-state model will reduce by calculation. Meanwhile, for the mismatch $k_{11} = -15.1$, the number of condition increases, so the former ill-condition is smaller and the control performance is also better. For most of the processes, the steady-state model can represent dynamic nature of the processes, so the ill-conditioned problem of dynamic model can be converted into steady state model to analyze, and the calculation method of condition number has been given above. The linear algebra tools singular value decomposition (SVD) will be used to analyze the model ill-conditioned problem further next.

For the description of steady-state relationship shown in formula (2), the singular value decomposition of steady-state matrix is shown in formula (8) as following.

$$K = U \Sigma V^T = \sum_{i=1}^{n} \sigma_i u_i v_i^T \tag{8}$$

If $K \in R^{p*m}$, $U = [u_1 \ \cdots \ u_p], U \in R^{p*p}$, $\Sigma = diag[\sigma_1...\sigma_p], \Sigma \in R^{p*m}$, $V = [v_1 \ \cdots \ v_m], V \in R^{m*m}$

The formula (9) can be gotten by formula (8),

$$KV = U\Sigma \tag{9}$$

where, U and V are two group of orthonormal basis vectors. A matrix represents a linear transformation from the view of linear algebra, and the significance of the linear transformation has two equivalent explanations as following. (1) Transforming a $x \in R^m$ vector to R^p space. (2) The linear transformation provides a representation with a same vector in two different spaces. For steady-state matrix K, singular value decomposition just provides a kind of geometric form. For every 2×2 matrix, singular value decomposition can always find a space to another space conversion and corresponding matrix transformation. The steady-state model of the process is $\Delta y_s = K \Delta u_s$, where, V is a set of orthonormal base of Δu_s, U is a set of orthonormal base of Δy_s, can be used as a set of orthonormal base of, singular matrix Σ represents the degree of stretch of steady-state gain matrix K. Specifically, u_i and v_i are a couple of one-to-one correspondence orthonormal base vector in the formula (8). The singular value describes stretch multiple in different directions, in other word, there are σ_i times stretch from direction of u_i to v_i. From the perspective of the steady state, the nature of control is steady state matrix inversion, the matrix is more close to singular matrix, the determinant value is more small, its inverse will be very big, that means the controller action might be very big, namely

$$\Delta u_s(k) = [U \Sigma V^T]^{-1} \times \Delta y_s(k) \tag{10}$$

For the model gain matrix of controller in this paper, its singular value decomposition results are written as following.

$$U = \begin{bmatrix} -0.8 & 0.6 \\ 0.6 & 0.8 \end{bmatrix}, \Sigma = \begin{bmatrix} 20 & \\ & 0.5 \end{bmatrix}, V = \begin{bmatrix} 0.8 & 0.6 \\ -0.6 & 0.8 \end{bmatrix}$$

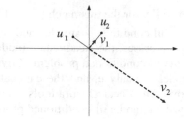

Fig. 9. Coordinate system consisting of base vectors of singular value decomposition

Where, the coordinate system which vertices are original point, are composed of base vector u_1, u_2. The imaginary lines coordinate system that vertices are allows, are composed of base vector v_1, v_2. The stretching amplitude σ_1 , σ_2 in the directions of u_1, u_2 are 0.5 times and 20 times respective, which means if movement output is existent in the direction of v_2, the direction of u_2 need will need twice the input amplitude. Otherwise, the direction of u_2 only will need 0.05 times the input amplitude. This example uses set point 1 and set point 2 which are used to simulate in previous section, the set point 1 $[-1.6, -1.2]^T$ is $(-0.56, -2)$ in the coordinate of which composed of base vector v_1, v_2, and set point 2 $[-1.6, 1.2]^T$ is $(-2, 0)$. Therefore, if the set point 1 is to be set point, the controller action must be more intense, and this analysis is also proved in the previous simulation.

5   Strategy of Modifying Ill-Conditioned Model

By simulation and analysis in previous section, the fundamental reason of ill-conditioned model is that the last singular value of model gain matrix is small and the solution are shown as following. (1) Make output not affected by direction movement of the model ill-condition influence, it means this strategy ensured magnitude of outputs in the minimum singular value corresponding to the projection of vector direction smaller, online strategy mentioned in introduction is used to this idea in essential. (2) Drastically reduce the model ill-condition by modifying model. The following strategy uses the second method, it is an off-line strategy. After finished identification, the identified models are modified immediately, and qualified modified models will be conFigured as a controller model.

The ill-conditioned model gain matrixes compared to ill-conditioned problem, the errors of model gain coefficient seem unimportant, and it is possible that identified models are different form actual process models. Therefore, a model modified strategy is presented, which means increasing the minimum singular value and reducing model ill-condition by modifying model gain matrixes [17, 18]. The method is shown as following.

1. After models are identified, the authors will examine the subunits of original model ill-condition by minimum scale, and inspection index is condition number.

2. Define the error tolerance factor α, it represents each model gain coefficient is allowed to change the amplitude size. The expression is $|\Delta k_{ij}| \leq |k_{ij}| \times \alpha$, it is noted that α can a s allowed factors for special gain, and also be a unified factor for every gain.

3. The initial value of minimum singular value $\sigma_{\min}$ is set $\sigma_{\max}$, and uses dichotomy to doing iterative search by golden section until the maximum allowable error range $\sigma_{\min}$ is found.

4. Update the models.

The error tolerance factor is set 0.1, the modified model was shown as following by application of above method.

$$K' = \begin{bmatrix} -12.3783 & 10.1623 \\ 10.1623 & -6.4503 \end{bmatrix}, \quad \Delta K = K' - K = \begin{bmatrix} 0.2417 & 0.3223 \\ 0.3223 & 0.4297 \end{bmatrix} \leq 0.1 \times K$$

K' are configured to controller model, and actual model of process are $\pm 20\%$ mismatched model.

$$K = \begin{bmatrix} -10.1 & 9.84 \\ 9.84 & -6.88 \end{bmatrix}, \quad K_1 = \begin{bmatrix} -15.1 & 9.84 \\ 9.84 & -6.88 \end{bmatrix}$$

Using set point 1 and set point 2 to simulate under model mismatch mode 1 and mode 2, Q, R are still unit matrixes, the control effect are shown as Fig.10 as below.

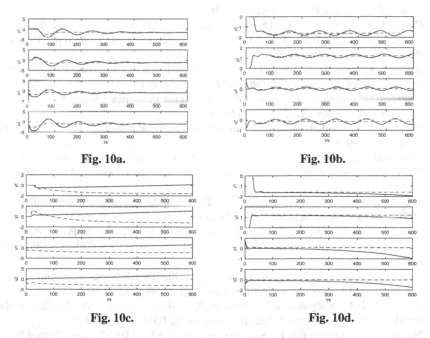

Fig. 10a.   Fig. 10b.

Fig. 10c.   Fig. 10d.

Fig. 10 are Control effect comparison between modified model 1, model 2 and unmodified model under mismatch mode 1, mode 2 and set point 1, set point 2.

In Fig.10, the solid lines represent input and output curves of unmodified models, and the dashed lines represent input and output curves of modified models. It is obviously seen that process shock amplitude and shock time are reduced significantly, while quickly reach the set point, and correct adverse effects for model mismatch Therefore, this strategy is feasible.

6   Conclusions

The ill-conditioned model will lead to control performance declining obviously, and changing of the direction of outputs movement will affect control performance too. Model mismatch problem is a common problem for control, but ill-conditioned model in case of model mismatch will have a greater impact for control performance. In this paper, geometry and linear algebra tools are used to analyze fundamental reason of ill-conditioned, and presents an offline modified strategy to solve model ill-conditioned problem well. Finally, the methods and solutions presented in this paper are proved by simulation.

References

1. Qin, S.J., Badgwell, T.A.: An overview of industrial model predictive control technology. AIChE Symposium Series. American Institute of Chemical Engineers, New York, pp. 1971--2002 (1997). 93(316): 232–256
2. Qin, S.J., Badgwell, T.A.: A Survey of Industrial Model Predictive Control Technology. Control Engineering Practice 11(7), 733–764 (2003)
3. Cutler, C.R., Ramaker, B.L.: Dynamic matrix control-A computer control algorithm. In: Proceedings of the Joint Automatic Control Conference, vol. 1, p. Wp5-B. American Automatic Control Council, Piscataway (1980)
4. Richalet, J., Rault, A., Testud, J.L., et al.: Model Predictive Heuristic Control: Applications to Industrial Processes. Automatica 14(5), 413–428 (1978)
5. Clarke, D.W., Mohtadi, C., Tuffs, P.S.: Generalized Predictive Control-Part I. The Basic Algorithm. Automatica 23(2), 137–148 (1987)
6. Skogestad, S., Postlethwaite, I.: Multivariable Feedback Control: Analysis and Design, New York (1996)
7. Skogestad, S., Morari, M., Doyle, J.C.: Robust Control of Ill-conditioned Plants: High-Purity Distillation. IEEE Transactions on Automatic Control 33(12), 1092–1105 (1988)
8. Grosdidier, P., Froisy, B., Hammann, M.: The Idcom-M controller. In: Proceedings of the 1988 IFAC workshop on Model Based Process Control, pp. 31–36 (1988)
9. Marroquin, J., Mitter, S., Poggio, T.: Probabilistic Solution of Ill-Posed Problems in Computational Vision. Journal of the American Statistical Association 82(397), 76–89 (1987)
10. Cai, J.F., Candès, E.J., Shen, Z.: A Singular Value Thresholding Algorithm for Matrix Completion. SIAM Journal on Optimization 20(4), 1956–1982 (2010)
11. Kassmann, D.E., Badgwell, T.A., Hawkins, R.B.: Robust Steady-State Target Calculation for Model Predictive Control. AIChE Journal 46(5), 1007–1024 (2000)
12. Nikandrov, A., Swartz, C.L.E.: Sensitivity Analysis of LP-MPC Cascade Control Systems. Journal of Process Control 19(1), 16–24 (2009)
13. Limebeer, D.J.N., Kasenally, E.M., Perkins, J.D.: On the Design of Robust Two Degree of Freedom Controllers. Automatica 29(1), 157–168 (1993)
14. Wu-zhong, L.: Singular Perturbation of Linear Algebraic Equations with Application to Stiff Equations. Applied Mathematics and Mechanics 8(6), 513–522 (1987)
15. Jun-liang, W., Fei, L.: Control Model of Ill-linear System and it Iterative Solution Method. Control and Decision 19(11), 1315–1317 (2004)

16. Gui, C., Jiang, Y., Lei, X., Rongjin, Z.: Research on Model-Plant Mismatch Detection Based on Subspace Approach. CIESC Journal 62(9), 2575–2581 (2011)
17. Sanliturk, K.Y., Cakar, O.: Noise Elimination from Measured Frequency Response Functions. Mechanical Systems and Signal Processing 19(3), 615–631 (2005)
18. Hu, S.-L.J., Bao, S., Li, H.: Model Order Determination and Noise Removal for Modal Parameter Estimation. Mechanical Systems and Signal Processing 24(6), 1605–1620 (2010)
19. Zou, T., Ding, B.-C., Zhang, R.: MPC-An Introduction to Industrial Applications (2010)
20. Ying, C., Zhen, C., Shu-liang, W.: Modified IMM Algorithm for Unmatched Dynamic Models. Systems Engineering and Electronics 33(12), 2593–2597 (2011)
21. Harrison, C.A., Qin, S.J.: Discriminating between Disturbance and Process Model Mismatch in Model Predictive Control. Journal of Process Control 19(10), 1610–1616 (2009)
22. Li, R., Fang, Y., Cai, W., et al.: Stability Analysis and a Fast Algorithm for the Predictive Control of Stochastic Systems. Information and Control 2, 000 (2013)

Quaternion-Based Adaptive Control
for Staring-Mode Observation Spacecraft

Chen Lei[1], Xiao Yan[2(✉)], Ye Dong[2], and Li Changjun[1]

[1] DFH Satellite Co., Ltd., Beijing 100094, People's Republic of China
[2] Research Center of Satellite Technology, Harbin Institute of Technology,
Harbin 150001, People's Republic of China
xiaoy@hit.edu.cn

Abstract. The spacecraft staring-mode observation is to render space-borne optical sensor an accurate and continuous pointing strategy at a specified ground target within a specified time span. First, the desired attitude reference trajectory are derived from the geographic information according to the need of the staring-mode. Then, an adaptive tracking control law is proposed to accomplish the staring-mode maneuver suffering from the inertia matrix uncertainty. In addition, a signum function is mixed in the switching surface in controller to produce a maneuver to the reference attitude trajectory in a shortest distance. The stability of the resulting system by the proposed controller is guaranteed by the Lyapunov-based method. Simulation results are presented to illustrate all the technical aspects of this work.

1   Introduction

Spacecrafts often have to undertake attitude maneuvers so as to render the space-borne optical sensor an accurate and continuous pointing at a ground target. In a spacecraft staring-mode observation, the onboard optical sensor should accurately fix its optical axis on the target steadily in a particular observation time span without relative rotation between sensor and target [1].

Based on the desired angular velocity reference for tracking, A quaternion-based controller for ground-target tracking of a three-axis stabilized Earth-pointing satellite is systematically analyzed [2]. In reference [3], attitude control problem of staring mode spacecraft in Low Earth Orbits is studied; moreover, a variable structure controller and a disturbance observer are proposed to reduce inherent chattering in order to improve the observation precision and prolong the observation time.

Quaternion-based feedback pointing control has been thoroughly discussed in a few papers [4, 5]. In the design process of practical attitude control systems, the dynamic model is mainly subject to the spacecraft inertia parameter uncertainty. The inertia uncertainty may be due to the measurement error during the pre-launch testing phase, changes in the overall spacecraft system configuration (such as retrieval of a spacecraft by the space shuttle), or fuel usage during the mission. Thus, the nominal inertia matrix used in the traditional controllers does not equal to the true inertia

© Springer International Publishing Switzerland 2015
H. Liu et al. (Eds.): ICIRA 2015, Part I, LNAI 9244, pp. 632–642, 2015.
DOI: 10.1007/978-3-319-22879-2_57

matrix, and therefore it largely reduces the performance of the control system. There are several robust control methods for spacecraft with model uncertainty [6, 7]. The sliding mode control approach to solve the attitude tracking problem especially in the presence of space disturbances and spacecraft model uncertainties is addressed in references [8, 9]. Another effective way to deal with the attitude control problem with the inertia parameter uncertainty is the adaptive control method [10-12]. An quaternion-based, attitude tracking control of rigid spacecraft without angular velocity measurements was designed in the presence of an unknown inertia matrix [13]. But, few papers about the adaptive attitude control method has considered the issue of attitude slewing unwinding problem due to the double values of unit quaternions, so in this paper this problem will be settled.

The paper is organized as follows. In Sec. 2, the desired attitude reference trajectory of the satellite including the desired attitude, the desired angular velocity and the desired angular acceleration, are derived from the geographic information according to the need of the staring-mode. In Sec. 3, we propose an adaptive controller to accomplish the staring-mode maneuver suffering from the inertia matrix uncertainty. In Sec. 4, numerical simulations are presented to describe the performance of the proposed theoretical results followed by the concluding statement in Sec. 5.

2   The Desired Motion Reference

In this section, the desired spacecraft attitude reference profiles are derived from the geographic information (the orientation between the spacecraft and the ground target) according to need of the staring-mode. Furthermore, the angular velocity reference profiles will also be derived from the kinematics of rigid body so that the onboard optical sensor observes the ground target without relative rotation. Subsequently, the desired angular acceleration will be deduced.

2.1   Desired Attitude

There are four coordinate frames of interest, denoted by "ORC" for the orbit referenced coordinates, "EIC" for the Earth Centered Inertial Coordinates, "EFC" for the Earth fixed coordinates, and "SBC" for the spacecraft body fixed coordinates, respectively. For the coordinates frame representation of a vector, a superscript indicates the frame of reference. For the attitude transformation matrix C, the subscript frame is transformed to the superscript frame.

The location of the ground target is given by

$$x_T^{EIC} = C_{EFC}^{EIC} x_T^{EFC} \tag{1}$$

where x_T represent the vector from the Earth center to the target.

If we assume the Earth has a constant angular rate $\boldsymbol{\omega}_E$ around its rotation axis during the tracking maneuver, then C_{EFC}^{EIC} can be represented by

$$C_{EFC}^{EIC} = \begin{bmatrix} \cos(\omega_E t + \phi) & -\sin(\omega_E t + \phi) & 0 \\ \sin(\omega_E t + \phi) & \cos(\omega_E t + \phi) & 0 \\ 0 & 0 & 1 \end{bmatrix} \tag{2}$$

where ϕ is the initial phase between the x-axes of EFC and the EIC coordinates at $t = 0$.

The vector from the satellite to the ground target can be described as

$$x_{s/T}^{ECI} = x_T^{EIC} - r = C_{EFC}^{EIC} x_T^{EFC} - r \tag{3}$$

where r denotes the vector from the center of the Earth to the satellite represented in the EIC coordinates. The unit direction vector of $x_{s/T}^{ECI}$ will be

$$u_{s/T}^{ECI} = x_{s/T}^{ECI} / \left\| x_{s/T}^{ECI} \right\| \tag{4}$$

The desired attitude should be such that the optical sensor points toward the target. Only one vector can't determine the orientation of the desired reference. So we need another vector such as x-axes of the ORC coordinates to completely specify the orientation. Then the corresponding Euler parameters q_d can be obtained.

2.2   Desired Angular Velocity

The desired angular velocity is not uniquely specified for this problem. A particular choice with no rotation component about $u_{s/T}^{ECI}$ is [14]

$$\omega_d^{ECI} = \left(u_{s/T}^{ECI}\right)^\times \dot{u}_{s/T}^{ECI} \tag{5}$$

The notation $\zeta^\times$ for $\forall \zeta = [\zeta_1, \zeta_2, \zeta_3]^T$ denotes the following skew-symmetric matrix:

$$\zeta^\times = \begin{bmatrix} 0 & -\zeta_3 & \zeta_2 \\ \zeta_3 & 0 & -\zeta_1 \\ -\zeta_2 & \zeta_1 & 0 \end{bmatrix} \tag{6}$$

To solve the vector $\dot{u}_{s/T}^{ECI}$, we take the time derivative of both sides of the Eq. (4), and obtain

$$\dot{u}_{s/T}^{ECI} = \frac{1}{\left\| x_{s/T}^{ECI} \right\|} \left[I_{3\times3} - u_{s/T}^{ECI} \left(u_{s/T}^{ECI} \right)^{T} \right] \dot{x}_{s/T}^{ECI} \tag{7}$$

From Eq. (3), the vector $\dot{x}_{s/T}^{ECI}$ can be computed by

$$\dot{x}_{s/T}^{ECI} = \left(\omega_{sid}^{EIC} \right)^{\times} C_{EFC}^{EIC} x_{T}^{EFC} - v \tag{8}$$

Furthermore, we can get the second derivative of $x_{s/T}^{ECI}$ as

$$\ddot{x}_{s/T}^{ECI} = \left[\left(\omega_{sid}^{EIC} \right)^{\times} \right]^{2} C_{EFC}^{EIC} x_{T}^{EFC} - \dot{v} \tag{9}$$

where v is the translational motion velocity of the satellite with respect to the EIC coordinates, and $\omega_{sid}^{EIC} = \begin{bmatrix} 0 & 0 & 2\pi/86164 \end{bmatrix}^{T}$ is the sidereal rotation rate of the Earth. Furthermore substitute Eq.(7) into Eq.(5), then [15]

$$\begin{aligned} \omega_{d}^{ECI} &= \frac{1}{\left\| x_{s/T}^{ECI} \right\|} \left(u_{s/T}^{ECI} \right)^{\times} \left[I_{3\times3} - u_{s/T}^{ECI} \left(u_{s/T}^{ECI} \right)^{T} \right] \dot{x}_{s/T}^{ECI} \\ &= \frac{1}{\left\| x_{s/I}^{ECI} \right\|} \left(u_{s/T}^{ECI} \right)^{\times} \dot{x}_{s/T}^{ECI} \end{aligned} \tag{10}$$

2.3   Desired Angular Acceleration

The desired angular acceleration follows by differentiating Eq.(10) with respect to time

$$\begin{aligned} \dot{\omega}_{d}^{ECI} &= \frac{-1}{\left\| x_{s/T}^{ECI} \right\|^{2}} \frac{d}{dt} \left(\left\| x_{s/T}^{ECI} \right\| \right) \left(u_{s/T}^{ECI} \right)^{\times} \dot{x}_{s/T}^{ECI} + \frac{1}{\left\| x_{s/T}^{ECI} \right\|} \dot{u}_{s/T}^{ECI\times} \dot{x}_{s/T}^{ECI} + \\ &\quad \frac{1}{\left\| x_{s/T}^{ECI} \right\|} \left(u_{s/T}^{ECI} \right)^{\times} \ddot{x}_{s/T}^{ECI} \end{aligned} \tag{11}$$

Using $a^{\times}b = -b^{\times}a$ and $a^{\times}a = 0$ and substitution of Eq.(7) goes here

$$\begin{aligned} \dot{\omega}_{d}^{ECI} &= \frac{1}{\left\| x_{s/T}^{ECI} \right\|^{3}} \left(x_{s/T}^{ECI} \right)^{T} \dot{x}_{s/T}^{ECI} \left(\dot{x}_{s/T}^{ECI} \right)^{\times} u_{s/T}^{ECI} - \\ &\quad \frac{1}{\left\| x_{s/T}^{ECI} \right\|^{2}} \left(\dot{x}_{s/T}^{ECI} \right)^{\times} \left[I_{3\times3} - u_{s/T}^{ECI} \left(u_{s/T}^{ECI} \right)^{T} \right] \dot{x}_{s/T}^{ECI} + \frac{1}{\left\| x_{s/T}^{ECI} \right\|} \left(u_{s/T}^{ECI} \right)^{\times} \ddot{x}_{s/T}^{ECI} \end{aligned} \tag{12}$$

Using $a^\times a = 0$ Eq.(4) yields

$$\dot{\omega}_d^{ECI} = \frac{2}{\left\| x_{s/T}^{ECI} \right\|^2} \left(\dot{x}_{s/T}^{ECI} \right)^\times u_{s/T}^{ECI} \left(u_{s/T}^{ECI} \right)^{\mathrm{T}} \dot{x}_{s/T}^{ECI} + \frac{1}{\left\| x_{s/T}^{ECI} \right\|} \left(u_{s/T}^{ECI} \right)^\times \ddot{x}_{s/T}^{ECI} \tag{13}$$

3   Adaptive Tracking Control Law

The unit quaternion $\tilde{q}(t) = \begin{bmatrix} \tilde{q}_o(t) & \tilde{q}_e(t) \end{bmatrix} = q_d^{-1} \otimes q$ describes the orientation of the body-fixed frame $\mathbb{B}$ with respect to the desired reference frame.

The rotation $C_{bd} = C(\tilde{q}) \in \mathrm{SO}(3)$ relating $\mathbb{B}$ to $\mathbb{D}$ is given by

$$C_{bd} = \left(\tilde{q}_0^2 - \tilde{q}_e^{\mathrm{T}} \tilde{q}_e \right) I_{3\times3} + 2\tilde{q}_e - 2\tilde{q}_0 \tilde{q}_e^\times \tag{14}$$

The kinematic equations are therefore

$$\delta \dot{\tilde{q}}_e = \tfrac{1}{2} \left(\tilde{q}_e^\times \tilde{\omega} + \tilde{q}_0 \tilde{\omega} \right)$$
$$\dot{\tilde{q}}_0 = -\tfrac{1}{2} \tilde{q}_e^{\mathrm{T}} \tilde{\omega} \tag{15}$$

The equations of motion of the spacecraft with respect to $\mathbb{D}$ are given by

$$J \dot{\tilde{\omega}} = -\omega^\times J\omega + J \left(\tilde{\omega}^\times C_{bd} \omega_d - C_{bd} \dot{\omega}_d \right) + u \tag{16}$$

where $\tilde{\omega} = \omega - C_{bd} \omega_d$ represent the angular velocity tracking error. We choose $s = \delta\omega + \lambda \mathrm{sgn}(\delta q_0) \delta q_e$ similar to the reference[16], and λ is a positive scalar constant. Then Eq. (16) can be rewritten as

$$J\dot{s} = -\omega^\times J\omega - J\beta_r + u \tag{17}$$

The parameters in Eq. (17) are $\beta_r = C_{bd} \dot{\omega}_d - \tilde{\omega}^\times C_{bd} \omega_d - \lambda \dot{\tilde{q}}_e$

From Eq. (16), we can observe that the inertia parameter $J_{i,j}$, where i, $j = 1, 2, 3$, appear linearly. For convenience, we defined a linear operator $L : \Re^3 \to \Re^{3\times6}$ acting on $a = \begin{bmatrix} a_1 & a_2 & a_3 \end{bmatrix}^{\mathrm{T}}$ by

$$L(a) = \begin{bmatrix} a_1 & 0 & 0 & 0 & a_3 & a_2 \\ 0 & a_2 & 0 & a_3 & 0 & a_1 \\ 0 & 0 & a_3 & a_2 & a_1 & 0 \end{bmatrix} \tag{18}$$

Letting

$$\boldsymbol{\alpha}(J) = \begin{bmatrix} J_{11} & J_{22} & J_{33} & J_{23} & J_{13} & J_{12} \end{bmatrix}^{\mathrm{T}} \tag{19}$$

It follows that

$$Ja = L(a)\boldsymbol{\alpha} \tag{20}$$

In this research work the inertia matrix is assumed fixed throughout the duration of the spacecraft attitude maneuver. The true spacecraft inertia matrix is given by $J = \hat{J} + \tilde{J}$, where $\hat{J}$ is the current best estimate of the inertia matrix and the $\tilde{J}$ is inertia matrix uncertainty.

Theorem 1. Assume that $\boldsymbol{\omega}_d$ and $\dot{\boldsymbol{\omega}}_d$ are bounded and let $\mathbf{K}_D \in \Re^{3\times3}$ and $\Gamma \in \Re^{6\times6}$ be positive definite. Then the parameters update law and the control algorithm

$$\begin{aligned} u &= -\mathbf{K}_D s + \hat{J}\boldsymbol{\beta}_r + \boldsymbol{\omega}^\times \hat{J}\boldsymbol{\omega} \\ \dot{\hat{\boldsymbol{\alpha}}} &= \Gamma\Phi(\boldsymbol{\omega} \quad \boldsymbol{\beta}_r)s \end{aligned} \tag{21}$$

where $\Phi(\boldsymbol{\omega} \quad \boldsymbol{\beta}_r) = -\left[\mathbf{L}(\boldsymbol{\beta}_r) + \boldsymbol{\omega}^\times \mathbf{L}(\boldsymbol{\omega})\right]^{\mathrm{T}}$, solves the tracking problem.

Proof. Consider the positive-definite Lyapunov candidate function V defined as

$$V = \frac{1}{2}s^{\mathrm{T}}Js + \frac{1}{2}\tilde{\boldsymbol{\alpha}}\Gamma^{-1}\tilde{\boldsymbol{\alpha}} \tag{22}$$

The time derivative of V along the trajectories of the system is given by

$$\dot{V} = s^{\mathrm{T}}J\dot{s} + \tilde{\boldsymbol{\alpha}}^{\mathrm{T}}\Gamma^{-1}\dot{\tilde{\boldsymbol{\alpha}}} \tag{23}$$

After substitution of the Eq. (17) and the control law (21), then

$$\begin{aligned} \dot{V} &= s^{\mathrm{T}}J\dot{s} - \tilde{\boldsymbol{\alpha}}^{\mathrm{T}}\Gamma^{-1}\dot{\hat{\boldsymbol{\alpha}}} \\ &= s^{\mathrm{T}}\left(-\boldsymbol{\omega}^\times\tilde{J}\boldsymbol{\omega} - \tilde{J}\boldsymbol{\beta}_r - \mathbf{K}_D s\right) + s^{\mathrm{T}}\left[\mathbf{L}(\boldsymbol{\beta}_r) + \boldsymbol{\omega}^\times\mathbf{L}(\boldsymbol{\omega})\right]\tilde{\boldsymbol{\alpha}} \end{aligned} \tag{24}$$

Using the definition of Eq. (20), we can get

$$\begin{aligned} \dot{V} &= s^{\mathrm{T}}\left(-\boldsymbol{\omega}^\times\tilde{J}\boldsymbol{\omega} - \tilde{J}\boldsymbol{\beta}_r - \mathbf{K}_D s\right) + s^{\mathrm{T}}\left(\tilde{J}\boldsymbol{\beta}_r + \boldsymbol{\omega}^\times\tilde{J}\boldsymbol{\omega}\right) \\ &= -s^{\mathrm{T}}\mathbf{K}_D s \end{aligned} \tag{25}$$

Since the semi-definite positive of the Eq. (25), applying the Invariant Set Theorems result in $s(t) \to \mathbf{0}_{3\times1}$ as $t \to \infty$ [17], which implies that $\lim_{t\to\infty} \tilde{\boldsymbol{\omega}} = \mathbf{0}_{3\times1}$ and

$\lim_{t\to\infty}\tilde{\boldsymbol{\omega}}=\mathbf{0}_{3\times 1}$, $\lim_{t\to\infty}\tilde{q}_v=\mathbf{0}_{3\times 1}$, and $\lim_{t\to\infty}\tilde{q}_0=\pm 1$ (depending on which equilibrium point is the minimum angular path) [18].

4   Simulation Results

In this section, we simulate the control law and the kinematics using Matlab in which the true inertia matrix is unknown. The goal presented here are of tracking of a ground target.

4.1   Simulation Model and Control Parameters

The radius vector and velocity vector of the spacecraft in Earth Centered Inertial Coordinates are [-798988.093, 2845623.429, 5958355.993] m and [106.021, 6990.439, -3324.417] m/s.

To illustrate the performance of the adaptive controller, the following inertia matrix $\boldsymbol{J}$ and the nominal inertia matrix $\boldsymbol{J}^*$ are chosen as [19]

$$\boldsymbol{J}=\begin{bmatrix} 95.0 & -0.69 & 0.18 \\ -0.69 & 190.0 & 0.12 \\ 0.18 & 0.12 & 142.5 \end{bmatrix}\mathrm{Kg\cdot m^2}, \boldsymbol{J}^*=\begin{bmatrix} 100 & 0 & 0 \\ 0 & 100 & 0 \\ 0 & 0 & 100 \end{bmatrix}\mathrm{Kg\cdot m^2} \quad (26)$$

The initial angular velocity is [0.02 -0.03 0.02] rad/s. The initial attitude is described by the quaternion [0.2795, 0.9424, -0.1820, -0.0191]. Besides, the initial reference attitude is described by the quaternion [0.3555, 0.9343, -0.0079, 0.0208].

The parameters for control law(19) are chosen as $\lambda=0.8$, $\mathbf{K}_D=200$, $\Gamma=50*\mathbf{I}_6$.

4.2   Results

We define tracking error angle θ_{error} to represent the angle between the vector $\boldsymbol{u}_{s/T}^{ECI}$ and the axis of actual optical sensor $\boldsymbol{u}_{s/T}$, which can be used to describe the tracking error. θ_{error} can be calculated as

$$\theta_{error}=\arccos\left(\boldsymbol{u}_{s/T}^{ECI}\bullet\boldsymbol{u}_{s/T}\right) \quad (27)$$

To illustrate the advantage of the term $\mathbf{sgn}(\delta q_0)$ in $\boldsymbol{s}$, a comparative simulation has been run under the same initial conditions, and the results are shown as follow. Fig. 1 depicts the attitude tracking error $\tilde{q}$ without $\mathbf{sgn}(\delta q_0)$ in $\boldsymbol{s}$. Meanwhile, Fig. 2 through Fig. 5 present the profiles of the control law(19).

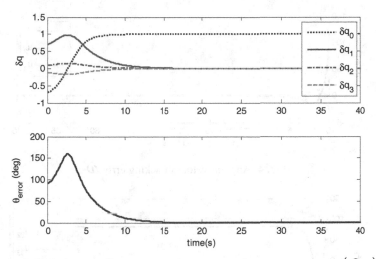

Fig. 1. Attitude tracking error $\tilde{q}$ and tracking error angle θ_{error} without $\mathbf{sgn}\left(\delta q_0\right)$ in s

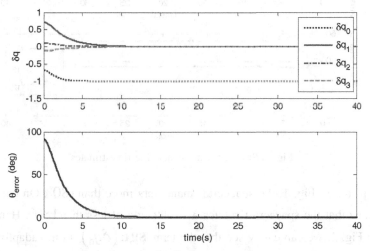

Fig. 2. Attitude tracking error $\tilde{q}$ and tracking error angle θ_{error}

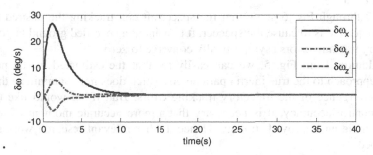

Fig. 3. Angular velocity tracking error $\tilde{\omega}$

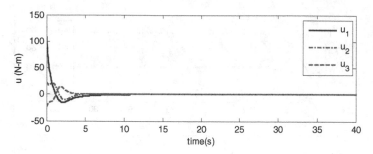

Fig. 4. Angular velocity tracking error $\tilde{\boldsymbol{\omega}}$

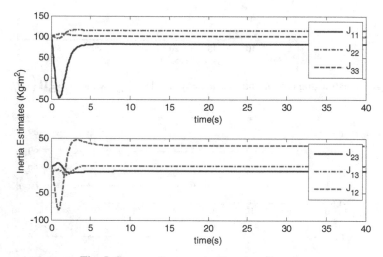

Fig. 5. Spacecraft moment of inertia estimates

As depicted in Fig. 1, the spacecraft maneuvers more than $180°$. On the contrary, Fig. 2 show that the spacecraft rotates a small angle of less than $180°$. Hence, From Fig. 1 and Fig. 2, we can easily see that the term $\mathbf{sgn}\left(\delta q_0\right)$ in the adaptive control law can produce a maneuver with respect to the reference attitude trajectory in the shortest distance.

Fig. 2 through Fig. 5 show that the spacecraft can tracking the desired trajectory whose objective is to cause the spacecraft to point at a specified ground target, and the trajectory tracking errors asymptotically converge to zero.

In addition, from Fig. 5, we can easily see that the estimated inertia parameters do not approach to the true inertia parameters, which does not contaminate the asymptotic convergence of the trajectory tracking errors. But, if we would like to get the higher control accuracy, it is necessary that a more accurate moment of inertia can be gotten via an improved strategy. Hence, further relevant research work should be developed.

5   Conclusions

This paper addressed the problem of spacecraft's accurately pointing at a specified ground target. The desired attitude reference profiles were derived from the geographic information expressed in the Earth Centered Inertial Coordinates according to need of the staring-mode. Then, a direct adaptive control strategy is proposed to achieve an accurate ground target tracking maneuver without the knowledge of the inertia parameters. From the theoretical proof and the simulation results, we can see that the proposed strategy can obtain an accurate tracking accuracy. Besides, simulation results indicated that the addition of the simple term in the control law always provides an optimal response so that the reference attitude motion is achieved in the shortest possible distance.

References

1. Lee, A.Y., Hanover, G.: Cassini spacecraft attitude control system flight performance. In: AIAA Guidance, Navigation, and Control Conference, August 15–18, 2005, San Francisco, CA, United states, pp. 4172–4254 (2005)
2. Goeree, B.B., Fasse, E.D.: Sliding mode attitude control of a small satellite for ground tracking maneuvers. In: 2000 American Control Conference, June 28–30, 2000 Chicago, IL, USA, pp. 1134–1138 (2000)
3. Sun, Z.-W., Wu, S.-N., Li, H.: Variable structure attitude control of staring mode spacecraft with disturbance observer. Harbin Gongye Daxue Xuebao/Journal of Harbin Institute of Technology 42, 1374–1378+1417 (2010)
4. Wu, S., Radice, G., Gao, Y., Sun, Z.: Quaternion-based finite time control for spacecraft attitude tracking. Acta Astronautica 69, 48–58
5. Song, Y.D., Cai, W.C.: Quaternion Observer-Based Model-Independent Attitude Tracking Control of Spacecraft. Journal of Guidance Control and Dynamics 32, 1476–1482 (2009)
6. Krstic, M., Tsiotras, P.: Inverse optimal stabilization of a rigid spacecraft. IEEE Transactions on Automatic Control 44, 1042–1049 (1999)
7. Luo, W., Chu, Y.-C., Ling, K.-V.: Inverse optimal adaptive control for attitude tracking of spacecraft. IEEE Transactions on Automatic Control 50, 1639–1654 (2005)
8. Robinett, R.D., Parker, G.G.: Spacecraft Euler parameter tracking of large-angle maneuvers via sliding mode control. Journal of Guidance, Control, and Dynamics 19, 702–703 (1996)
9. Jin, Y., Liu, X., Qiu, W., Hou, C.: Time-varying sliding mode controls in rigid spacecraft attitude tracking. Chinese Journal of Aeronautics 21, 352–360 (2008)
10. Wu, J., Liu, K., Han, D.: Adaptive sliding mode control for six-DOF relative motion of spacecraft with input constraint. Acta Astronautica 87, 64–76 (2013)
11. Ahmed, J., Coppola, V.T., Bernstein, D.S.: Adaptive Asymptotic Tracking of Spacecraft Attitude Motion with Inertia Matrix Identification. Journal of Guidance, Control, and Dynamics 21, 684–691 (1998)
12. Seo, D., Akella, M.R.: High-Performance Spacecraft Adaptive Attitude-Tracking Control Through Attracting-Manifold Design. Journal of Guidance, Control, and Dynamics 31, 884–891 (2008)
13. Costic, B.T., Dawson, D.M., de Queiroz, M.S., Kapila, V.: Quaternion-Based Adaptive Attitude Tracking Controller Without Velocity Measurements. Journal of Guidance, Control, and Dynamics 24, 1214–1222 (2001)

14. Shucker, G.B., Fasse, E.: Geometric attitude control of a small satellite for ground tracking maneuvers. In: Annual AIAA/Utah State University Conference on Small Satellites. United States, Logan (1999)
15. Chen, X., Steyn, W., Hashida, Y.: Ground-target tracking control of earth-pointing satellites. In: Proceedings AIAA-2000-4547, Guidance, Navigation, and Control Conference, Denver Colorado (2000)
16. Crassidis, J., Vadali, S., Markley, F.: Optimal Variable-Structure Control Tracking of Spacecraft Maneuvers. Journal of Guidance Control and Dynamics 23, 564–565 (2000)
17. Slotine, J.J., Li, W.: Applied Nonlinear Control. Prentice Hall Englewood Cliffs, New Jersey (1991)
18. Li, Z.X., Wang, B.L.: Robust Attitude Tracking Control of Spacecraft in the Presence of Disturbances. Journal of Guidance, Control, and Dynamics 30, 1156–1159 (2007)
19. Junkins, J.L., Akella, M.R., Robinett, R.D.: Nonlinear adaptive control of spacecraft maneuvers. Journal of Guidance Control and Dynamics 20, 1104–1110 (1997)

Model-Based Metacontrol for Self-adaptation

Carlos Hernández$^{(\boxtimes)}$, José L. Fernández, Guadalupe Sánchez-Escribano,
Julita Bermejo-Alonso, and Ricardo Sanz

Escuela Técnica Superior de Ingenieros Industriales, Universidad Politécnica de
Madrid, José Gutierrez Abascal 2, 28006 Madrid, Spain
{carlos.hernandez,mguadalupe.sanchez,ricardo.sanz}@upm.es,
{jlfdez,jbermejo}@etsii.upm.es
http://www.aslab.upm.es

Abstract. There is an increasing demand for more autonomous systems. Enhancing systems with self-aware and self-adaptation capabilities can provide a solution to meet resilience needs. This article proposes a general design solution to build autonomous systems capable of runtime reconfiguration. The solution leverages Model-Driven Engineering with Model-Based Cognitive Control. The key idea is the integration of a metacontroller in the control architecture of the autonomous system, capable of perceiving the dysfunctional components of the control system and reconfiguring it, if necessary, at runtime. At the core of the metacontroller's operation lies a model of the system's functional architecture, which can be generated from the engineering modeling of the system.

Keywords: Autonomy · Model-based control · Self-adaptation · Self-awareness

1 Introduction

There is a rising demand to build more resilient autonomous systems, capable of adaptation to uncertain circumstances in critical applications such as rescue robots, self-driving cars, Remotely Piloted Aircraft Systems (RPAS), smart grids, to mention some. From a control engineering perspective, adaptivity is limited by the knowledge explicitly exploitable by the system at runtime, as opposed to the knowledge that was embedded embedded in its own design during development. To design the control of autonomous system knowledge of the environment is used, typically in the form of mathematical models. Nevertheless, models are not perfect replicas of reality: unmodeled dynamics, model inaccuracies and disturbances produce uncertainty at runtime. The solution lies at the very core of control engineering: the feedback loop. For a system's capability to be resilient in the face of uncertainty, a control loop is designed that will leverage at runtime the knowledge in the model. This way, the control action executed at runtime to achieve it is not determined at design time, but computed on the fly by the loop using information of the instantaneous state, in addition to the model, and therefore accommodating the disturbances.

© Springer International Publishing Switzerland 2015
H. Liu et al. (Eds.): ICIRA 2015, Part I, LNAI 9244, pp. 643–654, 2015.
DOI: 10.1007/978-3-319-22879-2_58

However, control loops are only closed for certain capabilities, and not for every system trait. For example, the very control policy implemented in the controller itself is usually fixed at design time, based on the system model and the control technique selected. At most, modern control techniques such as adaptive control allow for a set or a range of control alternatives to be chosen at runtime. The capability to adapt at runtime is therefore limited by the knowledge incorporated in the controller's design during development. We say that these traits of the autonomous system, usually related to non-functional requirements such as resilience or performance, suffer the development-runtime gap: the systems does not have the explicit knowledge at runtime to adapt if they are compromised by unexpected events.

Engineering knowledge includes not only the models used to design the control algorithms, but also the infrastructure, components and technology used to integrate it in the system. Uncertainty has been traditionally regarded as affecting the environment, since the engineered control system was considered perfectly defined in both static structure and dynamic operation by design. However, this is far from true, especially as systems grow in complexity. Emergent failures and unexpected interactions may hamper mission fulfillment as environmental disturbances do. For these reasons, critical systems are supervised by human operators. They take corrective action "controlling the control system", from manual operation to system reconfiguration on the fly, if required by the situation. Their actions are based on all their knowledge about the system concerning physical architecture, capabilities, control actions, mission, etc. Structural and behavior models are not enough. The knowledge of these engineers about the system's mission and the mission-system relationship, i.e. *functional* knowledge, is crucial.

The work presented in this paper is based on the idea of making the system's control exploit this functional knowledge captured in the engineering models as self-knowledge to adapt at runtime. For this we have leveraged Model-Driven Engineering and Model?-Based Control in a metacontrol architecture capable of functional self-awareness, by monitoring the autonomous system controller, and adaptation, by reconfiguring it on the fly.

The rest of the paper is organized as follows. Section 2 discusses related approaches in the literature to the problem of autonomous systems self-awareness and adaptation. Section 3 analyzes the core idea behind our proposal. Section 4 presents the Operative Mind (OM) architectural framework for autonomous systems that we have developed. Section 5 describes as a case study how the OM framework has been applied to develop a control architecture with self-awareness and adaptivity for a mobile robot. Finally we conclude the paper with some conclusions in Section 6.

2   Self-adaptation and Autonomy

This work has been developed in the context of the ASys program [12], whose aim is to develop a domain-neutral technology to build custom autonomy in any kind of technical system. It is a multidisciplinary effort inspired by previous theories and solutions from different domains that address the engineering of systems that can self-manage and adapt autonomously: cognition, self-adaptive software, and fault-tolerant control. Our focus is to identify core architectural traits for building autonomy.

2.1   Cognition for Autonomy

The ASys strategy is to build more resilient autonomous systems by using cognitive control loops to make systems that can re-engineer themselves. Artificial Intelligence has been leading research on autonomous systems by providing systems with intelligent solutions for certain tasks, such as navigation or object recognition, to mention two examples. However there are still strong challenges concerning general solutions ready to reach the market. For such, cognitive architectures is the approach followed in our research. Cognitive Architectures provide mechanisms to integrate high-level intelligent capabilities and explore models of the cognitive phenomena. However, they remain experimental efforts that are in most cases far away from engineering applicability in industry, although there are some exceptions, such as the Real-time Control Systems Architecture [1], whose paradigmatic architectural approach and engineering methodology have been a reference for this work.

2.2   Self-adaptive Software

Self-adaptation has been an important area of research in software systems during the last decades. From early fault-tolerance at the code level, the solutions have moved to the architectural level. Services and components can be bound at runtime to result in new configurations, and new models for describing and reasoning about these dynamic architectures are being explored [7]. The autonomic computing approach [10] offers a roadmap for future computing systems to target self-management. It is bio-inspired by the homeostatic regulation maintained by the autonomic nervous system and the endocrine system. It proposes computing systems formed by a myriad of autonomic elements rendering services to each other but also capable of self- configuration, optimization, healing and protection. De Loach et al. [5] multiagent approach is especially relevant, because the system uses an explicit model of its organization and capabilities to reconfigure at runtime in order to maximize the achievement of its goals, and their framework is supported by a Model-Driven Development process.

2.3   Fault-Tolerant Control

Resilience issues have been addressed both in classical control theory -i.e. robust control- and computer-based systems -i.e. fault-tolerance- with domain specific methods. Fault-tolerant control [3] has arisen as a cross-domain approach to the fault-tolerant design of current complex and heterogeneous control systems.

It uses a components and services model, as well as control reconfiguration by exploiting analytical redundancy (as alternative to physical redundancy such as the classical triple modular redundancy) in the form of service versions. An example of it is the Autonomous Supervisor architecture proposed by Blanke et al. [3], capable of performing a limited controller re-design using fault-diagnosis information.

There is no yet universal engineering solution to design systems for resilient autonomy. AI more general techniques like cognitive architectures, are more focused on high-level cognitive capabilities, and the validation of scientific models of intelligence, rather than engineering solutions for robust autonomy. On the other hand, engineering approaches closer to industry, such as fault-tolerant control and autonomic-computing, are domain-dependent and they are still missing to incorporate the cognitive approach to exploit knowledge.

3   Bridging the Development-Runtime Gap

In the seek for architectural traits to build resilient autonomy, our solution aims to bridge the development-runtime gap discussed in the introduction by including a higher level control strategy that targets the functional issues of the autonomous system, as resident engineers do. Our work is based on the concept of function in the field of functional modeling [4]. Functional modeling is cross-disciplinary effort that focuses on representation formalisms and methods to incorporate knowledge about the intention and purpose of designers and the relation with the system physical structure and behavior, into the models of systems that are used in engineering activities. However, we claim that this functional models can also be exploited at runtime in autonomous systems.

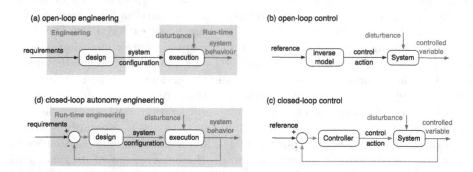

Fig. 1. We have defined the autonomy-loop as the application of the feedback control loop to the system run-time configuration.

Let's consider the analogy between control strategies and the engineering of autonomous systems depicted in Figure 1. The classical engineering process (Figure 1a) has a gap between development and runtime phases, making it impossible for the system to cope with situations not considered during development, in the same way that any disturbance invalidates open-loop control strategies (Figure 1b). The solution is to close what we could call the autonomy-loop shown in Figure 1d. The system requirements are the target reference of a loop that monitors the adequacy of system's behavior. When it is not met due to an unexpected disturbance, a re-design is performed to produces a new system configuration, which is the corrective control action by which the system adapts at runtime.

To implement the autonomy-loop, we propose to add a higher level controller to the control schema, a metacontroller, devoted to adapt the conventional controller to disturbances that deviate system behavior from its mission. For the metacontroller design, we have intertwined Model-Based Cognitive Control, functional modeling and Model-Driven Engineering (MDE). A conventional controller operates according to a model of the system that is based on variables of the system domain. The metacontroller shall use instead a model that represents the architecture of the system in terms of its components and functions, to assess the run-time configuration in relation to the mission requirements. Using MDE to build the autonomous system enables to obtain this run-time functional architecture model from the same engineering model used to build the system. The same knowledge is thus used for both development and runtime, effectively bridging the gap.

4   Metacontrol Engineering

Our metacontrol solution has been developed as an architectural framework to be of generic applicability. It is named the *OM Architectural Framework* and includes the following core elements: the *Self-awareness Patterns* [9] that reify the metacontroller design ideas as reusable design patterns, the *Teleological and Ontological Metamodel for Autonomous Systems* (TOMASys) that defines the run-time models, the *OM Architecture*, which is a reference architecture for the metacontroller development, and the *OM Engineering Process* (OMEP) to apply this solution assets to build autonomous applications.

4.1   Self-awareness Patterns

A pattern-based strategy [9] has been followed to flesh out the metacontrol ideas into reusable design solutions. Four patterns for self-awareness have been defined, and applied on the control architecture of an autonomous system to generate the metacontrol architecture.

Metacontrol pattern defines the integration of the metacontrol approach in the control architecture of an autonomous system. It separates the control system into two subsystems (see Figure 2 step 1): the domain controller, which consists of the traditional controller responsible for sensing and acting on the system to achieve a target reference that is typically the value of a variable in the system's domain of operation, e.g. a position or a velocity; and the metacontroller, which in turn controls the domain controller through an interface provided by its implementation component platform. The metacontroller's references are the system's mission requirements.

Functional Metacontrol pattern defines the metacontrol operation in order to explicitly target functional issues. It defines a layered structure for the metacontroller consisting of two loops (see Figure 2 step 2), one for controlling the configuration of the components of the controller, and another one on top of the former, controlling the functions performed; so the functional and structural concerns are explicitly represented.

Epistemic Control Loop pattern (ECL) defines a Model-Based Cognitive Control loop to build any controller. It formalizes the idea of the well-known sense-plan-act schema for a model-based controller. It prescribes the basics of the structure and behavior of a control loop that exploits explicit system knowledge –i.e. the model–, which is at the core of its operation (see Figure 2 step 3). It establishes an explicit separation of the operations in the loop by their nature regarding the exploitation of the information contained in the model.

Deep Model Reflection pattern proposes a MDE approach to produce the run-time executable model used by the metacontroller by model-to-model transformation from the engineering model (see Figure 2 step 4). For that the run-time model must conform to a metamodel for which a transformation exists from the engineering model.

4.2 TOMASys

The TOMASys [8] is a metamodel that has been developed to specify the functional model of the system used by the metacontroller at runtime, according the Deep Model Reflection pattern. TOMASys metamodel includes elements for representing the structure and its relation to the functional requirements of any autonomous system (Figure 3). A TOMASys-based model of an autonomous system contains two kinds of elements:

- Elements that capture the instantaneous state of the system (i.e. run-time information): the hierarchy of objectives that represent the mission at runtime, the set of components and their connections, the set of function groundings, which specify run-time realizations of function design by assigning specific components to roles to address the run-time objectives.
- Elements that capture the knowledge about the design of the control system and the properties of its components: the set of functions or systems capabilities in relation to the mission and requirements, the set of component classes in the system, the set of function designs, which represent the functional design as specifications of components (roles) whose joint behavior renders a certain function.

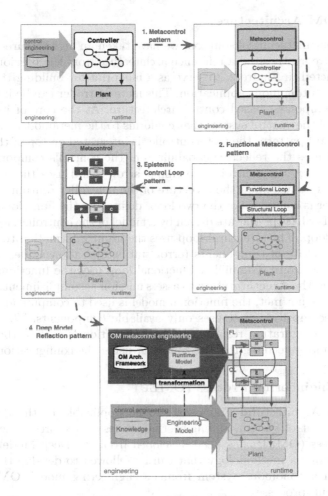

Fig. 2. The self-awareness patterns are combined resulting in the OM Architecture for metacontrollers and the OM Engineering Process to build them.

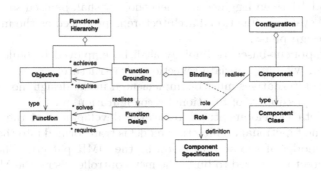

Fig. 3. Core elements in the TOMASys metamodel.

4.3   The OM Architecture

To develop our metacontrol solution, we have taken an architecture-driven approach. We have synthesized a reference architecture from the previous four patterns as depicted in Figure 2, to serve as a blueprint for building the metacontroller in any autonomous application. This metacontroller can be integrated on
top of any component-based control architecture. At the core of its operation
lies a model of the control system that conforms to the metamodel. According to
the Metacontrol pattern, the metacontroller is integrated on top of the standard
controller. It uses the reflective capabilities of the control's components to get
information about the control state (input arrow), and uses the exposed configuration API to act upon the structure of the controller (output arrow). The
metacontroller is composed of the two loops defined by the Functional Metacontrol pattern. Both loops operate driven by a model of the controller, as prescribed
by the ECL loop. The structural loop uses the input information to update the
instantaneous state of components (error status, operation mode, parameters)
and their connections. The higher functional loop uses the functional model of
the system and the structural state to assess mission accomplishment. If any mission objective is not met, the functional model is used to compute an alternative
functional decompsition that uses only available components. The alternative
components configuration resulting for this allocation is sent to the structural
loop, which finally reconfigures the controller using the configuration API.

4.4   OM Engineering Process (OMEP)

For the OM Architectural Framework to be applicable to the design of an
autonomous control system, an engineering process is required. The OM Engineering Process (OMEP) has been developed from the Deep Model Reflection
Pattern to define a methodology that can be followed to develop the metacontroller for an autonomous system from its engineering model. OMEP distin-
guishes two sub-processes:

1. OMEP Control Development: encompases the development of the control
   system, which shall meet the following requirements:
   - A Model-Driven Engineering methodology shall be used so that a model
     of the system's functional architecture is produced as the main result of
     the design phase.
   - A component-based technology shall be employed to build the control
     system, with appropriate reflection infrastructure.
   - Design alternatives must be implemented in the design model to provide
     for the possibility of functional reconfiguration.
2. OMEP Meta Development: addresses the development of the metacontrol
   system. The functional architecture model is transformed into the TOMASys
   run-time model of the system following the DMR pattern. The OM reference architecture is used to build the metacontroller using the Model-Driven
   Architecture process of progressive refinement up to the final implementation
   in the component platform of the control architecture.

5   Case Study: Mobile Robot

The OM Architectural Framework has been demonstrated by developing the control architecture of a mobile robot capable of autonomous navigation in the presence of disturbances. This application was selected because resilience issues are not yet fully solved in these applications, which involve sophisticated and heterogeneous components, both hardware and software. The application consists of a robot (a Pioneer 2AT8 platform instrumented with a laser sensor, a Kinect and a compass) navigating through an indoor environment to sequentially visit a number of waypoints in a given map. To implement the robot's control architecture the ROS navigation library and other ROS libraries for the sensory integration were used [11]. ROS is a component-based framework that exposes basic introspection and reconfiguration mechanisms as demanded by the Metacontrol pattern. Two failure scenarios were tested to validate the metacontrol: i) a temporary failure of the laser driver, ii) a permanent fault in the laser sensor. In both cases the metacontrol should be able to detect the problem, perform a functional assessment, and reconfigure the control in order to continue the mission.

5.1   Robot's Control Architecture Development

Following OMEP, a model-driven systems engineering process was used to develop the control architecture of the mobile robot and obtain the engineering model of the system from which to obtain the TOMASys functional model for the metacontroller. In this case we applied the ISE&PPOOA [6] process. The primary deliverable generated by the application of ISE&PPOOA process is the creation of the model of the functional architecture and its forward allocation to the robot physical architecture and its components, and its backward traceability to the capabilities related to its mission requirements, which included navigation and localisation amongst others. The functional architecture mappings include the contributions of all the components of the robot, including the hardware, the software and the personnel operating it. The main ISE&PPOOA process steps were followed:

Step 1. The system operational scenarios were specified, which included autonomous robot patrolling, autonomous navigation to a commanded location, and manual teleoperation.
Step 2. The system capabilities required were identified, which included the following: mission control, localization, teleoperation, navigation, motion and capabilities to acquire the different sensory inputs: ded reckoning about robot's motion, scans for localization landmarks, and 3D points information.
Step 3. The system functional architecture was created, to achieve the above capabilities. The key particularity in OMEP is that the functional architecture designed shall not be unique. Alternative functional breakdowns were defined for the range scan, dead reckoning, localisation and navigation capabilities, resulting in two alternative architectures (Figure 4 top).

Step 4. The system physical architecture was developed, representing the system decomposition into components and their allocation to the functional architecture alternatives, as schematized in Figure 4, where square shadow forms represent functions that fulfill capabilities and round shapes represent physical components.

5.2 Metacontrol Implementation

The metacontroller for the robot was built from the OM Architecture through a process of progressive platform-specific refinement, using the MDA model weaving pattern [2]. This way, the metacontroller that was finally obtained is applicable to any ROS-based system, and not just to our case study. Considering the MDA approach, TOMASys metamodel constitutes the computation-independent model and the OM Architecture the platform-independent model for our metacontrol solution. From them, we have developed libraries and components as platform-specific models of the OM metacontrol. Concretely, a Java library was built to provide a domain independent and multiplatform specific implementation of the OM Architecture. This library is the platform-specific model for Java as the metacontrol's implementation platform. Finally, a ROS library was developed to integrate the OMJava metacontroller in the ROS-based control architecture of the robot. It consist of ready-to-deploy metacontrol assets for any robotic application, and constitutes the platform-specific model for ROS as the platform for the complete control architecture.

Thanks to the application of the ECL pattern, which decouples the metacontroller from the run-time model that it uses, to implement the metacontroller for the patrolling robot we only needed to generate the TOMASys run-time model and deploy with the metacontrol implemented with the described libraries. This TOMASys model was obtained manually from the functional and structural atchitecture models generated with ISE&PPOOA in the development of the robot's control architecture.

5.3 Validation Scenarios

Let us analyze how the metacontrol operates using the robot's functional knowledge in each of the failure scenarios envisaged:

Failure Scenario 1: the Metacontroller detects the failure of the laser driver and repairs the component by re-launching the software process. Only the components loop intervenes, since the solution is achieved at the components level.

Failure Scenario 2: this time it is not possible to simply relaunch the laser driver because the error is due to a permanent fault in the laser device. The laser permanent failure scales-up to the functional loop. The impact of this failure on the robot objectives is assessed by bottom up propagation in the functional hierarchy. A new configuration is generated from the available functional alternatives and component classes (Figure 4). This new configuration uses the Kinect sensor instead of the laser. The new configuration is commanded to the structural

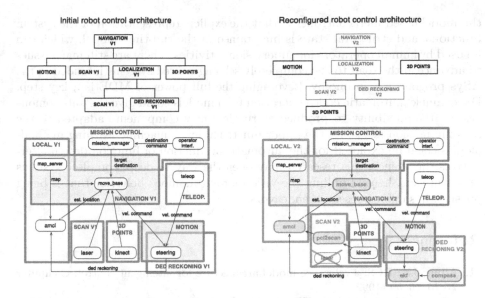

Fig. 4. The functional breakdown for the navigation capability and allocation of components in the functional architecture at runtime initially (left) and then after the laser permanent failure (right).

loop, which reconfigures the navigation system by deploying the new components needed, and re-connecting and re-parameterizing those that require it.

6   Concluding Remarks

The OM Architectural Framework is a tool to develop autonomous applications capable of run-time adaptation to unforeseen circumstances. Its novelty lies in the integration of two different approaches, MDE for leveraging engineering knowledge, and model-based cognition for intelligent behavior, to achieve self-aware and self-adaptation capabilities. These properties are provided by a metacontroller that operates at runtime as a resident engineering, using functional knowledge of the system to manage it. To materialize the solution we have synthesized a reference architecture using design patterns. It serves as a blueprint to build the metacontroller in any autonomous application. This patterned approach, involving progressive architectural refinement, allows to introduce improvements and novelties at any level, while still reusing the design assets of the framework. The OM Architectural Framerowrk offers more possibilities in addition to the construction of new autonomous applications. The OM metacontroller can also be added to the control architectures of extant autonomous systems. To implement it, appropriate reflection infrastructure to connect to the controller needs to be provided, and the TOMASys model of the system's architecture needs to be generated. Another important value of the model-based metacontroller is that the explicit representation of the system functional and structural state is maintained in the run-time model, which can be used by human operators for supervision activities. There are still many issues to advance in the road to the completely self-engineered systems intended in the ASys program for autonomy. Leveraging the full power of MDE is a key step. For example, automatic code generation from models could give the autonomous system the possibility to produce at runtime new components adapted to the ongoing mission needs. At the upper functional level, machine-learning methodologies could be explored for the automatic production of function design alternatives. In essence, this research may open the way to deep, complete systems self-awareness, hence fulfilling the vision for autonomous behavior that is being pursued in so many research endeavors.

References

1. Albus, J.S.: RCS: a reference model architecture for intelligent control. Computer **25**(5), 56–59 (1992)
2. Aßmann, U., Zschaler, S., Wagner, G.: Ontologies, meta-models, and the model-driven paradigm. In: Calero, C., Ruiz, F., Piattini, M. (eds.) Ontologies for Software Engineering and Technology. Springer (2006)
3. Blanke, M., Kinnaert, M., Lunze, J., Staroswiecki, M.: Diagnosis and Fault-Tolerant Control. Springer-Verlag, Berlin (2006)
4. Chittaro, L., Kumar, A.N.: Reasoning about function and its applications to engineering. Artificial Intelligence in Engineering **12**(4), 331–336 (1998)
5. DeLoach, S.A., Oyenan, W.H., Matson, E.T.: A capabilities-based model for adaptive organizations. Autonomous Agents and Multi-Agent Systems **16**, 13–56 (2008)
6. Fernandez-Sánchez, J.L.: ISE & Process Pipelines in OO Architectures (ISE&PPOOA). http://www.omgwiki.org/MBSE/doku.php?id=mbse:ppooa
7. Garlan, D.: Software architecture: a travelogue. In: Proceedings of the on Future of Software Engineering, FOSE 2014, pp. 29–39. ACM, New York (2014)
8. Hernández, C.: Model-based Self-awareness Patterns for Autonomy. Ph.D. thesis, Universidad Politécnica de Madrid, ETSII, Dpto. Automática, Ing. Electrónica e Informática Industrial, José Gutierrez Abascal 2, 28006 Madrid (SPAIN), October 2013
9. Hernández, C., Bermejo-Alonso, J., López, I., Sanz, R.: Three patterns for autonomous robot control architecting. In: Zimmermann, A. (ed.) The Fifth International Conferences on Pervasive Patterns and Applications, PATTERNS 2013, pp. 44–51. IARIA, May 27, 2013
10. Kephart, J., Chess, D.: The vision of autonomic computing. Computer **36**(1), 41–50 (2003)
11. Marder-Eppstein, E., Berger, E., Foote, T., Gerkey, B., Konolige, K.: The office marathon: robust navigation in an indoor office environment. In: 2010 IEEE International Conference on Robotics and Automation (ICRA), pp. 300–307, May 2010
12. Sanz, R., Rodríguez, M.: The ASys vision. Tech. Rep. ASLAB-R-2007-001, Autonomous System Laboratory, February 2007

Multi-core Control of the Switched Reluctance Drive with High-Performance Timing Analysis and Optimal Task Deployment

Ling Xiao, Gong Liang$^{(\boxtimes)}$, Wu Linlizi, Wang Bowen, Li Binchu,
Huang Yixiang, and Liu Chengliang

Department of Mechanical and Electronic Engineering,
Shanghai Jiao Tong University, Shanghai 200240, China
{lingxiao,gongliang_mi,wulinlizi,wangbowenmax,bcli,
huang.yixiang,chlliu}@sjtu.edu.cn

Abstract. The switched reluctance motor (SRM) is a promising direct drive unit for robot arms, however high performance control of the SRM faces the challenge of real-time constraint. A dual-core embedded control system, embracing a DSP ((Digital Signal Processor) and an FPGA (Field Programmable Gate Array) processor, is designed to reduce the computational burden. The hardware system is composed of three modules, i.e. the core processors module, the peripheral module and the drive module. The software system is correspondingly designed to match the hardware infrastructure to enhance the real-time performance. Timing analysis and task optimization show that the dual-core system is eligible for critical real-time drive applications.

Keywords: SRM · Dual-core · DSP · FPGA · Real-time analysis · Task deployment

1 Introduction

The switched reluctance motor (SRM) has received considerable attention for variable-speed drive applications in recent years, with the features of simple construction, rugged structure, robustness to operational conditions, and relatively high efficiency over a wide range of speed [1]. New structure of SRM has been designed for direct-drive applications [2]. Since a real-time control of rotational speed, current and position are needed, the real-time capability is one of primary criteria to evaluate a SRM control system. In practice, the MCU (Micro Controller Unit), FPGA, DSP are the three main choices for the SRM controller [3]. DSP has an obvious advantage over others at high-speed data processing and complex algorithm calculation. FPGA is prior to others for a high-speed parallel computing ability, rich digital interfaces and abundant on-chip storage resources and flexible programmability [4]. Considering the

G. Liang—This research was supported by the national natural science foundation of China under grant No. 11202125 and 51305258.

H. Liu et al. (Eds.): ICIRA 2015, Part I, LNAI 9244, pp. 655–666, 2015.
DOI: 10.1007/978-3-319-22879-2_59

advantages of each chip, a dual-core controller including DSP and FPGA is selected in this paper. DSP is treated as the primary controller to realize the high-level control algorithms, and FPGA as the auxiliary processor to process the peripheral sensor signals. The dual-core control system can realize advantageous complementarities [5], be suitable for modular design, and improve the efficiency of algorithm, hence provides an ideal platform for the real-time control of SRM. Software design for the dual-core controller is highlighted in the paper, as well as a real-time analysis of the system.

This paper is arranged as follows. Section 2 discusses the hardware structure of the control system.   Section 3 is about the realization of the software control system for SRM. Section 4 describes the real-time analysis of the system. Section 5 gives the conclusion.

2   The Hardware Structure of the System

The switched reluctance motor drive system consists of the motor, drive circuit, position detection circuit, current detector, and a dual-core controller. The hardware structure diagram is shown in Fig.1. The motor is a 4-phase 8/6 750W switched reluctance motor. The drive circuit uses asymmetric half bridge circuit power amplifier. Optocouplers combined with incremental decoder are adopted for position detection in the system. And Hall sensor is applied to measure current that flows through each stator winding. The main control chip DSP and auxiliary control chip FPGA used in the system are TMS320F2812 of TI® and EP2C8Q208C8 of Altera® respectively.

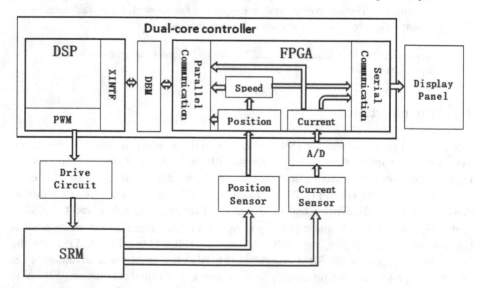

Fig. 1. SRM drive hardware

The SRD system processes as follows. When start the motor, firstly DSP chip carries out a series of initialization work, receives the current signal, position signal and speed signal from FPGA. According to these signals, DSP outputs PWM waveform to the drive circuit, and drive circuit supplies power to the motor, then the motor begins to spin. When the motor is rotating, based on the difference between measured speed signal and reference speed, a desired current derived through the control algorithm. Then detect the difference of expected current to the measured current to get a PWM waveform [6]. Outputs the PWM signal to drive circuit, and changes the average voltage loaded on winding to realize the speed regulation for SRM.

2.1 SRM Characteristic

The characteristic of the SRM used in this paper are given by Table 1.

Table 1. SRM parameters

number of phases, m	4
number of stator poles, N_S	8
number of rotor poles, N_R	6
max power	750W
rated speed	1500RPM
DC bus voltage	311V

2.2 Dual-Core Controller

Dual-core controller presented in this paper consists of DSP (TMS320F2812) and FPGA (EP2C8Q208C8). There are many communication mode between dual-core controller, such as UART, SPI, and Parallel mode [7]. To meet the demand of two-way data transmission and high speed requirements, this system select 16 bit parallel bus communication mode. TMS320F2812 has 16-bit bus XINTF module to connect to external memory. Considering FPGA as DSP's external memory, DSP communicates with FPGA by DSP's XINTF interfaces, which realize the data two-way communication between DSP and FPGA. The picture of this dual-core controller is showed in Fig. 2.

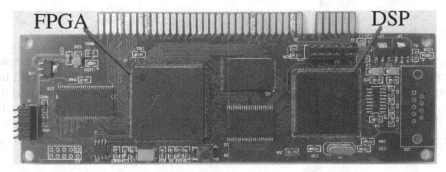

Fig. 2. Picture of the dual-core controller

FPGA Chip

FPGA has the rich on-chip resources and logical unit, users can customize the hardware circuit with FPGA. In this system, FPGA receives and pre-processes input control signals. Then it provide the feedback information of SRM to DSP, such as position, current and velocity signal. Besides, FPGA sends the operational parameters of SRM to the upper computer via UART and get target parameter from upper computer.

DSP Chip

TMS320F2812 has an EV (event manager) module which can produce up to 12 PWM outputs depending on GP (general timer) and compare registers, module by providing GP and compare registers. With DSP's characteristics of low power consumption, high performance processing capabilities, the control algorithm is realized in DSP, such as current controller and velocity controller.

2.3   Position (Velocity) Detection Module

Shaft position information is provided by the opto-coupler and incremental encoder. The opto-coupler detector uses a 6-slot, slotted disk connected to the rotor shaft and four opto-couplers mounted to the stator [8]. The opto-couplers located 67.5° apart from each other. This configuration generates an opto-coupler edge for every 7.5°. The incremental encoder used in the system has a revolution of 1024 which means it outputs 1024 edges for one circle.

2.4   Current Detection Circuit

The Hall current sensor has fine linearity, wide frequency response, high accuracy, good dynamic performance, anti-interference ability and current overload ability [9], this paper uses TBC15P to acquire four-phase winding currents of the SRM. Current signals are transferred to TLV2548 chip. TLV2548 is an A/D conversion chip, with 12 bits, multi-channel, small size, low power consumption, and high speed. The A/D output interface and controller interface chip are connected with FPGA. Finally, the converted digital signals are transferred to FPGA.

2.5   Power Electronics Hardware

Power electronics hardware included in drive circuits has numerous options in practical application [10]. And invariably the decision will come down to trading off the cost of the driver components against having enough control capability built into the driver. A popular configuration, and the one used in this article, asymmetrical half-bridge power inverter is depicted in Fig.5. With each phase is independent of each other, this structure can meet the requirements of multi-parameterized control of the system, and to ensure that the system is stable and reliable.

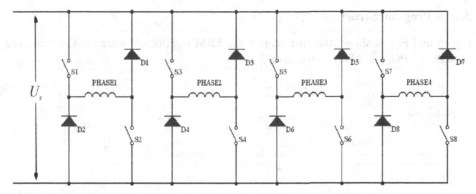

Fig. 3. Schematic diagram of asymmetrical half-bridge power inverter

3   Software Control System

The software described in this application report is written in C (for DSP) and Verilog (for FPGA) and is designed for operating the 4-phase 8/6 SRM in closed loop current control and closed loop speed control. A block diagram of the algorithms implemented is given in Fig. 4.

Velocity is estimated in FPGA by monitoring the elapsed time between position sensor edges, which are a known distance apart. A velocity compensation algorithm determines the torque required to bring the motor velocity to the commanded value. A commutation algorithm executed by DSP converts the torque command into a set of reference phase currents, and the current in each phase is individually regulated using a fixed-frequency PWM scheme. Further details on each of the algorithms are provided in subsequent sections of this paper.

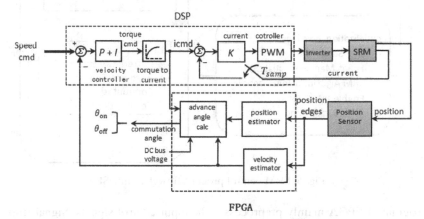

Fig. 4. Block diagram of the SRM Controller

3.1   Program Structure

Fig. 5 and Fig. 6 shows the structure of the SRM control software for the dual-core system in FPGA and DSP respectively.

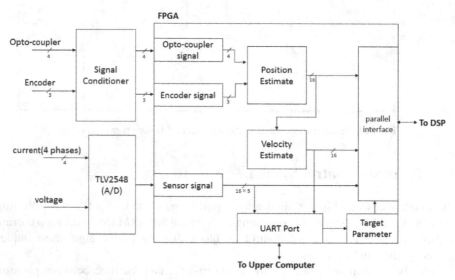

Fig. 5. SRM Control Program Structure in FPGA

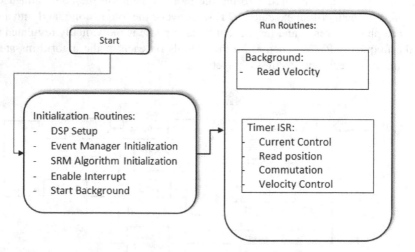

Fig. 6. SRM control program flowchart in DSP

Program in FPGA mainly preprocesses the input control signals. Signal from current and voltage sensor are conditioned and transformed to digital signal, and sent to FPGA in serial. Signals from Opto-coupler, Encoder are conditioned by hardware, and sent into FPGA. They are the basis of rotor Position Estimation and Velocity Estimation, which are vital feedback information for SRM closed-loop control. FPGA estimates the real-time position, rotate speed, and send all these feedback, as well as four phase current and voltage, to DSP via a parallel interface [11]. What's more, rotate speed and current are sent to upper computer via a UART port.

When DSP is powered on, initialization routines are executed, including DSP setup, event manager initialization and so on. In the run routines, DSP read velocity and position value from FPGA directly, which save resources and time to accomplish complex control algorithm.

3.2   A/D Controller

Signals collected by sensors are usually analog signal. A/D conversion module is necessary to convert the analog signals to digital ones. TLV2548 is used in this system to realize A/D conversion. Depending on its sequential logic, an A/D controller is customized by FPGA. Digital signal after the conversion is transferred to FPGA through SPI interface. With the FPGA's clk frequency at 50MHz, and conversion clk at 5MHz, it needs 94.4us to finish the conversion and transmission. Control signal monitored by oscilloscope is depicted in Fig. 7. Image marked by two black lines is a conversion and transmission process. Meaning of signals showed in the Fig. 7 are as follows: cs is chip select signal, cstart controls the start of A/D conversion process and sclk is the clock signal both for conversion and transmission process.

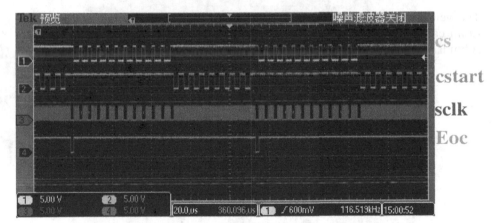

Fig. 7. Sequence diagram of A/D control signal

3.3   Position Estimation

As described in section 2.3, encoder provides increment of position counter, and opto-coupler provides the absolute position, which can be used for calibrating the accumulative position errors. Fig.8 shows principle of encoder and Fig.9 shows principle of opto-coupler.

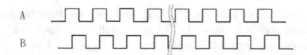

Fig. 8. Principle of incremental encoder

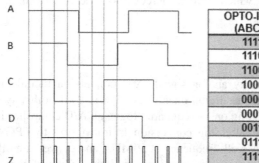

OPTO-Input (ABCD)	Mechanical Angle	Position Counter
1111	0	0
1110	7.5	85
1100	15	170
1000	22.5	256
0000	30	341
0001	37.5	426
0011	45	512
0111	52.5	597
1111	60	683

Fig. 9. Principle of opto-coupler

3.4   Velocity Estimation

The velocity is estimated based on the result of position estimation, with the algorithm called "Dynamic velocity point choosing based MT method" [12]. The algorithm is illustrated in Fig.10.

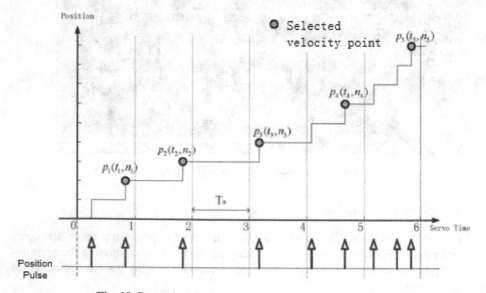

Fig. 10. Dynamic velocity point choosing based MT method

In the algorithm, a fixed servo time T_s is chosen. The arriving time and position counter of the last position pulse before each T_s ends is recorded, which are represented as $p_i(t_i,n_i)$ in Fig.10. The average velocity of the latest servo time is:

$$v_{avg} = \frac{K(n_i - n_{i-1})}{t_i - t_{i-1}}$$

where K is constant related to clock frequency and encoder resolution.

3.5   Current Controller

Current is regulated by a fixed-frequency PWM-signals with varying duty cycles [13]. The duty cycle is calculated by the current loop compensation algorithm. The model of current loop is given in Fig.11.

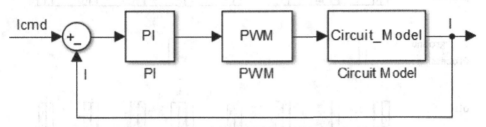

Fig. 11. Current control loop for SRM

3.6   Velocity Controller

Velocity is calculated in a closed-loop manner by comparing the desired shaft speed with to the calculated velocity and the compensating the error [14]. A PI controller is used in this paper to compensate speed error.

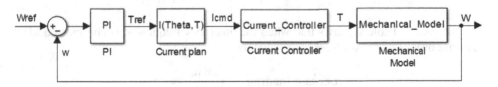

Fig. 12. Velocity loop for SRM

4   Program Timeline of the DSP Processor

At the highest level, the software consists of initialization routines and run routines. Upon completion of the necessary initialization, the background task is started as a lower priority processing. The background is simply an infinite loop, which includes the velocity estimation and the reference shaft speed acquiring. It costs DSP about 96 cycles to get the velocity estimation result from FPGA. All of the time critical motor control processing is done via timer interrupt service routines. The timer ISR is executed at each occurrence of the maskable CPU interrupt INT2. The shaft position is estimated by FPGA and the result is send to DSP at first in the interrupt service routines. The software decides which phase to turn on by the real-time feedback of the position. Besides the position estimation and control, a main part of SRM control algorithms which is implemented during the timer ISR are the current control. As illustrated in Fig.13, the current control and shaft position estimation are executed ate the frequency of 5KHz. Because of their lower bandwidth requirements, velocity estimation and velocity control are performed at a frequency of 1.25kHz.

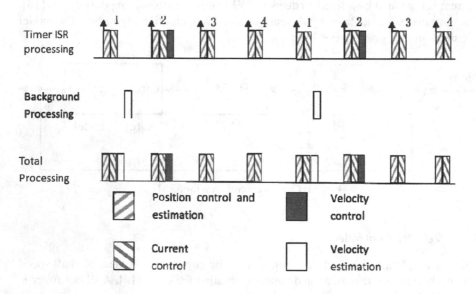

Fig. 13. Processor timeline showing typical loading and execution of SRM control algorithm

The data in Table 2 shows that when the timer ISR frequency, is chosen as 5 kHz, that the overall processor loading is equal to

$$\text{processor loading} = \frac{18 \ \mu s}{200.0 \ \mu s} = 9\%$$

The low-level data-acquiring and preprocessing routines are deployed in FPGA core, taking advantages of its extended input/output ports. Given 150MHz DSP frequency, the timing analysis shows that the proposed dual-core framework support the high speed control up to10000 RPM.

Table 2. Benchmark data for the various SRM drive software modules

S/W Block	Module	Number of Cycles	Execution Times @6.6ns	Execution Frequency	Relative Time @5kHz
Position Estimation and Control	Timer ISR	802	5294ns	5kHz	5294 ns
Current Control	Timer ISR	1642	10837 ns	5kHz	10837 ns
Velocity Control	Timer ISR	1022	6745ns	1.25kHz	1686 ns
Velocity Estimation	Background	96	634 ns	1.25kHz	160 ns
total					17977 ns

5   Conclusions

In traditional DSP-based control system, there are multiple interrupts to execute the program. It is hard to realize real-time control algorithm in a deterministic way due to the high complexity. The dual-core control system described in the paper outperforms other single-core systems in view of its high real-time performance. In the system, FPGA is used for the front-end signal processing, and DSP is used for the advanced control algorithm, which shares the computational load at different real-time scale in parallel.

References

1. Rallabandi, V., Fernandes, B.G.: Design procedure of segmented rotor switched reluctance motor for direct drive applications. Electric Power Applications, IET **8**(3), 77–88 (2014)
2. Krishnamurthy, M., Edrington, C.S., Fahimi, B.: Prediction of rotor position at standstill and rotating shaft conditions in switched reluctance machines. IEEE Trans. Power Electron. **21**(1), 225–233 (2006)
3. Cadambi, S., Durdanovic, I., Jakkula, V., Sankaradass, M., Cosatto, E., Chakradhar, S., Graf, H.P.: A massively parallel FPGA-based coprocessor for supportvector machines. In: Field Programmable Custom Computing Machines, pp. 115–122 (2009)
4. Zheng, Y., Zhao, D.: Study on Operation of Switched Reluctance Motor for Electric Vehicles Based on FPGA. Information Management, Innovation Management and Industrial Engineering **2**, 512–516 (2009)
5. Guojun, Y., He, C., Yingjie, H.: Research on the switched reluctance machine drive system for electric vehicles. In: Electrical Machines and Systems (ICEMS), pp. 192–197 (2014)
6. Stumpf, A., Elton, D., Devlin, J., Lovatt, H.: Benefits of an FPGA based SRM controller. In: Industrial Electronics and Applications (ICIEA), pp. 12–17 (2014)
7. Chang, T., Juang, K.-C.: CMOS SC-spinning, current-feedback hall sensor for high speed and low cost applications. In: SENSORS, pp. 527–530 (2014)

8. Hamouda, M., Blanchette, H.F., Al-Haddad, K., Fnaiech, F.: An Efficient DSP-FPGA-Based Real-Time Implementation Method of SVM Algorithms for an Indirect Matrix Converter. Industrial Electronics **58**(11), 5024–5031 (2011)
9. Tian, J., Petzoldt, J., Reimann, T., Scherf, M., Berger, G.: Control system analysis and design of a resonant inverter with the variable frequency variable duty cycle scheme. In: Power Electronics Specialists Conference, PESC 2006, pp. 1-5 (2006)
10. Cai, J., Deng, Z.: Switched-Reluctance Position Sensor **50**(11) (2014)
11. Fleming, F., Edrington, C.S: A comparison of machine modeling methods for real-time applications. In: IECON 2012-38th Annual Conference on IEEE Industrial Electronics Society, pp. 5346–5351 (2012)
12. Fang, W., Shanming, W., Shenghua, H., Yongjun, C.: Study on speed detection and control method of PMSM under ultra-low speed. In: Universities Power Engineering Conference, pp. 178–183 (2007)
13. Maswood, A.I., Al-Ammar, E.: Analysis of a PWM voltage source inverter with PI controller under non-ideal conditions IPEC. In: 2010 Conference Proceedings, pp. 193-198 (2010)
14. Ouddah, N., Boukhnifer, M., Chaibet, A., Monmasson, E.: Robust controller designs of switched reluctance motor for electrical vehicle. In: Control and Automation (MED), pp. 212–217 (2014)

A Position Domain Cross-Coupled Iteration Learning Control for Contour Tracking in Multi-axis Precision Motion Control Systems

Jie Ling, Zhao Feng, and Xiaohui Xiao[✉]

School of Power and Mechanical Engineering, Wuhan University, Wuhan 430072, China
{jamesling,fengzhaozhao7,xhxiao}@whu.edu.cn

Abstract. A novel cross-coupled iteration learning controller in position domain is presented to improve contour tracking performance for multi-axis micro systems executing repetitive tasks. The position domain iteration learning control (PDILC) is combined with position domain cross-coupled control (PDCCC) to develop a position domain cross-coupled iteration learning control (PDCCILC). The stability and performance analysis are given based on lifted system representation in time domain. To illustrate effectiveness and good tracking performance of the proposed control method, simulation studies are conducted based on an identified model of a three dimensional micro-motion stage.

Keywords: Position domain · CCILC · Contour tracking · Multi-axis motion · Precision motion control

1 Introduction

High precision manufacturing processes and micromanipulation have produced a need for increased research in precision motion control (PMC). In multi-axis PMC system like 3D printer, micro-assembling, nano-lithography and so on, contour tracking is one of the crucial control problems. Many control strategies have been developed to improve the tracking performance of each individual axis motion, such as proportional integral derivative (PID) controller [1], robust control [2], sliding-mode control [3], iterative control [4], repetitive control [5,6], polynomial-based pole placement control [7] and so on. For decoupled multi-input multi-output (MIMO) control systems, a traditional control regards MIMO system as many single-input single-output (SISO) systems designed each axis separately regardless of other axes. However, a good tracking performance for each individual axis does not guarantee the reduction of contour errors for a multi-axis motion system, as poor synchronization of relevant motion axes may result in diminished accuracy of the contour tracking performance [8].

Contour error is defined as an orthogonal component of the derivation of an actual contour from the desired one [8]. To improve the contour performance, cross-coupling control (CCC) was developed by Koren [9]. CCC utilizes coupling gains to

© Springer International Publishing Switzerland 2015
H. Liu et al. (Eds.): ICIRA 2015, Part I, LNAI 9244, pp. 667–679, 2015.
DOI: 10.1007/978-3-319-22879-2_60

couple the individual axis errors of SISO systems together and applies a controller to the combined signal. CCC has been used in multi-axis motions for manufacturing, especially in computerized numerical control (CNC). It should be mentioned that, in CCC, each axis still needs to be controlled, and there still have some tracking errors for each individual axis motion which will affect the final contour tracking performance [8]. Another method to decrease contour errors is event-driven control [10], which triggers the controller to update a control action through an event, such as a new measurement or distance information. The major drawback of event-driven control is the difficulty in developing system theory and performance sacrifice. Besides, Ouyang proposed a novel PID feedback controller based on the position domain (PDPID) which perceives motion system as a master-slave cooperative system to guarantee synchronization and improve the contour tracking performance. The method has been applied on CNC [11] and robotic system [12]. However, for those manufacturing systems to perform repetitive task, the feedback controller alone cannot decrease the repetitive contour error [13].

To improve repetitive performance, iterative learning control (ILC) can be implemented because of the repetitive feature. ILC allows the controller to learn from previous executions (trials, iterations, passes) to achieve better performance [14]. Barton [13], [15,16] combined cross-coupled iterative learning control (CCILC) with individual axis ILC to improve both individual axis and contour tracking performance. However, for multi-axis systems, no matter CCILC or individual axis ILC, the controllers are designed in time domain where reference trajectories are designed as a function of time, which may not achieve good motion synchronization.

Therefore the main goal of this paper is to provide a new method for multi-axis PMC to improve repetitive contour tracking performance by introducing position domain method into CCILC design. The novel position domain CCILC (PDCCILC) controller is advantageous on maintaining multi-axis synchronization with reducing individual axis and contour error simultaneously when compared with existing time domain CCILC.

The outline of this paper is as follows. Section II gives a brief introduction of the controller design background about ILC, CCC and PDC. In section III, PDCCILC control law is proposed and analyzed. The simulation results and comparison between time domain CCILC (TDCCILC) and PDCCILC are presented in Section IV. Conclusions are given in Section V.

2   Controller Design Background

Cross-coupled iteration learning control is a combination of traditional feedback CCC and feedforward ILC. Its main advantage is improving contour tracking performance in multi axis precision motion systems. Before discussing position domain CCILC, the following sections briefly introduce ILC, CCC and PDC.

2.1   ILC

ILC is firstly proposed by Uchiyama in 1978 [17] and widely discussed in [14], [18] and so on. As an intelligence algorithm, it is advantageous on achieve high tracking precision when a system executes repetitive tasks.

Considering a discrete LTI and SISO system

$$Y_j(z)=P(z)U_j(z)+D(z) \tag{1}$$

where z stands for the z-transformation of a system, j is the iteration index, $Y_j(z)$ is the output, $U_j(z)$ is the control signal, $D(z)$ is the exogenous signal and $P(z)$ is the transfer function of system.

A widely used control law of ILC is Eq.(2) and the ILC system is asymptotically stable (AS) if Eq.(3) can be satisfied.

$$U_{j+1}(z)=Q(z)[U_j(z)+L(z)E_j(z)] \tag{2}$$

$$\left\|Q(z)[1-L(z)P(z)]\right\|_\infty<1 \tag{3}$$

where $Q(z)$ is a filter, $L(z)$ is learning function, tracking error $E_j(z)=Y_d(z)-Y_j(z)$ ($Y_d(z)$is the desired output), and $\|\cdot\|_\infty$ is the infinite norm of the matrix.

2.2   CCC and CCILC

In some multi-axis systems, prime concern should be emphasized in contour error rather than separate axis tracking error [19]. Cross coupled control is a technique to reduce contour error by choosing appropriate coupling gains and coordinating the motion of two coupled axis.

Determining coupling gains is vital in CCC as they are used to calculate contour error and allocated control signal to individual axis. In linear XY plane contour tracking, contour error ε is defined as Eq.(4)

$$\varepsilon= -C_x e_x+C_y e_y \tag{4}$$

where $C_x=\sin\theta$, $C_y=\cos\theta$, θ is the angle between the x-axis and the desired linear trajectory, e_x and e_y are x-axis and y-axis tracking error. For a circular contour tracking, $C_x=\sin\theta-e_x/2R$, $C_y=\cos\theta+e_x/2R$, where R is the radius and θ is angle between the x-axis and the tangent of the desired tracking point in the circle. A complex contour can be achieved by combining a series of linear and circular parts together.

When associating ILC with CCC, the CCILC algorithm is approached. The general CCILC control structure is presented in Fig. 1 [13],[15]. A novel control law combining individual axis ILC algorithm for x-axis and y-axis with CCILC law is given in [16] as

$$\begin{bmatrix}U_x\\U_y\end{bmatrix}_{j+1}=Q\left(\begin{bmatrix}U_x\\U_y\end{bmatrix}_j+\begin{bmatrix}L_x & 0 & -C_x L_c\\0 & L_y & C_y L_c\end{bmatrix}\begin{bmatrix}E_x\\E_y\\E_c\end{bmatrix}_j\right) \tag{5}$$

where U, L, E, and C are control signal, learning function, individual axis tracking error and coupling gain matrix respectively, x, y and c represent x-axis, y-axis and contour.

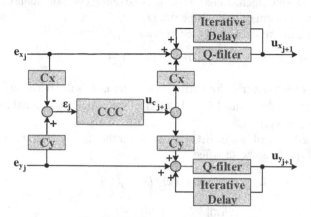

Fig. 1. CCILC control structure

2.3   PDC

Being different from the aforementioned control methods implemented in time domain, a position domain controller focuses on improving synchronization of relevant individual axis motion in multi-axis systems. A PD-type contour tracking control law established in position domain were proposed in [20].

In a two DOF decoupled parallel motion system, a PD-type feedback control signal $U_y(x)$ of y-axis (slave motion) in position domain is related to x-axis position (master motion), which can be expressed as

$$U_y(x)=K_{py}e_y(x)+K_{dy}e_y'(x) \tag{6}$$

where K_{py} and K_{dy} are proportional gain and differential gain, $e_y(x)=y_d(x)-y(x)$, $e_y'(x)=y_d'(x)-y'(x)$. It can be seen that position domain PD law uses x-axis position as a reference rather than time. Then, to achieve accurate contour performance, a high precision measurement is required in the master motion direction.

Convert Eq.(6) to time domain Eq.(8) using Eq.(7).

$$\dot{y}(t)=dy/dt=(dy/dx)\cdot(dx/dt)=y'(x)\dot{x}(t) \tag{7}$$

$$U_y(t)=K_{py}(y_d(t)-y(t))+K_{dy}(\dot{y}_d(t)-\dot{y}(t))/\dot{x}(t) \tag{8}$$

3   Position Domain CCILC

We here provide a novel method for multi-axis manufacturing systems to improve repetitive contour tracking performance by introducing position domain method into CCILC design.

3.1   Control Law

Considering a two input two output, linear time invariant (LTI) system, the PDCCILC control signal of x-axis and y-axis can be given as

$$u_{x_{j+1}}(t) = Q[u_x(t) + L_x e_x(t) - C_x L_\varepsilon \varepsilon(x)]_j \tag{9}$$

$$u_{y_{j+1}}(x) = Q[u_y(x) + L_y e_y(x) + C_y L_\varepsilon \varepsilon(x)]_j \tag{10}$$

where j is the iteration index, u is the control signal, Q is a filter, L is learning function and ε is contour error. Eq.(9) is a time domain law except the last term, which can be regard as a sacrifice of x-axis accuracy for contour tracking performance. We here mainly discuss the control law of y-axis as it is completely in position domain.

Applying PID type ILC and PD type CCC, Eq.(10) then can be written as

$$u_{y_{j+1}}(x) = Q[u_y(x) + K_{py} e_y(x) + K_{iy} \int_0^{\Delta s} e_y(x)dx + K_{dy} e_y{'}(x)$$

$$+ C_y(K_p^{xy} a_\upsilon(x) + K_d^{xy} a_\upsilon{'}(x))]_j \tag{11}$$

3.2   Stability

For linear trajectory, substituting Eq.(4) into Eq.(11), yielding Eq.(12).

$$u_{y_{j+1}}(x) = Q[u_y(x) + K_{py} e_y(x) + K_{iy} \int_0^{\Delta s} e_y(x)dx + K_{dy} e_y{'}(x)$$

$$+ C_y K_p^{xy} C_y e_y(x) - C_y K_p^{xy} C_x e_x(x) + C_y K_d^{xy} C_y e_y{'}(x) - C_y K_d^{xy} C_x e_x{'}(x)]_j \tag{12}$$

Convert Eq. (12) to time domain Eq. (13).

$$u_{y_{j+1}}(t) = Q[u_y(t) + K_{py} e_y(t) + K_{iy} \int_0^{\Delta t} e_y(t)\dot{x}(t)dt + K_{dy}(\dot{e}_y(t)/\dot{x}(t))$$

$$+ C_y K_p^{xy} C_y e_y(t) - C_y K_p^{xy} C_x e_x(t) + C_y K_d^{xy} C_y(\dot{e}_y(t)/\dot{x}(t)) - C_y K_d^{xy} C_x(\dot{e}_x(t)/\dot{x}(t))]_j \tag{13}$$

According to Eq. (13), numerical error of control signal may occur when x-axis under a low speed because the denominator $\dot{x}(t)$. To avoid this problem, an increment of x-axis position Δx in time Δt is applied.

$$\int_0^{\Delta t} e_y(t)\dot{x}(t)dt = \Delta x(e_y(t) + e_y(t - \Delta t))/2 \tag{14}$$

$$e_y'(t)/\dot{x}(t)= (e_y(t)-e_y(t-\Delta t))/\Delta x \qquad (15)$$

$$e_x'(t)/\dot{x}(t)= (e_x(t)-e_x(t-\Delta t))/\Delta x \qquad (16)$$

Substituting Eq.(14) - (16) into Eq.(13) yielding,

$$u_{y\,j+1}(t)=Q[u_y(t)+\alpha\cdot e_y(t)+\beta\cdot e_y(t-\Delta t)+\gamma]_j \qquad (17)$$

where

$$\alpha= K_{py}+K_{iy}\Delta x/2+K_{dy}/\Delta x+C_y^2 K_d^{xy}/\Delta x \qquad (18)$$

$$\beta=K_{iy}\Delta x/2-K_{dy}/\Delta x-C_y^2 K_d^{xy}/\Delta x \qquad (19)$$

$$\gamma=-C_y K_d^{xy} C_x(e_x(t)-e_x(t-\Delta t))/\Delta x-C_y K_p^{xy} C_x e_x(t) \qquad (20)$$

According to Eq. (17), there is time delay in control law. To analysis the definitive stability and convergence in time domain, the lifted-system representation is applied.

$$y_j(k)=P(q)u_j(k)+d(k) \qquad (21)$$

$$e_j(k)=y_d(k)-y_j(k) \qquad (22)$$

where k is the discrete time index, q is the forward time-shift operator ($qx(k)\equiv x(k+1)$). Using Eq.(21) and Eq.(22), Eq.(17) can be written as Eq.(23).

$$u_{y\,j+1}(k)=Q[(I-\alpha P_y-\beta P_y')u_{y\,j}(k)+(\alpha+\beta q^{-\Delta k})(y_d(k)-d(k))+\gamma'] \qquad (23)$$

where $P_y'(q)=q^{-\Delta k}\cdot P_y(q)$, $q^{-\Delta k}$ stands for time delay with discrete step length of Δk. γ' is independent of convergence of y-axis control signal because it is a term related to e_x rather than u_j.

Then, system is asymptotically stable (**AS**) if there exists appropriate parameters α and β satisfying

$$\rho(Q(I-\alpha P_y-\beta P_y'))<1 \qquad (24)$$

where ρ is the spectral radius of the matrix. The parameter α and β are bounded on condition that the trajectory and motion of x-axis are planned appropriate with the position increment Δx bounded.

3.3   Performance

If the AS condition Eq.(24) is satisfied, the performance of system under this law is based on the asymptotic value of the error.

$$u_{y\infty}(k)=Q[(\alpha+\beta q^{-\Delta k})(y_d(k)-d(k))+\gamma']/(I-Q(I-\alpha P_y-\beta P_y')) \qquad (25)$$

$$e_{y\infty}(k) = lim_{j\to\infty} (y_d(k)\text{-}P_y u_{y_j}(k)\text{-}d(k))$$

$$=y_d(k)\text{-}d(k)\text{-}P_y\cdot Q[(\alpha+\beta q^{-\Delta k})\cdot(y_d(k)\text{-}d(k))+\gamma']/(I\text{-}Q(I\text{-}\alpha P_y\text{-}\beta P_y')) \qquad (26)$$

where Q filter is designed to determine which frequencies are emphasized in the learning function. Term γ' indicates that tracking performance in y-axis is associated with x-axis position accuracy. The contour error ε can be calculated using Eq.(4).

4   Evaluation

4.1   Micro-motion Stage

For this paper, three-dimensional translational DOF micro-motion stage was used as a control case of multi-axis micromanipulator contour tracking. As shown in Fig. 2, each motion direction of the stage is driven by a piezoelectric actuator, and the motion is transferred by a compliant joint to the moving platform where an end-effector can be set up.

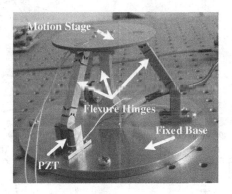

Fig. 2. Micro-motion stage

4.2   Control Structure

The control structure in this case was built as Fig. 3, a feedforward PDCCILC controller was adopted to decrease individual axis and contour errors, combining with a feedback H∞ controller to improve system robustness.

4.3   Simulation Cases and Results

In this part, comparison studies between time domain ILC (TDILC) and position domain ILC (PDILC), time domain CCILC (TDCCILC) and position domain CCILC (PDCCILC) were made to verify the effectiveness of position domain design method and the proposed PDCCILC controller.

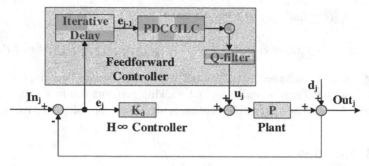

Fig. 3. Feedback and feedforward control structure

Two types of three dimension reference trajectories shown in Fig. 4 (zigzag motion and quadrangle motion) were adopted as examples to show contour tracking performance in linear motions. In ILC iterations, the initial conditions should be set the same, so the red marked point in Fig. 4 is set as the home position.

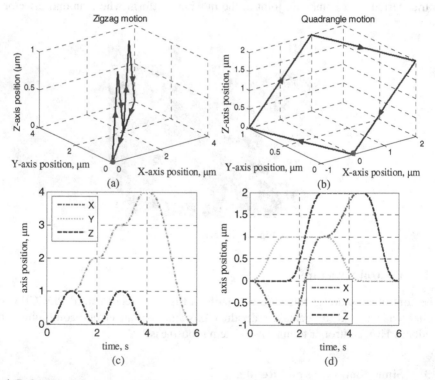

Fig. 4. Reference trajectories. (a) Zigzag trajectory. (b) Quadrangle trajectory. (c) Zigzag axis motion versus time. (d) Quadrangle axis motion versus time.

Base on system AS condition in Eq. (24), the PID-type ILC gains and PD-type CCC gains were selected. For comparison, all of the gains for the four type controllers were set the same (shown in Tab.1) and iteration was set 300 times.

Table 1. Control gains used in four controllers

	Axis gain in ILC			Contour gain in CCC	
Reference	K_p	K_i	K_d	K_p	K_d
Zigzag motion	1	0.3	0.3	0.4	0.4
Quadrangle motion	0.3	2	0.3	0.4	0.4

Zigzag Motion Tracking Case

As the x-axis and y-axis trajectory is the same, we here choose x-z plane to evaluate contour tracking performance for aforementioned four controllers. Fig. 5 shows MAX contour errors of x-z plane in iteration process. Errors under the four controllers are uniformly convergent from the 100th iteration to the last, which indicates the asymptotic stable condition is satisfied (Eq.(24) for the proposed PDCCILC).

Compared with time domain controllers, position domain controllers are superior to time domain controllers under same parameters and external disturbance.

For CCILC, the maximum steady error of PDCCILC is smaller (about 1.53nm), which is a 97% decrease of the initial error (58.36nm) and is 13% of MAX contour error of TDCCILC (11.9 nm).

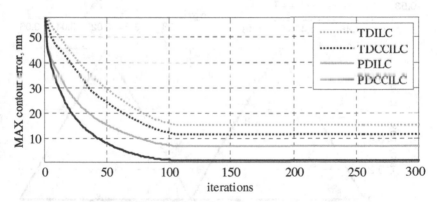

Fig. 5. MAX error of x-z plane zigzag contour tracking

Contour tracking in x-z plane for the four controllers is shown in Fig. 6. Part a and part b are respectively the amplified linear and angular contour of reference. In both parts, positon domain controllers prove to be better in following reference than time domain ones, and PDCCILC is better than TDCCILC, which keeps correspondence with analysis in Fig. 5. More specific supporting data about MAX and root mean square (RMS) contour errors can be found in Tab.2.

Table 2. MAX and RMS contour errors of the last iteration

Controller		Zigzag Motion (nm)			Quadrangle Motion (nm)		
		XY	XZ	YZ	XY	XZ	YZ
TDILC	RMS	0.01	6.21	6.21	3.41	9.02	7.58
	MAX	0.06	15.49	15.49	6.89	19.5	19.5
TDCCILC	RMS	0.01	1.43	1.43	1.62	3.77	3.26
	MAX	0.04	11.90	11.90	4.03	8.50	8.50
PDILC	RMS	0.00	0.11	0.11	1.60	1.97	2.23
	MAX	0.01	7.11	7.11	6.24	4.91	3.58
PDCCILC	RMS	0.00	0.02	0.02	0.86	1.02	1.17
	MAX	0.00	1.53	1.53	2.20	2.88	1.84

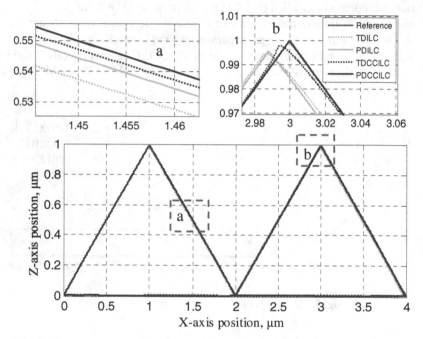

Fig. 6. Zigzag contour tracking in x-z plane of four controllers in the last iteration

Quadrangle Motion Tracking Case

In this example, 3-dimentional quadrangle motion was used to evaluate the four controllers. Fig. 7 shows the MAX contour error in iteration process. Fig. 10 shows contour tracking in x-z plane for the four controllers. (The x-y and y-z plane is omitted for brief.)

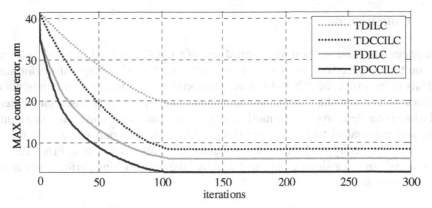

Fig. 7. MAX error of x-z plane quadrangle contour tracking

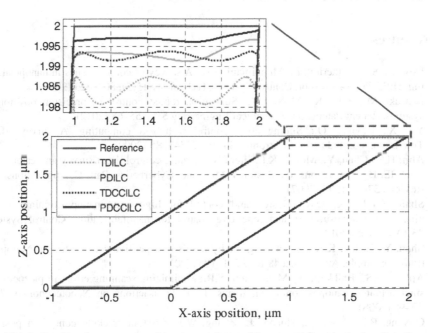

Fig. 8. Quadrangle contour tracking in x-z plane of four controllers in the last iteration

In Fig. 7, PDILC achieves better performance than TDILC (maximum steady error 4.91nm versus 19.5nm). On the other hand, MAX error under PDCCILC control is decreased by 93% from 41.6nm and keeps steady at 2.88nm, which is 34% of TDCCILC. Fig. 8 demonstrates the best contour tracking performance of the proposed PDCCILC, which is in agreement with Fig. 5 and Fig. 6.

5   Conclusions

This paper has presented a novel control law of CCILC designed in position domain for multi-axis precision motion control systems. The stability and performance analysis of the proposed PDCCILC were conducted using lifted system representation method. To evaluate the control law, simulations were performed based on an identified model of a three axis micro-motion stage. Four controllers under the same control gains and external disturbance were designed to make comparisons and to demonstrate the superiority of PDCCILC. Simulation results proved that PDCCILC decreased contour error significantly and achieved the best tracking performance among the four controllers.

Acknowledgment. This research is sponsored by National Natural Science Foundation of China (NSFC, Grant No.51375349 and No.51175383).

References

1. Devasia, S., Eleftheriou, E., Moheimani, S.R.: A survey of control issues in nanopositioning. IEEE Transactions on Control Systems Technology **15**(5), 802–823 (2007)
2. Iwasaki, M., Seki, K., Maeda, Y.: Survey on robust control of precision positioning systems. Recent Patents on Mechanical Engineering **5**(1), 55–68 (2012)
3. Yu, X., Kaynak, O.: Sliding-mode control with soft computing: A survey. IEEE Transactions on Industrial Electronics **56**(9), 3275–3285 (2009)
4. Ahn, H.S., Chen, Y., Moore, K.L.: Iterative learning control: brief survey and categorization. IEEE Transactions on Systems Man and Cybernetics Part C Applications and Reviews **37**(6), 1099 (2007)
5. Shan, Y., Leang, K.K.: Design and control for high-speed nanopositioning: serial-kinematic nanopositioners and repetitive control for nanofabrication. Control Systems **33**(6), 86–105 (2013)
6. Shan, Y., Leang, K.K.: Accounting for hysteresis in repetitive control design: Nanopositioning example. Automatica **48**(8), 1751–1758 (2012)
7. Aphale, S.S., Bhikkaji, B., Moheimani, S.R.: Minimizing scanning errors in piezoelectric stack-actuated nanopositioning platforms. IEEE Transactions on Nanotechnology **7**(1), 79–90 (2008)
8. Ouyang, P.R., Dam, T., Huang, J., Zhang, W.J.: Contour tracking control in position domain. Mechatronics **22**(7), 934–944 (2012)
9. Koren, Y.: Cross-coupled biaxial computer control for manufacturing systems. Journal of Dynamic Systems, Measurement, and Control **102**(4), 265–272 (1980)
10. Heemels, W.P.M.H., Sandee, J.H., Van Den Bosch, P.P.J.: Analysis of event-driven controllers for linear systems. International Journal of Control **81**(4), 571–590 (2008)
11. Dam, T., Ouyang, P.R.: Contour tracking control in position domain for CNC machines. In: IEEE International Conference on Information and Automation, Shenzhen, China (2011)
12. Ouyang, P.R., Pano, V., Acob, J.: Position domain contour control for multi-DOF robotic system. Mechatronics **23**(8), 1061–1071 (2013)
13. Barton, K., Alleyne, A.G.: Cross-coupled iterative learning control: design and implementation. In: Proceedings of IMECE (2006)

14. Bristow, D.A., Tharayil, M., Alleyne, A.G.: A survey of iterative learning control. Control Systems **26**(3), 96–114 (2006)
15. Barton, K.L., Alleyne, A.G.: A cross-coupled iterative learning control design for precision motion control. IEEE Transactions on Control Systems Technology **16**(6), 1218–1231 (2008)
16. Barton, K.L., Hoelzle, D.J., Alleyne, A.G., Johnson, A.J.W.: Cross-coupled iterative learning control of systems with dissimilar dynamics: design and implementation. International Journal of Control **84**(7), 1223–1233 (2011)
17. Uchiyama, M.: Formation of high-speed motion pattern of a mechanical arm by trial. Transactions of The Society of Instrument and Control Engineers **14**, 706–712 (1978)
18. Craig, J.J.: Adaptive control of manipulators through repeated trials. General Motors Research Laboratories (1983)
19. Koren, Y., Lo, C.C.: Variable-gain cross-coupling controller for contouring. CIRP Annals-Manufacturing Technology **40**(1), 371–374 (1991)
20. Ouyang, P.R., Dam, T.: Position domain PD control: stability and comparison. In: IEEE International Conference on Information and Automation, Shenzhen, China (2011)

Author Index

Printed in the United States
By Bookmasters